SURFACE WAVES AND FLUXES

ENVIRONMENTAL FLUID MECHANICS

Volume 8

The titles in this series are listed at the end of this volume.

Surface Waves and Fluxes

Volume II – Remote Sensing

edited by

G. L. GEERNAERT

Space Systems and Technology Department, Naval Research Laboratory, Washington D.C., U.S.A.

and

W. L. PLANT

Ocean Engineering Department, Woods Hole Oceanographic Institution, Massachusetts, U.S.A.

KLUWER ACADEMIC PUBLISHERS

DORDRECHT / BOSTON / LONDON

Library of Congress Cataloging in Publication Data

Surface waves and fluxes : current theory and remote sensing / edited
by G.L. Geernaert, W.J. Plant.
p. cm. -- (Environmental fluid mechanics)
Includes index.

1. Waves--Remote sensing. 2. Ocean-atmosphere interactions--Remote sensing. 3. Atmospheric physics. 4. Fluid mechanics.
I. Geernaert, G. L. II. Plant, William J. III. Series.
QC157.S94 1990
532'.0593--dc20 90-4725

ISBN-13: 978-94-010-6769-0 e-ISBN-13: 978-94-009-0627-3
DOI: 10,1007/978-94-009-0627-3

Published by Kluwer Academic Publishers,
P.O. Box 17, 3300 AA Dordrecht, The Netherlands.

Kluwer Academic Publishers incorporates
the publishing programmes of
D. Reidel, Martinus Nijhoff, Dr W. Junk and MTP Press.

Sold and distributed in the U.S.A. and Canada
by Kluwer Academic Publishers,
101 Philip Drive, Norwell, MA 02061, U.S.A.

In all other countries, sold and distributed
by Kluwer Academic Publishers Group,
P.O. Box 322, 3300 AH Dordrecht, The Netherlands.

Printed on acid-free paper

Softcover reprint of the hardcover 1st edition 1990

Foreword

During the Conference on Air-Sea Interaction in January 1986, it was suggested to me by David Larner of Reidel Press that it may be timely for an updated compendium of air-sea interaction theory to be organized, developed, and published. Many new results were emerging at the time, i.e., results from the MARSEN, MASEX, MILDEX, and TOWARD field projects (among others) were in the process of being reported and/or published. Further, a series of new experiments such as FASINEX and HEXOS were soon to be conducted in which new strides in our knowledge of air-sea fluxes would be made.

During the year following the discussions with David Larner, it became apparent that many of the advances in air-sea interaction theory during the 1970s and 1980s were associated with sponsor investments in satellite oceanography and, in particular, remote sensing research. Since ocean surface remote sensing, e.g., scatterometry and SAR, requires intimate knowledge of ocean surface dynamics, advances in remote sensing capabilities required coordinated research in air-sea fluxes, wave state, scattering theory, sensor design, and data exploitation using environmental models. Based on this interplay of disciplines, it was decided that this book be devoted to air-sea interaction and remote sensing as multi-disciplinary activities.

The book evolved into documenting a concise description of four important research arenas common to both the air-sea interaction and remote sensing disciplines: (1) marine atmospheric surface layer turbulence and fluxes; (2) surface wave statistics and dynamics; (3) theory of electromagnetic scattering off a roughened air-sea interface; and (4) the description and use of sensors for active and passive remote sensing. In most part, the first two research arenas are found in Volume 1 while the latter two are in Volume 2. The fourth arena of this book was decided early-on not to engage in the synthetic aperture radar problem, in most part because of the breadth of the problem and controversies still being resolved. Measurement techniques were additionally ignored as a focus for this book. It is hoped that this book merely acts as a state-of-the-art documentation of theory and practice in which a better understanding of remote sensing and air-sea interaction may be developed.

This book was prepared in camera-ready format with the patience and diligence of Dora Wilbanks and her staff at the Naval Research Laboratory. Since the compilation of all 18 chapters required extensive coordination among all authors and both editors, sincere gratitude is extended to all parties involved in the completion of this book.

Gary Geernaert
Naval Research Laboratory
April 1990

TABLE OF CONTENTS

VOLUME 2

TABLE OF CONTENTS

VOLUME 1

Chapter 10

QUASI-SPECULAR SCATTERING FROM THE AIR-SEA INTERFACE

GARY S. BROWN
Bradley Department of Electrical Engineering
Virginia Polytechnic Institute & State University
Blacksburg, Virginia 24061-0111

1. Introduction

Scattering of electromagnetic energy by an arbitrarily roughened surface comprises a very difficult boundary value problem (Beckmann and Spizzichino, 1963; Bass and Fuch, 1978; Desanto and Brown, 1986). When the surface shape changes with both time and space, this further complicates matters. Even describing the scattering as a random process does not simplify the analysis too much and the use of numerical methods has not proven to be very useful either. This state of affairs leads to a great deal of emphasis on approximate analysis methods to yield both quantitative predictions and physical insight (Bass, et al., 1968; Valenzuela, 1978; Brown, 1985). Chapters 11 and 12 both illustrate approximate scattering methodologies and this chapter comprises yet another, namely the quasi-specular scattering approximation.

The purpose of this chapter is to give an overview of the quasi-specular approximation with emphasis on its capabilities and limitations when applied to sea surface scattering. Since this topic is a relatively mature one, not much space will be devoted to those issues which are adequately covered elsewhere. Emphasis will be placed on the origin of the approximation and on those applications which have not received much development in the literature. The reader is cautioned that a great deal of literature exists on this topic but not all of it adequately states the *limitations* of the approximation (Barrick, 1970). This is due, in part, to the fact that hard and fast rules for its use have not yet been developed. On the other hand, there are some conservative guidelines (Barrick and Peake, 1968) which if followed will guarantee that the approximation will never be misapplied. This latter rationale is the one adopted in this chapter.

The quasi-specular scattering approximation is a method for predicting the average scattering from a randomly rough surface, such as the ocean surface, in and about a specific direction and for a certain range of surface roughness parameters. Nothing in the approximation itself *clearly* shows the range of surface parameters for which the technique is valid so this must be estimated from other considerations. In fact, the approximation can be applied to surfaces which do not satisfy the basic restrictions for the application of the technique. In this instance, effective surface parameters must be used in the resulting equations (Brown 1978) and the angular range over which the scattering is predictable is restricted. This approach is

G. L. Geernaert and W. J. Plant (eds.), Surface Waves and Fluxes, Volume II, 1–39.

also used in the analysis in Chapter 11. Thus, the key to the intelligent use of this approximation is knowing and understanding the assumptions that are contained within it.

1.1. REFLECTION FROM A FLAT SURFACE

A good starting point for introducing some of the concepts which are of use in the quasi-specular approximation is simple reflection of an incident field by a planar interface (see also Chapter 11). In this situation an arbitrarily polarized electromagnetic plane wave is incident upon the infinite planar interface; the interface represents a discontinuity in the electrical properties of the medium such as the permittivity (μ), conductivity (σ), and/or the permeability (ϵ). A portion, of the incident field is reflected from the interface while a portion may be transmitted through it and into the second medium. The electric field of the incident plane wave is of the form

$$\vec{E}_i = E_o \hat{e}_i \exp(-j\vec{k}_i \cdot \vec{r}) \tag{1}$$

where E_o is its amplitude, $\hat{e}_i$ is the polarization, and $\vec{k}_i = k_o \hat{k}_i$ where k_o is the wavenumber $(=2\pi/\lambda_o)$ of the upper medium. While this vector field "points" in the direction $\hat{e}_i$, it travels in the direction $\hat{k}_i$. Because the interface is perfectly flat, there is a plane wave *reflected* from the interface and back into the upper medium and one which is *transmitted* into the lower medium. These vector fields may be written as follows;

$$\vec{E}_r = E_{r_o} \hat{e}_r \exp(-j\vec{k}_r \cdot \vec{r}) \tag{2}$$

and

$$\vec{E}_t = E_{t_o} \hat{e}_t \exp(-j\vec{k}_t \cdot \vec{r}) \tag{3}$$

where the subscripts, r and t, denote reflected and transmitted quantities. The unknown field amplitudes, polarizations, and vector wavenumbers in (2) and (3) are determined by satisfying the appropriate boundary conditions on the *total* field across the interface at $z = 0$. For example, the tangential component of the electric field must be continuous across the interface so that

$$\hat{\tau} \cdot \left[\vec{E}_i + \vec{E}_r \right]_{z=0} = \hat{\tau} \cdot \vec{E}_t \,|_{z=0} \tag{4}$$

where the left side of (4) represents the total electric field above the interface and the right side is the total field below the interface. The unit vector $\hat{\tau}$ is tangent to the interface. Substituting (1)-(3) in (4) first of all shows that the reflected and transmitted fields must be coplanar with the incident field. This is a consequence of (4) implying that for all x and y on the interface

$$\vec{k}_i \cdot \vec{r}_T = \vec{k}_r \cdot \vec{r}_T = \vec{k}_t \cdot \vec{r}_T \tag{5}$$

where $\vec{r}_T = x\hat{x} + y\hat{y}$, the symbol "$T$" denotes *transverse* to z. Each of the wavevectors can be written in terms of their vector components as follows;

$$\vec{k}_i = -k_o \cos\theta_i \hat{z} + k_o \sin\theta_i \hat{k}_{i_T} \tag{6a}$$

$$\vec{k}_t = -k_1 \cos\theta_t \hat{z} + k_1 \sin\theta_t \hat{k}_{t_T} \tag{6b}$$

$$\vec{k}_r = k_o \cos\theta_r \hat{z} + k_o \sin\theta_r \hat{k}_{r_T} \tag{6c}$$

where $k_o = \omega\sqrt{\mu_o \epsilon_o}$ and $k_1 = \omega\sqrt{\mu_1 \epsilon_1}$ are the wavenumbers in the upper and lower media, respectively. The angles θ_i, θ_r, and θ_t are the smallest angles between the directions of propagation of the incident, reflected, and transmitted fields and the z-axis, respectively, which is the normal to the interface (see Figure 1). The unit vectors $\hat{k}_{i_T}$, $\hat{k}_{r_T}$, and $\hat{k}_{t_T}$ are the parts of $\vec{k}_i$, $\vec{k}_r$, and $\vec{k}_t$ in the x and y directions (transverse to z). Substituting (6) into (5) yields (from $\vec{k}_i \cdot \vec{r}_T = \vec{k}_r \cdot \vec{r}_T$)

$$\sin\theta_i = \sin\theta_r \tag{7}$$

and (from $\vec{k}_i \cdot \vec{r}_T = \vec{k}_t \cdot \vec{r}_T$)

$$k_o \sin\theta_i = k_i \sin\theta_t \tag{8}$$

The result in (7) is *Snell's law of reflection* which states that the angle of incidence is equal to the angle of reflection. The result in (8) is *Snell's law of refraction* which relates the angles of incidence and refraction to the contrast in electrical properties between the two media.

It should be noted that the angle of incidence and the angle of reflection are equal *regardless* of the details of the discontinuity at the interface. The angle of transmission is less than the angle of incidence when the lower medium is denser than the upper medium ($\sqrt{\mu_1 \epsilon_1} > \sqrt{\mu_o \epsilon_o}$). When the lower medium is lossy, i.e., ϵ_1 or μ_1 are complex, $\sin\theta_t$ becomes complex and this simply indicates the presence of phase progression and exponential attenuation in the lower medium.

1.1.1. *Fresnel Reflection Coefficients.* The previous section shows that the reflected and transmitted plane waves are coplanar with the incident plane wave and gives the directions in which they travel. To complete the reflection process description, it is necessary to determine how much of the incident field goes into the reflected and transmitted parts. This is accomplished by applying the boundary conditions on the fields at the interface, and these are that both the incident electric and magnetic fields are continuous across the interface. We will not go through the derivation here but will only outline it since it is a very basic result that is covered by just about any look on electromagnetics (Stratton, 1941; Harrington, 1961). (See also Chapter 11).

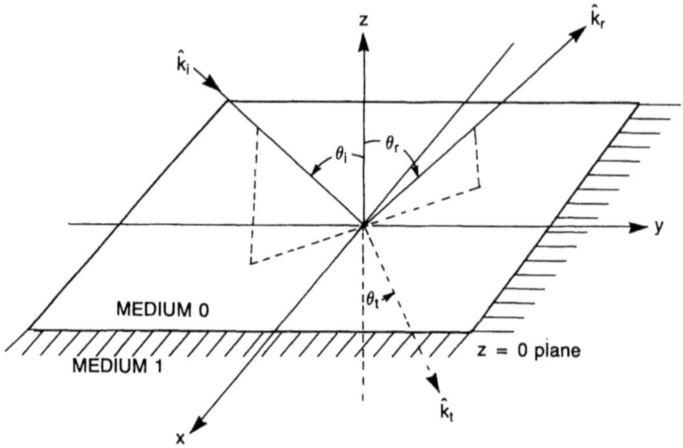

Fig. 1 — An illustration of the various directions of travel of the incident, reflected, and transmitted fields at the planar interface at $z = 0$

Because of the vector nature of electromagnetic waves, two reflection coefficients are required to completely describe the reflection process. One coefficient applies to that component of the electric field which is normal to the plane formed by incident wave vector ($\hat{k}_i$ or the direction of travel of the incident fields) and the normal to the surface. The reflection coefficient in this case relates the reflected electric field to the incident electric field. Its specific value is given by

$$\rho_h = \frac{\mu_1 k_o \cos\theta_i - \mu_o \sqrt{k_1^2 - k_o^2 \sin^2\theta_i}}{\mu_1 k_o \cos\theta_i + \mu_o \sqrt{k_1^2 - k_o^2 \sin^2\theta_i}} \tag{9}$$

The upper medium is denoted by the "o" subscript and the lower by the "1" subscript. The incident field comes from the upper medium. The second reflection coefficient applies to the component of the magnetic field which is normal to the plane formed by the incident wave vector ($\hat{k}_i$ or the direction of travel of the incident fields) and the normal to the interface. This reflection coefficient relates the reflected magnetic field to the incident field and its specific value is given by

$$\rho_v = \frac{\mu_o k_1^2 \cos\theta_i - \mu_1 k_o \sqrt{k_1^2 - k_o^2 \sin^2\theta_i}}{\mu_o k_1^2 \cos\theta_i + \mu_1 k_o \sqrt{k_1^2 - k_o^2 \sin^2\theta_i}} \tag{10}$$

The transmission coefficients are easily obtained from

$$t_h = 1 + \rho_h \tag{11a}$$

$$t_v = 1 + \rho_v \tag{11b}$$

but they are not normally of interest in ocean surface remote sensing.

Table 1 Alternate Descriptions of Polarization

I.	*CONDITION*: $\vec{E}_i$ is normal to the plane of incidence formed by $\vec{k}_i$ and the normal to surface *DEFINITION*: $\rho_h = \vec{E}_r \cdot \hat{e} / \vec{E}_i \cdot \hat{e}$ *POLARIZATION DESCRIPTION IN AREA OF x* EM THEORY — TE (Transverse Electric) Remote Sensing — *h* (Horizontal) Optics — *s*
II.	*CONDITION:* $\vec{H}_i$ is normal to the plane of incidence formed by k_i and the normal to the surface *DEFINITION*: $\rho_v = \vec{H}_r \cdot \hat{h} / \vec{H}_i \cdot \hat{h}$ *POLARIZATION DESCRIPTION IN AREA OF x* EM Theory — TM (Transverse Magnetic) Remote Sensing — v (Vertical) Optics — *p*

The coefficients in (9) and (10) are called, in a generic sense, the Fresnel reflection coefficients. However, the specific nomenclature for the individual coefficients is less universal because there is no unanimous agreement on how to refer to polarization. Table 1 is an attempt to clarify the situation. Polarization, as defined in many texts on electromagnetic waves (Harrington, 1961), refers to the direction of the electric field. However, in real-world situations, this is not always the best or least ambiguous method for defining the direction of the incident field. Table 1 gives the actual direction of either the incident electric field ($\vec{E}_i$) or the incident magnetic field ($\vec{H}_i$) along with the appropriate definition of the reflection coefficient and three (3) common usage descriptions of this particular polarizations. For example, if the $\vec{H}_i$ is normal to the plane of incidence then the reflection coefficient is defined in terms of the $\vec{H}$-fields, i.e.

$$\rho_v = \frac{\vec{H}_r \cdot \hat{h}}{\vec{H}_i \cdot \hat{h}} \tag{12}$$

where $\vec{h}$ is the direction of the $\vec{H}$-field. Furthermore, according to Table 1, this particular situation is called TM-polarization in the EM-theory literature, vertical or v-polarization in the remote sensing community, and p-polarization in optics. In view of the intended audience of this book, the remote sensing nomenclature will be used although it is potentially the most confusing.

Most texts on basic electromagnetics cover the behavior of the Fresnel reflection coefficients and there are some which present results for electrical properties common to salt water (Kerr, 1951), so we will only review the major behavior of ρ_v and ρ_h. In (9) and (10) it is assumed that the magnetic permeability of the two media are equal ($\mu_1 = \mu_o$) and that the upper medium is free space while the lower is a lossy dielectric such as sea water: thus, $|\epsilon_1| > \epsilon_o$ and $|k_1| > k_o$.

The horizontal polarization reflection coefficient cannot go to zero for $0 \leq \theta_i \leq 90°$ because the numerator in (9) cannot go to zero for $\mu_o = \mu_1$. It goes to -1 for grazing incidence ($\theta_i = 90°$) and the lower medium becoming a perfect conductor, $\epsilon_1 \rightarrow -j\infty$. For vertical polarization (ρ_v), the reflection coefficient goes to unity at grazing incidence and for a perfect conductor in the lower medium. It is possible to have the numerator in (10) go to zero and the reflection coefficient to consequently go to zero; this occurs when

$$\sin\theta_i = \sqrt{\frac{\epsilon_1}{\epsilon_o + \epsilon_1}} \tag{13}$$

This particular value of θ_i is called the Brewster angle and its occurrence implies that all of the incident energy is transmitted into the lower medium. With a lossy medium such as sea water, ϵ_1 is complex and a true Brewster's angle cannot be achieved. However, it is possible to obtain a minimum in $|\rho_v|$ near grazing incidence (θ_i near 90°) for sea water in the microwave frequency range (Kerr, 1951). The magnitude of the horizontal reflection coefficient does not exhibit this local minimum and, in fact, monotonically increases from a minimum value at normal incidence ($\theta_i = 0°$) to unity at grazing incidence.

At normal incidence, both $|\rho_v|$ and $|\rho_h|$ become equal, i.e.

$$|\rho_v| = |\rho_h| = \left|\frac{\sqrt{\epsilon_1} - \sqrt{\epsilon_o}}{\sqrt{\epsilon_1} + \sqrt{\epsilon_o}}\right| \tag{14}$$

and the polarization of the reflected electric field is exactly the same as the incident electric field for linear polarization. Although this is a trivial result, it is one that should be remembered for future use.

To close this discussion of Fresnel reflection coefficients, it should be noted that an incident plane wave field having an arbitrary polarization can always be decomposed into a TE part (to which ρ_h can be applied) and a TM part (to which ρ_v can be applied). Thus, the ρ_v and ρ_h coefficients are sufficient to deal with *any* incident plane wave although the algebra becomes very messy.

1.1.2. *Real Sources.* In the section above, we dealt exclusively with plane waves primarily because this is how the Fresnel coefficients are defined. From an analytical point of view, plane waves are convenient to deal with because three angles describe the entire situation. However, for finite size antennas, plane waves do not exist. That is, it truly takes an infinite antenna to produce a plane wave. However, if the point of observation of the field radiated by an antenna is further away than $2D^2/\lambda_o$ then it is in the so-called far-field region of the antenna and the radiated fields may be written as follows (Collin, 1985)*;

$$\vec{E} = f_E(\theta,\phi)\hat{e}(\theta,\phi)\frac{\exp(-jk_o r)}{4\pi r} \tag{15a}$$

$$\vec{H} = \frac{1}{Z_o}(\hat{k} \times \hat{e})\, f_E(\theta,\phi)\frac{\exp(-jk_o r)}{4\pi r} \tag{15b}$$

In these equations $k_o = 2\pi/\lambda_o$ is the free space wavenumber, r is the distance from the antenna to the point of observation of these fields, $\hat{e}$ is the polarization of the radiated fields (the direction of the $\vec{E}$ field, $Z_o = \sqrt{\mu_o/\epsilon_o} = 377\Omega$, and $\hat{k}$ is a unit vector in the direction of propagation of the fields ($\hat{k} = \hat{r}$). If we construct a spherical coordinate system (r,θ,ϕ) centered on the antenna and pointing to the point of observation of the field a distance r away, this defines the angles θ and ϕ, see Figure 2.

In (15) it should be noted that there is a distinct separation between the angle and distance dependent functions. The distance dependent part represents a spherical wave which is what a real source produces in the far-field. The angle dependent parts of $\vec{E}$ and $\vec{H}$ in (15) modulate the basic spherical wave nature of the far-fields and therefore give them angular structure. If the point of observation of the fields is sufficiently far removed from the antenna then it is possible that we can move over a portion of the r = constant sphere and approximate the behavior of the field in this region as a plane wave. This situation is illustrated in Figure 3. Clearly, the plane wave approximation holds over an area which increases as the curvature of the spherical wave decreases, i.e. the point of observation becomes further away from the source. Typically, this approximation should not be pushed beyond the point where $k_o d = \pi/2$ or $d = \lambda_o/4$. The area over which this condition holds is $\pi\rho_o^2$ and this can be shown to equal $2\pi dr$ or $\pi\lambda_o r/2$. The corresponding change in θ, i.e. $\delta\theta$, over this area is given by $\delta\theta \approx \lambda_o/2r$.

What the above material shows is that if the point of observation is taken to be sufficiently far from the source then the fields may be considered to be locally planar and the angle θ can be taken to be constant. The area over which this approximation is valid is given by $\pi\lambda_o r/2$ and the angular error at the boundaries of this area is $\lambda_o/2r$. In remote sensing we normally deal with wavelengths (λ_o) in the centimeter range and distances (r) in the km range. Consequently, the fields do indeed behave as local plane waves for most remote sensing applications.

*This criterion applies to a real aperture antenna having a maximum linear dimension D and λ_o is the electromagnetic wavelength.

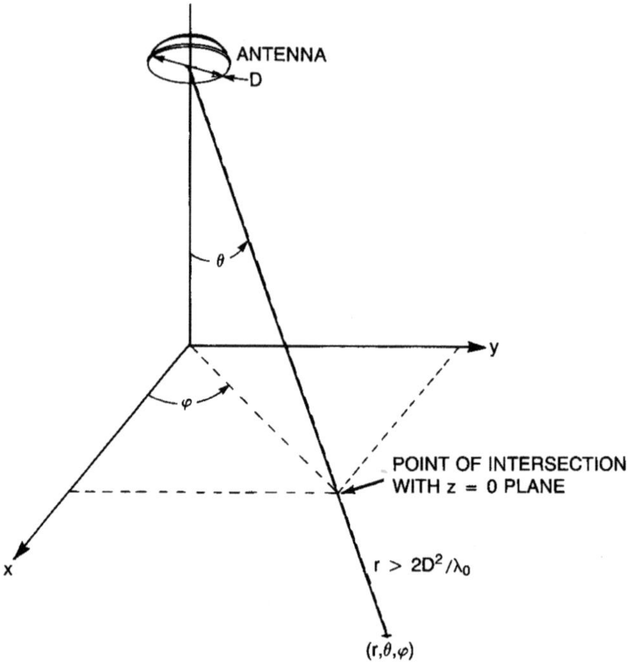

Fig. 2 — Geometry for a radiating antenna and the point (r, θ, ϕ) in the far field of the antenna

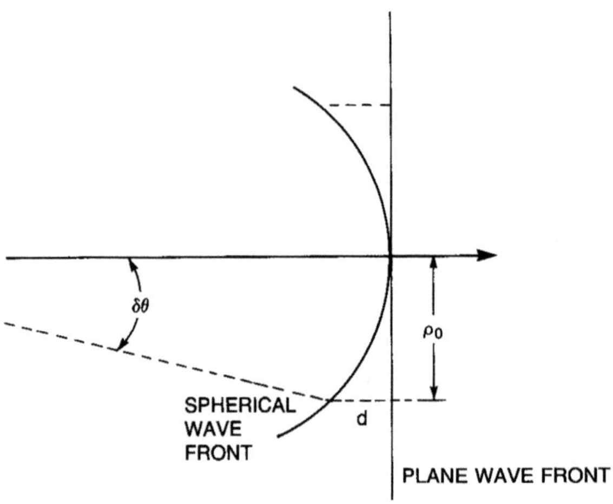

Fig. 3 — Approximation of a spherical wave by a plane wave. The phase $k_o d$ is the error for a distance ρ_o out from the center of the approximation

1.1.2.1. *Analysis Via Images.* In active remote sensing we generally transmit with one antenna and receive with either the same antenna or another one. When the same antenna is used to both transmit and receive, this is called a monostatic geometry. When the receiving antenna is displaced from the transmitter, this is called a bistatic geometry.

In remote sensing surfaces, we would like to be able to remove all of the sensor and geometrical factors in the data so that only information on the surface remains. To do this, we must be able to relate the received voltage or power to the transmitted voltage or power. When dealing with rough surfaces, power is normally used and the reason for this will be given later. For flat surfaces, we must deal initially with voltages.

Consider the situation in Figure 4 in which a transmitter illuminates a planar surface and the reflected energy is received by a receiver displaced from the transmitter, i.e. a bistatic situation. We know that the receiver only responds to the field propagating along the dashed path (from the transmitter to the receiver) because this is the only transmitted ray that is properly reflected to the receiver. However, instead of going through a complicated analysis involving the paths from the transmitter to ground to receiver, we can use *image theory* to relate transmitted and received voltages (Collin, 1985). In image theory, the transmitter is replaced by its image across the planar surface and a simple link equation between the image (in free space) and the receiver is used to describe the situation. The image takes on the same antenna pattern as the transmitter *but* it is weighted by the appropriate reflection coefficients for the surface. For example, if the field radiated by the transmitter has the form (see Figure 4)

$$\vec{E}_T = f_T(\theta_T, \phi_T)\hat{e}\frac{\exp(-jk_o R_T)}{4\pi R_T} \tag{16}$$

where $\hat{e}$ is normal to the plane of incidence then the image field at the location of the receiver is

$$\vec{E}_I = \rho_h(\theta_i) f_T(\theta_T, \phi_T)\hat{e}\ \frac{\exp(-jk_o[R_T + R_R])}{4\pi[R_T + R_R]} \tag{17}$$

The voltage at the output terminals of the receiving antenna is therefore given by

$$v_R = \rho_h(\theta_i) f_T(\theta_T,\phi_T) f_R(\theta_R,\phi_R)\frac{\exp(-jk_o[R_T + R_R])}{4\pi[R_T + R_R]} \tag{18}$$

where f_T and f_R are the radiation patterns of the transmitting and receiving antennas for the polarization direction $\hat{h}$. If there were a mixture of polarization states, this expression would become more complicated.

It is important to remember that this result applies only to reflection from a planar interface. In addition to the dependence on the reflection coefficient and the local angle of incidence, there is the appearance of the two antenna patterns as a product. However, for future purposes it is very important to note that the amplitude of the received voltage varies as the inverse of the distance from the transmitter to the surface and then to the receiver. Consequently, the received power varies as the inverse square of this distance. In future sections,

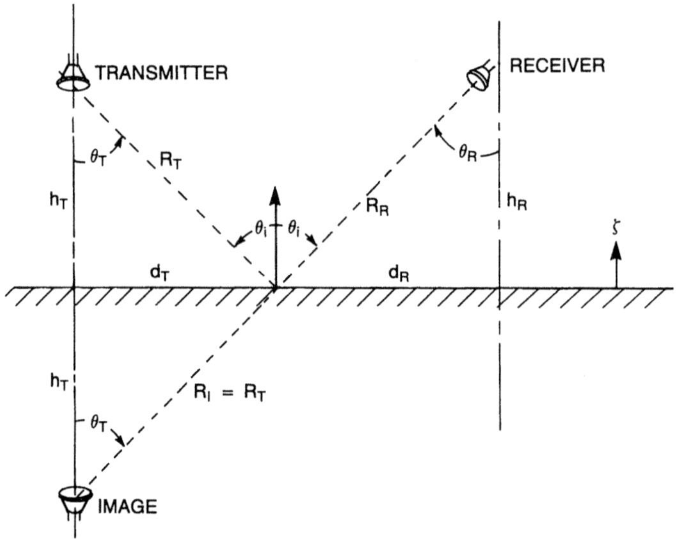

Fig. 4 — Bistatic measurement geometry and image location

it will be shown that the power can assume a range or distance dependence that is significantly different than this when the surface becomes rough.

1.2. REFLECTION FROM A RANDOMLY ELEVATED PLANAR SURFACE

We are now in a position to introduce a very simple randomness into the problem. For example, let the planar surface be displaced by a distance ζ from its position shown in Figure 4; ζ will be positive for an upward displacement of the surface. The new distance $R_T' + R_R'$ can then be written as

$$R_T' + R_R' = [d_T^2 + (h_T - \zeta)^2]^{1/2}$$

$$+ [d_R^2 + (h_R - \zeta)^2]^{1/2} \tag{19}$$

and for small height displacements relative to h_T, h_R, R_T, and R_R, there results

$$R_T' + R_R' \approx R_T + R_R - (\cos\theta_T + \cos\theta_R)\,\zeta \tag{20}$$

Substituting this result (for $R_T + R_R$) in the phase of (18) and ignoring the miniscule effect in the amplitude yields

$$v_R' = v_R \exp\,(jk_o\,[\cos\theta_T + \cos\theta_R]\,\zeta) \tag{21}$$

where v_R is the value given in (18). It is obvious from the geometry in Figure 4 that $\theta_T = \theta_R$ so (21) becomes

$$v_R' = v_R \exp(2jk_o \cos\theta_T \zeta) \tag{22}$$

This result merely states that the bulk of the surface displacement effect is a linear phase perturbation.

1.2.1. *The Average (Coherent) Reflected Field.* The average or coherent reflected field is obtained *experimentally* by measuring the reflected field for a number of different realizations of the surface displacement and then averaging these results, e.g.

$$\overline{v_R'} = \frac{1}{N}\sum_{n=1}^{N} v_R'(\zeta_n) \tag{23}$$

Because this result is based on a finite number of samples (N), it is subject to the usual statistical errors associated with a finite population mean. The ensemble or infinite sample average is given by

$$<v_R'> = \int_{-\infty}^{\infty} v_R' p(v_R') dv_R' \tag{24}$$

where $p(v_R')$ is the probability density function for v_R'. We frequently do not know the probability density function for the actual quantity to be averaged; however, we usually know the density function for the physical properties upon which the quantity depends. In the case of v_R' we may not know $p(v_R')$ but we do know $p(\zeta)$ so that

$$<v_R'> = \int_{-\infty}^{\infty} v_R' p(\zeta) d\zeta \tag{25}$$

There is a fundamental theorem in probability theory that guarantees that (24) and (25) are identical. Substituting from (22) into (25) yields

$$<v_R'> = v_R \int_{-\infty}^{\infty} p(\zeta) \exp(j2k_o \cos\theta_T \zeta) d\zeta \tag{26}$$

and we recognize this to be the Fourier transform of $p(\zeta)$. In probability theory the Fourier transform of the probability density function is called the characteristic function Φ so

$$<v_R'> = v_R \Phi(2k_o \cos\theta_T) \tag{27}$$

If the density function of ζ is Gaussian with zero mean, i.e.

$$p(\zeta) = \frac{1}{\sqrt{2\pi <\zeta^2>}} \exp(-\zeta^2/2<\zeta^2>), \tag{28}$$

the characteristic function is given by

$$\Phi(2k_o \cos\theta_T) = \exp(-2k_o^2 <\zeta^2> \cos^2\theta_i) \tag{29}$$

where we have used $\theta_T = \theta_i$. The quantity $k_o\sqrt{<\zeta^2>}\cos\theta_i$ is frequently called the Rayleigh parameter in honor of Lord Rayleigh's early work on rough surfaces. It is clear from the expressions in (29) and (27) that even a relatively small roughness can result in a severely attenuated average field. This is because the effect of the surface height variation appears in the phase of the reflected field and even relatively small effects can lead to large phase changes.

The average field reflected by a randomly elevated planar surface is always determined by the characteristic function for the random elevations. Of course, the specific functional dependence on the Rayleigh parameter is determined by the probability density function of the elevations.

1.2.2. *The Average Incoherent Reflected Power.* The square of (27) is proportional to the *coherent reflected power*, i.e.

$$P_c = \alpha \mid <v_R'> \mid^2 \tag{30}$$

or

$$P_c = \alpha \mid v_R \Phi(2k_o \cos\theta_i) \mid^2 \tag{31}$$

where α is a constant determined by the input resistance of the receiver. There is another descriptor which is useful when dealing with problems of this nature and this is the *incoherent power*. The incoherent power is defined as follows;

$$P_i = \alpha < \mid v_R' - <v_R'> \mid^2 > \tag{32}$$

where we first center v_R', i.e. make it zero mean, next square the centered quantity and finally average this result. The expression for P_i can be simplified to

$$P_i = \alpha \left[< \mid v_R' \mid^2 > - \mid <v_R'> \mid^2 \right] \tag{33}$$

Experimentally, this quantity would be determined by

$$\bar{P}_i = \alpha \frac{1}{N} \sum_{n=1}^{N} \mid v_R'(\zeta_n) \mid^2 - \alpha \mid \overline{v_R'} \mid^2 \tag{34}$$

where each measurement of v_R' for each realization of the surface is first squared and then averaged. The ensemble average is obtained from using (22) and (27) in (33), i.e.

$$P_i = \alpha \mid v_R \mid^2 \left[1 - \mid \Phi(2k_o \cos\theta_i) \mid^2 \right] \tag{35}$$

because

$$<|v_R'|^2> = |v_R|^2 \int_{-\infty}^{\infty} p(\zeta) d\zeta = |v_R|^2 \tag{36}$$

The result in (35) says that for small Rayleigh parameter, i.e. so that $|\Phi|^2 \approx 1$, the incoherent power is zero. When the parameter is large so that $|\Phi|^2 \approx 0$, the incoherent power represents *all* of the power reflected from the surface. These observations are just consequences of conservation of energy in that the total reflected power is given by

$$P_T = P_i + P_c = \alpha |v_R|^2 \tag{37}$$

Thus, as the coherent power decreases to zero, the incoherent power must increase to compensate for this decrease. Conversely, the coherent power must increase whenever the incoherent power decreases.

The randomly elevated plane is an interesting "first" example of random surface scattering because it illustrates the basic principles using relatively simple concepts. We can easily see the conservation of energy because *both* the coherent and incoherent powers propagate *only* in the specular direction, i.e. $\theta_R = \theta_i$. However, *this happens only for a randomly elevated planar surface.* When we start introducing horizontal structure on the surface, the incoherent power starts propagating in directions other than the specular direction. The coherent power, however, always propagates only in the specular direction.

2. Evolution of Quasi-Specular Scattering

When the surface can be represented by a randomly elevated plane, all of the incident power is reflected into the specular direction, i.e. $\theta_T = \theta_R = \theta_i$. The randomly elevated planar surface problem is therefore governed entirely by the laws of *specular reflection*. When the surface begins to exhibit some horizontal structure, the laws governing the process change significantly. That is, when the random surface elevation depends on the location of its point of measurement in the $\zeta = 0$ plane (see Fig. 5), the laws of reflection no longer hold. The relatively simple principles of reflection are replaced by the more complicated process of diffraction or scattering. In fact, diffraction or scattering may be defined as those processes in wave propagation which are not predictable by the laws of reflection and refraction. Diffraction and scattering are consequences of the wave nature of electromagnetic fields while reflection and refraction are more closely related to the ray-like behavior of fields in the high frequency limit.

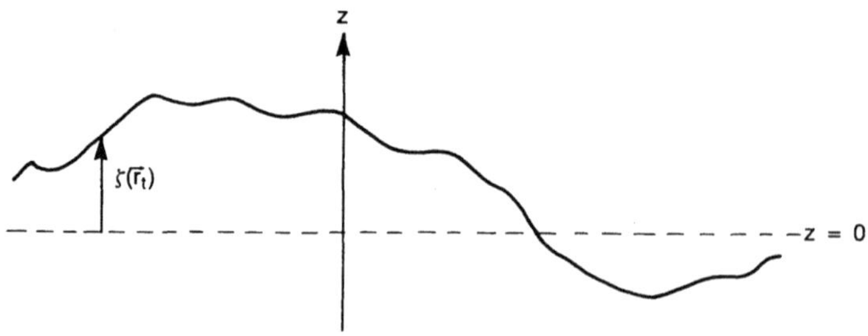

Fig. 5 — A roughened planar surface in which $\vec{r}_t$ denotes a dependence on x and y (the coordinates transverse to z)

2.1. THE TANGENT PLANE-PHYSICAL OPTICS (TP/PO) APPROXIMATION

In the planar reflection problem, we matched the fields (incident, reflected, and transmitted) across the interface at each point on the interface *independent* of any other part of the surface. When this kind of localized analysis can be used, we say that *single scattering* applies. However, once the surface departs from a plane, single scattering no longer applies and the field at any point on the surface depends on the behavior of both the fields *and* the surface in a *neighborhood* of the point in question (DeSanto and Brown, 1986). Mathematicians refer to this as multiple scattering even though it is not necessarily the same as rays bouncing about on a surface, i.e. the conventional physical picture of multiple scattering.

When we can no longer use the laws of reflection to obtain the reflected fields, it is necessary to go to an integral or superposition representation for the scattered fields.* One form of this representation is the set of vector Kirchhoff equations (Stratton, 1941). We give the one for $\vec{E}_s$ only (the one for $\vec{H}_s$ may be obtained by replacing $\vec{E}_T$ by $\vec{H}_T$ and $\vec{H}_T$ by $-\vec{E}_T$ and ϵ_o by μ_o in the $\vec{E}_s$ equation) as follows;

$$\vec{E}_s(\vec{r}) = \nabla \times \int_{s_0} (\hat{n}_o \times \vec{E}_T) g_o ds_o + \frac{1}{j\omega\epsilon_o} \left[\nabla(\nabla \cdot) + k_o^2 \right] \int_{s_o} (\hat{n}_o \times \vec{H}_T) g_o ds_o \tag{38}$$

where $\vec{E}_s(\vec{r})$ is the scattered electric field at the point $\vec{r}$ above the surface, $(\vec{E}_T, \vec{H}_T)$ are the total electric and magnetic fields on the surface at the point $\vec{r}_o = \vec{r}_{t_o} + \zeta_o \hat{z}$; $\hat{n}_0$ is the unit normal to the surface at $\vec{r}_o$, e.g.

$$\hat{n}_o = \left(-\zeta_{x_o}\hat{x} - \zeta_{y_o}\hat{y} + \hat{z} \right) \Big/ \sqrt{1 + \zeta_{y_o}^2 + \zeta_{y_o}^2} \tag{39}$$

where the slopes are

$$\zeta_{x_o} = \frac{\partial \zeta_o}{\partial x_o} \qquad \zeta_{y_o} = \frac{\partial \zeta_o}{\partial y_o}, \tag{40}$$

and g_o is the free space Green's function

$$g_o(|\vec{r} - \vec{r}_o|) = \frac{\exp\left(-jk_o |\vec{r} - \vec{r}_o| \right)}{4\pi |\vec{r} - \vec{r}_o|} \tag{41}$$

*We also use the nomenclature "scattered" rather than "reflected" since scattering is the dominant process.

Also in (38), ω is the angular frequency and k_o is the free space wavenumber $2\pi/\lambda_o$. The total fields ($\vec{E}_T$, $\vec{H}_T$) on the surface are the sum of the incident and the scattered fields, i.e.

$$E_T(\vec{r}_o) = \vec{E}_i(\vec{r}_o) + \vec{E}_s(\vec{r}_o) \tag{42a}$$

$$\vec{H}_T(\vec{r}_o) = \vec{H}_i(\vec{r}_o) + \vec{H}_s(\vec{r}_o) \tag{42b}$$

Thus, if we let the point of observation $\vec{r}$ move to the surface, we obtain integral equations for the scattered electric and magnetic fields *on the surface.* Once these fields are determined we can compute $\vec{E}_T$ and $\vec{H}_T$ from (42) and $\vec{E}_s$ from (38) *for any point above the surface.*

The accurate determination of $\vec{E}_s$ and $\vec{H}_s$ on the surface comprises a problem of continuing interest in rough surface scattering theory. Because it requires the solution of coupled, three dimensional, vector integral equations some degree of approximation is required. The simplest approximation that can be made is the tangent plane approximation. In this approximation, we made two fundamental assumptions. First, we assume that the only field striking the point in question is the incident field. This means that we do not allow any multiple scattering from any other point on the surface to the point in question. The second assumption is that at the point in question on the surface, the surface is so gently curving that we may replace the actual surface by a *plane* that is tangent to the surface at the point (see Fig. 6). This is a very crucial step because it means that we can now use the laws of reflection to estimate ($\vec{E}_s$, $\vec{H}_s$) and ($\vec{E}_T$, $\vec{H}_T$) on the surface.

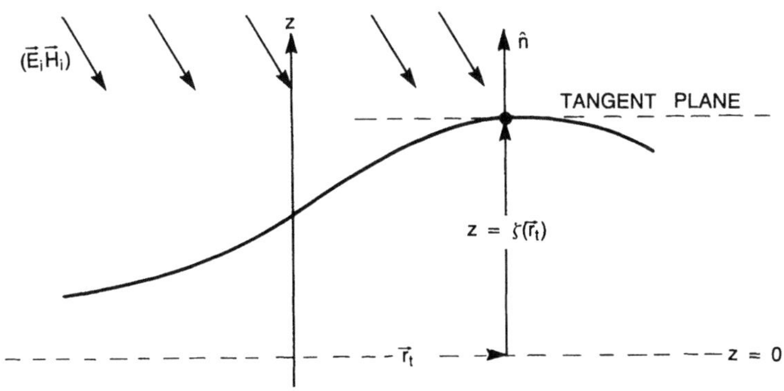

Fig. 6 — Replacing the actual surface at $(\vec{r}_t, \zeta)$ by a tangent plane

Assuming that the electric field incident on the surface is a plane wave of the form

$$\vec{E}_i = E_o \hat{a} \exp(-jk_o \hat{k}_i \cdot \vec{r}_o) \tag{43a}$$

and the corresponding $\vec{H}$-field is

$$\vec{H}_i = \frac{1}{Z_o}(\hat{k}_i \times \hat{a}) E_o \exp(-jk_o \hat{k}_i \cdot \vec{r}_o) \tag{43b}$$

these fields can be decomposed into locally horizontal and vertical components and the reflection coefficients of Section 1.1.1 can be used to obtain the scattered (reflected) fields on the surface. This leads to the following approximations for the fields that go into the vector Kirchoff formula (eq. 38) (Sancer, 1969).

$$(\hat{n}_o \times \vec{E}_T)_K = \Big[(1-\rho_v)(\hat{a}\cdot\hat{v}_1)(\hat{n}_o\cdot\hat{k}_i)\hat{h}_1 + (1+\rho_h)(\hat{n}_o \times \hat{h}_1)(\hat{a}\cdot\hat{h}_1)\Big] E_o \exp(-jk_o\hat{k}_i\cdot\vec{r}_o) \tag{44a}$$

$$(\hat{n}_o \times \vec{H}_T)_K = \Big[-(1-\rho_h)(\hat{a}\cdot\hat{h}_1)(\hat{n}_o\cdot\hat{k}_i)\hat{h}_1 + (1+\rho_v)(\hat{n}_o \times \hat{h}_1)(\hat{a}\cdot\hat{v}_1)\Big] \frac{E_o}{Z_o} \exp(-jk_o\hat{k}_i\cdot\vec{r}_o) \tag{44b}$$

where

$$\hat{h}_1 = \frac{\hat{k}_i \times \hat{n}_o}{|\hat{k}_i \times \hat{n}_o|} \quad , \quad \hat{v}_1 = \hat{h}_1 \times \hat{k}_i \tag{45}$$

and the K subscript denotes the Kirchoff approximation for the surface fields. If we now take the observation point $\vec{r}$ sufficiently far from the surface, we can simplify (38) to yield

$$\vec{E}_s(\vec{r}) \approx -jk_o g(r)(\overline{\overline{I}} - \hat{k}_s\hat{k}_s) \cdot \int_s \Big[\hat{k}_s \times (\hat{n}_o \times \vec{E}_T)_K + Z_o(\hat{n}_o \times \vec{H}_T)_K\Big] \cdot \exp(jk_o\hat{k}_s\cdot\hat{r}_o)ds_o \tag{46}$$

where $\overline{\overline{I}}$ is the unit dyad, $\hat{k}_s$ is a unit vector in the scattering direction, and

$$g(r) = \frac{\exp(-jk_o r)}{4\pi r} .$$

The operator $(\overline{\overline{I}} - \hat{k}_s\hat{k}_s)$ just selects the component of the integral that is *transverse* to the scattering direction. We can convert the integration over the rough surface to one over the $z = 0$ plane through

$$ds_o = \sqrt{1 + \zeta_{x_o}^2 + \zeta_{y_o}^2}\, dx_o dy_o . \tag{47}$$

The radical in the denominator of the expression for $\hat{n}_o$ cancels the one in (47) leaving

$$\vec{E}_s(\vec{r}) \approx -jk_o g(r)(\overline{\overline{I}} - \hat{k}_s\hat{k}_s) \cdot \int \left[\hat{k}_s \times (\vec{N}_o \times \vec{E}_T)_K + Z_o(\vec{N}_o \times \vec{H}_T)_K \right] \cdot \exp(jk_o\hat{k}_s\cdot\vec{r}_o)dx_o dy_o \tag{48}$$

where

$$\vec{N}_o = -\zeta_{x_o}\hat{x} - \zeta_{y_o}\hat{y} + \hat{z} \tag{49}$$

Equation (44) when substituted into (48) comprises the tangent plane approximation for the electric field scattered by the surface.

Now it may be that the incident field at a point on the surface is shadowed from that point by another part of the surface. In addition, a scattered ray drawn from the point on the surface in the direction $\hat{k}_s$ may also be blocked by yet another part of the surface. These two phenomena are called incident and scatter shadowing and we can crudely account for these by multiplying the integrand of (48) by the factor $I_2(\vec{r}_o)$ which is unity if neither the incident nor scattered rays are shadowed and zero if there is shadowing (Sancer, 1969), i.e.

$$\vec{E}_s(\vec{r}) \approx -jk_o g(r)(\overline{\overline{I}} - \hat{k}_s\hat{k}_s)\cdot \int \left[\hat{k}_s \times (\vec{N}_o \times \vec{E}_T)_K + Z_o(\vec{N}_o \times \vec{H}_T)_K \right] \cdot I_2(\vec{r}_o)\exp(jk_o\hat{k}_s\cdot\vec{r}_o)dx_o dy_o \tag{50}$$

Equation (50) contains what is frequently referred to as the physical optics approximation for the fields on the surface. It differs from the tangent plane approximation only in its use of a shadowing function.

2.1.1. *Capabilities and Limitations of the TP/PO Approximation.* Whenever we make as drastic an approximation as what leads to (50), we would certainly like to know its range of validity. Unfortunately, this is a formidable task in itself because it assumes that we know the *exact* answer. For the full three dimensional problem, we do not have an exact result and even numerical approaches are not yet up to dealing with this problem. Some progress has been made with the two-dimensional problem but this is still in an early stage of development (Thorsos, 1988). Experimentation does not seem to hold the key to resolving this uncertainty (at least completely) because of the difficulty in both generating surfaces of known statistics and enough of them.

If we look to deterministic problems to answer this question (Ruck, et al., 1970), we find some interesting results. First, we find that physical optics always gives the correct result in the optical limit, i.e. as $\lambda_o \rightarrow 0$ or $k_o \rightarrow \infty$. For frequencies where the size of the scatterer is not large compared to λ_0, the results are mixed, at best. That is, sometimes the approximation works very well and sometimes it fails completely. This is due largely to the inaccurate nature of the tangent plane approximation when the wavelength and the size of the scatterer become comparable.

This result essentially means that we can expect the TP/PO approximations to be accurate when all dimensions of the rough surface are large compared to a wavelength. For a rough surface which is a second order process, this means large rms height and correlation length. However, if the TP/PO approximation is to predict accurate scattering in *all* directions we must enforce an additional condition. In the spectral representation of the surface roughness, there can be no components whose spatial wavelengths are comparable to or less than the electromagnetic wavelength (Brown, 1985). This condition must be imposed because we know that the TP/PO approximation breaks down for these small wavelength features. Finally, in order that there be a very small possibility for rays from one part of the surface to reflect to another part (multiple scattering in the optical limit), it is necessary for the surface to have small rms slopes.

This is certainly a long list of conditions for the TP/PO approximation to yield an accurate result in *all scattering directions*. It is possible, however, to relax some of the conditions if we are willing to use the TP/PO approximation to only predict the scattering in a specific direction [Bass et al., 1968; Brown, 1978]. For example, we might expect that the TP/PO approximation should be valid in the specular direction even if there is some small scale structure on the surface. This is because of the rather strong effect of the phase (in the specular direction) contained in the TP/PO approximation relative to other competing effects. In this instance, however, the actual surface parameters that go into the TP/PO scattering model are those corresponding to a *filtered surface* (Brown, 1978; Hagfors, 1967), i.e. a surface formed by filtering out all of the surface features having a horizontal extent that is the order of or smaller than the electromagnetic wavelength. What this filtering action means is that in the specular direction these features have no effect on the scattering process. The reader will recognize that this is exactly the rationale for the composite surface model as it applies to scattering about the specular direction. (See Chapter 11.)

3. Specular Point Approach

We now have an expression for the scattered field, (50), and a reasonable idea of when the TP/PO approximation can be used with some degree of confidence. We are therefore in a position to compute the desired statistical moments of the scattered field.

3.1. SPATIAL INTEGRATION BEFORE AVERAGING

There are two ways to proceed at this point. We can either do the spatial integration first and then do the averaging or we can reverse this process. There are basic theorems in probability theory which, given the satisfaction of certain criteria, guarantee the equality of the two approaches, so why then would we proceed with the two approaches? The answer is that each method gives a little different insight into the problem and contains different levels of mathematical difficulty.

The first approach, in which we do the spatial integration first and then perform the averaging, starts out to be relatively simple from a mathematical point of view (Kodis, 1966).

It leads to a great deal of physical insight but ends up requiring averages which are very difficult to accomplish (Barrick, 1968). The technique was first developed by Kodis (Kodis, 1966) and then completed by Barrick (Barrick, 1969). Since k_o in the exponent of the integrand (50) is assumed to be very large, the major contributions to the integrand come from points where the phase is stationary, i.e.,

$$\nabla_{t_o}\left(\left[\hat{k}_s - \hat{k}_i\right]\cdot\vec{r}_o\right) = 0 \tag{51}$$

or

$$\zeta_{x_0} = \frac{\left(\hat{k}_s - \hat{k}_i\right)\cdot\hat{x}}{\left(\hat{k}_s - \hat{k}_i\right)\cdot\hat{z}} \tag{52a}$$

and

$$\zeta_{y_0} = \frac{\left(\hat{k}_s - \hat{k}_i\right)\cdot\hat{y}}{\left(\hat{k}_s - \hat{k}_i\right)\cdot\hat{z}} \tag{52b}$$

Although it is not immediately obvious, these are the values of slopes which will cause an incident ray traveling in the $\hat{k}_i$-direction to be reflected into the $\hat{k}_s$-direction. One way of demonstrating this is to substitute these values of slopes into the expression for the unit normal to the surface, i.e. (39), form $\hat{n}_o\cdot(-\hat{k}_i)$ and $\hat{n}_o\cdot\hat{k}_s$ and note that they are equal. These particular values of slopes (Barrick, 1968) will be called the specular slopes because they are properly oriented to specularly reflect the incident field in the $\hat{k}_s$ direction. This means that in the rest of the integrand in (50), all surface slopes are set equal to the specular values, as given by (52).

Completing the stationary phase integration yields

$$\vec{E}_s(\vec{r}) = 2\pi g(r)E_o(\bar{\bar{I}} - \hat{k}_s\hat{k}_s)\sum_{m=1}^{M}\left[\hat{k}_s \times \vec{p}_{e_m} + \vec{p}_{h_m}\right]\cdot$$

$$\frac{sgn\left(\zeta_{xx_{sm}}\zeta_{yy_{sm}} - \zeta^2_{xy_{sm}}\right)}{\sqrt{|\zeta_{xx_{sm}}\zeta_{yy_{sm}} - \zeta^2_{xy_{sm}}|}}\exp\left[jk_o(\hat{k}_s - \hat{k}_i)\cdot\vec{r}_m\right] \tag{53}$$

where the sum is over the m specular points on the surface; $\zeta_{xx_{sm}}$, $\zeta_{yy_{sm}}$, and $\zeta_{xy_{sm}}$ are the surface curvatures at the mth specular point, and $\vec{r}_m$ is the value of $\vec{r}_o$ at the mth specular point. The $\vec{p}$-functions are

$$\vec{p}_{e_m} = \left[(1-\rho_v)(\hat{a}\cdot\hat{v}_1)(\hat{N}_o\cdot\hat{k}_i)\hat{h}_1 + (1+\rho_h)(\hat{N}_o \times \hat{h}_1)(\hat{a}\cdot\hat{h}_1) \right]_m \tag{54a}$$

$$\vec{p}_{h_m} = \left[-(1-\rho_h)(\hat{a}\cdot\hat{h}_1)(\hat{N}_o\cdot\hat{k}_i)\hat{h}_1 + (1+\rho_v)(\hat{N}_o \times \hat{h}_1)(\hat{a}\cdot\hat{v}_1) \right]_m \tag{54b}$$

where the m-subscript denotes evaluation at the mth specular point The *sgn*-function is defined as 1 for a positive argument and -1 for a negative argument.

3.2. RELEVANT SURFACE STATISTICS

We can now average (53) to find the average scattered field or the square of the magnitude of (53) to find the incoherent power. However, before doing this it is worthwhile pointing out the physical implications of this result. Equation (53) is telling us that the contributions to the far field in the high frequency limit are coming from the specular points on the surface and the strength of the scattering is determined by the reflective properties of the flat surface and the curvature of the specular points (Kodis, 1966). If we were to square this and retain no cross terms, this would be what one would observe with sun light reflecting from a roughened sea surface. As we might expect, we are now back to a ray description of the process because we have gone to the high frequency or ray optic limit.

The random quantities in this result are the random surface heights ζ_m at the specular points, the number of specular points in the illuminated area, and the surface curvatures at the specular points. If we average the scattered field, we start by averaging over all surface heights at the specular points. These may not be distributed the same as ζ (Barrick, 1972), but it really does not matter because

$$\left[k_o \left(\hat{k}_s - \hat{k}_i \right) \cdot \hat{z} \right] \zeta_m >> 1 \tag{55}$$

and so the characteristic function for ζ_m is vanishingly small and thus

$$< \vec{E}_s(\vec{r}) > \approx 0 \tag{56}$$

Unfortunately, we cannot recover any better approximation than this and still be consistent with the high frequency nature of this technique.

If we square (53) we obtain

$$|\vec{E}_s(\vec{r})|^2 = \left(2\pi E_o \right)^2 |g|^2 \sum_{m=1}^{M} \sum_{n=1}^{M} \vec{F}_m \cdot \vec{F}_n^* \exp\left[jk_o \left(\hat{k}_s - \hat{k}_i \right) \cdot \left(\vec{r}_m - \vec{r}_n \right) \right] \tag{57}$$

where the $\vec{F}$-functions are obvious. Averaging this expression over the random heights results in the condition that we get a non-zero answer only when $m = n$; otherwise, we have the same type of situation that causes $< \vec{E}_s >$ to be zero. Thus,

$$< |\vec{E}_s(\vec{r})|^2 >_{\zeta_m} = (2\pi E_o)^2 |g|^2 \sum_{m=1}^{M} |\vec{F}_m|^2 \tag{58}$$

where

$$\vec{F}_m = \frac{\left(\bar{\bar{I}} - \hat{k}_s\hat{k}_s\right) \cdot \left(\hat{k}_s \times \vec{p}_{e_m} - \vec{p}_{h_m}\right)}{|\zeta_{xx_{sm}}\zeta_{yy_{sm}} - \zeta^2_{xy_m}|} \tag{59}$$

It can be shown that the denominator in (59) is equal to the magnitude of the reciprocal of the product of the principal radii of curvature at the specular points (Kodis, 1966), i.e.

$$|\zeta_{xx_{sm}}\zeta_{yy_{sm}} - \zeta^2_{sy_{sm}}| = |R_{1_s}R_{2_s}|^{-1} \tag{60}$$

Substituting this result in (58) and completing the average yields

$$< |\vec{E}_s|^2 > = (2\pi E_o)^2 |g|^2 M \, |\vec{\mathcal{R}}|^2 < |R_{1_s}R_{2_s}| > \tag{61}$$

where

$$|\vec{\mathcal{R}}|^2 = \left| \left(\bar{\bar{I}} - \hat{k}_s\hat{k}_s\right) \cdot \left(\hat{k}_s \times \vec{p}_{e_m} - \vec{p}_{h_m}\right) \right|^2 \tag{62}$$

The scattering cross section per unit scattering area is defined as

$$\sigma^0 = \lim_{\substack{A \to \infty \\ R \to \infty}} \left\{ \frac{4\pi R}{AE_o} < |\vec{E}_s|^2 > \right\} \tag{63}$$

so substituting (61) into (63) gives

$$\sigma^0 = \pi \left(\frac{M}{A} \right) < |R_{1_s}R_{2_s}| > |\mathcal{R}|^2$$

so defining m_A as the number of specular points per unit area yields

$$\sigma^0 = \pi m_A < |R_{1_s}R_{2_s}| > |\vec{\mathcal{R}}|^2 \tag{64}$$

which is the result obtained by Barrick (1968).

The difficulty comes in when attempting to find m_A and $< | R_{1_s} R_{2_s} | >$ because these are not trivial quantities to compute. Barrick (1968) calculates these quantities approximately to find*

$$\sigma^0 = \pi sec^2 \gamma \, p_{\nabla \zeta}(\zeta_{x_s}, \zeta_{y_s}) | \mathscr{R} |^2 \tag{65}$$

where $\tan^2\gamma = \zeta_{x_s}^2 + \zeta_{y_s}^2$ and γ is the angle between the normal to the mean surface and the normal at the specular point. The function $p_{\nabla\zeta}(\zeta_{x_s}, \zeta_{y_s})$ is the slope density function evaluated at the specular point slopes. We have chosen to ignore the shadowing effects, but these can be accounted for by the use of a shadowing function (Sancer, 1969).

The primary result here is that the scattering cross section of the rough surface is directly proportional to the number of specular points on the surface and the probability of occurrence of the specular point slopes. It should be noted that if the specular points had added coherently, σ^0 would have varied with $(m_A)^2$.

4. Correlation Function Approach

The specular point approach is very physical in its content but it leads us to needing averages of discrete events on the surface. There is, however, an alternate approach which appears to reverse this situation. That is, the physics are somewhat more obscure but the computations are generally easier to accomplish. This latter approach is usually called the correlation function approach (Hagfors, 1964; Barrick, 1969).

4.1. AVERAGE BEFORE SPATIAL INTEGRATION

The expression for $\vec{E}_s$ in (50) may be rewritten as follows;

$$\vec{E}_s(\vec{r}) = -jk_o g(r) E_o \left(\bar{\bar{I}} - \hat{k}_s \hat{k}_s \right)$$

$$\cdot \int \left[\hat{k}_s \times \vec{p}_e + \vec{p}_h \right] I_2(\vec{r}_o) \exp \left(jk_o \left[\hat{k}_s - \hat{k}_i \right] \cdot \vec{r}_o \right) dx_o dy_o \tag{66}$$

where p_e and p_h are defined as in (54) but with the m-subscript. The average of this quantity requires an average of the quantity

$$I = < \left[\hat{k}_s \times \vec{p}_e + \vec{p}_h \right] I_2(\vec{r}_o) \exp \left(j \left[k_{sz} - k_{iz} \right] \zeta \right) > \tag{67}$$

*There is a $\sec^2\gamma$ contained in $| \mathscr{R} |^2$ so this is why this result is not identical in appearance to Barrick's result.

If the height and slopes are independent, this averaging can be done rather straightforwardly. If independence is not the case, the averaging becomes more difficult. For independent height and slopes, we first average over the height to give

$$< \exp(j(k_{sz} - k_{iz})\zeta)p > \; = \Phi\left(\left[k_{sz} - k_{iz}\right]\right) \tag{68}$$

which is just the characteristic function we encountered with the randomly elevated plane. We now have

$$I = < \left[\hat{k}_s \times \vec{p}_e + \vec{p}_h\right] I_2(\vec{r}_o)> \; \Phi\left(k_{sz} - k_{iz}\right) \tag{69}$$

It will be recalled that I is unity if neither an incident ray is blocked or a scattered ray in the direction $\hat{k}_s$ is blocked after the incident ray strikes the surface at $\vec{r}_o$. Thus, to complete the average of (69) we need the density $p(\zeta_x, \zeta_y, I_2)$ which we write as (after Sancer, 1969)

$$p(\zeta_x, \zeta_y, I_2) = p(\zeta_x, \zeta_y)p(I_2 \mid \zeta_x, \zeta_y) \tag{70}$$

where $p(I_2 \mid \zeta_x, \zeta_y)$ is the conditional probability density. This conditional density can be written as (Saucer, 1969),

$$p(I_2 \mid \zeta_x, \zeta_y) = P_2(\hat{k}_i, \hat{k}_s \mid \zeta_x, \zeta_y)\delta(I_2 - 1)$$

$$+ [1 - P_2(\hat{k}_i, \hat{k}_s \mid \zeta_x, \zeta_y)]\delta(I_2) \tag{71}$$

where $P_2(\hat{k}_i, \hat{k}_s \mid \zeta_x, \zeta_y)$ is the probability that a point having slopes ζ_x and ζ_y will be illuminated by an incident ray having the direction $\hat{k}_i$ and the resulting scattered ray in the direction $\hat{k}_s$ will not strike the surface. Substituting (71) in (70) and then into (69) yields

$$I = \Phi(k_{sz} - k_{iz}) \cdot \iint [\hat{k}_s \times \vec{p}_e + \vec{p}_h] P_2(\hat{k}_i, \hat{k}_s \mid \zeta_x, \zeta_y)$$

$$\cdot p(\zeta_x, \zeta_y) d\zeta_x d\zeta_y \tag{72}$$

If the slopes are small, the slope density function is highly peaked about $\zeta_x = 0$ and $\zeta_y = 0$ the approximate evaluation of (72) yields

$$I \approx \Phi(k_{sz} - k_{iz})[\hat{k}_s \times \vec{p}_e + \vec{p}_h]_{\substack{\zeta_x \to 0 \\ \zeta_y \to 0}} P_2(\hat{k}_i, \hat{k}_s \mid 0, 0) \tag{73}$$

and so $< \vec{E}_s >$ becomes

$$<\vec{E}_s(\vec{r})> \; = \; -jk_o g(r) E_o (\bar{\bar{I}} - \hat{k}_s \hat{k}_s) \cdot \left[\hat{k}_s \times \vec{p}_e + \vec{p}_h \right]_{\substack{\zeta_x = 0 \\ \zeta_y = 0}} P_2(\hat{k}_i, \hat{k}_s \,|\, 0, 0)$$

$$\cdot \Phi(k_{sz} - k_{iz}) \delta(k_{sx} - k_{ix}) \delta(k_{sy} - k_{iy}) \tag{74}$$

The delta functions are one way of indicating that the average scattered field is specularly reflected from the rough surface (Brown, 1983). It should be noted that if the slopes are not small, the integrations in (72) over ζ_x and ζ_y will have to be carried out more rigorously.

The second moment of $\vec{E}_s$ may be written as follows;

$$< |\vec{E}_s|^2 > \; = \; k_o^2 |g|^2 E_o^2 \iint \; < \; \left[\hat{k}_s \times \vec{p}_e + \vec{p}_h - (\hat{k}_s \cdot \vec{p}_h) \hat{k}_s \right] \cdot \tag{75}$$

$$\left[\hat{k}_s \times \vec{p}_e^{\,'} + \vec{p}_h^{\,'} - (\hat{k}_s \cdot \vec{p}_h^{\,'}) \hat{k}_s \right]^* I_2(\vec{r}_o) I_2(\vec{r}_o^{\,'}) \exp[jk_o(\hat{k}_s - \hat{k}_i)(\vec{r}_o - \vec{r}_o^{\,'})\,] > dx_o dy_o dx_o^{'} dy_o^{'}$$

There are *eight* random variables contained within the average symbol e.g. $\zeta_{xo}, \zeta_{yo}, \zeta_{xo}^{'}, \zeta_{yo}^{'}, I_2(\vec{r}_o), I_2(\vec{r}_o^{\,'}), \zeta$, and $\zeta^{'}$. We first assume that the two points $\vec{r}_o$ and $\vec{r}_o^{\,'}$ are not too far apart when contributing to the integral so that

$$I_2(\vec{r}_o) I_2(\vec{r}_o^{\,'}) \approx I_2^2(\vec{r}_o) \tag{76}$$

so this reduces the content of the integral to one shadowing function. In principle, the remaining averages in (75) can be accomplished assuming that we know the joint probability density function for the heights, slopes, and shadowing function, $p(\zeta_o, \zeta_o^{'}, \nabla_t \zeta_o, \nabla_t \zeta_o^{'}, I_2)$. With

$$\tilde{\tilde{p}}\,([k_{sz} - k_{iz}], -[k_{sz} - k_{iz}], \nabla_t \zeta_o, \nabla_t \zeta'_o, I_2)$$

$$= \iint p(\zeta_o, \zeta_o^{'}, \ldots) \exp(j(k_{sz} - k_{iz})(\zeta_o - \zeta_o^{'})) d\zeta_o \, d\zeta_o^{'} \tag{77}$$

the average in (75) becomes

$$< |\vec{E}_s|^2 > = k_o^2 |g|^2 E_o^2 \int \ldots \int \left[\hat{k}_s \times \vec{p}_e + \vec{p}_h - (\hat{k}_s \cdot \vec{p}_h) \hat{k}_s \right] \cdot \left[\hat{k}_s \times \vec{p}_e^{\,'} + \vec{p}_h^{\,'} - (\vec{k}_s \cdot \vec{p}_h^{\,'}) \hat{k}_s \right]^*$$

$$I_2^2(\vec{r}_o^{\,'}) \tilde{\tilde{p}}\,(\Delta k_z, -\Delta k_z, \nabla_t \zeta_o, \nabla_t \zeta_o^{'}, I_2) \exp \left[j[\vec{k}_s - \vec{k}_i] \cdot (\vec{r}_{t_o} - \vec{r}_{t_o}^{\,'}) \right]$$

$$dx_o dy_o dx_o^{'} dy_o^{'} d\nabla_t \zeta_o d\nabla_t \zeta_o^{'} dI_2 \tag{78}$$

where $\nabla_t \zeta_o$ and $\nabla_t \zeta_o'$ denote the surface slopes and

$$\Delta k_z = k_{sz} - k_{iz}$$

$$(\vec{r}_{t_o} - \vec{r}_{t_o}') = (x_o - x_o')\hat{x} + (y_o - y_o')\hat{y}$$

When $|\vec{r}_{t_o} - \vec{r}_{t_o}'| \to \infty$, all the stochastic variables decorrelate and we assume that this is equivalent to statistical independence so that

$$\lim_{|r_{t_o} - r_{t_o'}| \to \infty} \overset{\approx}{p} = \Phi(\Delta k_z) \cdot \Phi(-\Delta k_z) p(\nabla_t \zeta_o, I_2) p(\nabla_t \zeta_o', I_2) \tag{79}$$

and we recognize (78) to be the square of the coherent field in this limit. Thus, we can write

$$< |\vec{E}_s|^2 > - |<\vec{E}_s>|^2 = k_o^2 |g|^2 E_o^2 \int \ldots \int \vec{\mathcal{R}}(\nabla_t \zeta_o) \cdot \vec{\mathcal{R}}(\nabla_t \zeta_o')^* I_2^2(\vec{r}_o')$$

$$\left\{ \overset{\approx}{p} (\Delta k_z, -\Delta k_z, \nabla_t \zeta_o, \nabla_t \zeta_o', I_2) - \lim_{|\Delta \vec{r}_{t_o}| \to \infty} \overset{\approx}{p} \right\} \exp \left[j(\vec{k}_s - \vec{k}_i) \cdot \Delta \vec{r}_{t_o} \right]$$

$$dx_o dy_o dx_o' dy_o' d \nabla_t \zeta_o d \nabla_t \zeta_o' dI_2 \tag{80}$$

where $\Delta \vec{r}_{t_o} = \vec{r}_{t_o} - \vec{r}_{t_o}'$. The remaining slope integrations are very difficult to accomplish because the $\vec{\mathcal{R}}$ functions are complicated functions of slopes; the $\vec{\mathcal{R}}$ are

$$\vec{\mathcal{R}}(\nabla_t \zeta_o) = \left[\hat{k}_s \times \vec{p}_e + \vec{p}_h - (\hat{k}_s \cdot \vec{p}_h) \hat{k}_s \right] \tag{81a}$$

$$\vec{\mathcal{R}}^*(\nabla_t \zeta_o') = \left[\hat{k}_s \times \vec{p}_e' + p_h' - (\hat{k}_s \cdot \vec{p}_h') \hat{k}_s \right]^* \tag{81b}$$

There are a number of different approximations that have appeared in the literature for how to deal with the slopes (Beckmann and Spizzichino, 1963). However, the one which can be rigorously justified in the high frequency limit is the one which essentially replaces the slopes in the $\vec{\mathcal{R}}$-functions by their value which reflects an incident ray traveling in the direction $\hat{k}_i$ to one traveling in the $\hat{k}_s$ direction, i.e. their specular reflection value (Sancer, 1969). That is, (80) can be written as

$$< |\vec{E}_s|^2 > - |<\vec{E}_s>|^2 = k_o^2 |g|^2 E_o^2 |\vec{\mathcal{R}}(\nabla_t \zeta_{sp})|^2 P_2(\hat{k}_i, \hat{k}_s | \nabla_t \zeta_{sp})$$

$$\cdot \int \ldots \int \left\{ \Phi_2(\Delta k_z, -\Delta k_z; \Delta \vec{r}_{t_o}) - \Phi_1(\Delta k_z) \Phi_1(-\Delta k_z) \right\} \cdot \exp \left[j(\vec{k}_s - \vec{k}_i) \cdot \Delta \vec{r}_{t_o} \right] dx_o dy_o dx_o' dy_o' \tag{82}$$

where Φ_2 is the joint characteristic function for two surface heights separated by a distance $\Delta\vec{r}_{t_o}$ and Φ_1 is the single point characteristic function. Converting to a difference coordinate permits the reduction of two spatial integrations, i.e.

$$<|\vec{E}_s|^2> - |<\vec{E}_s>|^2 = k_o^2 A\,|g\,|^2 E_o^2|\,\mathcal{R}(\nabla_t\zeta_{sp})|^2\,P_2(\hat{k}_1,\hat{k}_s\,|\,\nabla_t\zeta_{sp})$$

$$\cdot\iint\left\{\Phi_2(\Delta k_z, -\Delta k_z;\Delta\vec{r}_{t_o}) - \Phi_1(\Delta k_z)\Phi_1(-\Delta k_z)\right\}\cdot\exp\left[j(\vec{k}_s-\vec{k}_i)\cdot\Delta\vec{r}_t\right]d\Delta x_o\,d\Delta y_o \quad (83)$$

where A is the illuminated area.

4.2. HIGH FREQUENCY LIMIT AND THE SURFACE STATISTICS

The evaluation of (83) has led to a great deal of controversy in the literature (Peake, et al., 1970). The primary point of contention centers on whether an evaluation of (83) in *other than* the high frequency limit has any real meaning. For example, we know that the TP/PO approximation is strictly only valid for a very gently undulating surface, the high frequency limit, or in the specular direction. Furthermore, we have already replaced the slope-dependent functions by their values at the specular slopes which is certainly a high frequency approximation. Thus, it would seem that to be self consistent, we should either

(a) go to the high frequency limit, or

(b) not go to this limit but restrict the observation region to the vicinity of the specular direction.

In the high frequency limit,

$$\Phi_1(\Delta k_z) \rightarrow 0$$

$$|<\vec{E}_s>|^2 \rightarrow 0$$

and Barrick (1968) has shown that

$$\lim_{k_o\to\infty}\iint\Phi_2(\Delta k_z, -\Delta k_z;\Delta\vec{r}_t)\exp\left[j\Delta\vec{k}\cdot\Delta\vec{r}_t\right]d\Delta\vec{r}_t = \frac{4\pi^2}{(\Delta k_z)^2}p_{\nabla\zeta}(\zeta_{x_s},\zeta_{y_s}) \quad (84)$$

where

$$\zeta_{x_s} = \frac{\Delta k_x}{\Delta k_z} \qquad \zeta_{y_s} = \frac{\Delta k_y}{\Delta k_z} \quad (85)$$

and

$$\Delta k_x = k_o (k_{sx} - k_{ix})$$

$$\Delta k_y = k_o (k_{sy} - k_{iy})$$

$$\Delta k_z = k_o (k_{sz} - k_{iz})$$

Substituting these results into (83) and then converting these results to σ^o yields

$$\sigma^o (\hat{k}_i , \hat{k}_s) = \pi \, sec^2 \gamma \, p_{\nabla \zeta}(\zeta_{x_s}, \zeta_{y_s}) \, | \vec{\mathcal{R}} |^2 \tag{86}$$

which is the same result as obtained with the specular point approach.

5. Comparisons Between the Two Approaches

The above analysis showns that both methods agree in the high frequency limit and this is as it should be because this is when the TP/PO approximation must hold. There is no way to back-off from this limit in the specular point approach except to use "effective" surface parameters and limit the direction of observation to the vicinity of specular scattering. For example, rather than use the actual slope variances of a surface which may not obey the TP/PO approximations, we could use "filtered" values and restrict the use of (65) to the near specular region. That is, we filter the surface with a low pass filter until the filtered surface does satisfy the approximations of TP/PO and we use the slope variances of the filtered surface.

We can do exactly the same thing with the correlation function approach. However, it is also tempting to do more! That is, it is tempting to go back to (83) and perform this integration (perhaps numerically) for other than a large Rayleigh parameter. It should be noted that this has been done in the literature (Fung and Moore, 1966; Fung and Chan, 1969). The important question here is whether or not this result has meaning! For example, it is certainly at odds with the assumptions in the TP/PO approximation and it is not consistent with the way the slope dependent term was evaluated (Peake, et al., 1970). However, there may be specific instances when a non-asymptotic evaluation of (83) does indeed produce an accurate result. The problem is that we do not know when to trust this result!

As far as the coherent or average scattered field goes, the specular point theory predicts that it is vanishingly small in the high frequency limit. The correlation function method agrees with this limiting result but also provides non-asymptotic results. In fact, (74) is essentially the same as the result for reflection from a randomly elevated plane, i.e. (27). The only difference is the appearance of the shadowing function in (74) and it is negligible for small slopes except near grazing incidence. What this means is that (74) ignores the horizontal surface structure and treats the surface as a randomly elevated plane. Experimental results tend to indicate that (74) is reasonably accurate for sea surface scattering as long as (Beard, 1961).

$$k_o \sqrt{<\zeta^2>}(\cos\theta_s + \cos\theta_i) \lesssim 1 \qquad (87)$$

That is, once (87) is violated, the horizontal surface structure appears to become important.

6. Application of Quasi Specular Scatter to the Ocean

Unlike wide angle scattering from the ocean surface which appears to be highly dependent on the small scale surface waves (comparable in wavelength to the electromagnetic wavelength), quasi-specular scatter is more dependent on the larger scale waves (Brown, 1978). This means that quasi specular scatter provides a more direct measurement of the large scale structure on the surface.

6.1. COHERENT SCATTERING

Not much has been done with using coherent scattering as a remote sensing tool because of the difficulty of maintaining phase coherency in the transmitter-path-receiver chain. For example, it is possible, in principle, to measure the rms surface height of the ocean by making a number of measurements at different frequencies of the average reflected field. These measurements could then be fitted to some *a priori* form of Φ, as in (74), to obtain estimates of $<\zeta^2>$. The difficulties with this method are, however, very significant. Complete phase coherence in the system and the path* must be maintained. Thus, any movement of the sensors must be avoided. Unless the measurement is done at normal incidence, separate receivers and transmitters must be used. The dynamic range of such a technique is severely limited unless the sensors have the ability to go through some fairly large frequency changes.

For shipborne tracking radars, coherent reflections can lead to very severe problems. Generally, when the target gets very close to the water surface but at a fair distance from the radar, the radar can actually lock up on the image of the target. This is due to the radar energy striking the target, reflecting off the target and then coherently reflecting off the ocean surface back to the radar. If we trace the reflected ray back through the water it will appear to come from an image of the target (through the water surface). Thus, the radar never appears to receive energy directly from the target but rather from its image.

6.2. INCOHERENT SCATTERING

Probably the most interesting application to make use of quasi-specular scattering is short pulse radar altimetry (Barrick, 1972; Brown, 1977). Such radars usually operate within one or two degrees of normal incidence, over the ocean, and in the GHz frequency band. Except for glassy calm seas, all the assumptions in the TP/PO approximation are or can be satisfied. That is, even though the ocean surface contains waves which have wavelengths smaller than the electromagnetic wavelength, we can deal with a filtered surface because we are only

---

*By the term path, we mean the electrical distance from the transmitter to the *mean* surface and then to the receiver.

interested in the normal incidence backscatter (note that at normal incidence, the specular direction is the backscatter direction).

A remote sensing, short pulse, radar altimeter measures three independent quantities-the backscattered power at normal incidence, the shape of the average backscattered pulse, and the altitude from the radar to the sea surface. Each of the measurements, within the framework of quasi-specular scattering, provides information on a different surface characteristic. The power measurement leads to an inferrence of the surface wind speed. The pulse shape measurements lead to estimates of the wave height on the sea surface. Finally, the altitude measurement has the potential of providing estimates of large scale ocean circulation patterns.

Of the three measurement techniques, the two most closely coupled to the quasi-specular scattering theory are the power and waveform measurements, and they will be discussed below.

6.2.1. *Cross Section Dependence on Wind Speed.* According to the result in (86), the normal incidence value of the backscattering cross section per unit area for the sea surface is given by

$$\sigma^o(0°) = \alpha \frac{|\rho(0)|^2}{<(\nabla\zeta)^2>_f} \tag{88}$$

where $\rho(0)$ is the Fresnel reflection coefficient for normal incidence*, $<(\nabla\zeta)^2>_f$ is the mean square slope of a filtered version of the true surface, and α is a constant (Brown, 1979). The filtering is necessary to remove any structure on the surface which is not amenable to the TP/PO assumption. The constant α is dependent on the form of the slope density function; for a Gaussian density $\alpha = 1$ while for a Laplacian or double-sided exponential $\alpha = 3$. If we assume that the filtered *surface* has Gaussian slopes then only the reflection coefficient and the filtered surface mean square slope remain unknown. To date, all spaceborne radar altimeters have operated in the vicinity of 14 GHz and at this frequency (Matthews, 1975).

$$-2.37 \text{ dB} \lesssim 10\log_{10}|\rho(0)|^2 \lesssim -2.08 \text{ dB}$$

so the reflection coefficient is essentially known. Thus, combining the power measurements from the radar along with the radar equation, we can obtain estimates of $\sigma^o(0°)$ and, hence, $<(\nabla\zeta)^2>_f$.

Figure 7 is a plot of $<(\nabla\zeta)^2>_f$ versus wind speed as inferred from 184 data points obtained from the GEOS-3 radar altimeter (Brown, et al., 1981). Also shown are linear and logarithmic curve fits to the Cox and Munk [1954] measurements of mean squared slope using optical glitter photography. It is interesting to note that microwave measured slopes are greater than the optically measured results. Of course, one would expect that this situation should be reversed because of the effective filtering with the microwave-derived values. We should note that this same situation was observed with SEASAT and GEOSAT data. The discrepancy is too large to be due to an error in $|\rho(0)|^2$ but it might possibly be due to the different sensitivities of the two techniques to the presence of foam and surfactants.

Figure 8 backs the process up one step and plots $\sigma^o(0°)$ vs. wind speed at the standard 10 m anemometer height. These data were obtained from overflights of NOAA data buoy (Brown, et al., 1981) and the figure shows that a measurement of $\sigma^o(0°)$ can be used to infer

*We make no distinction between vertical or horizontal polarization because at normal incidence $|\rho_v| = |\rho_h|$.

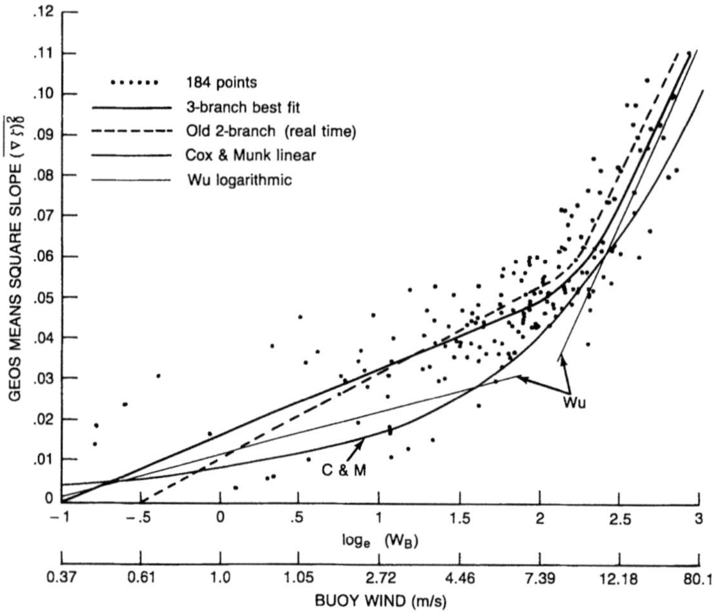

Fig. 7 — Radar derived mean square slope as a function of wind speed

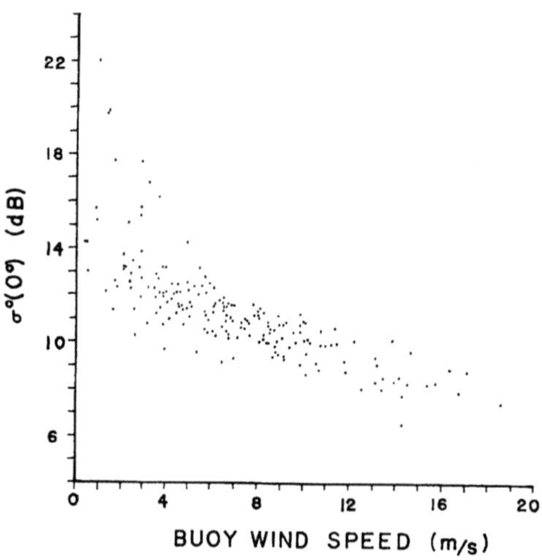

Fig. 8 — Normal incidence cross section as a function of wind speed

surface wind speed beneath the altimeter. The precision of this estimation technique appears to be about 1.7 m/s (Brown, et al., 1981; Dobson, et al., 1987).

6.2.2. *Surface Roughness Estimation.* There are a number of ways for describing how a short pulse radar can measure the rms height of ocean waves. We shall present a relatively simple mathematical derivation because it emphasizes the physics of the actual process.

Figure 9 illustrates a radar at an altitude h above the surface transmitting a pulse of width T. In (a) the pulse is just striking the surface; in (b) it is illuminating a circular area; in (c) it is illuminating an *annular* area. For schematic purposes, we have shown the incident pulse penetrating the surface while the scattered pulse is the dashed curved line. For incoherent scattering, the backscattered power increases linearly with illuminated area [see Eq. (83)]. Thus, a plot of illuminated area, corresponding to the situation in Figure 9, is shown in Figure 10 along with its value corresponding to (a), (b), and (c) in Figure 9. If there were no other effects due to, say, antenna pattern, σ^o variation, surface roughness, and range-to-surface increase, Figure 10 would be a plot of the average backscattered waveform as a function of time but with h/c replaced by $2h/c$ (to account for the two-way travel of the pulse). That is, the radar would record the illuminated area. Now what we need to do is to go back and include the effects which were omitted from Figure 10. We will do this by starting with the radar range equation for the received power because we are dealing with incoherent scattering and, hence, can add powers rather than voltages.

The radar range or power balance equation describes the power received back at the radar antenna after being scattered by the surface. It may be written as follows:

$$P_r(t) = \sum_{m=1}^{M'} A \cdot B \cdot C \cdot D \tag{89}$$

where

$$A = \frac{P_t(t - 2r_m/c)G(\theta_m, \phi_m)}{4\pi \, r_m^2}$$

$$B = \sigma_m$$

$$C = \frac{1}{4\pi \, r_m^2}$$

$$D = \frac{G(\theta_m, \phi_m)\lambda^2}{4\pi}$$

The upper limit M' on the summation is determined by the number of facets or specular points illuminated by the pulse at time t. The various terms in (89) have been lettered in order to explain how each one arises. The factor $P_t(t - 2r_m/c)$ is the transmitted modulation power delayed by the time it takes for the pulse to propagate to the mth facet and back to the radar antenna. The quantity $G(\theta_m, \phi_m)$ is the gain of the radar antenna at $\theta = \theta_m$ and

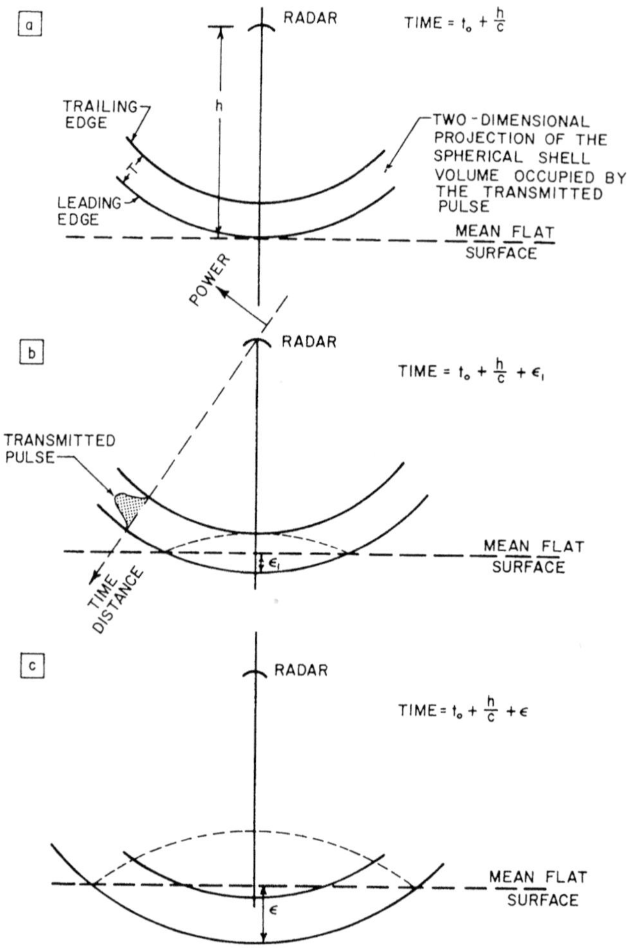

Fig. 9 — Two-dimensional view of pulse scattering

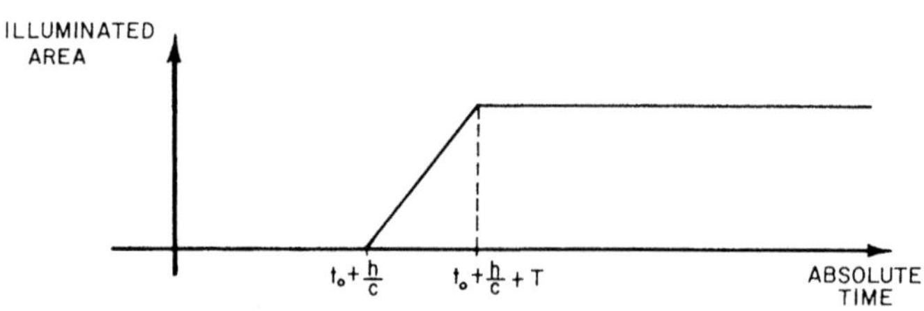

Fig. 10 — Illuminated area as a function of absolute time for a flat-top transmitted pulse of width T

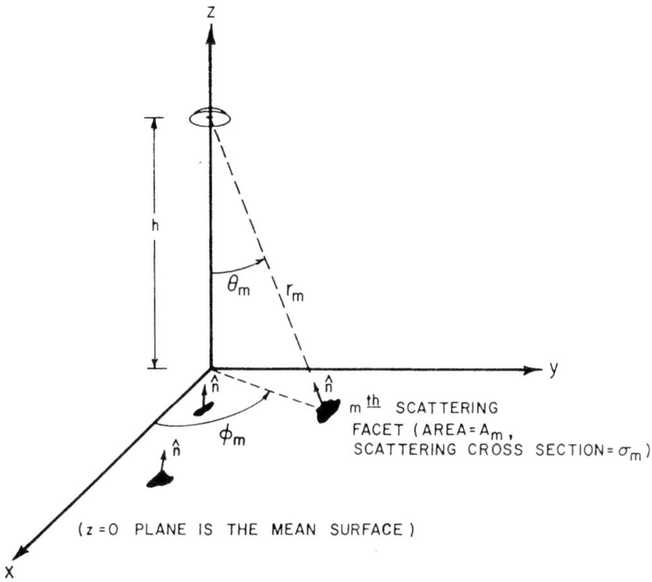

Fig. 11 — Facet or specular point scattering geometry

and we have assumed that the same antenna is used for transmitting and receiving (see Fig. 11). The term A is just the power per unit area incident upon the mth facet (with the appropriate time delay to account for the two-way travel path). The term B is the effective scattering cross section (or scattering area) of the mth facet. Note that the product of A and B yields the power incident upon the mth facet. The term C is just the free-space radiation loss incurred in the propagation from the mth facet back to the antenna. The last term D is the "capture" area of the radar antenna in the direction (θ_m, ϕ_m). In essence then, equation (89) is more or less just a power balance equation. Those who are used to dealing with deterministic (non-random) scattering problems may find (89) somewhat shocking in that we are adding power rather than voltage, e.g. if we add voltages and then square the result to get power we would have to consider the relative phases between the component voltages. However, it must be remembered that we are trying to obtain the *average* return power and since this is an incoherent scattering process, we can add powers because phase is not preserved.

Rearranging the terms in (89), we have

$$P_r(t) = \frac{\lambda_o^2}{(4\pi)^3} \sum_{m=1}^{M'} \frac{P_t\left(t - \dfrac{2r_m}{c}\right) G^2(\theta_m, \phi_m)}{r_m^4} \left(\frac{\sigma_m}{A_m}\right) A_m \tag{90}$$

where we have introduced the area A_m of the mth elemental scattering facet. Note that the quantity (σ_m / A_m) is the scattering cross section per unit scattering area. Using the assumption that there are a sufficiently large number of scatterers or facets (independent of the illuminated area), we can let the summation go to an integral, i.e.

$$P_r(t) = \frac{\lambda_0^2}{(4\pi)^3} \int_{\substack{Illum.\\ Area}} \frac{P_t\left[t - \frac{2\tilde{r}}{c}\right] G^2(\theta, \phi)}{\tilde{r}^4} \left[\frac{\sigma}{\delta A}\right] dA \tag{91}$$

where $\tilde{r}$ is the distance from the radar antenna to the point on the true surface at an angel (θ, ϕ). The term $(\sigma/\delta A)$ is the scattering cross section per unit scattering area of the elemental surface area dA; it is a function of the angle (θ, ϕ) and the local surface slope. From Figure 12, we can see that

$$\tilde{r} = r_o - \frac{\zeta(x, y)}{\cos\theta} \tag{92}$$

where $\zeta(x, y)$ is the height of the surface at (x, y) and r_o is the distance from the antenna to the mean flat surface. For θ small,

$$\tilde{r} \approx r_o - \zeta(x, y)$$

and

$$P_r(t) \approx \frac{\lambda_o^2}{(4\pi)^3} \int_{\substack{Illum.\\ Area}} \frac{P_t\left[t - \frac{2r_o}{c} + \frac{2}{c}\zeta(x, y)\right] G^2(\theta, \phi)}{r_o^4} \left[\frac{\sigma}{\delta A}\right] dA \tag{93}$$

where we have ignored the $\zeta(x, y)$ term in the denominator, i.e. $\tilde{r}^{-4}$, since r_o is so much more dominant. We now average over the random quantities in (93), i.e. $\zeta(x, y)$ and $(\sigma/\delta A)$. To the order of approximation used in this model, these quantities are independent. That is, we are essentially assuming that the surface height and slope are independent random variables. Thus,

$$< P_t\left[t - \frac{2r_o}{c} + \frac{2}{c}\zeta(x, y)\right] \left[\frac{\sigma}{\delta A}\right] > \, = \, < P_t\left[t - \frac{2r_o}{c} + \frac{2}{c}\zeta(x, y)\right] > < \frac{\sigma}{\delta A} >$$

The quantity $<\sigma/\delta A>$ is the average scattering cross section per unit scattering area of the surface and it is a function of the surface and the angle of incidence (θ, ϕ). The first bracketed quantity on the r.h.s. is given by

$$< P_t\left[t - \frac{2r_o}{c} + \frac{2}{c}\zeta(x, y)\right] > \, = \int_{-\infty}^{\infty} P_t\left[t - \frac{2r_o}{c} + \frac{2}{c}\zeta\right] p_\zeta(\zeta) d\zeta \tag{94}$$

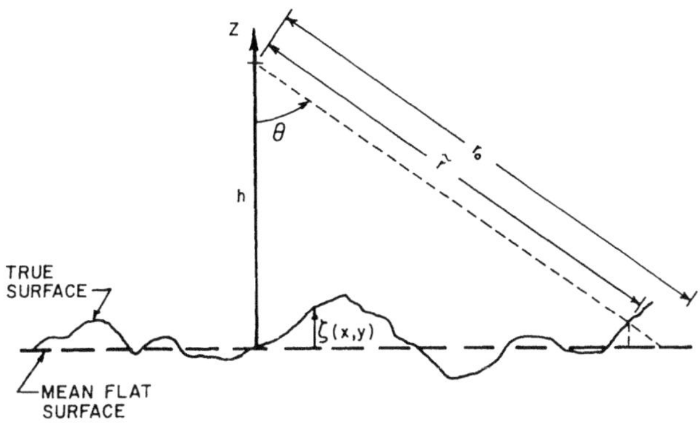

Fig. 12 — Local surface height geometry

where $p_\zeta(\zeta)$ is the surface height probability density function. We note from the form of (94), that this is just a convolution integral (with a sign reversal), thus we can write

$$<P_r(t)> = \frac{\lambda_o^2}{(4\pi)^3} \int_{\substack{Illum.\\ Area}} \frac{\left[P_t\left[t - \frac{2r_o}{c} + \frac{2}{c}\zeta \right] cp_\zeta(\zeta) \right]}{r_o^4} \cdot G^2(\theta,\phi)\sigma^o(\theta,\phi)dA \qquad (95)$$

where $<P_r(t)>$ is the average return power* as a function of time, the symbol © denotes convolution, and $\sigma^o(\theta,\phi)$ is the average scattering cross section per unit scattering area (of the surface) at an angle of incidence (θ,ϕ). Note that the integration is now over the mean flat surface rather than the random surface. For the flat surface, the element of area can be taken equal to $r_o dr_o d\phi$ and the above expression reduces to the following form;

$$<P_r(t)> = \frac{\lambda_o^2}{(4\pi)^3} \int_{\substack{Illum\\ r_o}} \left[P_t\left[t + \frac{2\zeta}{c} - \frac{2r_o}{c} \right] ©p_\zeta(\zeta) \right] \cdot F(r_o)dr_o \qquad (96)$$

where

$$F(r_o) = \frac{1}{r_o^3} \int_o^{2\pi} G^2(\theta,\phi)\sigma^o(\theta,\phi)d\phi \qquad (97)$$

Using the property of the delta function ($\delta(x)$ or unit impulse) that

$$\int_{-\infty}^{\infty} q(x)\delta(x - x_o)dx = q(x_o) \ ,$$

*Averaged over all possible configurations of the random surface.

we can rewrite (97) in the following form;

$$< P_r(t) > = \frac{\lambda_o^2}{(4\pi)^3} \int_{t'} \int_{r_o} \left[P_t \left(t + \frac{2\zeta}{c} - t' \right) p_\zeta(\zeta) \right]$$

$$\left[\delta \left(t' - \frac{2r_o}{c} \right) F(r_o) \right] dr_o dt' \tag{98}$$

If we define $P_{FS}(t')$ such that

$$P_{FS}(t') = \frac{\lambda_o^2}{(4\pi)^3} \int \delta \left(t' - \frac{2r_o}{c} \right) F(r_o) dr_o \tag{99}$$

then equation (98) becomes a double convolution, i.e.

$$< P_r(t) > = P_t(t) © p_\zeta(\zeta) © P_{FS}(t) \tag{100}$$

Equation (100) is a complete description of the average return power in terms of the transmitted waveform $P_t(t)$, the surface height density function $p_\zeta(\zeta)$, and the function $P_{FS}(t)$. Combining equations (99) and (97), we see that the equation for $P_{FS}(t)$ is given by the following:

$$P_{FS}(t) = \frac{\lambda_o^2}{(4\pi)^3} \iint \frac{\delta \left(t - \frac{2r_o}{c} \right)}{r_o^4} \cdot G^2(\theta, \phi) \sigma^o(\theta, \phi) dA \tag{101}$$

Comparing this result with equation (95), we see that an apt description for $P_{FS}(t)$ might be the *flat surface impulse response*. That is, if we ignored the surface roughness and transmitted an impulse, (101) would describe the average return power as a function of delay. Implicit in (101) is the fact that $P_{FS}(t) = 0$ for $t < \frac{2h}{c}$ since $t = 2h/c$ represents the shortest possible round trip delay between the antenna and the flat surface. If $\sigma^o(\theta, \phi)$ were independent of the angle coordinates, i.e. a constant, and if we used a very broadbeam antenna, then $P_{FS}(t)$ would be a unit step function (such as shown in Figure 13) with an eventual decay due to the r_o^{-4} term. When antenna and/or σ^o effects are included, the behavior of $P_{FS}(t)$ is somewhat like the dashed line in Figure 13. Apart from the contribution of the surface height density function, the average return power is determined by the surface area as weighted by the pulse shape, the antenna pattern, and the per unit area scattering cross section.

It is now possible to see how we might "measure" the probability density function of the surface height. For example, if we consider the case of a broadbeam antenna pattern and short enough pulse so that σ^o does not vary over the spatial extent of the pulse, the flat surface impulse response would be the step function shown in Figure 13. If the transmitted pulse is rectangular with width T, the convolution of it and the flat surface impulse response is the

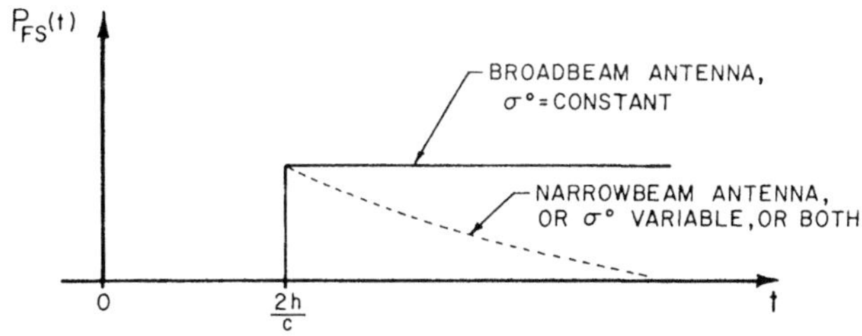

Fig. 13 — Flat surface impulse response

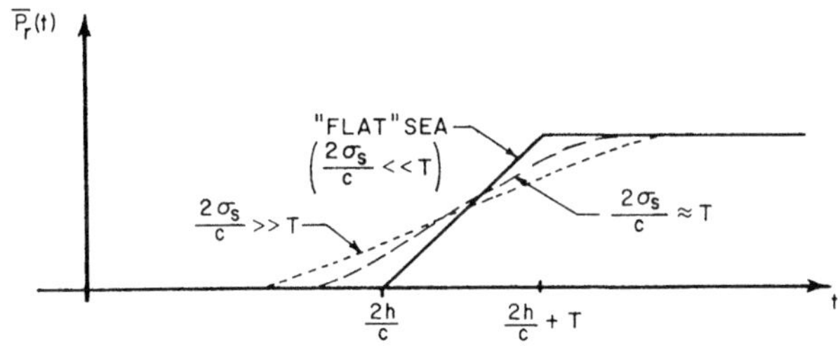

Fig. 14 — Average return power waveform as a function of rms surface height

solid curve shown in Figure 14. Note that the plot of $< P_r(t) >$ versus time essentially portrays how the illuminated area varies with time (see Figure 10). If we now include the surface height probability density function, the character of the leading edge of the return will change. In particular, the rise time of the leading edge will increase. For example, if $<\zeta^2>^{1/2}/c$ is much smaller than the pulse width we will see little change in the leading edge ($<\zeta^2>^{1/2}$ = rms height of the surface). When $<\zeta^2>^{1/2}/c \approx T$, the leading edge rise time will increase to something on the order of the large dashed curve in Figure 14. When $<\zeta^2>^{1/2}/c >> T$, the small dashed curve applies. Thus we note that the leading edge of the average return power as a function of time contains information on the probability density function of the surface height. As should be obvious, the plateau portion of the return contains information on the variation of $\sigma^o(\theta)$.

Thus, by measuring the stretching of the leading edge of the average return waveform, it is possible to directly infer the rms wave height. The precision of the measurement of wave height using this method is generally better than 10% of the significant wave height (Dobson, et al., 1987).

7. REFERENCES

Barrick, D.E. and W.H. Peake, ''A Review of Scattering from Surfaces with Different Roughness Scales,'' *Radio Science,* **3**, pp. 865-868, 1968.

Barrick, D.E., "Relationship Between Slope Probability Density Function and the Physical Optics Integral in Rough Surface Scattering," *Proc. IEEE* (Letters), **56**, pp. 1728-1729, 1968. (also see correction, vol. 57, pg. 256, 1969).

Barrick, D.E., "Remote Sensing of Sea State by Radar," in *Remote Sensing of the Troposphere,* ed. by V.F. Derr, Ch. 12, U.S. Govt. Printing Office, Washington, DC, 1972.

Barrick, D.E., "Rough Surface Scattering Based on the Specular Point Theory," *IEEE Trans. Antennas and Propagation,* **AP-16**, pp. 449-454, 1968.

Barrick, D.E., "Unacceptable Height Correlation Coefficients and the Quasi-Specular Component in Rough Surface Scattering," *Radio Science,* **5**, pp. 647-654, 1970.

Bass, F.G. and I.M. Fuchs, *Wave Scattering from Statictically Rough Surfaces,* transl. and edited by C.B. Vesecky and J.F. Vesecky, Pergamon Press, New York, 1978.

Bass, F.G., I.M. Fuks, A.I. Kalmykov, I.E. Ostrosky, and A.D. Rosenberg, "Very High Frequency Radiowave Scattering by a Disturbed Sea Surface, Parts I and II," IEEE Trans. Antennas and Propag., **AP-16**, pp 554-568, 1968.

Beard, C.I., "Coherent and Incoherent Scattering of Microwaves from the Ocean," *IRE Trans. Antennas and Propagation,* AP-9, pp. 470-483, 1961.

Beckmann, P. and A. Spetzichino, *The Scattering of Electromagnetic Waves From Rough Surfaces,* Pergamon Press, New York, 1963.

Brown, G.S., H.R. Stanley, and N.A. Roy, "The Wind Speed Measurement Capability of Spaceborne Radar Altimeters," *IEEE J. of Oceanic Engrg.*, **OE-6**, pp. 59-63, 1981.

Brown, G.S., "A Comparison of Approximate Theories for Scattering from Rough Surfaces," *Wave Motion,* **7**, pp. 195-205, 1985.

Brown, G.S., "Backscattering from a Gaussian Distributed Perfectly Conducting Rough Surface," *IEEE Trans. Antennas and Propagation*, AP-26, pp 472-482, 1978.

Brown, G.S., "Estimation of Surface Wind Speeds Using Satellite-Borne Radar Measurements at Normal Incidence," *J. Geophys. Res.*, **B8**, pp. 3974-3978, 1979.

Brown, G.S., "Scattering from Randomly Rough Surfaces and the Far Field Approximation," *Radio Science,* **18**, pp. 71-81, 1983.

Brown, G.S., "The Average Impulse Response of a Rough Surface and Its Applications," *IEEE Trans. Antennas and Propagation*, AP-25, pp. 67-77, 1977.

Collin, R.E., *Antennas and Radiowave Propagation,* McGraw-Hill Book Co., New York, 1985.

Cox, C.S. and W.H. Munk, "Statistics of the Sea Surface Derived from Sun Glitter," *J. Mar. Res.*, **13**, pp. 198-227, 1954.

DeSanto J.A. and G.S. Brown, "Analytical Techniques for Multiple Scattering from Rough Surfaces," in *Progress in Optics,* Vol XXIII, edited by E. Wolf, 1986.

Dobson, E., F. Monaldo, J. Goldhirsh, and J. Wilkerson, "Validation of GEOSAT Altimeter-Derived Wind Speeds and Significant Wave Heights Using Buoy Data," *Johns Hopkins APL Technical Digest,* **8**, pp. 222-233, 1987.

Fung, A.K. and H.L. Chan, "Backscattering of Waves by Composite Rough Surfaces," *IEEE Trans Antennas and Propagation,* AP-17, pp. 590-597, 1969.

Fung, A.K. and R.K. Moore, "The Correlation Function in Kirchhoff's Method of Solution for Scattering of Waves from Statistically Rough Surfaces," *J. Geophys. Res.*, **71**, pp. 2939-2943, 1966.

Hagfors, T., "Backscattering from an Undulating Surface with Application to Radar Returns from the Moon," *J. Geophys. Res.*, **69**, pp. 3779-3783, 1964.

Hagfors, T., "Scattering and Transmission of Electromagnetic Waves at a Statistically Rough Boundary Between Two Dielectric Media," in *Electromagnetic Wave Theory* edited by J. Brown, part 2, pg. 997, Pergamon Press, New York, 1967.

Harrington, R.F., "Time-Harmonic Electromagnetic Fields," McGraw-Hill Book Co., New York, 1961.

Kerr, D.E. (Ed.), *Propagation of Short Radio Waves,* McGraw-Hill Book Co., New York, 1951.

Kodis, R.D., "A Note on the Theory of Scattering from an Irregular Surface," *IEEE Trans. Antennas and Propagation,* **AP-14** pp. 77-82, 1966.

Matthews, R.E. (Ed.), "Active Microwave Workshop Report," NASA Special Publication SP-376, pp. 171-172, 1975.

Peake, W.H., D.E. Barrick, A.K. Fung, and H.L. Chan, "Comments on Backscattering of Waves by Composite Rough Surfaces," *IEEE Trans. Antennas and Propagation*, AP-18, pp. 716-726, 1970.

Ruck, G.T., D.E. Barrick, W.D. Stuart, and C.K. Krichbaum, *Radar Cross Section Handbook, Vol. I,* Plenum Press, New York, 1970.

Sancer, M.I., "Shadow-Corrected Electromagnetic Scattering from a Randomly Rough Surface," *IEEE Trans. Antennas and Propagation,* AP-17, pp. 577-584, 1969.

Stratton, J.A., *Electromagnetic Theory,* McGraw-Hill Book Co., New York, 1941.

Thorsos, E.I., "The Validity of the Kirchhoff Approximation for Rough Surface Scattering Using a Gaussian Roughess Spectrum," to be published in *J. Acoust. Soc. Am.*, Jan, 1988.

Valenzuela, G.R., "Theories for the Interaction of Electromagnetic and Oceanic Waves-A Review," *Boundary Layer Met.*,**13**, pp. 61-85, 1978.

Chapter 11

BRAGG SCATTERING OF ELECTROMAGNETIC WAVES FROM THE AIR/SEA INTERFACE

WILLIAM J. PLANT
Woods Hole Oceanographic Institution
Woods Hole, Massachusetts

1. Introduction

1.1. A HISTORICAL PERSPECTIVE

The advent of radar during World War II had consequences for research in air/sea interactions which were far from obvious in the face of the immediate wartime need to detect enemy aircraft. When the newly-discovered tool was put to use to detect targets on or near the surface of the sea, these targets were often obscured by strong echos from the ocean itself. This "sea echo" constituted a considerable nuisance to those engaged in locating enemy vessels so investigations into the nature of this unwanted return were launched. Early theoretical work concentrated on explaining sea echo in terms of either return from the sea surface itself or from the spray and bubbles above the surface. Methods of modelling electromagnetic scattering which had proved useful to the characterization of many types of targets were adapted for use modelling sea return. Specular reflection seemed to explain the characteristics of sea echo for small incidence angles but not for larger ones. The standard method of carrying out such calculations utilized the "Kirchoff principle," also known as physical optics or the tangent plane method, which assumes that the surface is smooth in the sense that the radius of curvature is large compared to the electromagnetic wavelength. In addition to yielding disturbingly low return at large incidence angles, the method was unable to account for the observation that vertically polarized return was generally stronger than horizontal at such angles. Since interference between radiation directly incident on bubbles above the surface and radiation incident on the bubbles following reflection from the surface could in principal yield such polarization differences, theories based on Rayleigh scattering by the bubbles were also proposed. If the postulated bubble sizes were much smaller than the electromagnetic wavelength, however, these models were unable to explain the observed frequency dependence of sea echo. Observed return decreased with electromagnetic wavelength, λ, at rates slower than λ^{-4} while the theory predicted rates between λ^{-4} and λ^{-6}. Explanations based on larger bubbles, though overcoming this difficulty, seemed unlikely due to the fact that large polarization differences were present in the echo even at very low wind speeds and sea states (Goldstein 1946, 1951). Thus by 1951 no consistent picture of sea echo at large incidence angles had emerged and its importance for air/sea interaction research was not yet realized.

The situation began to change as the decade of the fifties progressed. In 1951, Rice published a theoretical investigation of the scattering of electromagnetic waves from what he called "slightly rough surfaces" (Rice 1951). Applying essentially the same method that Rayleigh had first applied to the scattering of sound from rough surfaces, Rice characterized the surface roughness as a small perturbation of a smooth surface (Rayleigh 1929). The result

G. L. Geernaert and W. J. Plant (eds.), Surface Waves and Fluxes, Volume II, 41–108.

called "slightly rough surfaces" (Rice 1951). Applying essentially the same method that Rayleigh had first applied to the scattering of sound from rough surfaces, Rice characterized the surface roughness as a small perturbation of a smooth surface (Rayleigh 1929). The result of this calculation indicated that the scattered return was proportional to the Fourier transform of the surface roughness at a wavenumber which depended on radar parameters. Thus the amplitude and frequency dependence of the return signal depended on those of a particular Fourier component of the surface roughness. Furthermore, if this Fourier component was in motion, the theory showed that a Doppler shift equal to its frequency would occur. Since the perturbation expansion required that surface displacements be small compared to the electromagnetic wavelength and that all surface slopes be small, its applicability to scattering from the sea surface was not immediately obvious. The first evidence that such diffraction-type scattering was in fact applicable to sea echo came from the work of Crombie with electromagnetic waves with wavelengths of 22 meters (Crombie 1955). He showed that the most prominent line in the Doppler spectrum of low-grazing-angle backscattered sea return occurred at exactly the frequency of an ocean wave whose wavelength was 11 meters or one half that of the electromagnetic waves. Thus at some electromagnetic wavelengths perturbation theory seemed to explain one striking characteristic of sea echo.

For the electromagnetic wavelengths used by Crombie, the approximations of perturbation theory were well satisfied. Wait (1966) demonstrated that the theory could quantitatively explain HF ground wave backscatter from sea waves, the situation Crombie had observed. In 1959, Peak had shown that the method also yielded results which compared well with experiment at higher electromagnetic frequencies in the microwave range when applied to slightly rough stationary surfaces such as concrete or asphalt driveways (Peak 1959). The applicability of perturbation theory to sea return at microwave frequencies where sea surface displacements are large compared to electromagnetic wavelengths remained questionable, however, until the late sixties. At that time, apparently independently, researchers in both the United States and the Soviet Union published results of theoretical and laboratory experimental studies which showed the applicability of perturbation theory to very small disturbances of a water surface (Wright 1966, Bass et al. 1968a). Although the Soviet study utilized sound scattering while the Americans used microwaves, both studies investigated scattering from artificially-generated sinusoidal surface waves to show the resonant nature of the backscatter and its proportionality to surface wave height. The same investigators followed these papers with ones proposing and offering support for composite surface theory to explain the scattering of microwaves from a realistic ocean surface (Bass et al. 1968b, Wright 1968). Basically, they postulated that the longer waves on the surface could be treated using a tangent plane approximation while shorter waves did the scattering. Thus the large-scale ocean surface was approximated by an array of plane surfaces each of which was displaced and tilted due to long waves and each of which had dimensions which were a fraction of the long wavelength. First-order perturbation theory was then applied to each small plane, or facet, to yield the scattered signal from it. Summing over all facets provided the net sea echo. Both studies stressed that short wave structure must decorrelate in distances short compared to long wavelengths so that total return power could be calculated by incoherently summing return from individual facets; both presented evidence that this was indeed the case. These studies showed that the frequency, incidence angle, and polarization dependence of sea echo at microwave frequencies could be rather well explained by the theory. The stage was now set for the application of electromagnetic backscatter to research in air/sea interactions.

1.2. PURPOSE OF THIS CHAPTER

In the nearly two decades since the development of composite surface theory, evidence for its applicability has accumulated and it has been utilized to obtain information on air/water interactions which is difficult to obtain by alternate means. Perturbation-type scattering has come to be called "Bragg scattering" after similar diffraction-type scattering which occurs when solids are illuminated by X-rays, a process first explained by W. L. Bragg (Bragg 1913). Periodically, Bragg and composite surface theories have been reviewed and evidence for their applicability has been presented (Barrick and Peak 1968, Hasselmann and Schieler 1970, Valenzuela 1978a&b, Bass and Fuks 1979). Yet it is difficult to find a treatment which provides full details of the derivation of Bragg backscatter employing realistic antenna systems as applied to sea echo. Furthermore much of the experimental evidence supporting Bragg and composite surface theories is widely scattered in the literature and not easily available to either students or researchers in other fields. In this chapter we shall attempt to bring together theoretical details, realistic antenna systems, and experimental evidence. We develop the concept of Bragg scattering from Maxwell's equations of electromagnetism, couple these results with the large-scale surface in composite surface theory, and present evidence that this approach indeed explains most of the characteristics of the scattering of electromagnetic waves from air/water interfaces at intermediate incidence angles.

We will not attempt to present detailed mathematical discussions of all aspects of scattering from rough surfaces at intermediate incidence angles. Rather we will try to present a simple, yet hopefully complete, development of first-order Bragg scattering and composite surface theory. Higher-order scattering will not be discussed in any detail although it is being utilized today to obtain directional surface wave spectra with HF radars operating in the megahertz range. This omission is necessary to keep the size and complexity of the chapter within manageable limits. Readers interested in higher-order Bragg scattering are referred to the papers listed in the bibliography (Valenzuela 1967 & 1974, Hasselmann and Schieler 1970, Barrick 1972). Much of the development to follow involves standard electromagnetic concepts and may be found in greater detail in books such as those by Stratton (1949), Silver (1941), and Jackson (1962).

2. Electromagnetic Waves in Air and Water

2.1. ELECTROMAGNETIC WAVES

The fundamental equations of electromagnetic theory are Maxwell's equations. In rationalized mks units these are,

$$\nabla \times \vec{E} = -\frac{\partial \vec{B}}{\partial t} \tag{2.1}$$

$$\nabla \times \vec{H} = \frac{\partial \vec{D}}{\partial t} + \vec{J} \tag{2.2}$$

$$\nabla \cdot \vec{B} = 0 \tag{2.3}$$

$$\nabla \cdot \vec{D} = \rho \tag{2.4}$$

where $\vec{E}$ is electric field intensity, $\vec{H}$ is magnetic field intensity, $\vec{D}$ is electric displacement, $\vec{B}$ is magnetic induction, $\vec{J}$ is current density and ρ is charge density. The following relationships exist between vectors in isotropic media:

$$\vec{B} = \mu\vec{H} \tag{2.5}$$

$$\vec{D} = \epsilon'\vec{E} \tag{2.6}$$

$$\vec{J} = \sigma\vec{E} \tag{2.7}$$

where μ is permeability, ϵ' is permittivity, and σ is conductivity.

Since Maxwell's equations are linear, we may consider individual Fourier components, producing general solutions by superposition. Thus we assume that the time dependence of all fields may be represented by $e^{-i\omega t}$. Maxwell's equations then become,

$$\nabla \times \vec{E} = i\omega\mu\vec{H} \tag{2.8}$$

$$\nabla \times \vec{H} = (-i\omega\epsilon' + \sigma)\vec{E} = -i\omega\epsilon\vec{E} \tag{2.9}$$

$$\nabla \cdot \vec{H} = 0 \tag{2.10}$$

$$\nabla \cdot \vec{E} = \rho/\epsilon' \tag{2.11}$$

where we let

$$\epsilon = \epsilon' + i\sigma/\omega \tag{2.12}$$

which is a complex permittivity.

Now, elementary vector analysis gives the identity,

$$\nabla \times \nabla \times \vec{E} = \nabla(\nabla \cdot \vec{E}) - \nabla^2\vec{E}$$

Thus the curl of Eq. (2.8) after substitution of Eq. (2.9) and (2.11) yields,

$$\nabla^2\vec{E} + \omega^2\mu\epsilon\vec{E} = 0 \tag{2.13}$$

if charge density and permittivity are constant. This equation is satisfied by $\vec{E} = \vec{E}_c e^{i\vec{k}\cdot\vec{r}}$, where $\vec{E}_c$ is constant, $\vec{k}$ is a wavevector, and $\vec{r}$ is a position vector, if

$$k = \omega\sqrt{\mu\epsilon}. \tag{2.14}$$

An identical equation may be derived for $\vec{H}$ so $\vec{H} = \vec{H}_c\, e^{i\vec{k}\cdot\vec{r}}$. Thus we see that Maxwell's equations require electromagnetic fields to propagate through isotropic media as waves with phase speed $c_o = \omega/k = 1/\sqrt{\mu\epsilon}$. In free space, $\mu = 4\pi \times 10^{-7}$ henry/meter, $\epsilon = 8.854 \times 10^{-12}$ farad/meter, and $c_o = 2.998 \times 10^8$ meters/sec.

For each Fourier component, Maxwell's equations require that in the absence of charge,

$$\nabla \cdot \vec{E} = i\vec{k} \cdot \vec{E} = 0 \tag{2.15}$$

$$\nabla \cdot \vec{H} = i\vec{k} \cdot \vec{H} = 0 \tag{2.16}$$

and

$$\vec{H} \cdot \vec{E} = \frac{i}{\omega\mu} (\nabla \times \vec{E}) \cdot \vec{E} = 0. \tag{2.17}$$

Thus the vectors $\vec{E}$, $\vec{H}$, and $\vec{k}$ form a mutually orthogonal set. The sense of the vectors may be obtained from Eq. (2.8) which yields for any component,

$$\vec{k} \times \vec{E} = \omega\mu\vec{H}. \tag{2.18}$$

Solution of Eq. (2.13) in spherical coordinates yields

$$\vec{E} = \vec{E}_c'\, e^{ikr}/r \tag{2.19}$$

where now $\vec{E}_c'$ is always tangential to a sphere centered on the origin. Equation (2.18) also holds for these spherical waves.

It is often desirable to know the power per unit area contained in an electromagnetic field. This quantity is given by the Poynting vector

$$\vec{S} = \vec{E} \times \vec{H}. \tag{2.20}$$

In general the mean power is more relevant than the instantaneous power. Indicating the time dependence explicitly and averaging we find,

$$\vec{S} = \overline{\mathrm{Re}(\vec{E}e^{-i\omega t}) \times \mathrm{Re}(\vec{H}e^{-i\omega t})} = \frac{1}{2}\,\mathrm{Re}(\vec{E} \times \vec{H}^*) \tag{2.21}$$

where the overbar represents a temporal average, Re indicates the real part of a complex quantity, and the asterisk indicates complex conjugation. The mean energy flowing across a surface is the scalar product of $\vec{S}$ with that area.

2.2. BOUNDARY CONDITIONS

When the electromagnetic waves of the previous section encounter interfaces where discontinuities of ϵ or μ occur, boundary conditions exist at the interface which assure that fields in both media satisfy Maxwell's equations. Consider first a small pill-box-shaped volume whose two flat surfaces are in the two different media and whose cylindrical surface intersects the interface (Fig. 1a). Integration of Maxwell's divergence equations, (2.3) and (2.4), throughout this volume yield,

$$\int_V \nabla \cdot \vec{D}\, dV = \int_s \vec{D} \cdot \hat{n}\, dS = q$$

$$\int_V \nabla \cdot \vec{B}\, dV = \int_s \vec{B} \cdot \hat{n}\, dS = 0$$

where q is the total charge in the volume, S is the surface of the volume, $\hat{n}$ is a unit vector normal to the surface directed out of the volume and we have employed Gauss' divergence theorem. If we imagine that the sides of the pill box become infinitesimally small, then the surface integrals are only over the flat surface areas which are equal in size. Thus we find that

$$\vec{D}_o \cdot \hat{n} - \vec{D} \cdot \hat{n} = q/A \tag{2.22}$$

and

$$\vec{B}_o \cdot \hat{n} - \vec{B} \cdot \hat{n} = 0 \tag{2.23}$$

where $\vec{D}$ and $\vec{D}_o$ are electric displacements in the two media, $\vec{B}$ and $\vec{B}_o$ are magnetic inductions in the two media, and A is the area of either flat surface. The unit vector $\hat{n}$ is now normal to the interface. Thus normal components of $\vec{B}$ are continuous across the interface but normal components of $\vec{D}$ may be discontinuous if charge resides on the surface.

Next we consider a rectangular area whose short sides intersect the interface (Fig. 1b). Integration of Maxwell's curl equations, (2.1) and (2.2), yields

$$\int_s (\nabla \times \vec{E}) \cdot \hat{n}\, dS = \int_l \vec{E} \cdot d\vec{l} = \int_s i\omega\vec{B} \cdot \hat{n}\, dS$$

$$\int_s (\nabla \times \vec{H}) \cdot \hat{n}\, dS = \int_l \vec{H} \cdot d\vec{l} = \int_e \vec{H} \cdot d\vec{l} = \int_s (\vec{J} - i\omega\vec{D}) \cdot \hat{n}\, dS$$

where S is the rectangular area, l is its perimeter, and $\hat{n}$ is a unit vector perpendicular to S. As we let the lengths of the short sides of the rectangle become infinitesimal, the surface integrals on the right must go to zero unless the conductivity of one medium is infinite. Excluding this case, and recalling that the long sides are of equal length, we have,

$$\vec{E}_o \cdot \hat{l} - \vec{E} \cdot \hat{l} = 0 \tag{2.24}$$

$$\vec{H}_o \cdot \hat{l} - \vec{H} \cdot \hat{l} = 0 \tag{2.25}$$

where $\hat{l}$ is a unit vector parallel to the interface, $\vec{E}$ and $\vec{E}_o$ are electric field intensities in the two media, and $\vec{H}$ and $\vec{H}_o$ are magnetic field intensities in the two media. Thus tangential components of $\vec{E}$ are continuous across the interface but tangential components of $\vec{H}$ are continuous only if both conductivities are finite.

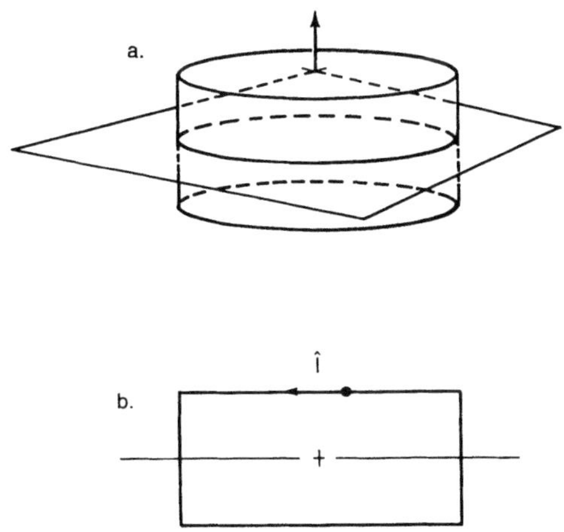

Figure 1. (a) Pill box used to determine normal electromagnetic boundary conditions. Vertical sides of box intersect the boundary between two media. The unit vector $\hat{n}$ is normal to the surface of the box. (b) Rectangle used to determine tangential boundary conditions. Vertical sides intersect the boundary. The unit vector $\hat{l}$ is parallel to the side of the rectangle.

2.3. PLANE WAVES INCIDENT ON A SMOOTH WATER SURFACE

When plane electromagnetic waves are incident on a flat water surface, some energy is generally transmitted across the interface while some is reflected. We may determine characteristics of these refracted and reflected waves by satisfying boundary conditions at the interface. We have seen that $\vec{E}$, $\vec{H}$, and $\vec{k}$ are mutually orthogonal. Thus we can always describe an incident wave having an arbitrary electromagnetic field orientation as a sum of waves having $\vec{E}$ parallel and perpendicular to the plane of incidence, *i.e.*, the plane containing $\vec{k}$ and the vertical to the surface. We refer to waves with $\vec{E}$ perpendicular to the plane of incidence as "horizontally polarized" and those with $\vec{E}$ in the plane of incidence as "vertically polarized."

We shall consider reflection and refraction of plane waves by a smooth water surface below, beginning with the case of horizontal polarization. The conductivity of the water will be considered to be large but finite and no charge will be present on the surface. As far as possible, we shall subscript quantities in air with "o" or "a" and those in water with "w". On occasion, however, we shall indicate a reflected wave in air with a subscript "r". Figure 2 defines the coordinate system and notation to be used. The notation $\hat{a}$ always refers to unit vectors; $\hat{a}_x$,$\hat{a}_y$, and $\hat{a}_z$ are orthogonal vectors in a right-handed Cartesian coordinate system with $\hat{a}_z$ directed vertically upward. Position vectors will be denoted by $\vec{r}$ and will be directed away from the origin. Thus

$$\vec{r} = x\hat{a}_x + y\hat{a}_y + z\hat{a}_z.$$

Throughout this chapter, the plane of incidence will be taken to be the y-z plane and θ will define the acute angle between the z-axis and the direction of wave propagation.

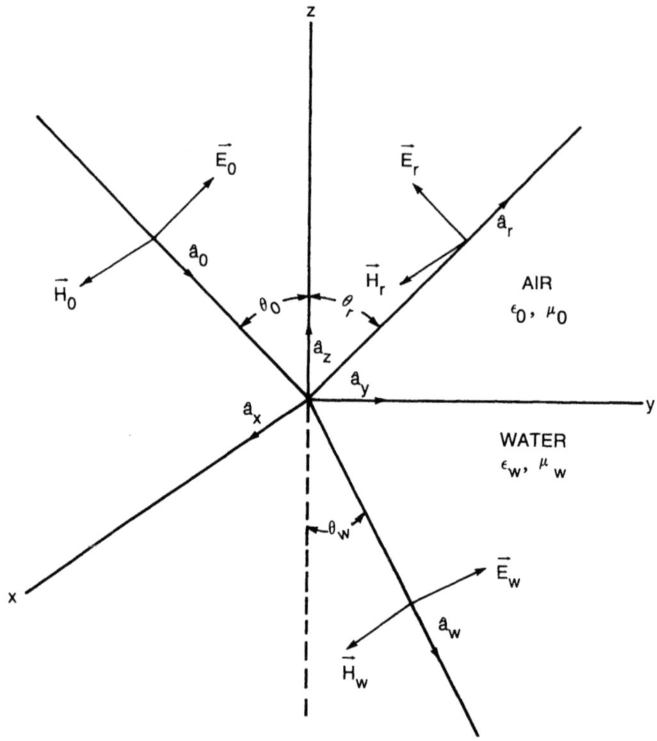

Figure 2. Coordinate system used to calculate reflection and refraction from a smooth air/water interface.

Horizontal Polarization: From the properties of plane electromagnetic waves previously noted, we may write equations for $\vec{E}$ and $\vec{H}$ in air and water:

$$\vec{E}_a = E_o \hat{a}_x e^{ik_o \hat{a}_o \cdot \vec{r}} + E_r \hat{a}_x e^{ik_o \hat{a}_r \cdot \vec{r}}$$

$$\vec{H}_a = H_o(- \cos \theta_o \ \hat{a}_y - \sin \theta_o \ \hat{a}_z) e^{ik_o \hat{a}_o \cdot \vec{r}} + H_r(\cos \theta_r \hat{a}_z - \sin \theta_r \hat{a}_z) e^{ik_o \hat{a}_r \cdot \vec{r}}$$

$$\vec{E}_w = E_w \ \hat{a}_x \ e^{ik_w \hat{a}_w \cdot \vec{r}}$$

$$\vec{H}_w = H_w(-\cos \theta_w \ \hat{a}_y - \sin \theta_w \ \hat{a}_z) e^{ik_w \hat{a}_w \cdot \vec{r}}.$$

Since boundary conditions at $z = 0$ must be satisfied at all values of x and y, we require that all phases be equal at $z = 0$. Thus

$$\sin\theta_o = \sin\theta_r \tag{2.26}$$

$$k_o \sin\theta_o = k_w \sin\theta_{w.} \tag{2.27}$$

These are Snell's laws of reflection and refraction.

In order to satisfy boundary conditions on $\vec{E}$ and $\vec{H}$ (Eqs. (2.24) and (2.25)), we require that

$$E_o + E_r = E_w \tag{2.28}$$

$$(H_o - H_r)\cos\theta_o = H_w \cos\theta_w. \tag{2.29}$$

But Eq. (2.18) requires the following relationship between the amplitudes of electric and magnetic fields in plane waves:

$$H = \left(\frac{k}{\omega\mu}\right) E. \tag{2.30}$$

Using this relation and Snell's laws, we may solve Eqs. (2.28) and (2.29) for E_r and E_w in terms of E_o:

$$R_H \equiv \frac{E_r}{E_o} = \frac{Z\cos\theta_o - \cos\theta_w}{Z\cos\theta_o + \cos\theta_w} \tag{2.31}$$

$$T_H \equiv \frac{E_w}{E_o} = \frac{2Z\cos\theta_o}{Z\cos\theta_o + \cos\theta_w} \tag{2.32}$$

where

$$Z = \frac{k_o\mu_w}{k_w\mu_o} = \sqrt{\frac{\epsilon_o\mu_w}{\epsilon_w\mu_o}} = \frac{\epsilon_o k_w}{\epsilon_w k_o} \tag{2.33}$$

is the normalized surface impedance.

These equations define the Fresnel reflection and transmission coefficients, R_H and T_H, for horizontal polarization. They imply that

$$T_H - R_H = 1 \tag{2.34}$$

and

$$\cos\theta_w = \left[\frac{(1 - R_H)Z}{T_H}\right]\cos\theta_o. \tag{2.35}$$

We may now easily write down expressions for the fields in air and water in terms of R_H and T_H. They are

$$\vec{E}_a = E_o e^{ik_o y \sin\theta_o}\{e^{-ik_o z\cos\theta_o} + R_H e^{ik_o z\cos\theta_o}\}\hat{a}_x \tag{2.36}$$

$$\vec{H}_a = -H_o e^{ik_o y\sin\theta_o}\{\cos\theta_o [e^{-ik_o z\cos\theta_o} - R_H e^{ik_o z\cos\theta_o}]\hat{a}_y$$

$$+ \sin\theta_o [e^{ik_{oz}\cos\theta_o} + R_H e^{ik_{oz}\cos\theta_o}]\hat{a}_z\} \tag{2.37}$$

$$\vec{E}_w = T_H E_o e^{ik_o y\sin\theta_o}\{e^{-ik_w z\cos\theta_w}\}\hat{a}_x \tag{2.38}$$

$$\vec{H}_w = -H_o e^{ik_o y\sin\theta_o}\{(1 - R_H)\cos\theta_o \hat{a}_y + \left(\frac{\mu_o}{\mu_w}\right) T_H \sin\theta_o \hat{a}_z\Big\} z^{-ik_w z\cos\theta_w}. \tag{2.39}$$

These equations show that tangential components of $\vec{E}$ and $\vec{H}$ are continuous at $z = 0$ and that normal components of $\vec{D}$ and $\vec{B}$ are continuous there as required by boundary conditions.

A reflected wave may fail to exist if

$$Z\cos\theta_o = \cos\theta_w$$

as shown by the form of the reflection coefficient. Using Snell's law and noting that $Z = k_o/k_w$ if $\mu_w = \mu_o$ which is a good approximation for water, we find that

$$Z\cos\theta_o = \sqrt{1 - Z^2\sin^2\theta_o}$$

which is satisfied only if $Z = 1$, *i.e.*, if no interface exists. Thus a reflected wave must always exist when horizontally polarized waves are reflected from the air-water interface.

Vertical Polarization: Turning to the case of vertical polarization, we have the following fields in air and water:

$$\vec{E}_a = E_o(\cos\theta_o\ \hat{a}_y + \sin\theta_o\ \hat{a}_z)e^{ik_o\hat{a}_o\cdot\vec{r}} - E_r(\cos\theta_r\ \hat{a}_y - \sin\theta_r\ \hat{a}_z)e^{ik_o\vec{a}_r\cdot\vec{r}}$$

$$\vec{H}_a = H_o\ \hat{a}_x\ e^{ik_o\hat{a}_o\cdot\vec{r}} + H_r\ \hat{a}_x\ e^{ik_o\hat{a}_r\cdot\vec{r}}$$

$$\vec{E}_w = E_w(\cos\theta_w\ \hat{a}_y + \sin\theta_w\ \hat{a}_z)e^{ik_w\hat{a}_w\cdot\vec{r}}$$

$$\vec{H}_w = H_w\hat{a}_x e^{ik_w\vec{a}_w\cdot\vec{r}}.$$

Once again Snell's laws follow from matching phases.

Boundary conditions on $\vec{E}$ and $\vec{H}$ require that

$$(E_o - E_r)\cos\theta_o = E_w \cos\theta_w \tag{2.40}$$

$$H_o + H_r = H_w. \tag{2.41}$$

In a manner identical to the case of horizontal polarization, these equations lead to Fresnel reflection and transmission coefficients for vertical polarization. In this case, however, we define the coefficients as ratios of magnetic fields:

$$R_V = \frac{H_r}{H_o} = \frac{\cos\theta_o - Z\cos\theta_w}{\cos\theta_o + Z\cos\theta_w}. \tag{2.42}$$

$$T_V = \frac{H_w}{H_o} = \frac{2\cos\theta_o}{\cos\theta_o + Z\cos\theta_w}. \tag{2.43}$$

Once again we see that $T - R = 1$ but now

$$\cos\theta_w = \left[\frac{1 - R_V}{ZT_V}\right]\cos\theta_o. \tag{2.44}$$

Writing out the expressions for fields in air and water, we have

$$\vec{E}_a = E_o e^{ik_o y\sin\theta_o}\{\cos\theta_o\,[e^{-ik_o z\cos\theta_o} - R_V\, e^{ik_o z\cos\theta_o}]\hat{a}_y \tag{2.45}$$

$$+ \sin\theta_o[e^{-ik_o z\cos\theta_o} + R_V e^{ik_o z\cos\theta_o}]\hat{a}_z\}$$

$$\vec{H}_a = H_o e^{ik_o y\sin\theta_o}\{e^{-ik_o z\cos\theta_o} + R_V e^{ik_o z\cos\theta_o}\}\hat{a}_x \tag{2.46}$$

$$\vec{E}_w = E_o e^{ik_o y\sin\theta_o}\{(1 - R_V)\cos\theta_o\,\hat{a}_y + \left(\frac{\epsilon_o}{\epsilon_w}\right) T_V\sin\theta_o\,\hat{a}_z\}e^{-ik_w z\cos\theta_w} \tag{2.47}$$

$$\vec{H}_w = T_V\, H_o\, e^{ik_o y\sin\theta_o}\{e^{-ik_w z\cos\theta_w}\}\hat{a}_x. \tag{2.48}$$

Once again we see that boundary conditions are satisfied at $z = 0$.

The reflection coefficient for vertical polarization is zero if

$$\cos \theta_o = Z \cos \theta_w .$$

Again letting $\mu_w = \mu_o$ and using Snell's law, this equation becomes

$$\cos \theta_o = Z\sqrt{1 - Z^2 \sin^2 \theta_o}$$

which is satisfied if $Z = 1$ or $Z = \cot \theta_o$. Because $\sigma \neq 0$, the dielectric constant of water is complex so that k_w and Z are also complex. The real part of Z defines Brewster's angle of zero reflection while the imaginary part represents phase shifts between the waves in air and water and attenuation of the waves in water.

We shall see shortly that for the air/water interface, Z is small. Thus we may approximate the zero-reflection conditions by

$$\cos \theta_o \doteq Z$$

which gives

$$\sin \theta_o \doteq 1 - Z^2/2$$

the electric field above the interface then is given by

$$\vec{E}_a = E_o [Z\hat{a}_y + (1 - Z^2/2)\hat{a}_z] e^{ik_o[(1-Z^2/2)y - Zz]} \tag{2.49}$$

$$\vec{E}_w = E_o [Z\hat{a}_y + Z^2 \hat{a}_z] e^{ik_o[(1-Z^2/2)y - z/Z]} \tag{2.50}$$

to second order in Z. This is the surface, or ground, wave mode of propagation. It is characterized by an electric field in air which tilts in the plane of incidence and by phase shift and attenuation of fields in both air and water. This mode of propagation is quite important at low radio frequencies but attenuates rapidly with increasing frequency.

2.4. ELECTROMAGNETIC PROPERTIES OF WATER

The ground wave mode is a special case since waves in both air and water are rather rapidly attenuated. In all cases, however, electromagnetic waves in water of finite conductivity are attenuated as a result of the imaginary part of the wavenumber in water. That is, fields in water are proportional to

$$e^{ik_w r} = e^{-k_i r} e^{ik_r r}$$

where k_r and k_i are the real and imaginary parts of k_w and are given by

$$k_r = \omega\sqrt{\mu_w |\epsilon|} \cos(\delta/2) \quad (2.51)$$

$$k_i = \omega\sqrt{\mu_w |\epsilon|} \sin(\delta/2) \quad (2.52)$$

where

$$\tan \delta = \epsilon_i / \epsilon_r$$

defines the ''loss tangent.''

Real and imaginary parts of the dielectric constant of pure, fresh, and sea water, ϵ_i and ϵ_r, as given by Saxton and Lane (1952) are shown in Figure 3 normalized by ϵ_o. Very pure water has little conductivity at the lower frequencies so that ϵ_i drops rapidly to zero there. On the other hand, sea water becomes nearly a perfect conductor at low frequencies. Furthermore, for frequencies above about 3 GHz, the difference in the dielectric properties of fresh and sea water is generally very small. Thus laboratory studies of microwave backscatter from the air/sea interface do not normally require salt water facilities.

To a very good approximation $\mu_w = \mu_o$ in water so Eqs. (2.51) and (2.52) indicate that k_r/k_o and k_i/k_o for pure, fresh, and sea water vary with frequency as shown in Figure 4. Since the normalized surface impedance of the air/sea interface, Z, is $k_o/\sqrt{k_r^2 + k_i^2}$, this figure shows that Z is less than 0.3 for frequencies below 100 GHz thereby justifying the approximation made to obtain the ground wave expressions. Fields in water are attenuated by e^{-1} in a distance k_i^{-1}, the skin depth. This depth is shown in Figure 5 for pure, fresh, and sea water as a function of frequency. We see that the penetration of electromagnetic radiation into sea water for frequencies between 1 MHz and 100 GHz, which are those of interest in Bragg scattering, is generally quite negligible.

3. Radiation of Electromagnetic Waves

3.1. THE STRATTON-CHU EQUATION

Up to this point, we have primarily considered electromagnetic fields to be plane waves, infinite in extent. In fact, however, all electromagnetic fields must have their sources in charge or current distributions and only approach plane waves at very large distances from these sources. In this section we shall deal with electromagnetic waves more realistically in terms of their sources. Our approach will be a special case of that of Stratton and Chu (Stratton 1941) in which we shall consider all sources to be outside of the volume of space in which we are interested. Fields at any point in the volume will then be expressed in terms of field distributions on the surface of the volume.

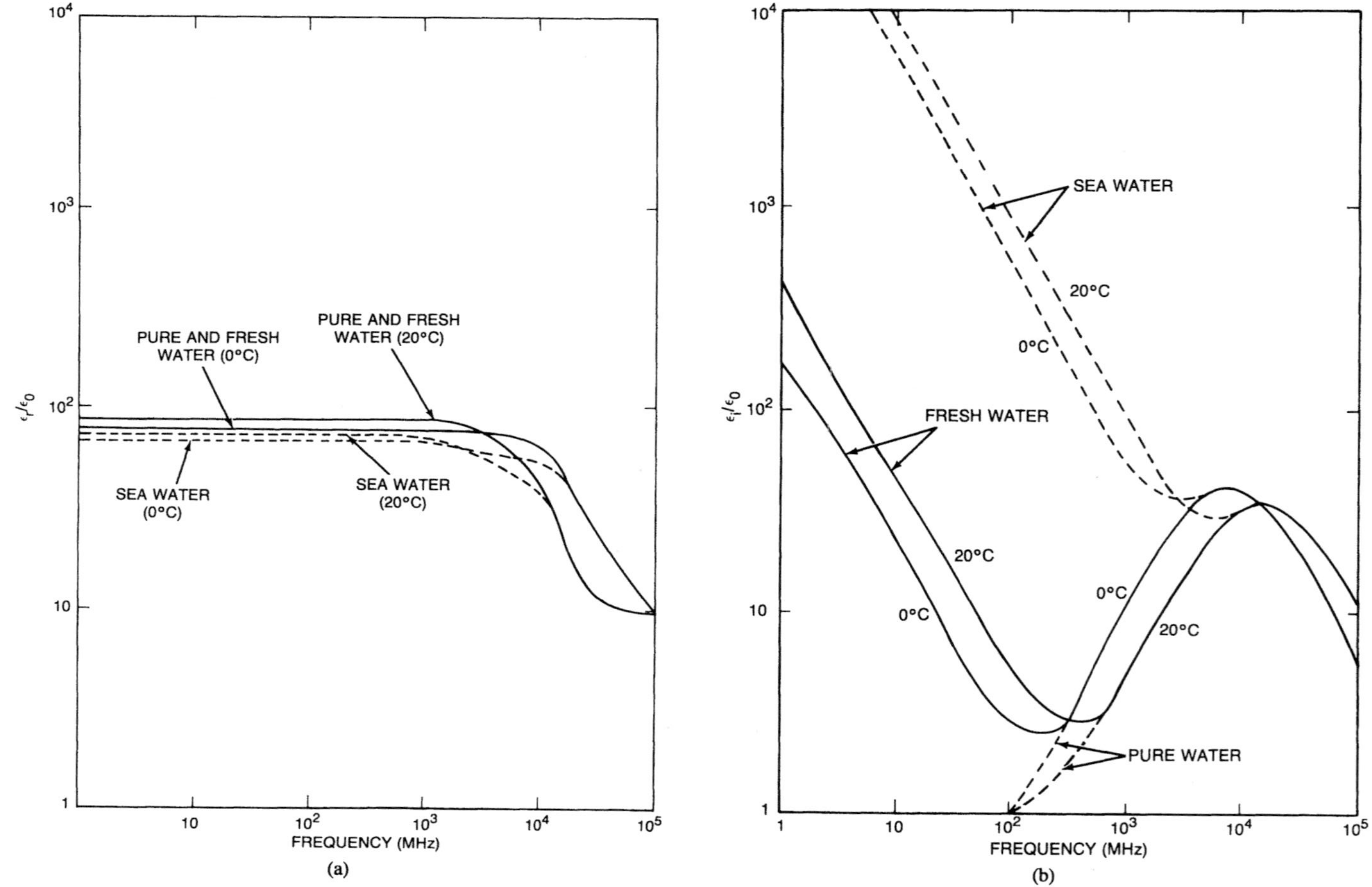

Figure 3. a) Real part of the dielectric constant of water as a function of frequency. b) Same plot of the imaginary part. The dielectric constant of free space is ϵ_o. Shown are curves for pure, fresh, and sea water at temperatures of 0°C and 20°C .

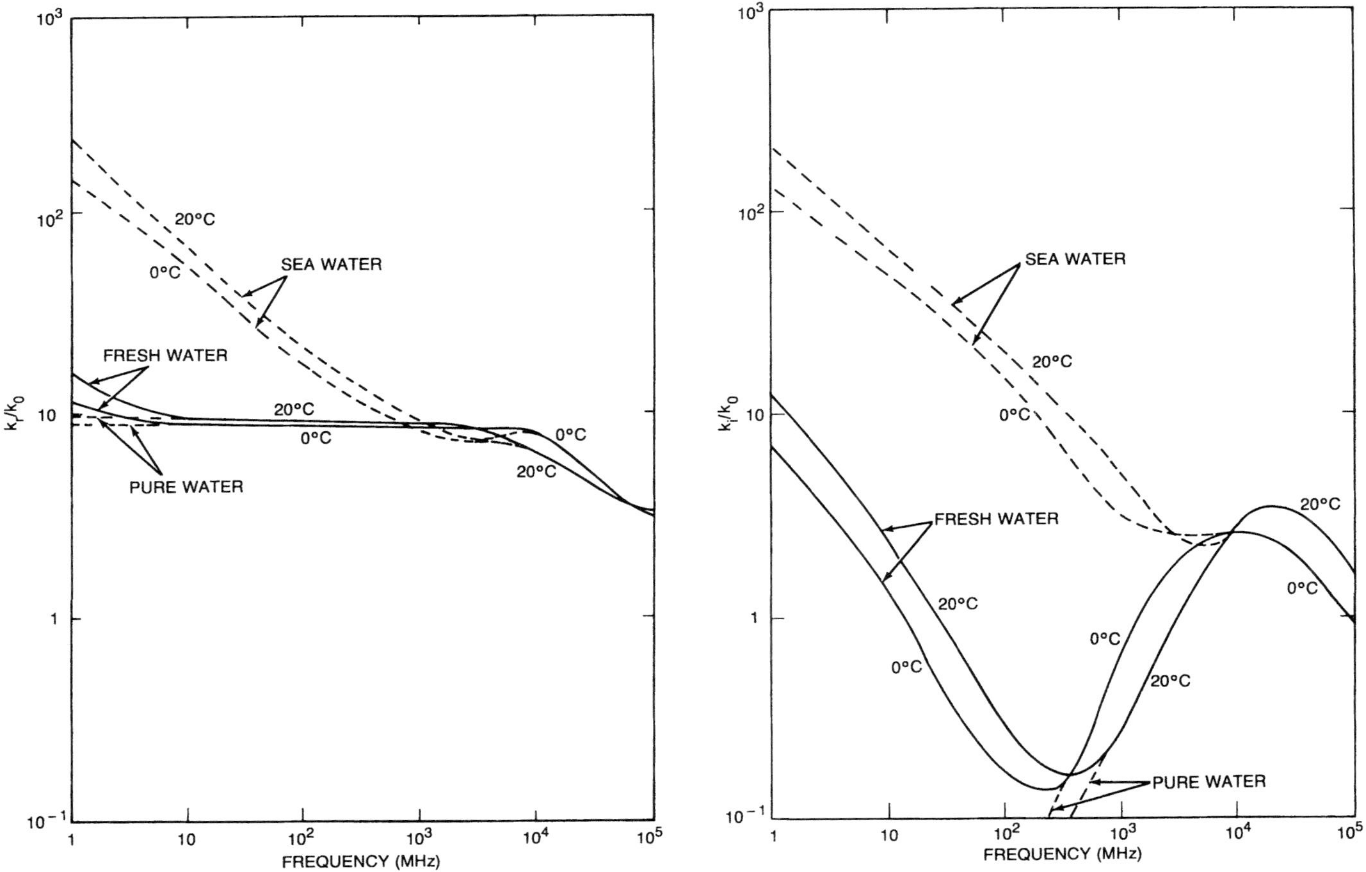

(a) (b)

Figure 4. a) Real part of the wavenumber in water as a function of frequency. b) Imaginary part of the wavenumber in water. The wavenumber in air is represented by k_o. Values for pure, fresh, and sea water are shown at temperatures of 0°C and 20°C.

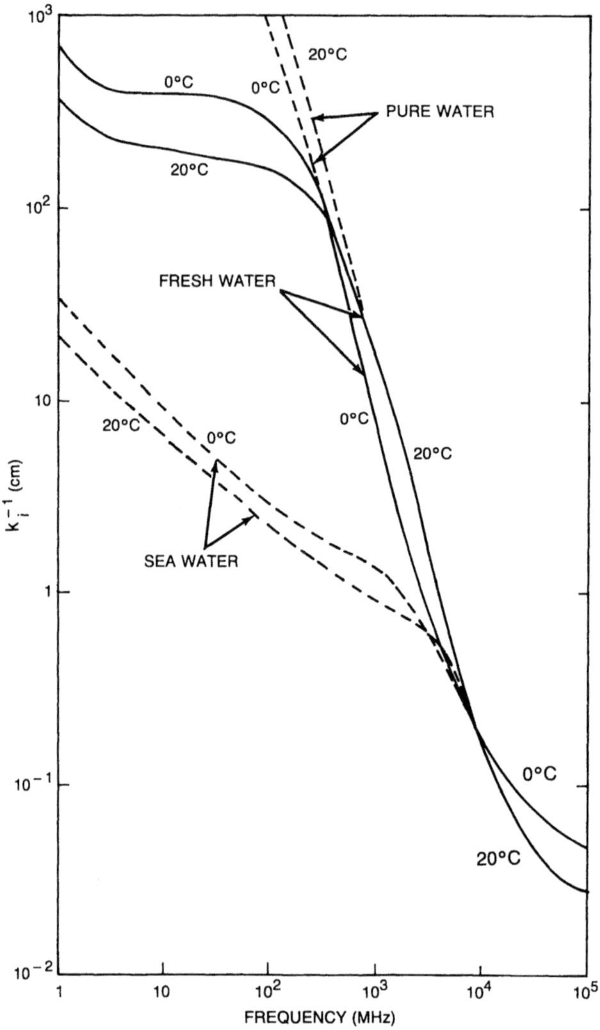

Figure 5. Skin depth, k_i^{-1}, in water as a function of frequency. Values are shown for pure, fresh, and sea water at 0°C and 20°C.

To begin this analysis, we need to establish a vector identity. As given in the last section, Gauss' divergence theorem is

$$\int_V \nabla \cdot \vec{A}\, dV = \int_s \vec{A} \cdot \hat{n}\, dS.$$

Let

$$\vec{A} = \vec{P} \times \nabla \times \vec{Q}$$

where $\vec{P}$ and $\vec{Q}$ are any vector quantities. Then

$$\nabla \cdot \vec{A} = \nabla \cdot (\vec{P} \times \nabla \times \vec{Q}) = \nabla \times \vec{P} \cdot \nabla \times \vec{Q} - \vec{P} \cdot \nabla \times \nabla \times \vec{Q}$$

and

$$\int_V (\nabla \times \vec{P} \cdot \nabla \times \vec{Q} - \vec{P} \cdot \nabla \cdot \nabla \times \vec{Q}) dV = \int_s (\vec{P} \times \nabla \times \vec{Q}) \cdot \hat{n}\, dS \qquad (3.1)$$

Interchanging P and Q in this equation and subtracting the resulting equation from Eq. (3.1) yields,

$$\int_V (\vec{Q} \cdot \nabla \times \nabla \times \vec{P} - \vec{P} \cdot \nabla \times \nabla \times \vec{Q})\, dV$$

$$= \int_s (\vec{P} \times \nabla \times \vec{Q} - \vec{Q} \times \nabla \times \vec{P}) \cdot \hat{n}\, dS. \qquad (3.2)$$

Now let $\vec{P} = \vec{E}$ and $\vec{Q} = \phi\hat{a}$ where $\vec{E}$ is an electric field, $\hat{a}$ is an arbitrary constant unit vector, and $\phi = e^{ik|\vec{r}-\vec{r}'|}/|\vec{r} - \vec{r}'|$. The position vector $\vec{r}$ defines a fixed point at which we wish to know the electromagnetic fields while $\vec{r}'$ defines the position variable over which the integrations of Eq. (3.2) are taken. Then we have

$$\nabla \times \vec{Q} = \nabla\phi \times \hat{a}, \quad \nabla \times \nabla \times \vec{Q} = \hat{a}k^2\phi + \nabla(\hat{a} \cdot \nabla\phi)$$

$$\nabla \times \vec{P} = i\omega\mu\vec{H}, \quad \nabla \times \nabla \times \vec{P} = k^2E.$$

Inserting these expressions in Eq. (3.2), we find

$$-\int_V \vec{E} \cdot \nabla(\hat{a} \cdot \nabla\phi) dV = \int_s (\vec{E} \times \nabla\phi \times \hat{a} - i\omega\mu\phi\hat{a} \times \vec{H}) \cdot \hat{n}\, dS. \qquad (3.3)$$

Since

$$\vec{E} \cdot \nabla(\vec{a} \cdot \nabla\phi) = \nabla \cdot [(\hat{a} \cdot \nabla\phi)\vec{E}]$$

$$- (\hat{a} \cdot \nabla\phi)\nabla \cdot \vec{E} = \nabla \cdot [(\hat{a} \cdot \nabla\phi)\vec{E}]$$

in the volume of interest, Eq. (3.3) may be written

$$\hat{a}\int_s [\nabla\phi(\vec{E}\cdot\hat{n}) + (\hat{n}\times\vec{E})\times\nabla\phi + i\omega\mu\phi(\hat{n}\times\vec{H})]\, dS = 0. \tag{3.4}$$

Here we have used the following vector identity:

$$\vec{A}\cdot(\vec{B}\times\vec{C}) = \vec{B}\cdot(\vec{C}\times\vec{A}) = \vec{C}\cdot(\vec{A}\times\vec{B}).$$

Since the vector $\hat{a}$ is arbitrary, Eq. (3.4) implies that the integral must be zero. Furthermore, since the derivation of Eq. (3.4) is strictly valid only if $\vec{P}$ and $\vec{Q}$ along with their first and second derivatives are continuous throughout the volume V, we must exclude the point $\vec{r}$, where $\vec{Q}$ is singular, from this volume. This is easily accomplished by encircling this point by a sphere whose surface is part of the surface S. For this sphere, $\hat{n}$ is directed radially toward the point $\vec{r}$, i.e., $\hat{n} = (\vec{r} - \vec{r}')/|\vec{r} - \vec{r}'|$. Thus on the sphere we find

$$\nabla\phi = \hat{n}(1/\rho - ik)e^{ik\rho}/\rho \tag{3.5}$$

where we have let $\vec{\rho} = \vec{r} - \vec{r}'$ and $\vec{k}\,\|\,\vec{\rho}$. Since as $\rho \to 0$ the surface area of the sphere vanished as $4\pi\rho^2$, the only terms of the integral over this sphere in Eq. (3.4) which do not vanish in the limit are those that vary as ρ^{-2}. Thus the integral over this sphere yields in the limit,

$$4\pi[(\hat{n}\times\vec{E}(\vec{r}))\times\hat{n} + (\hat{n}\cdot\vec{E}(\vec{r}))\cdot\hat{n}] = 4\pi\vec{E}(\vec{r})$$

where $\vec{E}(\vec{r})$ denotes the electric field at the point $\vec{r}$. Finally then, Eq. (3.4) yields

$$\vec{E}(\vec{r}) = \frac{-1}{4\pi}\int_s [i\omega\mu\phi(\hat{n}\times\vec{H}) + (\hat{n}\times\vec{E})\times\nabla\phi + (\vec{E}\cdot\hat{n})\nabla\phi]\, dS \tag{3.6}$$

where $\vec{r}$ is any interior point of S and

$$\phi = e^{ik\rho}/\rho.$$

Equation (3.6) is the Stratton-Chu integral equation for electromagnetic fields inside a volume enclosed by a known surface distribution of fields.

3.2 FIELDS RADIATED BY ANTENNAS

Our aim in this chapter is to apply the Stratton-Chu equation to the scattering by the air/sea interface of electromagnetic fields produced by antennas. Thus we choose our surface S to be the surface of a radar system and the surface of the interface connected by a hemisphere as shown in Figure 6. If we let the radius of the hemisphere approach infinity, the fields on its surface approach zero and contribute nothing to the fields observed at a point $\vec{r}$ above the air/water interface. The electric field at $\vec{r}$ may then be written

$$\vec{E}(\vec{r}) = \vec{E}_o + \vec{E}_s \tag{3.7}$$

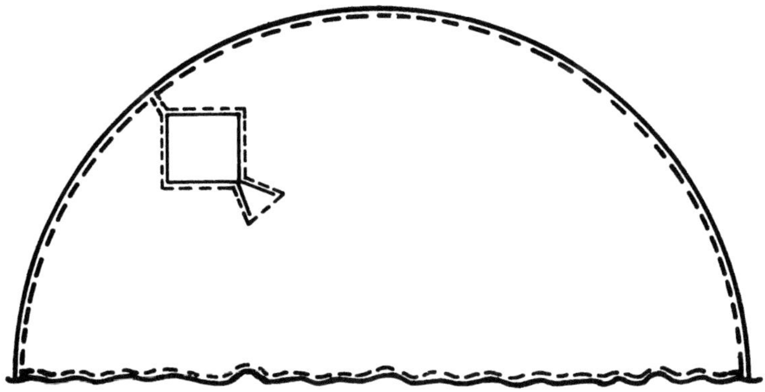

Figure 6. The surface over which the integral in the Stratton-Chu Equation is to be evaluated to determine backscatter from the air/sea interface.

where $\vec{E}_o$ is the field radiated by the antenna and $\vec{E}_s$ is the field scattered by the interface. These two fields may be determined from the Stratton-Chu equation with the surface being that of the radar and that of the interface, respectively. We shall obtain $\vec{E}_s$ in three steps. In this section, we first obtain an expression for the field $\vec{E}_o$ produced by an antenna in the Fraunhofer, or far-field, region. Next we develop an expression for the pattern of illumination produced by an antenna on a flat interface. Finally, in the following section we evaluate the field $\vec{E}_s$ scattered from a slightly rough air/water interface.

Consider now the Stratton-Chu equation applied to the surface of the radar. We shall assume that the radar system is constructed of a highly conductive material and is sufficiently well grounded that the only non-zero fields which exist on the surface are those at the aperture of the antenna. Thus we need only integrate over this aperture. Once again, Eq. (3.5) holds for $\nabla\phi$ except that in this case, the vector $\vec{\rho}$ is not necessarily parallel to the surface normal $\hat{n}$. In most cases of interest, however, the angle between these two vectors is small enough that they may be considered to be parallel. Then we have

$$\vec{E}_o(\vec{r}) = \frac{-1}{4\pi}\int_{\text{aperture}}\left[\left(\frac{1}{\rho} - ik_o\right)\left(\frac{e^{ik_o\rho}}{\rho}\right)\{(\hat{n}\times\vec{E})\times\hat{n}\right.$$

$$\left. + (\hat{n}\cdot\vec{E})\hat{n}\} + i\omega\mu\left(\frac{e^{ik_o\rho}}{\rho}\right)(\hat{n}\times\vec{H})\right] dS. \tag{3.8}$$

where we let $k = k_o$ to indicate that the field is in air. If the fields across the aperture have nearly uniform phase then the quantity in braces is simply $\vec{E}$ and

$$\hat{n}\times\vec{H} \doteq \vec{k}_o\times\vec{H}/k_o = -\omega\epsilon\,\vec{E}/k_o.$$

Finally, let us assume that $k_o >> \rho^{-1}$. Then using these approximations in Eq. (3.8) we have

$$\vec{E}_o(\vec{r}) = \frac{ik_o}{2\pi} \int_{\text{aperture}} \vec{E} \left(\frac{e^{ik_o \rho}}{\rho} \right) dS. \tag{3.9}$$

This is the expression for the field radiated by an antenna which we shall use for our further developments. The approximations leading to this equation are usually valid but should be carefully noted. For further discussion the reader is referred to Silver (1949).

3.3. SURFACE ILLUMINATION PATTERNS

Consider now a directional antenna pointed at the air/water interface as shown in Figure 7. We would like to calculate the fields produced by the antenna on flat surfaces $z = z_o$ near this interface. Thus the point $\vec{r}$ is located in the plane $z = z_o$ while $\vec{r'}$ is the integration variable located on the antenna aperture. From Figure 7, we see that

$$\rho = \{(x - n)^2 + (y + r_o \sin\theta_o - \xi \cos\theta_o)^2 + (z_o - r_o \cos\theta_o - \xi \sin\theta_o)^2\}^{1/2} \tag{3.10}$$

since the x and η axes are parallel and ξ makes an angle θ_o with the y axis in the yz plane.

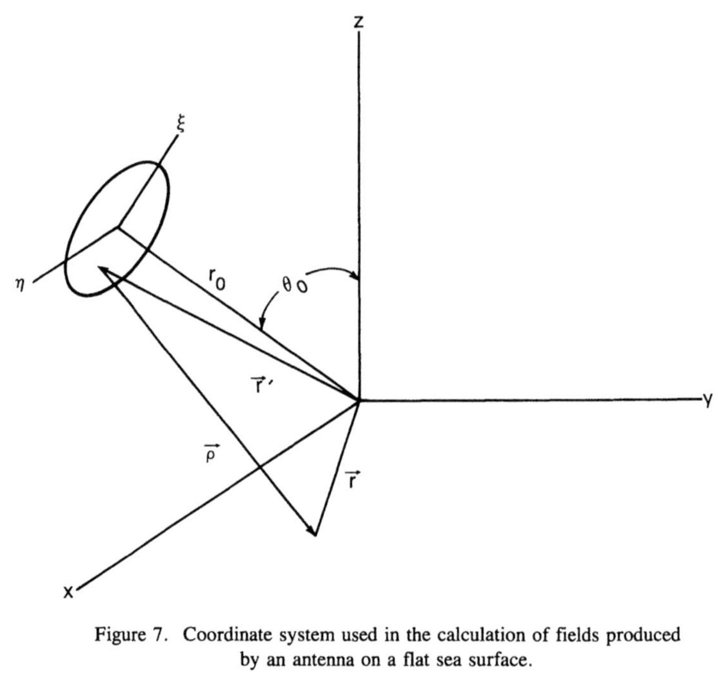

Figure 7. Coordinate system used in the calculation of fields produced by an antenna on a flat sea surface.

We now assume that r_o is much larger than any other variable and expand ρ in powers of the ratios of these variables to r_o. To second order, this yields

$$\rho = r_o + y \sin\theta_o - z_o \cos\theta_o + \frac{1}{2r_o}\left[x^2 + \nu^2 + \eta^2 + \xi^2 - 2x\eta - 2\nu\xi\right] \qquad (3.11)$$

where $\nu = y \cos\theta_o + z_o \sin\theta_o$. In the far field we may approximate ρ to first order in the denominator of Eq. (3.9) and omit those second-order terms involving η^2 and ξ^2 in the exponent. Noting that $dS = d\eta d\xi$, this yields

$$\vec{E}_o(\vec{r}) = \left\{\frac{ik_o(r_o - y\sin\theta_o + z_o\cos\theta_o)}{2\pi r_o^2}\right\} e^{ik_o[r_o + y\sin\theta_o - z\cos\frac{\theta_{o+1}}{2r_o}(x^{2+nu^2})]}$$

$$\int \vec{E}(\xi,\eta) e^{-ik_o(x\eta + \nu\xi)/r_o} d\xi d\eta. \qquad (3.12)$$

The omission of η^2 and ξ^2 terms is only valid if they induce very small phase shifts in the incident field. If we require that these be less than $\pi/8$ then the antenna diameter D_o must satisfy the following relationship:

$$\frac{k_o(D_o/2)^2}{2r_o} < \frac{\pi}{8}.$$

If $\lambda = 2\pi/k_o$ is wavelength, this implies that in the far field,

$$r_o > \frac{2D_o^2}{\lambda}. \qquad (3.13)$$

Actual fields $E(\xi,\eta)$ at the antenna aperture vary with antenna type. Often, however, Gaussian distributions are good approximations to these fields so we let

$$\vec{E}(\xi,\eta) = \vec{E}_c e^{-2\pi(\eta^{2+rho^2})/A_e} \qquad (3.14)$$

where A_e is the effective area of the aperture. This area does not necessarily coincide with the geometric area of the aperture. It is defined by

$$A_e = \frac{|\int \vec{E}(\xi,\eta) d\xi d\eta|^2}{\int |\vec{E}(\xi,\eta)|^2 d\xi d\eta} \qquad (3.15)$$

where integration is taken over the entire $\xi\eta$ plane. Obviously, for constant aperture illumination, A_e is identical to the geometric area of the aperture. An equation similar to Eq. (3.14) may be written for the magnetic field at the aperture so the mean transmitted power is [see Eq. (2.21)]

$$P_t = \frac{1}{2}\int \mathrm{Re}(EH^*)d\,\xi d\eta = A_e E_c H_c/8 \tag{3.16}$$

or simply half the effective area times the magnitude of the product of the weighted mean fields over the aperture.

Substituting Eq. (3.14) into Eq. (3.12), we find that the field produced at $z = z_o$ by this aperture illumination pattern is

$$\vec{E}_o(\vec{r}) = \left\{ \frac{ik_o A_e \vec{E}_c (r_o - y\ \sin\theta_o + z_o\ \cos\ \theta_o)}{4\pi r_o^2} \right\} \tag{3.18}$$

$$e^{\frac{-A_e k_o^2(x^2+\nu^2)}{8\pi r_o^2}}\ e^{\frac{ik_o(x^2\ +\ \eta^2)}{2r_o}}\ e^{-ik_o(r_{o+y}\sin\theta_o - z_o\cos\theta_o)} \tag{3.18}$$

The standard measure of the angular spread of an antenna is the full width at half the maximum power, Φ. Since power is proportional to the square of the modulus of the electric field, this angle may be obtained by noting the value of x at which

$$e^{-A_e k_o^2 x^2/4\pi r_o^2} = 0.5.$$

Thus,

$$\Phi \doteq \frac{2x}{r_o} = \sqrt{\frac{4\ \ln 2}{\pi}}\ \frac{\lambda}{\sqrt{A_e}} \doteq \frac{\lambda}{\sqrt{A_e}}. \tag{3.19}$$

Since $\sqrt{A_e}$ may be approximated by the antenna dimension in a particular direction, a close approximation for the beamwidth in that direction is

$$\Phi \sim \lambda/D_o. \tag{3.19a}$$

The gain G of an antenna is defined as (see Silver, 1949),

$$G = \frac{k_o^2 A_e}{\pi}. \tag{3.20}$$

In terms of Φ this is,

$$G \doteq \frac{4\pi}{\Phi^2} \tag{3.20a}$$

thus the gain of an antenna is approximately equal to the ratio of the total solid angle to that subtended by the antenna's beam.

Using Eq. (3.20), Eq. (3.18) for the field produced on a flat surface in the far field becomes,

$$\vec{E}_o(\vec{r}) = \left[\frac{i\vec{E}_c Ga(x,y,z_o)}{4k_o r_o}\right] e^{ik_o(r_{o+y}\sin\theta_o - z_o\cos\theta_o)} \tag{3.21}$$

where

$$a(x,y,z_o) \equiv \left(1 - \frac{y}{r_o}\sin\theta_o + \frac{z_o}{r_o}\cos\theta_o\right) e^{\frac{-G(x^2+\eta^2)}{8r_o^2}} e^{\frac{-ik_o(x^2+\eta^2)}{2r_o}}. \tag{3.22}$$

and, as before,

$$\nu = y\cos\theta_o + z_o\sin\theta_o.$$

4. Bragg Scattering from a Slightly Rough Surface

4.1. SCATTERED FIELDS RECEIVED BY AN ANTENNA

We are now in a position to calculate fields scattered from a slightly roughened air/water interface. Since the most important case for air/sea interaction research is backscatter, we shall consider only this case and let the transmitting antenna also be the receiving antenna. Then the scattered electric field is given by the Stratton-Chu equation, Eq. (3.6), but now with $\vec{r}$ being a position vector to the interface over which the integral is taken and $\vec{r}'$ being a position vector to a location very near the antenna aperture. The unit vector normal to the rough interface is $\hat{n}$; its primary component is vertical but it tilts depending on the local surface slope. The fields $\vec{E}$ and $\vec{H}$ on the interface are unknown. The perturbation approach is to replace these unknown fields on the wavy surface by fields on a flat surface whose values can be calculated from expansions of the boundary conditions. Thus, letting $\vec{E}'$ and $\vec{H}'$ be the flat surface fields, we write the Stratton-Chu equation as

$$\vec{E}_s(\vec{r}') = \frac{-1}{4\pi}\int \left[i\omega\mu(\hat{a}_z \times \vec{H}')\left(\frac{e^{ik_o\rho}}{\rho}\right) + (\hat{a}_z \times \vec{E}') \times \nabla \left(\frac{e^{ik_o\rho}}{\rho}\right) + (\vec{E}'\cdot\hat{a}_z)\nabla\left(\frac{e^{ik_o\rho}}{\zeta}\right)\right] dxdy. \tag{4.1}$$

Ultimately, however, we would like to obtain the mean power scattered by the interface to the receiving antenna. To this end, it is convenient to calculate the average scattered fields over the antenna aperture weighted by the appropriate field pattern for the aperture. Then the received power is simply half the magnitude of the product of average electric and magnetic fields and the effective area of the aperture. For the average electric field, we have

$$\langle \vec{E}_s \rangle = \frac{2}{A_e} \int \vec{E}_s(\vec{r}) e^{\frac{-2\pi(\eta^2+\xi^2)}{A_e}} d\eta d\xi. \tag{4.2}$$

Since the only quantity in Eq. (4.1) which depends on $\vec{r}\,'$, and therefore on η and ξ, is ρ, Eq. (4.2) may be evaluated in a manner identical to our evaluation of Eq. (3.9) to yield the following result:

$$\langle \vec{E}_s \rangle = -\frac{e^{ik_o r_o}}{4\pi r_o} \int \Big\{ i\omega\mu(\hat{a}_z \times \vec{H}\,')ae^{ik_o y \sin\theta_o} + (\hat{a}_z \times \vec{E}\,') \times \nabla(ae^{ik_o \hat{a}_o \cdot \vec{r}})\big|_{z=o}$$

$$+ (\vec{E}\,' \cdot \hat{a}_z)\nabla(ae^{ik_o \hat{a}_o \cdot \vec{r}})\big|_{z=o} \Big\} dxdy \tag{4.3}$$

where a is given by Eq. (3.22) and $\hat{a}_o = \sin\theta_o \hat{a}_y - \cos\theta_o \hat{a}_z$ is a unit vector in the direction of incidence. Now,

$$\nabla a \big|_{z=o} \doteq -\left(\frac{G}{4r_o} - ik_o\right)\frac{ax}{r_o}\hat{a}_x$$

$$- \left[\frac{1}{r_o} + \left(\frac{G}{4r_o} - ik_o\right)\frac{y}{r_o}\right] a \sin\theta_o \hat{a}_y + \frac{\cos\theta_o}{r_o}\hat{a}_z. \tag{4.4}$$

If antenna beamwidths are not too broad, we may safely assume that

$$|\nabla a|_{z=o}/a << k_o \tag{4.5}$$

i.e., that a is a slowly varying function of position. Then Eq. (4.3) may be written more simply as

$$\langle \vec{E}_s \rangle = \frac{-ik_o e^{ik_o r_o}}{4\pi r_o} \int a \Big\{ \frac{\omega\mu}{k_o}(\hat{a}_z \times \vec{H}\,')$$

$$+ (\hat{a}_z \times \vec{E}\,') \times \hat{a}_o + (\vec{E}\,' \cdot \hat{a}_z)\hat{a}_o \Big\} e^{ik_o y \sin\theta_o} dxdy. \tag{4.6}$$

Thus, $\langle \vec{E}_s \rangle$ is the two-dimensional Fourier transform of a times the quantity in brackets evaluated at $k_x = 0$, $k_y = -k_o \sin\theta_o$. Representing Fourier transforms with a "$\sim$" and noting that the Fourier transform of a product equals the convolution of the Fourier transforms, we have,

$$\langle \vec{E}_s \rangle = \frac{-ik_o e^{ik_o r_o}}{4\pi r_o} \left\{ \tilde{a} \otimes \left[\frac{\omega\mu}{k_o} (\hat{a}_z \times \tilde{\vec{H}}') + (\hat{a}_z + \tilde{\vec{E}}') \times \hat{a}_o + (\tilde{\vec{E}}' \cdot \hat{a}_z)\hat{a}_o \right\} \right. \tag{4.7}$$

where $\otimes$ represents convolution.

Now the Fourier components in (4.7) are by definition plane waves and only those components propagating from the surface toward the antenna will have $k_y = -k_o \sin\theta_o$. For these plane waves, we have from Eq. (2.18),

$$-\hat{a}_o \times \tilde{\vec{E}}' = \frac{\omega\mu_o}{k_o} \tilde{\vec{H}}'. \tag{4.8}$$

Substituting this in Eq. (4.7) and noting the vector identities

$$\hat{a}_o \times (\hat{a}_o \times \tilde{\vec{E}}') = (\hat{a}_z \cdot \tilde{\vec{E}}')\hat{a}_o - (\hat{a}_z \cdot \hat{a}_o)\tilde{\vec{E}}'$$

and

$$(\hat{a}_z \times \tilde{\vec{E}}') \times \hat{a}_o = (\hat{a}_z \cdot \hat{a}_o)\tilde{\vec{E}}' - (\hat{a}_o \cdot \tilde{\vec{E}}')\hat{a}_z$$

along with the fact that $\hat{a}_o \cdot E = 0$ from Eq. (2.15), we find,

$$\langle \vec{E}_s \rangle = \frac{-ik_o e^{ik_o r_o} \cos\theta_o}{2\pi r_o} [\tilde{a} \otimes \tilde{\vec{E}}']. \tag{4.9}$$

But the total field on our imaginary flat surface may be written

$$\vec{E}' = \vec{E}_a + \vec{e}_a \tag{4.10}$$

where $\vec{E}_a$ represents the incident and reflected fields if the surface were truly flat while $\vec{e}_a$ represents the scattered field caused by the surface roughness. Only the latter field has a Fourier component directed toward the antenna, so we have finally

$$\langle \vec{E}_s \rangle = \frac{-ik_o e^{ik_o r_o} \cos\theta_o}{2\pi r_o} [\tilde{a} \otimes \tilde{\vec{e}}_a] \tag{4.11}$$

and we are left only with the task of evaluating $\vec{e}_a$ from boundary conditions.

At this point, we must make the essential assumptions of slightly rough surface theory. We assume that surface slopes are small and that the projection of surface displacement on the incident ray is small compared to the electromagnetic wavelength. More precisely, if γ is surface displacement and γ_x, γ_y its derivatives in the x and y directions, we require that

$$|\gamma_x|, |\gamma_y| < 1 \tag{4.12}$$

and

$$|k_o \gamma \cos \theta_o| < 1. \tag{4.13}$$

Then we have for the normal vector $\hat{n}$,

$$\hat{n} = \frac{-\gamma_x \hat{a}_x - \gamma_x \hat{a}_y + \hat{a}_z}{\sqrt{1 + \gamma_x^2 + \gamma_y^2}} \doteq -\gamma_x \hat{a}_x - \gamma_y \hat{a}_y + \hat{a}_z \tag{4.14}$$

to first order in surface slope.

In Eq. (4.10) $\vec{E}_a$ is the sum of the incident and reflected fields $\vec{E}_o$ and $\vec{E}_r$ at a flat interface; below the interface, $\vec{E}_w$ is the refracted field. Since boundary conditions require that these fields be evaluated at $z = \gamma(x,y,t)$, the perturbation approach is to expand these fields around $z = 0$. Thus using Eq. (3.21) for $\vec{E}_o$ we have

$$\vec{E}_o = \frac{i\vec{E}_c G}{4k_o r_o} e^{ik_o r_o} e^{ik_o y \sin\theta_o} \left\{ a(x,y)[1 - ik_o \gamma \cos \theta_o] + \frac{\gamma \cos \theta_o}{r_o} \right\} \tag{4.15}$$

to first order in $k_o \gamma \cos \theta_o$ and γ / r. The latter term results from the expansion of $a(x,y,z)$ and is extremely small in cases of interest. Neglecting it yields a plane incident wave and the results of Section II may be used for incident, reflected, and refracted fields by letting

$$E_o = \frac{iE_c G a e^{ik_o r_o}}{4k_o r_o} \tag{4.16}$$

and

$$H_o = \frac{k_o E_o}{\omega \mu_o}.$$

At the slightly-rough interface, boundary conditions require that tangential components of electric and magnetic fields be continuous. Thus we require that

$$\vec{E}_a + \vec{e}_a - \hat{n}[\hat{n} \cdot (\vec{E}_a + \vec{e}_a)] = \vec{E}_w + \vec{e}_w - \hat{n}[\hat{n}(\vec{E}_w + \vec{e}_w)] \tag{4.17}$$

and

$$\vec{H}_a + \vec{h}_a - \hat{n}[\hat{n} \cdot (\vec{H}_a + \vec{h}_a)] = \vec{H}_w + \vec{h}_w - \hat{n}[\hat{n} \cdot (\vec{H}_w + \vec{h}_w)]$$

where $\vec{E}_{a,w}$ and $\vec{H}_{a,w}$ represent fields generated by a flat interface but evaluated at $z = \gamma(x,y,t)$ while $\vec{e}_{a,w}$ and $\vec{h}_{a,w}$ represent scattered electric and magnetic fields.

Since Eq. (4.11) shows that we are ultimately interested only in a single Fourier component of these fields, we may write

$$\vec{k}_s \times \vec{e}_a = \omega\mu_o\vec{h}_a$$

and (4.18)

$$\vec{k}_{sw} \times \vec{e}_w = \omega\mu_o\vec{h}_w$$

where $\vec{k}_s$ and $\vec{k}_{sw}$ are the wavevectors of scattered fields in air and water, respectively. These are given by

$$\vec{k}_s = -k_o \sin\theta_o\hat{a}_y + k_o \cos\theta_o\hat{a}_z$$

$$\vec{k}_{sw} = -k_o \sin\theta_o\hat{a}_y - k_w \cos\theta_w\hat{a}_z \tag{4.19}$$

for the Fourier components of interest. From the orthogonality of wavevector and electric field for any Fourier component,

$$\sin\theta_o e_{ay} = \cos\theta_o e_{az}$$

$$k_o \sin\theta_o e_{wy} = -k_w \cos\theta_w e_{wz} \tag{4.20}$$

where subscripts y and z indicate vector components along those directions. Thus we may write the scattered magnetic fields as

$$\vec{h}_a = \frac{k_o}{\omega\mu_o}\left\{-\frac{e_{ay}}{\cos\theta_o}\hat{a}_x + \cos\theta_o e_{ax}\hat{a}_y + \sin\theta_o e_{ax}\hat{a}_z\right\}$$

$$\vec{h}_w = \frac{k_o}{\omega\mu_o}\left\{\left[\frac{k_w e_{wy}}{k_o\cos\theta_w}\right]\hat{a}_x - \left[\frac{k_w\cos\theta_w e_{wx}}{k_o}\right]\hat{a}_y + \sin\theta_o e_{wx}\hat{a}_z\right\}. \tag{4.21}$$

Since $\vec{e}$ and $\vec{h}$ are first-order or higher in surface roughness, we may set $\hat{n} = \hat{a}_z$ when it multiplies these fields. Thus we may write the boundary conditions in the form

$$\vec{e}_a - \vec{e}_w - (e_{az} - e_{wz})\hat{a}_z = \vec{E}_w - \vec{E}_a - \hat{n}[\hat{n} \cdot (\vec{E}_w - \vec{E}_a)]$$

$$\vec{h}_a - \vec{h}_w - (h_{az} - h_{wz})\hat{a}_z = \vec{H}_w - \vec{H}_a - \hat{n}[\hat{n} \cdot (\vec{H}_w - \vec{H}_a)] \tag{4.22}$$

It is convenient now to consider horizontal and vertical polarization separately.

Horizontal Polarization: Using Eq. (4.14) for $\hat{n}$ and expanding the fields $\vec{E}_a$, $\vec{E}_w$, $\vec{H}_a$, and $\vec{H}_w$ given in Eqs. (2.36) - (2.39) to first order in $k_o \gamma \cos \theta_o$, we find the following relationships among the vector components:

$$e_{ax} - e_{wx} = 0$$

$$e_{ay} - e_{wy} = 0$$

$$h_{ax} - h_{wx} = 0$$

$$h_{ay} - h_{wy} = -4i(R_H/T_H)H_o k_o \gamma \cos^2 \theta_o e^{ik_o y \sin\theta_o}.$$

Here we have used Eq. (2.34) and (2.35) to obtain the result. Substituting for $\vec{h}$ components in the last two expressions using Eq. (4.21), we obtain

$$e_{ay} = 0$$

and

$$e_{ax} = 2iR_H E_o k_o \gamma \cos \theta_o e^{ik_o y \sin\theta_o}.$$

Furthermore, Eq. (4.20) shows that $e_{az} = 0$. Thus we have for horizontally polarized transmission and reception

$$\tilde{\vec{e}}_a = 2iR_H k_o \cos \theta_o \hat{a}_x \int E_o \gamma(x, y, t) e^{i2k_o y \sin\theta_o} dxdy. \tag{4.23}$$

Substituting this expression and Eq. (4.16) into Eq. (4.11) we obtain,

$$\langle \vec{E}_s \rangle = \frac{iE_c k_o G R_H \cos^2\theta_o e^{i2k_o r_o} \hat{a}_x}{4\pi r_o^2} \int a^2(x, y)\gamma(x, y, t) e^{i2k_o y \sin\theta_o} dxdy. \tag{4.24}$$

where we have again used the equivalence of a convolution of transforms with a transform of a product. Equation (4.24) is the expression for the mean electric field received by a horizontally polarized antenna when a horizontally polarized transmitted field is backscattered from a slightly rough surface.

Vertical Polarization: Again using $\hat{n}$ to first order in wave slope and expanding $\vec{E}_a$, $\vec{E}_w$, $\vec{H}_a$, and $\vec{H}_w$ to first order in $k_o \gamma \cos\theta_o$ using Eqs. (2.45) to (2.48) we find

$$e_{ax} - e_{wx} = E_o e^{ik_o y \sin\theta_o} \left(\frac{k_o^2 - k_w^2}{k_w^2} \right) T_V \sin\theta_o \gamma_x$$

$$e_{ay} - e_{wy} = E_o e^{ik_o y \sin\theta_o} \left(\frac{k_o^2 - k_w^2}{k_w^2} \right) T_V \sin\theta_o \{ik_o \gamma \sin\theta_o + \gamma_y\}$$

$$h_{ax}\ h_{wx} = H_o e^{ik_o y \sin\theta_o} \left(\frac{k_o^2 - k_w^2}{k_o^2} \right) (ik_o \gamma \cos\theta_o)(1 - R_V)$$

$$h_{ay} - h_{wy} = 0.$$

Here we have obtained the result using Eqs. (2.34) and (2.44) and the fact implicit in Eq. (4.19) that $k_w \sin\theta_w = -k_o \sin\theta_o$ for the Fourier components of interest.

The last equation yields a relationship between e_{ax} and e_{wx} which can be used to solve the first equation for e_{ax}. The result is that e_{ax} is proportional to $a\gamma_x$ since E_o depends on a. Now within the approximation of Eq. (4.5) $a\gamma_x$ may be replaced by $\partial(a\gamma)/\partial x$. But in order to obtain $\tilde{\tilde{e}}_a$ to use in Eq. (4.11) we must integrate $\partial(a\gamma)/\partial x$ over x from $-\infty$ to $+\infty$ which simply yields the difference in $a\gamma$ evaluated on the lines $(-\infty, y)$ and $(+\infty, y)$. Thus since $a \to o$ as $x \to \pm\infty$, e_{ax} contributes nothing to the final result.

In order to evaluate the remaining two components of $\tilde{\tilde{e}}_a$, we use Eq. (4.21) to convert the above equation for $h_{ax} - h_{wx}$ into one relating e_{ay} and e_{wy}. This can then be combined with the above equation for $e_{ay} - e_{wy}$ to obtain an expression for e_{ay}. After some algebraic manipulation, we find

$$e_{ay} = 2ik_o E_o e^{ik_o y \sin\theta_o} \left\{ \gamma R_V \cos^2\theta_o \quad - \gamma_y T_V^2 \left(1 - \frac{\epsilon_o}{\epsilon_w} \right) \sin\theta_o / 4ik_o \right\}. \qquad (4.25)$$

In obtaining $\tilde{\tilde{e}}_a$, the y transform yields the following integral which may be evaluated by parts:

$$\int \frac{\partial(a\gamma)}{\partial y} e^{i2k_o \sin\theta_o} dy = a\gamma e^{i2k_o y \sin\theta_o} \Big|_{-\infty}^{\infty} - i2k_o \sin\theta_o \int a\gamma e^{i2k_o y \sin\theta_o} dy.$$

The first term of the right hand side yields zero by the same arguments as applied above to e_{ax}. Thus Eq. (4.25) yields

$$\tilde{e}_{ay} = 2ik_o \left[R_V \cos^2\theta_o + \frac{1}{2} T_V^2 \left(1 - \frac{\epsilon_o}{\epsilon_w} \right) \sin^2\theta_o \right] \int E_o \gamma e^{i2k_o y \sin\theta_o} dy.$$

But Eq. (4.20) shows that,

$$\tilde{e}_{az} = \frac{\sin\theta_o \tilde{e}_{ay}}{\cos\theta_o}.$$

Thus we finally have for vertical polarization

$$\tilde{\vec{e}}_a = 2ik_o \left[R_V \cos^2\theta_o + \frac{1}{2} T_V^2 \left(1 - \frac{\epsilon_o}{\epsilon_w} \right) \sin^2\theta_o \right]$$

$$\left[\hat{a}_y + \frac{\sin\theta_o \hat{a}_z}{\cos\theta_o} \right] \int E_o \gamma e^{i2k_o y \sin\theta_o} dydx. \tag{4.26}$$

Substituting this equation and Eq. (4.16) into Eq. (4.11) yields

$$\langle \vec{E}_s \rangle =$$

$$\frac{iE_c k_o G \left[R_V \cos^2\theta_o + \frac{1}{2} T_V^2 \left(1 - \epsilon_o/\epsilon_w \right) \sin^2\theta_o \right] [\cos\theta_o \hat{a}_y + \sin\theta_o \hat{a}_z] e^{i2k_o r_o}}{4\pi r_o^2}$$

$$\int a^2(x,y)\gamma(x,y,t) e^{i2k_o y \sin\theta_o} dxdy. \tag{4.27}$$

This is the expression for the mean electric field received by a vertically polarized antenna when a vertically polarized transmitted field is backscattered from a slightly rough surface.

4.2. NORMALIZED RADAR CROSS SECTION FOR BRAGG BACKSCATTER

Equations (4.24) and (4.27) show that the electric field backscattered from a slightly rough air/sea interface is proportional to the two-dimensional Fourier transform of the surface displacement $\gamma(x,y,t)$ evaluated at $k_x = 0$, $k_y = 2k_o \sin\theta_o$. The area over which the transform is taken is limited by the tapering function $a^2(x,y)$. Thus backscatter is caused only by surface waves which lie within a restricted wavelength range and travel nearly toward or away from the antenna. We will discuss these features of the backscatter later in some detail. First, however, we obtain an expression for the normalized radar cross section σ_o of the air/sea interface.

The power received by the antenna may be written

$$P_r = \frac{1}{2} A_e \overline{\langle \vec{E}_s \rangle \times \langle \vec{H}_s \rangle} = \frac{A_e k_o}{2\omega\mu_o} \overline{|\langle \vec{E}_s \rangle|^2} \tag{4.28}$$

where the overbar indicates a time average. For horizontal polarization, we substitute Eq. (4.24) to obtain,

$$P_r = \left[\frac{A_e k_o E_c^2}{8\omega\mu_o} \right] \left[\frac{k_o^2 G^2 |R_H|^2 \cos^4 \theta_o}{2\pi^2 r_o^4} \right] \overline{|\int a^2 \gamma e^{i2k_o y \sin\theta_o} dxdy|^2} \tag{4.29}$$

From Eq. (3.16) we see that the first quantity in brackets is simply the transmitted power, P_t. Thus Eq. (4.29) may be written in the standard form of the radar equation

$$P_r = \frac{P_t G^2 \lambda^2 \sigma}{(4\pi)^3 r_o^4} \tag{4.30}$$

if the cross section σ is defined as

$$\sigma = \frac{4k_o^4 |R_H|^2 \cos^4 \theta_o}{\pi} \overline{|\int a^2 \gamma e^{i2k_o y \sin\theta_o} dxdy|^2}. \tag{4.31}$$

Because this expression for σ increases in direct proportion to the illuminated surface area, it is convenient to normalize by this area and define the result as σ_o. We may rewrite Eq. (4.31) as,

$$\sigma = \frac{4k_o^4 |R_H|^2 \cos^4\theta_o}{\pi} \int a^{2*}(x,y) a^2(x+u, y+v)$$

$$\overline{\gamma^*(x,y,t)\gamma(x+u,y+v,t)}\, e^{i2k_o v \sin\theta_o} dudvdxdy. \tag{4.32}$$

If $a(x,y)$ varies only slowly in distances over which the surface displacement decorrelates, then we may set u and v equal to zero in a^2. Furthermore, if $\gamma(x,y,t)$ is homogeneous, then $\overline{\gamma\gamma^*}$ does not depend on x and y so we may write

$$\sigma_o = \frac{\sigma}{A} = 16\pi k_o^4 |g_H(\theta_o)|^2 \psi(0, 2k_o \sin\theta_o) \tag{4.33}$$

where

$$g_H(\theta_o) = R_H \cos^2 \theta_o \tag{4.34}$$

and

$$\psi(0, 2k_o \sin\theta_o) = \frac{1}{4\pi^2}\int \overline{\gamma^*(x, y, t)\gamma(x + u, y + v, t)}\, e^{i2k_o v \sin\theta_o} dudv \qquad (4.35)$$

is the variance spectrum of surface displacement and

$$A = \int |a^2(x, y)|^2\, dxdy\ , \qquad (4.36)$$

is the illuminated area. If the function $a^2(x,y)$ varies more rapidly, it is convenient to retain the definition of A but generalize the definition of the variance spectrum to

$$\psi(0, 2k_o\ \sin\theta_o) = \frac{1}{4\pi^2 A}\int a_1(u,v)\overline{\gamma\gamma^*}\ e^{i2k_o v \sin\theta_o} dudv \qquad (4.37)$$

where

$$a_1(u,v) = \int a^{2*}(x,y) a^2(x + u, y + v)\, dxdy \qquad (4.38)$$

In this case, Eq. (4.33) still holds but the radar will respond to surface waves having a range of wavenumbers determined by the shape of the Fourier transform of $a_1(u,v)$.

Using Eq. (3.22) we may evaluate the expression for the illuminated area, Eq. (4.36). Thus,

$$A = \int \left[1 - \frac{y}{r_o}\sin\theta_o\right]^4 e^{\frac{-G(x^2 + y^2\cos^2\theta_o)}{2r_o^2}}\, dxdy.$$

To be consistent with the approximations of our scattering model, we need only retain first-order terms in y/r_o. Thus we find

$$A = \frac{2\pi r_o^2}{G\cos\theta_o}.$$

Using Eq. (3.20a) for the antenna gain G, we find

$$A \doteq \frac{1}{2} r_o{}^2 \Phi^2/\cos\theta_o. \qquad (4.39)$$

The point of maximum illuminated intensity within the illuminated area A need not coincide with $y = 0$. We may find the true point of maximum intensity by finding the point at which $|a^2|^2$ maximizes as a function of y. Setting the y derivative of $|a^2|^2$ equal to zero yields y_m, the point of maximum illuminated intensity:

$$y_m \doteq -\frac{r_o \Phi^2 \sin\theta_o}{\pi \cos^2\theta_o}. \tag{4.40}$$

We may define an effective incidence angle θ_e to be the angle to this point. It is given by

$$\sin\theta_e \doteq \sin\theta_o \left[1 - \frac{\Phi^2}{\pi \cos^2\theta_o}\right]. \tag{4.41}$$

Obviously θ_e deviates from θ_o by larger amounts as θ_o approaches 90°. In fact we must omit large incidence angles from this analysis in order to retain the approximation $y/r << 1$.

The fact that Φ^2 appears in all three of the last equations is a consequence of our assumptions of equal vertical and horizontal beamwidths for the antenna. In the more general case of unequal beamwidths, Φ^2 in Eq. (4.39) must be replaced by $\Phi_V \Phi_H$ while that in the last two equations becomes Φ_V^2.

Turning now to vertical polarization, we substitute Eq. (4.27) into Eq. (4.28) to obtain

$$P_r = \left[\frac{A_e k_o E_c^2}{8\omega\mu_o}\right] \left[\frac{k_o^2 G^2 |R_V \cos^2\theta_o + \frac{1}{2} T_V^2 (1 - \epsilon_o/\epsilon_w) \sin^2\theta_o|^2}{2\pi r_o^4}\right]$$

$$\overline{|\int a^2 \gamma e^{i2k_o y \sin\theta_o} dx dy|^2} \tag{4.42}$$

Thus we have for σ_o,

$$\sigma_o = 16\pi k_o^4 |g_V(\theta_o)|^2 \psi(0, 2k_o \sin\theta_o) \tag{4.43}$$

where

$$g_V(\theta_o) = R_V \cos^2\theta_o + \frac{1}{2} T_V^2 (1 - \epsilon_o/\epsilon_w) \sin^2\theta_o. \tag{4.44}$$

We may express σ_o for either polarization as

$$\sigma_o = 16\pi k_o^4 |g_p(\theta_o)|^2 \psi(0, 2k_o \sin\theta_o) \tag{4.45}$$

where p may be either H or V to indicate polarization. We have for horizontal polarization

$$g_H(\theta_o) = \frac{(\epsilon - 1)\cos^2\theta_o}{\left[\cos\theta_o + \sqrt{\epsilon - \sin^2\theta_o}\right]^2} \tag{4.46}$$

and for vertical polarization

$$g_V(\theta_o) = \frac{(\epsilon - 1)[\epsilon(1 + \sin^2\theta_o) - \sin^2\theta_0]\cos^2\theta_o}{\left[\epsilon\cos\theta_o + \sqrt{\epsilon - \sin^2\theta_o}\right]^2}. \tag{4.47}$$

Here $\epsilon = \epsilon_w/\epsilon_o$. The quantities $|g_H|^2$ and $|g_V|^2$ are plotted as functions of incidence angle in Figure 8 for a frequency of 10 GHz.

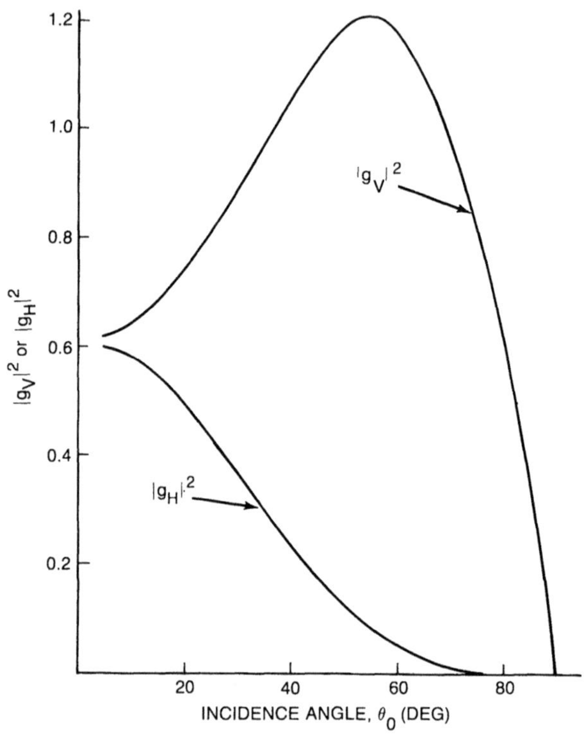

Figure 8. Bragg scattering geometric coefficients $|g_V(\theta_o)|^2$ and $|g_H(\theta_o)|^2$ at 10 GHz plotted as a function of incidence angle, θ_o. Note that the coefficient $|g_V|^2$ for vertically polarized transmission and reception is generally much larger and has less dependence on incidence angle than that for horizontal polarization.

5. Basic Surface Wave Theory

Before considering the experimental evidence supporting Bragg scattering from slightly rough air/sea interfaces, we need to understand some basic characteristics of surface waves on such interfaces. Detailed discussions of the properties of such waves and of the turbulent wind fields which generally produce them are presented in other chapters of this book. Here we shall develop only those properties necessary for understanding Bragg scattering.

The equations governing the motions of the fluids near the air/sea interface are the Navier-Stokes equations along with the condition of incompressibility. If we neglect viscous effects, these may be written

$$\frac{\partial \vec{u}_t}{\partial t} + (\vec{u}_t \cdot \nabla)\vec{u}_t = -\nabla P \tag{5.1}$$

$$\nabla \cdot \vec{u}_t = o \tag{5.2}$$

Here $\vec{u}_t$ is the total fluid velocity and P is the normal stress to density ratio

$$P = p/\rho + gz \tag{5.3}$$

where p is pressure, ρ is density, and g is the gravitational acceleration.

Equation (5.1) is non-linear in the fluid velocity. We may, however, consider this velocity to consist of a mean $\vec{U}$ and a turbulent part $\vec{u}$:

$$\vec{u}_t = \vec{U} + \vec{u}. \tag{5.4}$$

For simplicity we shall consider the turbulent fluctuations of velocity to be small and work only to first order in the fluctuations. Furthermore, we shall consider $\vec{U}$ to be horizontal vector which is a function of z only. Then Eq. (5.1) becomes

$$\frac{\partial \vec{u}}{\partial t} + (\vec{U} \cdot \nabla)\vec{u} + (\vec{u} \cdot \nabla)\vec{U} = -\nabla P. \tag{5.5}$$

This equation is now linear in $\vec{u}$ so we may consider individual Fourier components of the fluctuations separately. Again for simplicity we shall assume that the component of interest travels in the y direction and let $\vec{U}$ be in that direction also. Then

$$\vec{u} = [v_o(z)\hat{a}_y + w_o(z)\hat{a}_z]e^{i(k_w y - \omega t)} \equiv v\hat{a}_y + w\hat{a}_z \tag{5.6}$$

and Eqs. (5.2) and (5.5) may be written in component form as

$$ik_w(U - c)v + wU' = -\frac{\partial P}{\partial y} \tag{5.7}$$

$$ik_w(U - c)w = -P' \tag{5.8}$$

$$ik_w u + w' = 0 \tag{5.9}$$

where $c = \omega/k_w$ and primes represent differentiation with respect to z.

Differentiating Eq. (5.8) by z and Eq. (5.9) by y and eliminating the resulting $\partial P'/\partial y$ yields,

$$ik_w[U'v + (U - c)v'] + w'U' + [U'' + k_w^2(U - c)]w = 0.$$

Using Eq. (5.9) to eliminate v and v' we obtain the Rayleigh Equation (Rayleigh, 1876):

$$w'' - \left[\frac{U''}{(U - c)} + k_w^2\right] w = 0. \tag{5.10}$$

Closed form solutions of this equation cannot be obtained for most functions $U(z)$. In the particular case $U(z) = 0$, however, the solution is simple:

$$w(z) = Ae^{-k_w z} + Be^{k_w z}. \tag{5.11}$$

In air, we require that $w \to 0$ as $z \to \infty$. Thus

$$w(z) = Ae^{-k_w z} \quad , z > 0. \tag{5.12}$$

If the water has a finite depth d, however, we must require that $w(-d) = 0$ or that $A/B = -e^{-2k_w d}$, $z < 0$. Thus we may write

$$w(z) = A \sinh k_w(z + d) \quad , z < 0 \tag{5.13}$$

Furthermore, if the surface is to remain connected, we must have

$$w(z = \gamma) = \frac{\partial \gamma}{\partial t} + \vec{u} \cdot \nabla \gamma. \tag{5.14}$$

Linearizing this equation and working with a single Fourier component of γ then yields for $U(z) = 0$,

$$w(0) = -i\omega\gamma. \tag{5.15}$$

Thus from Eq. (5.12),

$$A = -i\omega\gamma \quad , z > 0$$

and

$$w(z) = -i\omega\gamma e^{-k_w z} \quad , z > 0. \tag{5.16}$$

From Eq. (5.9),

$$v(z) = \omega\gamma e^{-k_w z} \quad , z > 0. \tag{5.17}$$

Similarly, using Eq. (5.13),

$$A = \frac{-i\omega\gamma}{\sinh k_w d} \quad , z < 0$$

and

$$w(z) = \frac{-i\omega\gamma \sinh k_w(z + d)}{\sinh k_w d} \quad , z < 0. \tag{5.18}$$

Again using Eq. (5.9) we find

$$v(z) = \frac{\omega\gamma \cosh k_w(z + d)}{\sinh k_w d} \quad , z < 0. \tag{5.19}$$

Equations (5.16) to (5.19) give the fluid velocity components which accompany a Fourier component of surface displacement with wavenumber k_w and radian frequency ω . Note that horizontal components of velocity do not match at $z = 0$ due to our neglect of molecular viscosity. If viscous effects were included, a very thin viscous boundary layer near the interface would provide the transition between v above and below the surface. When $z = 0$, Eqs. (5.18) and (5.19) describe velocity components just below the surface itself. These are usually referred to as components of the orbital velocity and are given by

$$w(0) = i\omega\gamma \tag{5.20}$$

$$v(0) = \omega\ \gamma \coth k_w d.$$

Thus the vertical velocity component is 90° out of phase with the surface displacement and upward in front of the wave crest while $v(0)$ is in phase with the displacement and in the direction of wave propagation at the crest. Wave orbital velocities are illustrated in Figure 9.

One additional condition must be satisfied at the interface. The difference in normal stress across the interface must be due only to surface tension. To first order, surface tension is proportional to surface curvature so we may write

$$p_w + \rho_w g\gamma - (p_a + \rho_a g\gamma) = \rho_w T \left(\frac{\partial^2\gamma}{\partial x^2} + \frac{\partial^2\gamma}{\partial y^2} \right) \tag{5.21}$$

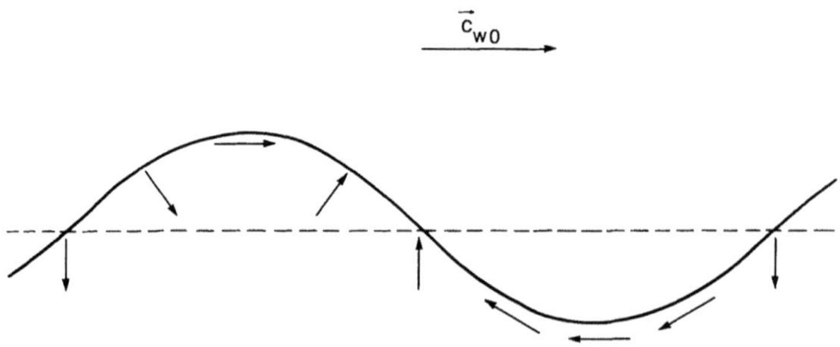

Figure 9. Diagram of orbital velocities at the surface of water disturbed by a sinusoidal wave. The wave is moving to the right as indicated by the phase speed c_{wo}.

at $z = 0$. Here T is the ratio of surface tension to water density. Substituting Eqs. (5.20) into Eq. (5.7) yields, for $U(z) = 0$ and $z = 0$,

$$p = -\rho c^2 k_w \gamma \coth k_w d \tag{5.22}$$

where p and ρ may refer to either air or water. Substituting this into Eq. (5.21) and neglecting air density compared to water density, we find

$$c_{wo} = \sqrt{(g/k_w + Tk_w) \tanh k_w d}. \tag{5.23}$$

This is the phase speed of a surface wave of wavenumber k_w in the absence of wind. The corresponding wave frequency is

$$f_{wo} = \frac{1}{2\pi} \sqrt{(gk_w + Tk_w^3) \tanh k_w d} \tag{5.24}$$

The group speed of a set of water waves is given by $c_{go} = 2\pi df/dk$. The three quantities c_{wo}, c_{go}, and f_{wo} are plotted in Figure 10 as functions of k_w for $d = \infty$.

If a current U_s exists in the water, then the wave frequency observed from a stationary platform, f_w, is the above f_{wo} plus $k_w U_s/2\pi$ while phase and group speeds, c_w and c_g, are changed by the addition of U_s. Thus in general

$$\begin{aligned} f_w &= f_{wo} + k_w U_s/2\pi \\ c_w &= c_{wo} + U_s \\ c_g &= c_{go} + U_s \end{aligned} \tag{5.25}$$

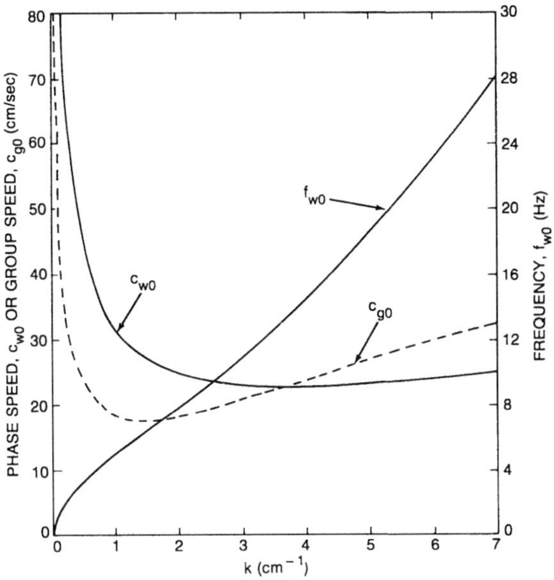

Figure 10. Phase speed, group speed, and frequency of surface water waves in the absence of wind and currents plotted as a function of wavenumber for deep water.

The above expressions for wave orbital velocity, phase speed, and frequency are valid only in the case of no mean air flow and low velocity fluctuations. Since the equations were linearized in $\vec{u}$ in order to obtain these expressions, they hold as long as γ may be represented as a linear superposition of Fourier components. When $U(z) \neq 0$ but the turbulence is low, the same procedures may be used to evaluate these quantities but in most cases, closed form solutions cannot be obtained. If the turbulence of the wind field becomes large enough, then the above first-order procedures are no longer valid and the original equations must be solved (numerically) to higher order in $\vec{u}$.

6. Experimental Support for Slightly Rough Surface Theory

As pointed out in the introduction, the first evidence that perturbation theory might explain some aspects of sea echo was obtained at HF, or megahertz frequencies, utilizing the ground wave mode of propagation. Such scattering still provides some of the clearest evidence for Bragg scattering through its Doppler spectrum. This spectrum is defined in general as

$$F(f_d) = \int_{-T/2}^{T/2} \overline{E_s^*(t)E_s(t+\tau)}\, e^{i2\pi f_d \tau} d\tau \tag{6.1}$$

where $E_s(t)$ is the magnitude of the electric field received by the antenna. If this field is given by either Eq. (4.24) or (4.27) then

$$F(f_d) = \frac{E_c^2 k_o^2 G^2 |g_p|^2}{(4\pi)^2 r_o^4} \int_{-T/2}^{T/2} \overline{\tilde{\gamma}^*(0,k_B,t)\tilde{\gamma}(0,k_B,t+\tau)}\, e^{i2\pi f_d \tau} d\tau \tag{6.2}$$

where $k_B = 2k_o \sin\theta_o$ and $\tilde{\gamma}$ represents the generalized Fourier transform given by the integral in Eq. (4.24) or (4.27). Here we have omited the microwave frequency which is typically removed from the signal through heterodyning, or beating with a similar frequency sine wave. It is customary to remove range, transmitted power, and antenna gain effects from $F(f_d)$ so that

$$\sigma_o = \int_{-\infty}^{\infty} F(f_d)\, df_d \tag{6.2a}$$

Thus we may write in place of Eq. (6.2),

$$F(f_d) = \frac{\sigma_o}{\psi(0,k_B)} \int_{-T/2}^{T/2} \overline{\tilde{\gamma}^*(0,k_B,t)\tilde{\gamma}(0,k_B,t+\tau)}\, e^{2\pi f_d \tau} d\tau \tag{6.2b}$$

Equation (6.2) shows that if the spectrum of surface displacements contains components

$$\gamma(x,y,t) = Ae^{i(k_{By} \pm 2\pi f_w t)} \tag{6.3}$$

then

$$F(f_d) = \sigma_o \delta(f_d \pm f_w) \tag{6.4}$$

if we assume that the time T is long enough that the Fourier transform may be represented as a pseudo δ-function. Equation (6.4) says that the Doppler spectrum of a signal Bragg-backscattered from the sea will contain two sharp lines whose Doppler shift, or frequency, f_d is plus and minus the wave frequency:

$$f_d = \pm f_w \tag{6.5}$$

Figure 11 shows a Doppler spectrum of HF (9.4 MHz) ground wave sea echo taken from the paper of Barrick et. al. (1974). The ordinate is $F(f_d)$ in decibels or 10 log $F(f_d)$ while the abcissa is f_d/f_{wo}. Clearly, prominent peaks are observed at nearly ± 1. The small shift Δ is due to surface currents as discussed in the last section; Bragg scattering provides a method for measuring ocean surface currents (Barrick et. al. 1977). The peak near $+1$ is by far the dominant feature of the Doppler spectrum being nearly 15 dB above, or 32 times as strong as, the largest second-order peak. The electromagnetic wave frequency of 9.4 MHz corresponds to an electromagnetic wavelength, $\lambda = c_o/f$, of 31.9 meters. The Bragg resonance condition $k_w = 2k_o \sin\theta_o$ may be written

$$\lambda_w = \frac{\lambda_o}{2\sin\theta_o} \tag{6.6}$$

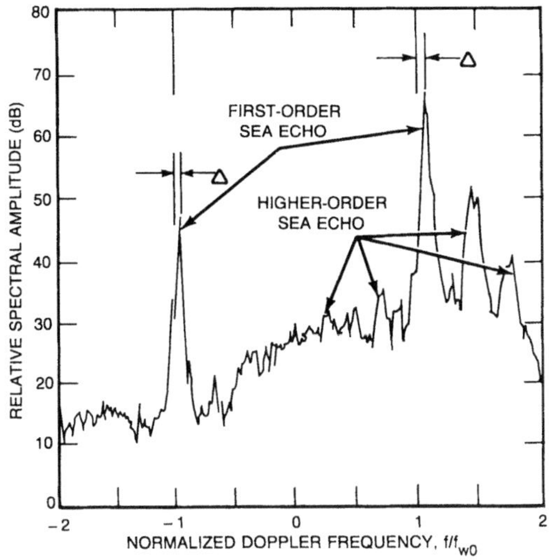

Figure 11. Doppler spectrum of HF backscatter (9.4 MHz) from the sea surface in the ground wave mode. The small shift Δ of the first-order peaks from the wave frequency f_{wo} is due to surface currents. From Barrick et. al. (1974).

where $\lambda_w = 2\pi/k_w$ is water wavelength. Since in this case, $\theta_o \approx 90°$, this implies that the Bragg-resonant surface wave, or Bragg wave, is about 16 meters long. Thus the strong first-order peak in Figure 11 corresponds to a 16 meter wave traveling downwind while the much weaker first-order peak corresponds to a wave of the same wavelength traveling upwind. Such upwind-traveling waves are nearly always observed in HF ground wave backscatter from the sea. Whether it's Doppler shift is assigned a positive or negative sign depends on whether the frequency used to beat out the HF frequency, ie, the local oscillator frequency, is higher or lower than the transmitted frequency. Thus although the Bragg resonance condition requires that the Bragg wave be traveling nearly toward or away from the antenna, the sign of the Doppler shift does not by itself determine the direction of wave travel.

At microwave frequencies, i.e., above about 1 GHz, evidence for first-order Bragg scattering is more difficult to obtain in the field; we will explore the reasons for this in the following two sections. In laboratory wind-wave tanks, however, the viability of first-order Bragg theory is relatively easy to investigate. The equations developed above show that according to this theory, the backscatter is caused only by surface waves satisfying the resonance condition Eq. (6.6) and traveling nearly toward or away from the antenna. Both of these predictions may be checked by generating small sinusoidal waves on the water surface with a paddle or plunger. Figure 12 shows the results of experiments of this type performed in the Naval Research Laboratory wind-wave tank. Figure 12 was obtained by pointing the antenna of an X-Band (10.0 GHz, vertical polarization) microwave system along the tank axis at a 48° incidence angle and varying the frequency of a mechanically-generated sinusoidal surface wave. The amplitude of this surface wave was monitored with a laser and kept constant to

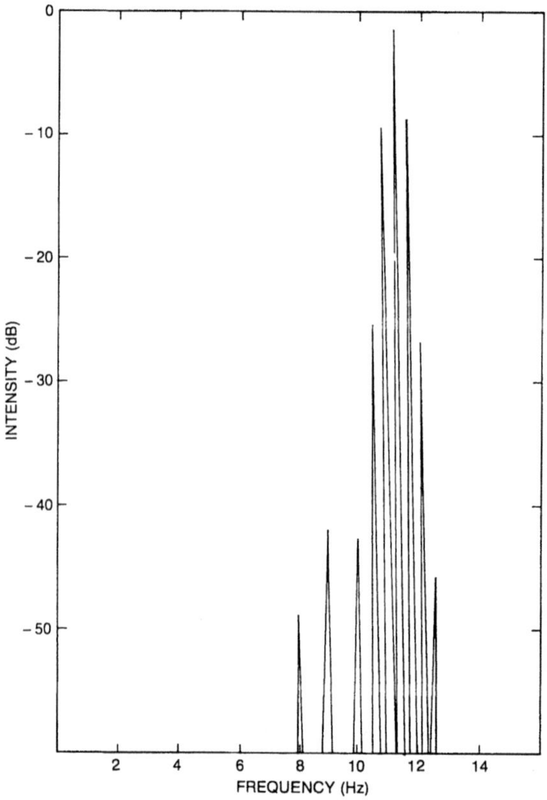

Figure 12. Doppler spectra of microwave backscatter from sinusoidal surface waves of variable frequency. Data were taken in an wavetank with an X-Band system (10.0 GHz) which was vertically polarized and directed into the wave propagation direction. Doppler frequency is exactly the frequency of the sinusoidal wave. Note that the antenna responded most readily to the 11.4 Hz waves which satisfy the Bragg resonance condition Eq. (6.6).

within 3 dB. As in Figure 11, the Doppler spectrum of the system output is displayed in Figure 12 with the ordinate in decibels. In Figure 12, however, the true Doppler frequency is given on the abcissa and spectra from surface waves of several frequencies are superimposed. Obviously, this microwave system responds most strongly to surface waves with a frequency of 11.4 Hz. From Figure 9, this corresponds to a water wavenumber k_w of 3.06 cm^{-1} or a wavelength of 2.05 cm, almost exactly those which satisfy the resonance condition.

Figure 12 shows that the microwave system responds not only to waves of exactly the Bragg wavenumber but also to those waves in a small wavenumber range around k_B. For scattering from the far field of the antenna, we may calculate the expected width of this wavenumber region by computing the Fourier transform of the tapering function $a_1(u,v)$ of Eq. (4.36a). After some calculation, we find that

$$\tilde{a}_1(k_x, 2k_o \sin\theta_o - k_y) \propto \exp\left\{-\frac{k_x^2}{K^2}\right\} \exp\left\{\frac{-(k_y - 2k_o \sin\theta_e)^2}{K^2 \cos^2\theta_o}\right\} \tag{6.7}$$

where

$$K = \sqrt{\frac{2\pi}{(r_o \Phi)^2} + \frac{2k_o^2 \Phi^2}{\pi}} \tag{6.8}$$

Note that $\tilde{a}_1$ is maximized by $k_y = 2k_o \sin\theta_e$ where the effective incidence angle θ_e is given by Eq. (4.39). The width of the wavenumber resonance region in the y direction at the e^{-1} point is simply $K \cos\theta_o$. For the X-Band system used to obtain Figure 12, $r_o = 3.7$ m, $k_o = 1.96$ cm^{-1} and $\Phi = \lambda/D = 0.05$. Thus Eq. (6.8) gives $K \cos\theta_o = 0.105$ cm^{-1}. From Figures 10 and 12 the half width to the e^{-1}, or -4.3 dB, point is about 0.094 cm which is satisfactory agreement since far-field conditions were not really satisfied in this wavetank situation. Phase differences across the antenna footprint had been minimized by focusing the antenna, however. For a discussion of this focusing technique, see Larson and Wright (1974).

Figure 13 shows the effect of rotating the antenna in azimuth when a mechanically-generated sinusoidal wave of exactly the Bragg wavelength is on the surface. In this case a vertically polarized L-Band system with a frequency of 1.85 GHz and a 32° incidence angle was used. This yielded a Bragg wavelength of 15.3 cm, a wave which has a frequency of 3.2 Hz. In Figure 12b, Doppler spectra are again displayed with a logarithmic ordinate showing intensity and a linear abcissa showing Doppler frequency. Zero Doppler frequency is shifted right as the antenna is rotated so that the abcissa also represent angle as indicated on the lower scale. We may find the expected angular half-width to the e^{-1} point easily from Eq. (6.7) for a far-field situation. It is

$$\Delta\phi \doteq \frac{K}{2k_o \sin\theta_o} = \sqrt{\frac{2\pi}{(k_B r_o \Phi)^2} + \frac{\Phi^2}{2\pi \sin^2\theta_o}}. \tag{6.9}$$

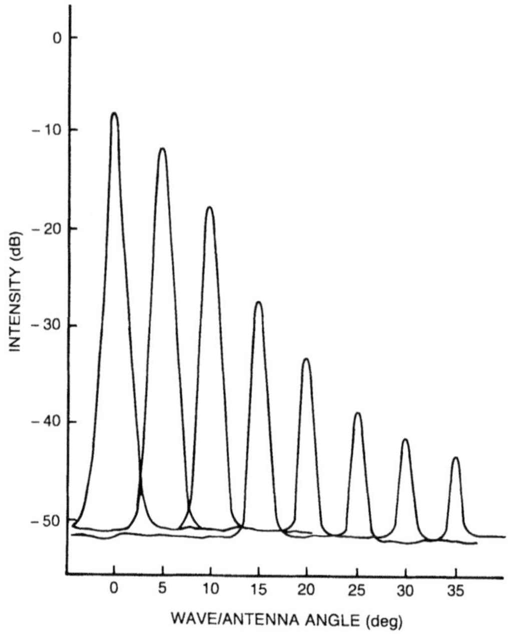

Figure 13. Doppler spectra of microwave backscatter from a sinusoidal surface wave making various angles with the horizontal antenna-look direction. Data were taken in a wavetank with an L-Band system (1.85 GHz) which was vertically polarized and operated in a bistatic mode. The Bragg wavelength was 15.3 cm so that all peaks shown are at 3.2 Hz.

For our L-Band antenna, $\Phi = 0.133$ and $r_o = 2.4$ m so Eq. (6.9) yields $\Delta\phi = 12.4°$ while the figure indicates a value closer to 7 or 8° for the focussed antenna system which, of course, did not satisfy far-field conditions.

Wright (1966) extensively examined the magnitude of backscatter from mechanically-generated sinusoidal surface waves both to see if it was proportional to water wave amplitude as Eqs. (4.24) and (4.27) indicate and to determine whether Eqs. (4.45) to (4.47) gave the proper dependence on incidence angle. He used an X-Band system with a transmit-receive horn located about five feet from the surface. Rather than focus the antenna to a small spot on the surface, he masked all but a portion of the surface with absorbing material. His results are summarized in Figures 14 to 18. Figure 14 shows that the square root of cross section is linearly dependent on water wave amplitude as expected. These results were obtained at 30° and 60° incidence angles and horizontal polarization. Figure 15 shows the dependence of horizontal and vertical cross sections on depression angle (90° - incidence angle) while Figure 16 shows their ratio. Also plotted in these figures are the theoretical values expected from Eqs. (4.45) to (4.47). The agreement between experimental and theoretical ratios of σ_V/σ_H is striking although each of the individual cross sections show rather large deviations from theoretical curves in some places. To check these angular dependences for a fluid with a very

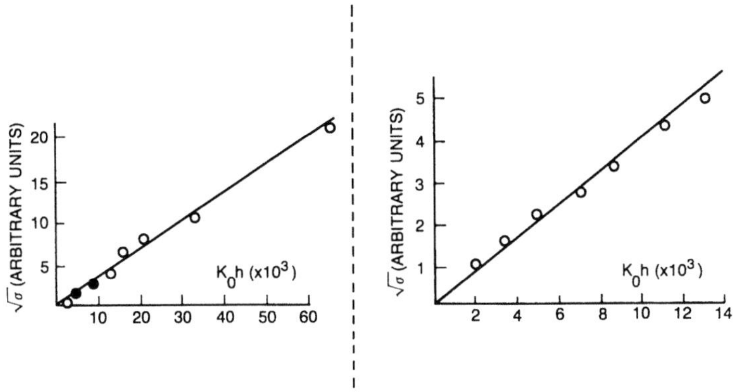

Figure 14. Dependence of the backscattering cross section, σ, at X-Band on the amplitude of a mechanically-generated surface wave. a) $\theta_o = 30°$ and horizontal polarization. b) $\theta_o = 60°$ and horizontal polarization. From Wright (1966).

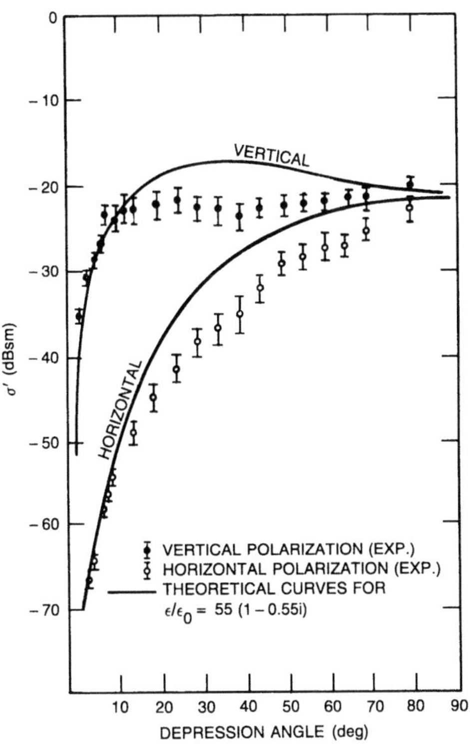

Figure 15. Vertical and horizontal backscattering cross sections of mechanically-generated surface waves on water versus depression angle, $90 - \theta_o$. Circles are data collected with an X-Band system. Lines indicate the predictions of Bragg scattering theory. From Wright (1966).

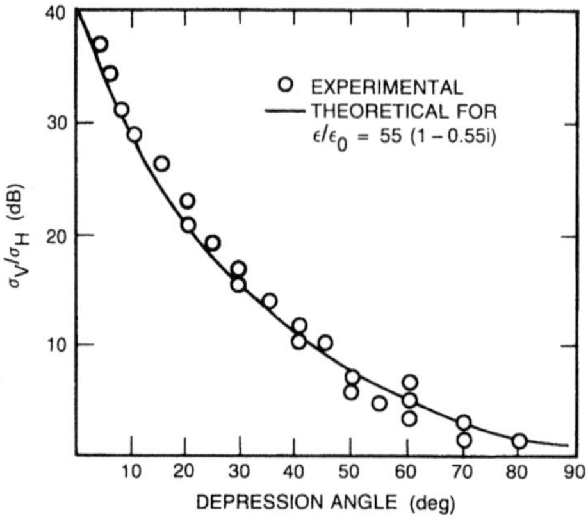

Figure 16. Ratio of vertical to horizontal cross sections versus depression angle. Circles are data of Figure 15 while the solid line is the prediction of Bragg scattering theory. From Wright (1966).

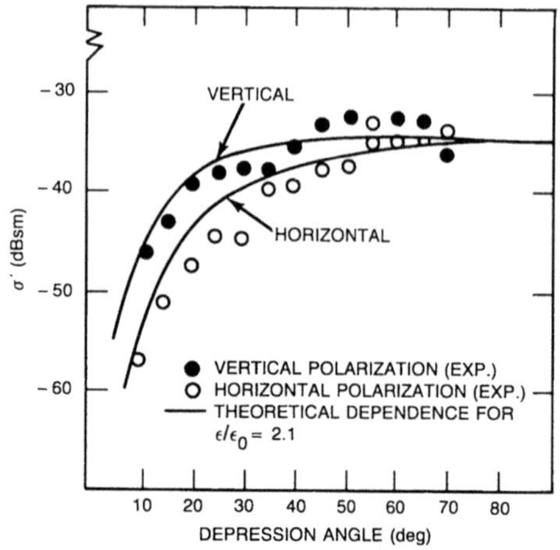

Figure 17. Vertical and horizontal backscattering cross sections of mechanically-generated surface waves on JP-5 jet fuel versus depression angle. Data were collected with an X-Band system; lines indicate theoretical predictions. From Wright (1966).

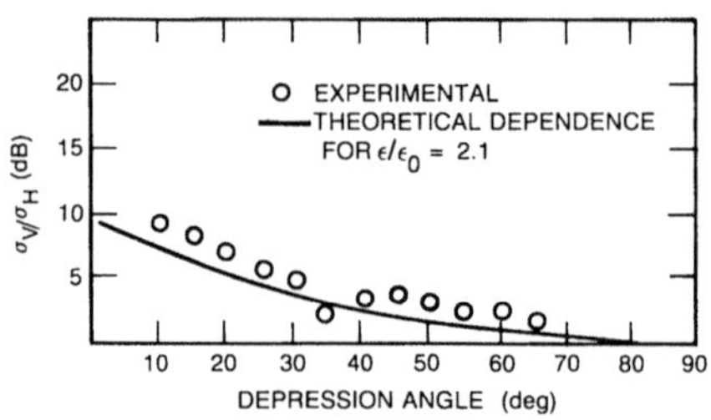

Figure 18. Ratio of vertical to horizontal cross sections versus depression angle. Circles are data of Figure 17 while the solid line is the prediction of Bragg scattering theory. From Wright (1966).

different dielectric constant, Wright also performed backscattering experiments with his tank full of U.S. Navy JP-5 jet fuel. Figure 17 shows that the individual cross sections agreed with theory somewhat better than for the case of a water surface. Again Figure 18 shows that the ratio of cross sections closely followed the expected curve. These results indicate that first-order Bragg theory rather well explains the dependence of backscatter on wave amplitude and incidence angle although the deviations from the expected curves for scattering from water remain unexplained.

Taken together, these wavetank results on mechanically-generated waves indicated that first-order Bragg theory will explain backscatter from an air/water interface when the basic assumptions of Eq. (4.12) and (4.17) are well satisfied. When the surface waves are generated by wind, however, these equations cannot always be satisfied. We will examine this problem further in the following section. Conditions can be established in wind/wave tanks, however, which allow the assumptions of Bragg theory to be satisfied most of the time. In such situations, first-order Bragg scattering can be demonstrated. Furthermore, the limitations of the theory can be examined by increasing wind speed or fetch in the tank.

Figures 19 and 20 show two series of wind-wave Doppler spectra obtained at various wind speeds in the wind/wave facility at the Naval Research Laboratory. Figure 19 is taken from Larson and Wright (1974) and shows Doppler spectra taken with a C-Band system (4.3 GHz) at a 60° incidence angle. This yields a Bragg wavelength of 4.05 cm. Friction velocity, u_*, (See Chapter 5) is indicated in the figure as a measure of wind speed. A very sharp first-order Bragg peak is clearly seen up to friction velocities of about 34 cm/sec. At this point, the assumptions of first-order theory begin to break down and a second-order peak at a higher frequency begins to appear. Further increases in wind speed cause the spectra to broaden greatly indicating that first-order Bragg theory alone is not a satisfactory explanation of the scattering. These spectra were taken at very short fetch, about 2 meters, so that the wavelength of the dominant surface wave was shorter than the Bragg wavelength until about $u_* = 26$ cm/sec. It thus appears that first-order Bragg theory is viable until the dominant wind wavelength slightly exceeds the Bragg wavelength.

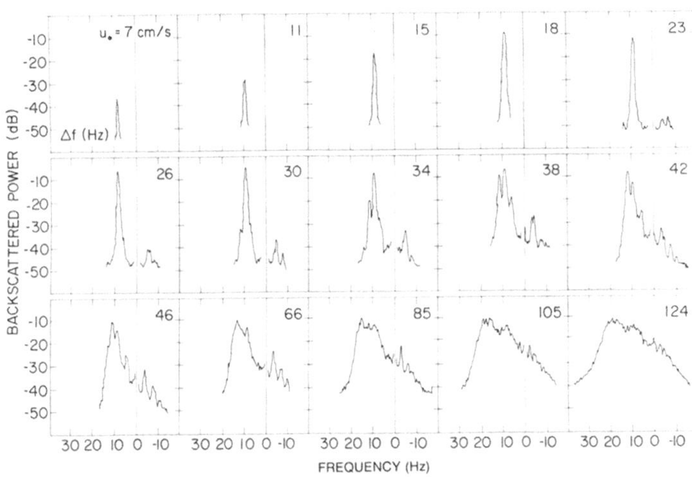

Figure 19. Microwave Doppler spectra of wind-generated water waves for various wind speeds indicated by the friction velocity, u_*. Data were collected with a C-Band system at a 60° incidence angle. The Bragg wavelength was 4.05 cm, accounting well for the sharp first-order Bragg line observed at low wind speeds. From Larson and Wright (1974).

This effect can also be observed in Figure 20 which is taken from Plant and Wright (1977). In this case, the data were also taken with a C-Band system operating at 4.375 GHz but were taken in a bistatic mode in which the receiving antenna was arranged to receive forward scatter. For this arrangement, the Bragg condition must be written

$$k_B = |k_o[\sin \theta_t + \sin \theta_r]| \tag{6.10}$$

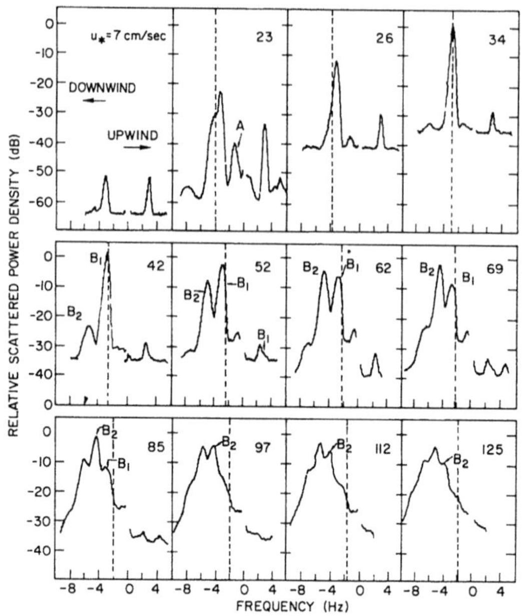

Figure 20. Microwave Doppler spectra of wind-generated water waves for various wind speeds. Data were collected with a bistatic C-Band system with a transmit incidence angle of 30° and a receive incidence angle of $-65°$. The Bragg wavelength of 16.5 cm accounts for the first-order Bragg peaks, B_1. Second order peaks are clearly evident for friction velocities above 42 cm/sec. Dashed lines indicate the wavelength of the dominant wind-generated wave on the surface. First-order theory holds until this wavelength exceeds the Bragg wavelength. From Plant and Wright (1977).

where θ_t is the incidence angle of the transmitting antenna and θ_r is the incidence angle of the receiving antenna. In the case of backscatter, $\theta_t = \theta_r = \theta_o$ and the equation yields our standard result. For the case of forward scatter, however, θ_r is negative and the Bragg wavelength can be made much longer than that for backscatter. For the case of Plant and Wright, $\theta_t = 30°$, $\theta_r = -65°$ and $\lambda_B = 16.5$ cm. The spectra of Figure 19 were taken at an 8.4 meter fetch so that the dominant surface wavelength reached the Bragg wavelength at about the same wind speed as in the earlier study. Note the upwind-traveling wave similar to those observed in HF backscatter from the ocean which is visible at low wind speeds. Dashed lines in this figure show the frequency of the dominant wind wave. Bragg wavelength and dominant wavelength are equal at $u_* = 34$ cm/sec. Above this wind speed the conditions for first-order scattering again begin to break down. Lines indicated by B_1 and B_2 designate Bragg lines of first and second order. The line indicated by A is unexplained but quite small; the misshapen Bragg peak in this same panel can be explained as a dominant wave effect. Essentially, the k-space antenna pattern is broad enough in this case to include both the Bragg wave and the dominant wave.

7. Composite Surface Theory

The last two figures clearly show that when the dominant wind-generated surface wave becomes longer than the Bragg wavelength, first-order Bragg theory begins to lose its viability. This means that one of the conditions given in Eq. (4.12) and (4.13) is being violated. Since these two conditions are limitations on acceptable surface slopes and displacements, respectively, we can determine which is most likely to be violated by looking at Figures 21 and 22 which show probability distributions of wave slope and height on the ocean. Figure 21 is taken from Tang and Shemdin (1983) and shows probability distributions of downwind and crosswind slopes obtained using their laser wave follower. The figure plots the probability distributions of surface slopes versus the ratio of the slope to the standard deviation of slope. The wind speed when these data were taken was 11.3 m/s, the downwind standard deviation was 0.185 radians and that for crosswind was 0.176 radians. The system was capable of measuring waves up to 100 Hz in frequency. The solid curve in Figure 21 shows that the data fit a Gaussian distribution rather well. For such a distribution, 95% of all slopes lie within ±2 standard deviations. Thus the figure indicates that there is only about a 5% probability of the slopes of ocean waves being larger than 0.37 radians or 21.2°. To a very high level of probability then, Eq. (4.12) is satisfied by surface wave slopes.

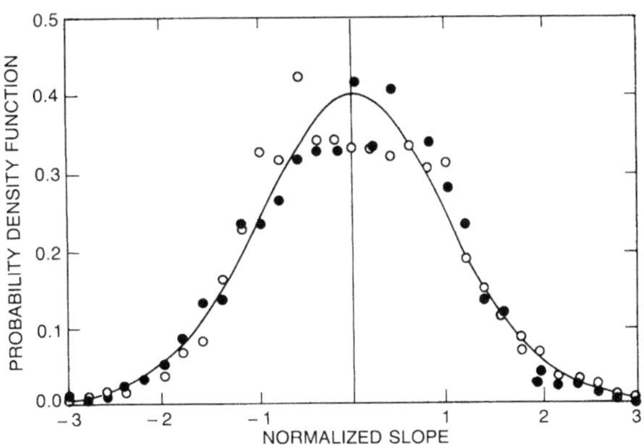

Figure 21. Distribution of surface wave slopes on the ocean obtained with a laser wave follower. Abcissa is slope normalized by the standard deviation of the sample. Solid circles represent downwind slopes with a standard deviation of 0.185 radians. Open circles indicate cross wind slopes whose standard deviation is 0.176 radians. The wind speed was 11.4 m/sec. The solid curve represents a Gaussian distribution. From Tang and Shemdin (1983).

The condition which is obviously much more likely to be violated by wind-generated waves is that requiring small heights compared to the microwave length. This is illustrated by Figure 22 taken from Longuet-Higgins et. al. (1963). The figure shows the probability distribution of surface wave heights measured with one of the earliest pitch-roll buoys. The buoy was about 1.7 meters in diameter so only heights of longer waves were measured. This figure indicates that a Gaussian distribution is also a close approximation to that of the height

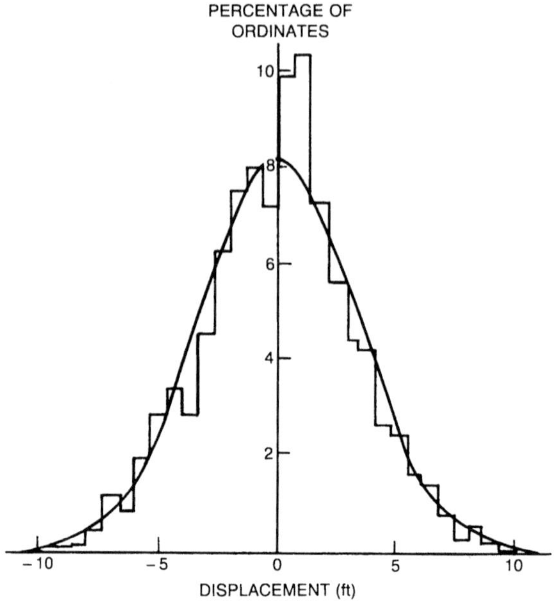

Figure 22. Distribution of surface wave displacements on the ocean obtained with an early pitch-roll buoy. The buoy was 1.7 meters in diameter so only displacements due to long waves were measured. The wind speed was 11.8 m/sec. The solid curve shows a Gaussian distribution. From Longuet-Higgins et. al. (1963).

distribution of these long waves. Now, however, the probability that Eq. (4.13) is satisfied at microwave lengths is almost zero.

Composite surface theory was developed in order to deal with the large surface displacements found in ocean waves within the framework of first-order Bragg scattering. The theory postulates that the surface may be divided into a large-scale part which has small curvature and a small-scale part which backscatters microwave radiation according to first-order Bragg theory. The large-scale surface is divided into patches which may be approximated by plane surfaces. Slightly-rough surface scattering theory is applied to each patch individually with the first-order expansion being made with respect to the plane, mean surface of the patch. The effect of the large-scale surface is therefore to simply tilt and advect the small-scale structure which backscatters according to first-order Bragg theory. Total received power is obtained by adding the power received from all patches illuminated by the radar system.

Several conditions must be met by the surface waves in order that composite surface theory be viable. First, higher-order Bragg scattering must be small compared to first-order scattering within each patch. Second, a cutoff wavelength must exist between the large and small-scale waves such that the small-scale surface satisfies Eq. (4.13) while at the same time the curvature of the large scale surface can be ignored over distances in which the small-scale structure decorrelates. That is, if γ_s represents the small-scale displacement, the correlation function, $\overline{\gamma_s(x,y,t)\gamma_s(x+u,y+v,t)}$, must approach zero as u and v increase. The distance $\xi \equiv |u\hat{a}_x + v\hat{a}_y|$ at which the correlation function falls to e^{-1} times its maximum value is the decorrelation length. Only this decorrelation of the small-scale structure allows powers from various patches to be added to obtain the total power. If backscattered return from individual patches were correlated, then the total backscattered field would first have to be determined and power obtained from its squared modulus. In this case, interference between fields scattered from individual patches would affect the total power received.

We shall examine the viability of these requirements presently. First, let us establish the approximate range of incidence angles over which we expect this composite surface theory to be valid. At low incidence angles, we must require that the slope be less than the tangent of the incidence angle to insure that no portion of the surface is oriented perpendicular to the incoming radiation so that specular scattering can occur. That is, we require that

$$|\gamma_x|, |\gamma_y| \lesssim \tan\theta_o. \tag{7.1}$$

This is satisfied 95% of the time if $\tan\theta_o$ exceeds 0.37 when the wind speed is 11.3 m/s. Thus an approximate lower limit on θ_0 is

$$\theta_o \gtrsim 20° \tag{7.2}$$

Similarly, if the large-scale surface tilts too far away from the antenna at large incidence angles, portions of the surface will not be fully illuminated. Thus we also require that

$$|\gamma_x|, |\gamma_y| \gtrsim \cot\theta_o. \tag{7.3}$$

Again, with 95% probability this is satisfied at 11.3 m/sec if

$$\theta_o \lesssim 70°. \tag{7.4}$$

These limits on θ_o are necessarily approximate since they must be specified only on a probability basis, must depend on wind speed and sea state, and must vary with antenna beamwidth. Nevertheless they give us a basis for consideration of the other requirements of composite surface theory.

We turn now to the question of whether we may safely ignore second and higher-order Bragg scattering. Figure 20 shows that second-order scattering equals or exceeds first-order in some wavetank situations. The reason for this is that the dominant surface wave has grown to be twice as long as the Bragg wave and quite steep. For such steep waves, Stokes (1947) showed that higher harmonics are generated whose amplitudes are proportional to the slope of the first harmonic raised to the power of the harmonic. These harmonics move with the first harmonic, or dominant wave, at the same phase speed. In our wavetank situation, then, the second spatial harmonic of the dominant wave at some point has the same wavelength as the Bragg wave and therefore backscatters according to first-order theory. Its frequency, f_2, however, is twice that of the dominant wave. Thus it may be calculated from the Bragg wavenumber according to

$$f_2 = \frac{1}{2\pi}\sqrt{2gk_B + \frac{1}{2}Tk_B^3} \tag{7.5}$$

if we ignore wind drift effects. For the situation of Figure 20, this predicts the position of the second-order Bragg peak to be at 4.35 Hz, a value in good agreement with observation. We conclude that the peak B_2 in Figure 20 is primarily the result of scattering from the second harmonic of the dominant wave. Figure 23 shows frequency spectra of wind-generated waves obtained in the same tank as the Doppler spectra of Figure 20. The figure is taken from Plant (1980) and plots the spectral density of wind-wave displacements, $E(f)$, times the dominant wave frequency, f_m, to the fifth power versus f/f_m. The figure shows that dominant wind waves in the tank have spectral densities given by $E(f_m) = 100 f_m^{-5}$. Frequency spectra from the ocean, however, exhibit spectral densities given approximately by $0.5 f^{-5}$ in the high frequency range (Phillips, 1977). Thus dominant waves in wavetanks are much larger in amplitude than waves of comparable frequencies on the ocean and should be expected to generate much steeper harmonics. In fact, for the above values of spectral densities, the second harmonic of the oceanic wave would be only 0.005 times as large as in the tank. In such a situation, the second-order spectral line would be reduced below that of Figure 20 by 23 dB making it negligible compared to the first-order one. On this basis it appears that the neglect of higher-order Bragg scattering by composite surface theory is justified.

We must now examine the question of whether a cutoff wavelength λ_c can exist such that the small-scale structure satisfies Eq. (4.13) while the large-scale surface has small curvature over the small-scale decorrelation length. Assuming Gaussian statistics for the small-scale displacements, Eq. (4.13) will be satisfied 95% of the time for this scale if

$$2k_o \sqrt{\overline{\gamma_s^2}} \cos\theta_o < 1 \tag{7.6}$$

Phillips (1977) shows that short wave spectral densities may be approximated by $F(k) \approx 0.005\, k^{-4}$ although the constant must vary somewhat with wind speed. Thus

$$\overline{\gamma_s^2} \doteq \int_{2\pi/\lambda_c}^{\infty} F(k) k dk \doteq \frac{0.0025\, \lambda_c^2}{(2\pi)^2}$$

Substituting into Eq. (7.6), we find that the requirement on the amplitude of the small-scale structure will be met 95% of the time if

$$\lambda_c < 20\lambda_B \tan\theta_B \tag{7.7}$$

The long-wave displacement, γ_L, may be expanded about any point in a Taylor series. We require that terms in this series beyond the linear one which defines the tangent plane be small compared to one when multiplied by the vertical projection of the wavenumber. Thus we require that

$$\frac{1}{2} \left| \frac{\partial^2 \gamma_L}{2x^2} \right| \left(\frac{\xi}{2} \right)^2 k_o \cos\theta_o < 1. \tag{7.8}$$

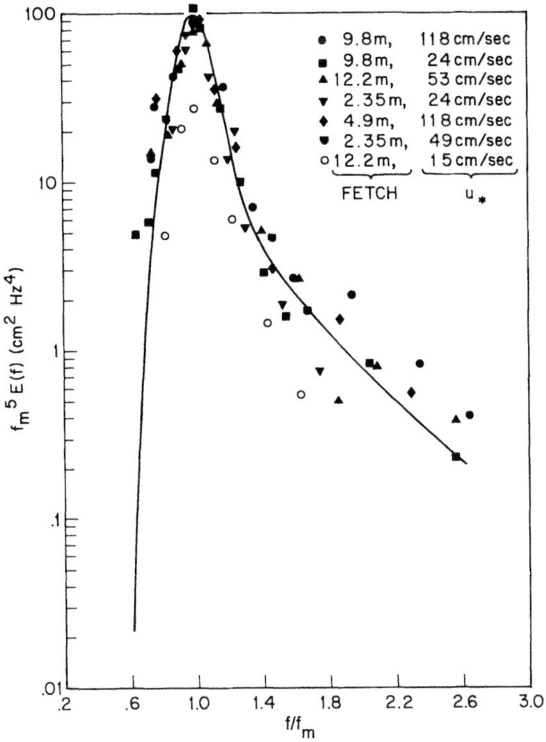

Figure 23. Self-similar spectra of wind-generated waves in a wavetank. Plotted as dominant-wave frequency, f_m, to the fifth power times spectral density versus frequency normalized by f_m, spectra at various fetches and wind speed collapse to a single curve. Spectral densities of dominant waves in the wavetank are much larger than those of waves of the same frequency on the ocean. From Plant (1980).

where the decorrelation length ξ, is the expansion parameter of the Taylor series. For a long wave of the cutoff wavelength, we have

$$\frac{\partial^2 \gamma_L}{2x^2} = \frac{2\pi \gamma_{cx}}{\lambda_c} \tag{7.9}$$

where γ_{cx} defines the long-wave slope at the cutoff wavelength. Substituting Eq. (7.9) into (7.8) yields

$$\lambda_c > \frac{|\gamma_{cx}|\pi^2 \xi^2}{4\lambda_B \tan\theta_o}. \tag{7.10}$$

But a patch size is defined by the decorrelation length and if it becomes small compared to λ_B, Bragg scattering can no longer hold for the patch. Therefore if composite surface theory is to be viable, ξ must be at least a few Bragg wavelengths. Let $\xi = 3\lambda_B$. Then, at the 95% confidence level, the cutoff wavelength must satisfy

$$\lambda_c > \frac{8.3\ \lambda_B}{\tan\theta_o}. \tag{7.11}$$

If this lower limit on λ_c is too low, however, the mean surface $z = 0$ itself satisfies the requirement of Eq. (4.13) and the wave of length λ_c need not be included with the long waves according to composite surface theory. The condition is that

$$k_o \gamma_c \cos\theta_c < 1$$

where γ_c is the surface displacement of the wave of length λ_c. Writing this in terms of the slope of this wave and the Bragg wavelength we obtain

$$\lambda_c < \left| \frac{2}{\gamma_{cx}} \right| \lambda_B \tan\theta_o \tag{7.12}$$

At the 95% confidence level, this is

$$\lambda_c < 6\ \lambda_B \tan\theta_o \tag{7.13}$$

If λ_c satisfies this relationship, then the surface displacement of long waves of length λ_c can be ignored on a composite surface theory.

Combining Eqs. (7.11) and (7.13) with (7.7), we find that a cutoff wavelength must exist which satisfies

$$\max\left\{6\ \lambda_B \tan\theta_o,\ \frac{8.3\ \lambda_B}{\tan\theta_o}\right\} < \lambda_c < 20\ \lambda_B \tan\theta_o \tag{7.14}$$

where max ( ) indicates that the maximum value of the two expressions in parenthesis should be used.

This condition is well satisfied at large incidence angles but it can be met only down to a lower incidence angle of about 35°. If we allow ourselves to work at confidence levels of 70%, however, the condition can be satisfied down to the lowest incidence angles at which composite surface theory can be expected to hold. Thus the theory should be a rather good approximation to the actual backscatter for incidence angles between about 20° and 70°.

We may develop equations appropriate to composite surfaces directly from the basic equations of first-order theory, Eqs. (4.24) and (4.27). These may be written

$$E_s = \frac{iE_c k_o G g_p(\theta_o)}{4\pi r_o^2} e^{i2k_o r_o} \int a_1^2 \gamma e^{i2k_o y \sin\theta_o} dxdy \tag{7.15}$$

where $g_p(\theta_o)$ represents Eq. (4.34) or (4.44) depending on polarization and we have written only the magnitude of $\vec{E}_s$. We may apply this equation to each patch of the large-scale surface if we remember that we set z_o of Eq. (3.10) equal to zero in obtaining Eq. (7.15) while now z_o must equal γ_L at the center of the patch. Also, the vector normal to any patch may now be tilted both in and out of the plane of incidence so that we must alter $g_p(\theta_o)$ slightly (Wright 1968, Valenzuela 1968). Tilting in the plane of incidence simply changes the incidence angle slightly so we must replace θ_o in g_p by θ' where

$$\theta' = \theta_o - \gamma_{Ly}. \tag{7.16}$$

Tilting out of the plane of incidence, however, is equivalent to rotating the surface by an angle ϕ', say, about the direction of propagation. Thus the incident field is a combination of vertical and horizontal polarization with respect to the tilted surface. For a vertically polarized incident field,

$$\vec{E}_o = E_o \sin \phi' \hat{a} + E_o \cos \phi' \hat{b}$$

where $\hat{a}$ is a unit vector perpendicular to $\vec{k}_o$ and lying in the tilted surface and $\hat{b}$ is a unit vector perpendicular to $\vec{k}_o$ and in the local plane of incidence. Thus the component of the scattered field which is vertically polarized is

$$E_s = E_{SV} \cos^2 \phi' + E_{SH} \sin^2 \phi'$$

where E_{SH} and E_{SV} are scattered fields polarized horizontally and vertically with respect to the tilted surface. These latter fields are each given by Eq. (7.15) with the appropriate $g_p(\theta')$. Thus $g_V(\theta', \phi')$ for the tilted surface is given by

$$g_V(\theta', \phi') = g_V(\theta') \cos^2 \phi' + g_H(\theta') \sin^2 \phi'. \tag{7.17}$$

In a similar manner, we find

$$g_H(\theta', \phi') = g_H(\theta') \cos^2 \phi' + g_V(\theta') \sin^2 \phi'. \tag{7.18}$$

For small angles ϕ', $\cos \phi' \approx 1$ and $\sin \phi' \approx \gamma_{Lx}/\sin \theta'$. Furthermore, Figure 8 indicates that for most incidence angles $g_H << g_V$. Thus we may write approximately

$$g_V(\theta', \gamma_{Lx}) \doteq g_V(\theta') \tag{7.19}$$

$$g_H(\theta', \gamma_{Lx}) \doteq g_H(\theta') + [\gamma_{Lx}/\sin \theta']^2 g_V(\theta').$$

Tilting the normal to a patch out of the plane of incidence is more important for horizontal polarization than for vertical.

We may now write down the field scattered from each small patch according to composite surface theory. We have

$$E_s = \frac{iE_c k_o G g_p(\theta', \gamma_{Lx})}{4\pi r_o^2} e^{i2k_o r_o} e^{-i2k_o \gamma_L \cos\theta_o} \int_{-\xi/2}^{\xi/2} a_1^2 \gamma_s e^{i2k_o y \sin\theta_o} dxdy \tag{7.20}$$

where a_1^2 is evaluated at the center of the patch since it varies little across the patch. The small-scale displacement, γ_s, is now expanded about the plane tangent to the surface at the center of the patch. Note that the entire time-dependent structure γ_s is being advected at the velocity of the large-scale patch on which it resides.

Adding fields from all patches together (recall that all vector fields are in the same direction), we obtain

$$E_s(t) = \frac{iE_c k_o G}{4\pi r_o^2} e^{i2k_o r_o} \int a_1^2 g_p(\theta', \gamma_{Lx})$$

$$e^{-i2k_o \gamma_L \cos\theta_o} \gamma_s e^{i2k_o y \sin\theta_o} dxdy. \tag{7.21}$$

Here we have taken $g_p(\theta', \gamma_{Lx})$ and the exponential containing γ_L inside the integral since θ' and γ_L depend now on x and y. The integral is taken over the entire horizontal plane and the time-dependence of E_s is due to that of both γ_L and γ_s. This form of E_s is useful but is only valid on the assumption that γ_s decorrelates very rapidly with position. We may understand the effect of the exponential containing γ_L by expanding $\gamma_L(x,y,t)$ about any point $(x_o, y_o, t_o,)$ and neglecting spatial terms above the linear one describing the tangent plane

$$e^{-i2k_o \gamma_L(x,y,t)\cos\theta_o} = e^{-i2k_o \gamma_L^o \cos\theta_o} e^{-i2k_o \gamma_{Lt}^o t \cos\theta_o}$$

$$e^{-ik_o \gamma_{Ltt}^o t^2 \cos\theta_o} e^{-i2k_o \gamma_{Lx}^o x \cos\theta_o} e^{-i2k_o \gamma_{Ly}^o y \cos\theta_o}. \tag{7.22}$$

Superscripts "o" indicate quantities evaluated at (x_o, y_o, t_o). The first exponential on the left of Eq. (7.22) merely adds a random phase to the already random phases of the components of γ_s. Thus it contributes nothing to the resulting backscatter. The next two exponentials containing time derivatives of γ_L account for the vertical components of velocity and acceleration of the large-scale patch. (Note that horizontal components are intrinsic to the time-dependence of γ_s.) Finally, the last two exponentials change the resonance condition slightly due to the tilting of the surface. Thus the Bragg resonance condition for the patch centered at (x_o, y_o) is

$$\vec{k}_w = (2k_o \gamma_{Lx}^o \cos\theta_o, 2k_o \sin\theta') \tag{7.23}$$

where we make the approximation using Eq. (7.16) that

$$\sin\theta' \doteq \sin\theta_o - \gamma_{Ly}^o \cos\theta_o$$

We may now easily write down the composite surface expression for the normalized radar cross section for backscatter from the sea surface. Let $\alpha \doteq \gamma_{Ly}$ and $\phi \doteq \gamma_{Lx}$. Then the cross section of each patch may be obtained from Eq. (7.20) in a manner identical to that of Section IV. We obtain the normalized radar cross section of the entire illuminated area by summing the cross sections of all patches weighted by the probability $P(\alpha,\phi)$ of finding a particular large-scale slope and dividing by the area. This yields

$$\sigma_o = 16\pi k_o^4 \int |g_p(\theta_o,\alpha,\phi)|^2 \psi(2k_o\phi \cos\theta_o,\, 2k_o \sin\theta')\, P(\alpha,\phi) d\alpha d\phi. \qquad (7.24)$$

It is usually acceptable to approximate $2k_o\phi \cos\theta_o$ by zero. Equation (7.19) with this value set to zero was first derived by Wright (1968).

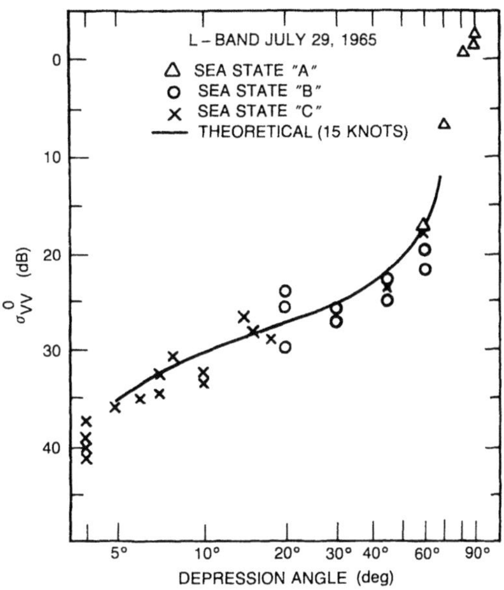

Figure 24. The normalized radar backscatter cross section of the sea looking upwind versus depression angle for L-Band, vertically polarized data of Macdonald (1956). The line indicates the prediction of composite surface theory. From Wright (1968).

8. Experimental Support for Composite Surface Theory

The earliest, and still some of the most compelling evidence that composite surface theory accounts for microwave backscatter from the real sea surface came from measurements of σ_o versus incidence angle. Such measurements date from the mid 1950s (See Macdonald 1956, Wiltse et al., 1957) and some of these early data were used by Wright (1968) to check the predictions of the theory. Figure 24 is taken from Wright's paper and compares the L-Band (1.225 GHz) data of Macdonald with composite surface theory predictions of σ_o as a function of depression angle. The agreement between these vertical polarization data and predictions is striking. Further measurements of $\sigma_o(\theta_o)$ were made by the Naval Research Laboratory in the 1960s using four different transmitted frequencies (Daley et. al. 1968, 1970, 1971, 1973). These data were summarized for both vertical and horizontal polarization by Guinard et al. (1971) for cases where the antenna was directed upwind. These authors chose to normalize the cross section of the sea by an area computed using one-way antenna beamwidths. Thus σ_o presented in this paper should be increased by 3 dB to agree with the more standard definition based on two-way antenna beamwidths presented in this chapter. Figure 25 shows an example of these data at C-Band (4.455 GHz) which is taken from Valenzuela (1978). Valenzuela has renormalized σ_o to agree with the standard definition. As pointed out by Wright (1968), $g_V(\theta',\phi)$ varies sufficiently slowly with incidence angle that composite surface theory predictions of σ_o for vertical polarization are essentially the same as those of slightly rough surface theory. Hence the solid curve in Figure 25a is labeled ''slightly rough'' and the assumed form of the short wave spectral density, here called $W(k)$, is given. This curve follows the data rather well for the vertical polarization case but obviously fails for horizontal polarization at large incidence angles. Assuming large-scale surfaces with various rms slopes brings the curve into somewhat better agreement with the data, however, as Figure 25b shows. Nevertheless, composite surface theory appears to be less than adequate for horizontal polarization at large incidence angles and other forms of scattering must be included here (See Chapter 12). More recent comparisons of composite surface predictions of σ_o with those measured by the K_u-Band (14.6 GHz) scatterometer flown on SEASAT indicate the same trend. Figure 26 shows such a comparison taken from Plant (1986) for the cross section averaged over all antenna azimuth angles at a wind speed of 10 m/s. Open circles in this figure show the SASS I model function used to produce wind fields from the scatterometer (See Chapter 13, Schroeder et. al. 1982). Squares are the best-fit model function obtained by Wentz et. al. (1984) who utilized all three months of SEASAT data. The solid circles are composite surface theory values computed from a short-wave spectrum calculated in the paper. Except for high incidence angles at horizontal polarization, agreement between the theory and SEASAT curves is rather good.

The primary uncertainty involved in making comparisons such as these is the uncertainty of the form of the short wave spectrum as a function of wavenumber to be used in the composite surface calculations. Such spectra are difficult to measure at sea and theoretical spectra have been used to date in most comparisons. Wright and Keller (1971) circumvented this problem in a series of wavetank experiments by photographically imaging the wind-roughened water surface at the same time that radar backscatter measurements were made. Fourier transforms of these images were used to obtain short-wave spectra for comparison with those deduced from the radar backscatter using composite surface theory. Measurements were made with vertically-polarized microwave systems operating at X-Band (9.375 GHz) and K-Band (24 GHz). Figure 27 shows the results of these experiments plotted as 10 log $(k^2\psi(k))$ versus k. At the low wind speed, both microwaves and photographs showed a dip in the wavenumber spectrum near $k = 3\ \text{cm}^{-1}$. At the higher wind speed, this dip had vanished according to both instruments. The wind blew over the water for about three meters in these experiments so that even at low wind speeds, the dominant wavelength in the tank was more than three times the wavelength where the spectral dip occurred. Thus most of the microwave backscatter in these experiments occurred in situations where composite surface theory, not slightly rough theory, should be applied. Even though the microwave data are rather scattered (as were the photographic data shown in the paper), these experiments provide persuasive evidence for composite surface scattering theory.

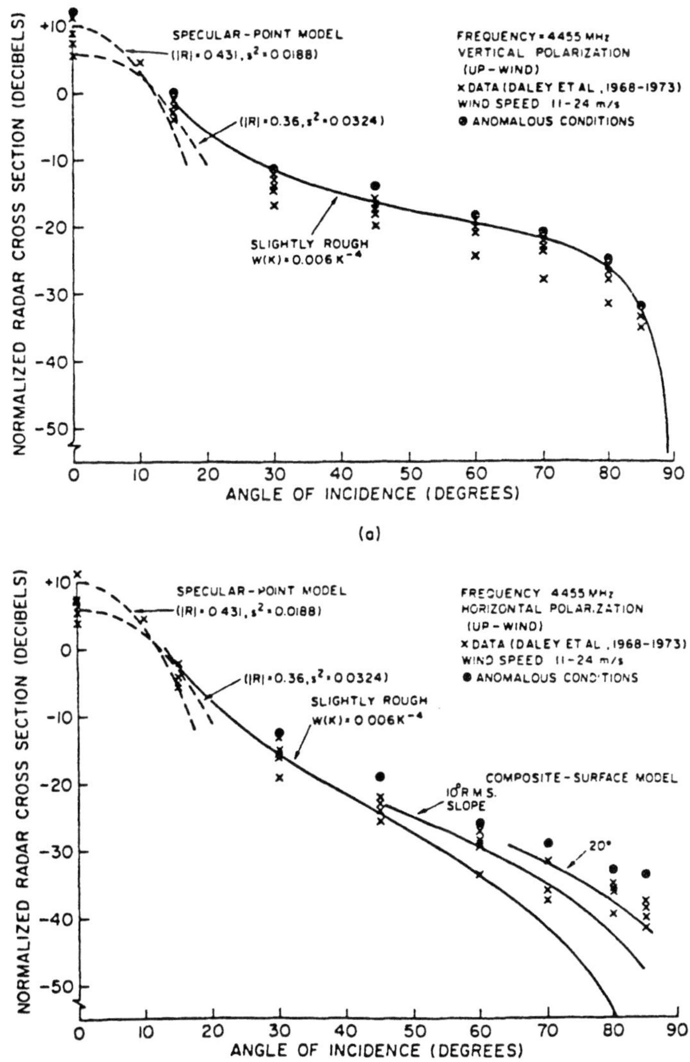

(b)

Figure 25. The normalized radar backscatter cross section of the sea looking upwind versus incidence angle for C-Band data of Daley et. al. (1968-1973). Solid curves indicate the predictions of composite and slightly-rough surface theory while the dashed lines indicate predictions of specular point theory (See Chapter 13). Note that the rms slope of the long waves severely affects the cross section at high incidence angles for horizontal polarization according to composite surface theory. From Valenzuela (1978).

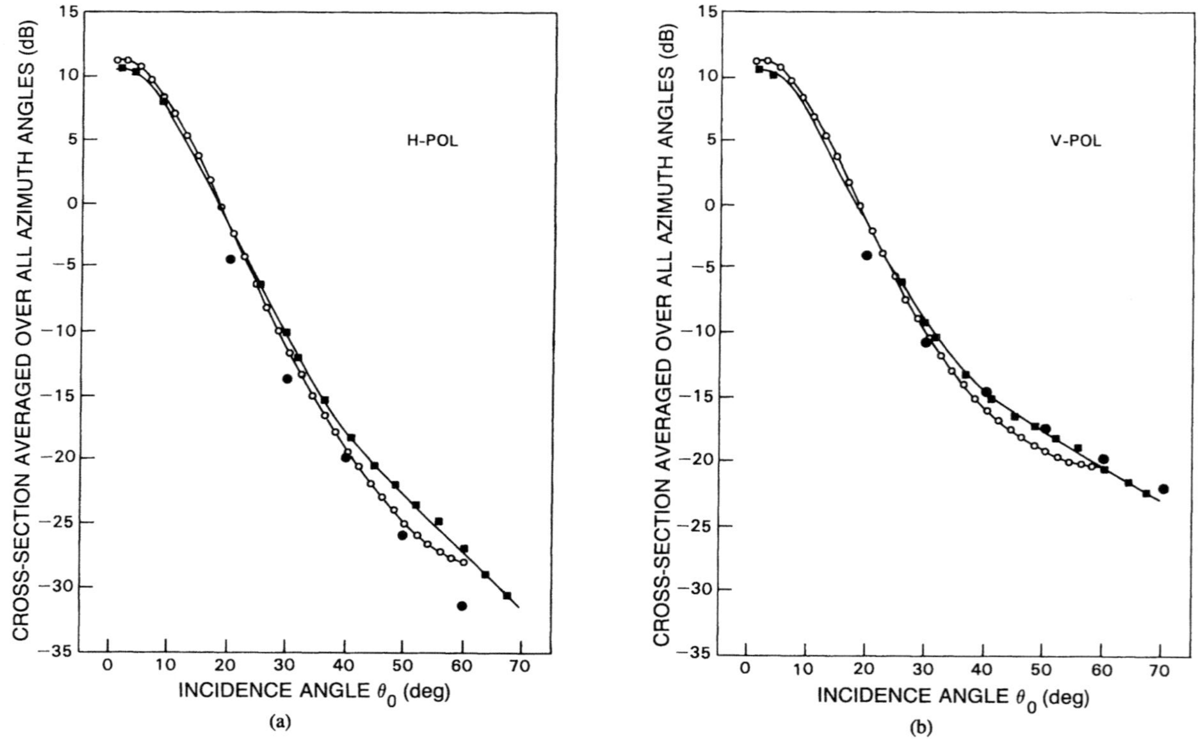

Figure 26. The normalized radar backscatter cross section of the sea averaged over all azimuth angles versus incidence angle for K_u-Band. Open circles indicate the SASS I model function used to obtain wind speed and direction from the SEASAT scatterometer. Squares show the best-fit model function of Wentz et. al. (1984) obtained using the entire SEASAT data set. Solid circles show the prediction of composite surface theory according to Plant (1986). From Plant (1986).

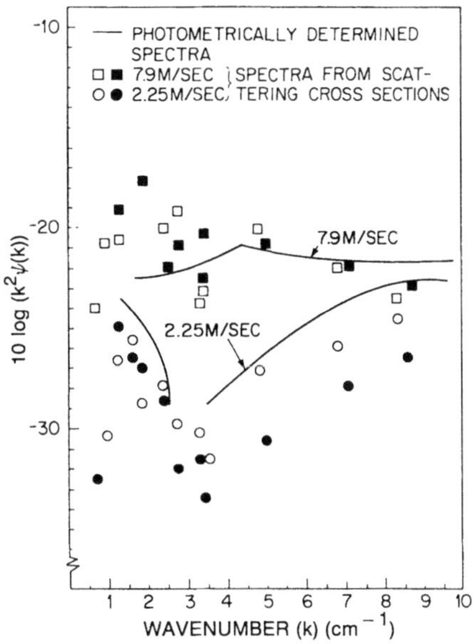

Figure 27. Plot of $k^2\Psi(k)$ versus k from Wright and Keller (1971). Circles and squares indicate values deduced from microwave backscatter in a wind wavetank using composite surface theory. Lines show the corresponding spectral densities measured simultaneously by Fourier transforming photographs of the surface. Note that both instruments show the spectral "dip" at a wind speed of 2.25 m/s and indicate that it fills at 7.9 m/sec.

One of the more easily investigated predictions of composite surface theory is that the small-scale structure will be advected by the larger-scale waves. Thus if one illuminates an area of sea surface which is small compared to dominant ocean wavelengths and forms Doppler spectra of the received signal in times short compared to dominant wave periods, these Doppler spectra should show affects attributable to the long wave orbital velocities (See Section V). Figure 28 shows Doppler spectra of vertically polarized, X-Band sea return obtained in this manner. The figure is taken from Plant and Keller (1983) and shows spectra obtained in 0.5 seconds from the output of a microwave system operated at a 45° incidence angle which illuminated a 2 × 3 meter area of sea surface. Successively lower spectra were obtained from consecutive 0.5 second segments of the time record. Two features of these spectra are of particular interest in judging the merit of composite surface theory. First, the widths of individual spectra indicate that the scattering remained coherent for about 20 milliseconds (the inverse of the bandwidth). In this length of time, the small-scale structure would travel less than 20 centimeter even if advected at the enormous rate of 10 m/sec. Thus the assumption of rapid spatial decorrelation of the small-scale structure appears to be warranted. Second, the centroids of the Doppler spectra shift away from zero by a maximum of about 50 Hz as long waves pass through the illuminated area on the surface. Such shifts imply variable surface currents up to about 1 m/s, about the size of orbital velocities of dominant waves implied by Eq. (5.20). Thus the small-scale structure does indeed appear to be advected by the large.

This conclusion may be checked more closely by arranging for the antenna of the microwave system to point into the direction from which the large waves are coming. Measurement of the time variation of the centroid of the Doppler spectrum then yields a time record of line-of-sight surface velocities, V_{los}, using the standard equation for Doppler shifts:

$$f_d = \frac{2V_{los}}{\lambda_o}. \tag{8.1}$$

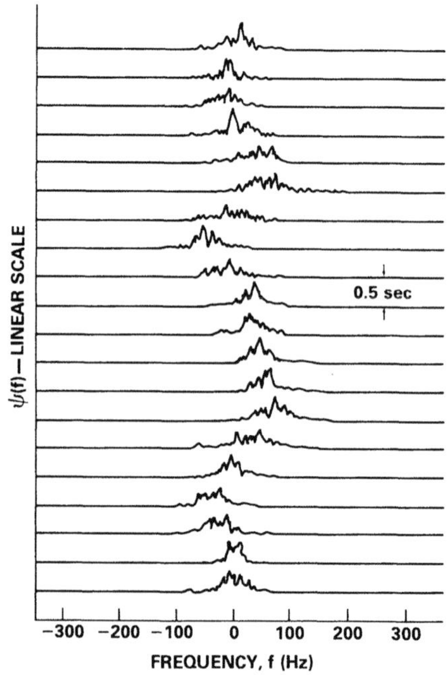

Figure 28. X-Band Doppler spectra of backscatter from the ocean. Each spectrum was obtained from 0.5 seconds of data. Consecutively lower spectra were calculated from consecutive 0.5 second time records. Data were taken at a 45° incidence angle and with vertical polarization. Note the variation in the position of the centroid of the spectra caused by long-wave orbital velocities.

From this time record, variance spectra of long wave heights may be computed using Eq. (5.20). A wave height spectrum obtained in this manner using an X-Band microwave system during the MARSEN experiment is shown in Figure 29 (Wright et. al. 1980). Also shown in the figure is a spectrum of the waves during the same period obtained from a standard wave height gauge. Over 900 individual spectra have been averaged to obtain the two mean spectra shown. Thus the 95% confidence interval is ±7%. Within this uncertainty, the two spectra agree well out to frequencies at which the spatial resolution of the radar begins to degrade its spectrum. Obviously, the motion of the Doppler spectrum centroid is caused by orbital velocities of the long waves as required by composite surface theory.

This advection of the small-scale structure by the large is responsible for the fact that sharp first-order Bragg lines are not commonly observed in microwave Doppler spectra of sea return as they are in HF backscatter (see Figure 11). Occasionally, however, it is possible to observe two broad, first-order Bragg peaks in microwave Doppler spectra of sea return. This is illustrated in Figure 30 which is taken from Plant and Keller (1988). This figure shows Doppler spectra obtained from an L-Band (1.5 GHz) microwave system mounted on the Coastal Engineer's research pier at Duck, N.C. The antenna was set at a 45° incidence angle and was directed 20° west of north (340°), parallel to the shoreline. The figure shows Doppler spectra obtained under several different wind conditions. The vertical dashed lines show the expected positions of first-order Bragg peaks for this particular arrangement. Clearly, such peaks are visible either when the wind speed is very low or when the wind is blowing offshore, ie, from about 250°. Since the microwave system was mounted only about 400 meters offshore, these wind conditions describe situations which produce short dominant wavelengths and, therefore, small orbital velocities. Thus Figure 30 strongly supports the

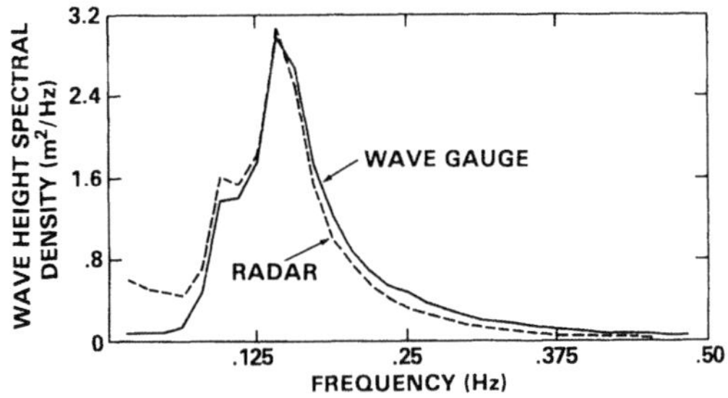

Figure 29. Comparison of wave height variance spectra obtained from a radar and a wave gauge. Spectra were obtained from the X-Band radar by converting time variations of the Doppler centroid to wave height using linear wave theory (See Section V). 960 individual spectra were averaged to produce each mean spectrum shown.

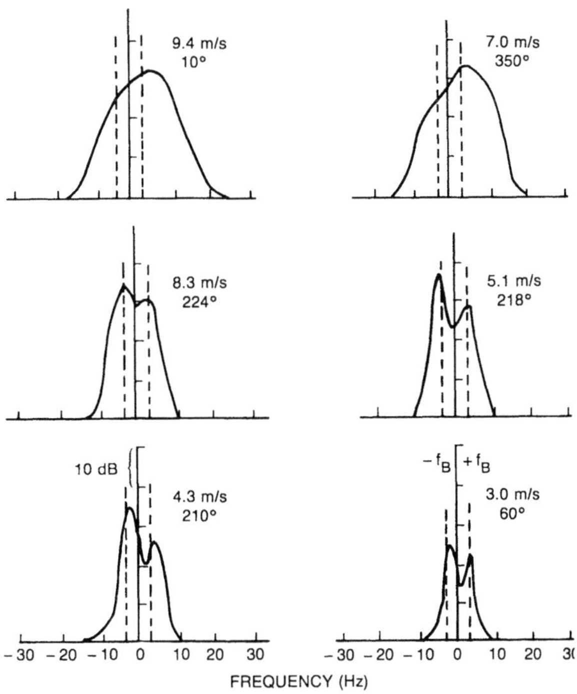

Figure 30. L-Band Doppler spectra obtained from the Coastal Engineer's research pier at Duck, North Carolina under various wind conditions. The radar operated at a 45° incidence angle in vertical polarization and was directed toward 340°, along the coastline. When winds were low or blowing offshore, dominant waves were small and first-order Bragg peaks were seen in the spectra. Dashed lines indicate the expected position of these peaks according to Bragg theory.

contention of composite surface theory that microwave backscatter is primarily first-order Bragg scattering from a small-scale surface riding on a larger-scale one.

Much other evidence exists which supports the composite surface theory of microwave backscatter from the sea surface at intermediate incidence angles. Generally, however, this evidence requires lengthy explanation and will not be presented here in detail. Thus the wind-speed dependence of σ_o can be rather well explained using composite surface theory but requires a development of the wind speed dependence of $\psi(k)$ (See Chapter 13, Plant 1986, Donelan and Pierson 1987). Scattering of two coherent microwave signals at different frequencies from the sea surface can produce return from which long-wave dispersion relations may be deduced. Explanations of such "Delta-K" radars based on composite surface theory have proven to be in good agreement with experiment (Alpers and Hasselmann 1978, Schuler et. al. 1982, Plant and Schuler 1980). The operation of synthetic aperture radars when observing the sea surface can be understood in terms of composite surface theory (Alpers and Rufenach 1979, Plant and Keller 1983, Hasselmann et. al. 1986). Finally, the modulation of microwave backscatter from the sea by internal waves has been explained quantitatively by composite surface theory to within a few dB (Alpers and Hennings 1984, Thompson 1988).

9. Summary

The intent of this chapter was to develop the theoretical equations describing Bragg scattering from a roughened air/sea interface using realistic electromagnetic fields as transmitted by antennas and to present experimental evidence supporting those equations. We began with the basic equations of electromagnetism, Maxwell's equations, and developed the laws of reflection and refraction at a flat air/water interface. These are summarized in Eqs. (2.36) - (2.39) and (2.45) - (2.48) using the Fresnel reflection and transmission coefficients of Eqs. (2.31), (2.32), (2.42), and (2.43). We then developed the Stratton-Chu Equation for fields at a point in space produced by known fields on the surface of a volume containing the point. This is Eq. (3.6). From the Stratton-Chu Equation, we derived the field transmitted by an antenna, Eq. (3.9), and the illumination pattern produced by this field on a flat surface in the far-field of the antenna, Eq. (3.21). In the process, we determined expressions for the range requirement for the far-field, Eq. (3.13), and for the beamwidth of an antenna, Eq. (3.19a). We then proceeded to obtain fields backscattered to an antenna from a slightly-rough air/water interface located in the far-field. These are given in Eqs. (4.24) and (4.27). An expression for the normalized radar cross section of a slightly-rough surface was derived from these fields and is given in Eq. (4.45) with geometrical coefficients for horizontal and vertical polarization given in Eqs. (4.46) and (4.47) respectively. An expression for the illuminated surface area used to normalize the radar cross section is given in Eq. (4.41). The latter expressions are for the case of beam-limited scattering and change slightly for a pulse-limited case.

Then followed a brief introduction to the theory of surface waves on water beginning with the Navier-Stokes equations. Here we developed the Rayleigh Equation for small-amplitude waves on water without viscosity, Eq. (5.10), and obtained expressions for phase speed, group speed, and frequency of these waves in Eqs. (5.22) - (5.24). Expressions for the orbital velocities of the waves are given in Eqs. (5.21).

We next presented evidence for the applicability of slightly rough surface theory. We showed that when the basic assumptions the theory imposes on surface wave height and slope,

Eqs. (4.12) and (4.13), are satisfied, Doppler spectra consist of two lines offset from zero frequency by plus and minus the frequency of the wave satisfying the Bragg resonance condition, Eq. (6.6). We showed that a microwave radar responds only to waves near the Bragg wavelength which are traveling nearly toward or away from the antenna and gave far-field expressions for the wavenumber and azimuth-angle resolution, Eqs. (6.7) and (6.9). Finally, we showed that wind waves can easily grow to such a size that first-order Bragg theory (slightly-rough theory) is no longer valid.

This led to the development of composite surface theory. We showed that the theory could be expected to hold in a mid-range of incidence angles between approximately 20 and 70 degrees if a cutoff wavelength could be found which satisfied Eq. (7.14). If this is the case then expressions for the backscattered field and normalized cross section are given by Eqs. (7.21) and (7.24), respectively. For this composite theory, we showed that the geometric coefficients had to be modified slightly to account for tilting in and out of the plane of incidence. The modified expressions are given in Eq. (7.19).

The last section presented experimental evidence supporting composite surface scattering theory. Normalized radar cross sections of the sea were shown to follow predictions of the theory well at intermediate incidence angles except possibly above 60° for horizontal polarization. We showed that the composite surface theory led to the picture of small patches of Bragg waves being advected about by orbital velocities of long waves and that this picture accounted well for centroid shifts of microwave Doppler spectra. Furthermore, widths of these spectra support the rapid spatial and temporal decorrelation of the small-scale waves required by composite theory. Finally, recent evidence of Bragg splittings in microwave Doppler spectra of the sea obtained under conditions of low long-wave orbital velocities was presented.

Thus Bragg scattering theories, slightly-rough and composite, explain most of the overall properties of microwave backscatter from the air/sea interface at intermediate incidence angles. The theories are approximations which, as shown here, can be derived directly from Maxwell's Equations by matching boundary conditions at the interface. As approximations, these theories cannot be expected to account for backscatter from every individual realization of the random rough surface which is the air/sea interface. Information obtained from a single realization of a random process is not generally relevant to understanding the process as a whole, however. Since we have shown that Bragg scattering theories explain the ensemble-averaged properties of electromagnetic backscatter from the air/sea interface rather well, its utility for studies of the interface is apparent. Thus a microwave radar can act as a contactless probe for the interrogation of the air/sea interface. Using the known characteristics of first-order Bragg scattering, interfacial parameters affecting wind-wave dispersion relations, growth rates, interactions, and equilibrium states can be investigated remotely without disturbing interface. Many of these applications are described in other chapters of this book. Electromagnetic Bragg scattering from the air/sea interface has given, and promises to give much more, valuable information about air/sea interactions.

10. References

Alpers, W. and K. Hasselmann, "The Two-Frequency Microwave Technique for Measuring Ocean-Wave Spectra from an Airplane or Satellite," Bdry Layer Met., 13, p. 215, 1978.

Alpers, W. and C.L. Rufenach, "The Effect of Orbital Motions on Synthetic Aperture Radar Imagery of Ocean Waves," IEEE Trans. Antennas Propag., AP-27, p. 685, 1979.

Alpers, W. and I. Hennings, "A Theory of the Imaging Mechanism of Underwater Bottom Topography by Real and Synthetic Aperture Radar," J. Geophys. Res., 89, p. 10529, 1984.

Barrick, D.E., "A Review of Scattering from Surfaces with Different Roughness Scales," Radio Sci., 3, p. 865, 1968.

Barrick, D.E., "Remote Sensing of Sea State by Radar," in *Remote Sensing of the Troposphere,* edited by V.F. Derr, Chap. 12, U.S. Government Printing Office, Washington, D.C., 1972.

Barrick, D.E., J.M. Headrick, R.W. Bogle, and D.D. Crombie, "Sea Backscatter at HF: Interpretation and Utilization of the Echo," Proc. IEEE, 62, p. 673, 1974.

Barrick, D.E., M.W. Evans, and B.L. Weber, "Ocean Surface Currents Mapped by Radar," Sci, 198, p. 138, 1977.

Bass, F.G. and I.M. Fuks, Wave Scattering from Statistically Rough Surfaces," trans. and edited by C.B. Vesecky and J.F. Vesecky, Pergammon Press, Oxford, England, 1979.

Bass, F.G., I.M. Fuks, A.I. Kalmykov, I.E. Ostrovsky, and A.D. Rosenberg, "Very High Frequency Radiowave Scattering by a Disturbed Sea Surface Part I: Scattering from a Slightly Disturbed Boundary," IEEE Trans. Antennas and Prop., AP-16, p. 554, 1968.

Bass, F.G., I.M. Fuks, A.I. Kalmykov, I.E. Ostrovsky, and A.D. Rosenberg, "Very High Frequency Radiowave Scattering by a Disturbed Sea Surface Part II: Scattering from an Actual Sea Surface," IEEE Trans. Antennas and Prop., AP-16, p. 560, 1968.

Bragg, W.L., "The Diffraction of Short Electromagnetic Waves by a Crystal," Proc. Cambridge Phil. Soc., 17, p. 43, 1913.

Crombie, D.D., "Doppler Spectrum of Sea Echo at 13.56 Mc./s.," Nature, 175, p. 681, 1955.

Daley, J.C., J.T. Ransone, J.A. Burkett, and J.R. Duncan, "Sea-Clutter Measurements on Four Frequencies," NRL Report 6806, 1968.

Daley, J.C., W.T. Davis, and N.R. Mills, "Radar Sea Return in High Sea States," NRL Report 7142, 1970.

Daley, J.C., J.T. Ransone, and J.A. Burkett, "Radar Sea Return - JOSS I," NRL Report 7268, 1971.

Daley, J.C., J.T. Ransone, and W.T. Davis, "Radar Sea Return - JOSS II," NRL Report 7534, 1973.

Donelan, M.A. and W.J. Pierson Jr., "Radar Scattering and Equilibrium Ranges in Wind-Generated Waves with Application to Scatterometry," J. Geophys. Res., 92, p. 4971, 1987.

Goldstein, H., "Frequency Dependence of the Properties of Sea Echo," Phys. Rev., 70, p. 938, 1946.

Goldstein, H., "Sea Echo," in *Propagation of Short Radio Waves,* edited by D.E. Kerr, p. 481, MIT Radiation Laboratory Series, 13, McGraw-Hill, New York, 1951.

Guinard, N.W., J.T. Ransone, Jr., and J.C. Daley, "Variation of the NRCS of the Sea with Increasing Roughness," J. Geophysic. Res., 76, p. 1525, 1971.

Hasselmann K., and M. Schieler, "Radar Backscatter from the Sea Surface," in Eighth Symposium, Naval Hydrodynamics, editor by M.S. Plesset, T.Y.T. Wu, and S.W. Doroff, Office of Naval Research, Arlington, Virginia, 1970.

Hasselmann, K., R.K. Raney, W.J. Plant, W. Alpers, R.A. Shuchman, D.R. Lyzenga, C.L. Rufenach, and M.J. Tucker, "Theory of Synthetic Aperture Radar Ocean Imaging: A MARSEN View," J. Geophys. Res., 90, p. 4659, 1985.

Jackson, J.D., *Classical Electrodynamics*, John Wiley and Sons, New York, New York, 1962.

Larson, T.R. and J.W. Wright, "Wind Wave Studies Part 2 — The Parabolic Antenna as a Wave Probe," NRL Report 7850, December 31, 1974.

Longuet-Higgins, M.S., D.E. Cartwright, and N.D. Smith, "Observations of the Directional Spectrum of Sea Waves Using the Motions of a Floating Buoy," in Ocean Wave Spectra, Prentice-Hall, Englewood Cliffs, N.J., p. 111, 1963.

Macdonald, F.C., "The Correlation of Radar Sea Clutter on Vertical and Horizontal Polarization with Wave Height and Slope," IRE Conv. Rec., pt. 1, p. 29, 1956.

Peake, W.H., "Theory of Radar Return from Terrain," IRE Intl. Conf. Record, I, pt. 1, p. 27, 1959.

Phillips, O.M., *The Dynamics of the Upper Ocean,* Cambridge University Press, Cambridge, England, 1977.

Plant, W.J. and J.W. Wright, "Growth and Equilibrium of Short Gravity Waves in a Wind-Wave Tank," J. Fluid Mech., 82, p. 767, 1977.

Plant, W.J., "On the Steady-State Energy Balance of Short Gravity Wave Systems," J. Phys. Ocean., 10, p. 1340, 1980.

Plant, W.J. and D.L. Schuler, "Remote Sensing of the Sea Surface Using One- and Two-Frequency Microwave Techniques," Radio Science, 15, p. 605, 1980.

Plant, W.J. and W.C. Keller, "The Two-Scale Radar Wave Probe and SAR Imagery of the Ocean," J. Geophys. Res., 88, p. 9776, 1983.

Plant, W.J., "A Two-Scale Model of Short Wind-Generated Waves and Scatterometry," J. Geophys. Res., 91, p. 10735, 1986.

Plant, W.J. and W.C. Keller, "Evidence of Bragg Scattering in Microwave Doppler Spectra of Sea Echo". To be published, J. Geophys. Res., 1989.

Rayleigh, Lord, "On Waves," Philosphical Mag., 5, p. 257, 1876.

Rayleigh, Lord, *Theory of Sound, Vol. II*, Macmillan, London, England, p. 89, 1929.

Rice, S.O., "Reflection of Electromagnetic Waves from Slightly Rough Surfaces," Comm. Pure Appl. Math, 4, p. 351, 1951.

Saxton, J.A., and J.A. Lane, "Electrical Properties of Sea Water," Wireless Engineer, 29, p. 269, 1952.

Schroeder, L.C., D.H. Boggs, G. Dome, I.M. Halberstam, W.L. Jones, W.J. Pierson, and F.J. Wentz, "The Relationship Between Wind Vector and Normalized Radar Cross Section Used to Derive SEASAT-A Satellite Scatterometer Winds," J. Geophys. Res., 87, p. 3318, 1982.

Schuler, D.L., W.J. Plant, W.P. Eng, W. Alpers, and F. Schlude, "Dual-Frequency Microwave Backscatter from the Ocean at Low Grazing Angles: Comparison with Theory," Intl. J. Remote Sens., 3, p. 363, 1982.

Silver, S., ed., *Microwave Antenna Theory and Design,* MIT Radiation Laboratory Vol. 12, McGraw-Hill, New York, New York, 1949.

Stokes, G.G., ''On the Theory of Oscillatory Waves,'' Trans. Cambridge Phil. Soc., 8, p. 441, 1847.

Stratton, J.A., *Electromagnetic Theory,* McGraw-Hill, New York, New York, 1941.

Tang, S. and O.H. Shemdin, ''Measurement of High Frequency Waves Using a Wave Follower During Marsen,'' J. Geophys. Res., 88, p. 9832, 1983.

Thompson, D.R., ''Calculation of Radar Backscatter Modulations from Intternal Waves,'' J. Geophys. Res., 93, p. 12371, 1988.

Valenzuela, G.R., ''Depolarization of EM Waves by Slightly Rough Surfaces,'' IEEE Trans. Antennas and Prop., AP-15, 1967.

Valenzuela, G.R., ''Scattering of Electromagnetic Waves from a Tilted Slightly Rough Surface,'' Radio Sci., 3, p. 1057, 1968.

Valenzuela, G.R., ''The Effect of Capillarity and Resonant Interactions on the Second-Order Doppler Spectrum of Radar Sea Echo,'' J. Geophys. Res., 79, p. 5031, 1974.

Valenzuela, G.R. ''Theories for the Interaction of Electromagnetic and Oceanic Waves A Review,'' Bdry-Layer Met., 13, p. 61, 1978.

Valenzuela, G.R., ''Scattering of Electromagnetic Waves from the Ocean,'' in *Surveillance of Environmental Pollution and Resources by Electromagnetic Waves*, D. Reidel, Dordrecht, Holland, p. 199, 1978.

Wait, J.R., ''Theory of HF Ground Wave Backscatter from Sea Waves,'' J. Geophys. Res., 71, p. 4839, 1966.

Wentz, F.J., S. Peteherych, and L.A. Thomas, ''A Model Function for Ocean Radar Cross Sections at 14.6 GHz,'' J. Geophys. Res., 89, p. 3689, 1984.

Wiltse, J.C., S.P. Schlesinger, and C.M. Johnson, ''Backscattering Characteristics of the Sea in the Region from 10 to 50 kMc,'' Proc. IRE, 45, p. 220, 1957.

Wright, J.W., ''Backscattering from Capillary Waves with Application to Sea Clutter,'' IEEE Trans. Antennas and Prop., AP-14, p. 749, 1966.

Wright, J.W., ''A New Model for Sea Clutter,'' IEEE Trans. Antennas and Prop., AP-16, p. 217, 1968.

Wright, J.W. and W.C. Keller, ''Doppler Spectra in Microwave Scattering from Wind Waves,'' Phys. Fluids, 14, p. 466, 1971.

Wright, J.W., W.J. Plant, W.C. Keller, and W.L. Jones, ''Ocean Wave-Radar Modulation Transfer Functions from the West Coast Experiment,'' J. Geophys. Res., 85, p. 4957, 1980.

Chapter 12

ELECTROMAGNETIC SCATTERING FROM THE SEA AT LOW GRAZING ANGLES

LEWIS B. WETZEL
Naval Research Laboratory
Washington, DC 20375

1. Introduction

Radars operating from ships, shore installations, and low-flying aircraft inevitably view the sea at low grazing angles from zero to a few degrees. Under these conditions it is expected that the surface might often be deeply shadowed, yet the experimental evidence is unclear, sometimes even contradictory. It is obvious that this is a very special scattering regime, in which the conventional ideas of sea scatter are quite possibly of little use. Any approach to understanding this regime must be guided by two basic questions:

1. In the most general case of bistatic scattering, how much, and what part of the sea surface is actually illuminated and observed simultaneously in a scattering interaction?
2. What are the special structural characteristics of the scattering elements in such regions, and what theoretical models are to be used.

Attempts to answer these questions run into severe problems in conceptualization (visualizing the physical processes involved), and modeling (finding useful predictive analytical descriptions of these processes). While shadowing of some sort must certainly take place at extreme grazing angles, radars operate at finite wavelengths, so the effects of diffraction cannot be ignored. But a diffraction-corrected shadowing theory does not exist, and the task of constructing one appears formidable. Moreover, it can be expected that modulation of the surface illumination by shadowing will preferentially highlight special regions of the surface—the raised wave peaks. Yet their morphology cannot be described with any confidence, and their scattering properties are largely unexplored and imperfectly understood. Present approaches to modeling in this area rest mainly on analogical guesswork and uncritical application of models that appear to work in other angular regimes.

The purpose of this Chapter will be to define these problems with greater precision, see how far existing knowledge can be carried, describe some new models and new approaches, and identify those topics requiring further work.

2. Shadowing

There are two ways to approach the problem of surface illumination at low angles: as a diffraction problem for which the usual geometrical optics shadowing theory is considered a lim-

G. L. Geernaert and W. J. Plant (eds.), Surface Waves and Fluxes, Volume II, 109–171.

iting case, or as a problem in geometrical optics in which there might be need for some diffraction corrections. It is the second approach that has been exclusively pursued [Bass and Fuks, 1963; Beckmann, 1965; Wagner, 1967; Smith, 1967; Sancer, 1969; Lynch and Wagner; 1970; etc.]. The reason, of course, is because it is mathematically manageable, albeit under certain severe assumptions: two-dimensional geometry; uncorrelated Gaussian surface; geometrical optics transition across the shadow line. Many useful statistical expressions are derived during the course of these treatments, but the end result is usually a "Shadowing Function," which gives, essentially, the percentage of surface area, on the average, that will be illuminated at a given grazing angle. Asymptotically, it becomes a linear function of angle for small grazing angles. Some form of this crude shadowing function is used whenever a shadowing theory is called for, yet there is little evidence that these results are of any value when applied to a real sea at low grazing angles.

At extreme grazing angles the standard geometrical shadowing theory described above develops some interesting implications, which were explored heuristically by Wetzel [1977]. It was found that at grazing angles below about 1 degree, the sea should begin to resemble a "dark" plane, populated by illuminated "islands," whose size and spacing are determined by the mean grazing angle and the wind speed. This conceptual model has a strong intuitive appeal, and bears a closer relation to visual observation and to radar events such as "sea spikes" than does the diffuse "vanishing scatter" picture suggested by the "shadowing function" described above. Nevertheless, this is still a geometrical theory, and it will be necessary to take a close look at the diffraction problem in order to find when, if ever, such models may be used. But first some general relations among illumination probabilities will be reviewed.

2.1. SOME BASIC SHADOWING PROBABILITIES

Associated with the illumination and observation of points on a surface, there are several probabilities that can be discussed without reference to the details of any particular shadowing theory. In Fig. 1, the most general case of bistatic scattering from an element of the surface is shown. The plane of incidence contains the illumination path at grazing angle ψ_I, while observation takes place at grazing angle ψ_0 in the plane of observation, which is rotated from the plane of incidence through the azimuthal angle ϕ_0. These planes intersect at a point P on the surface, whose scattering properties are left unspecified.

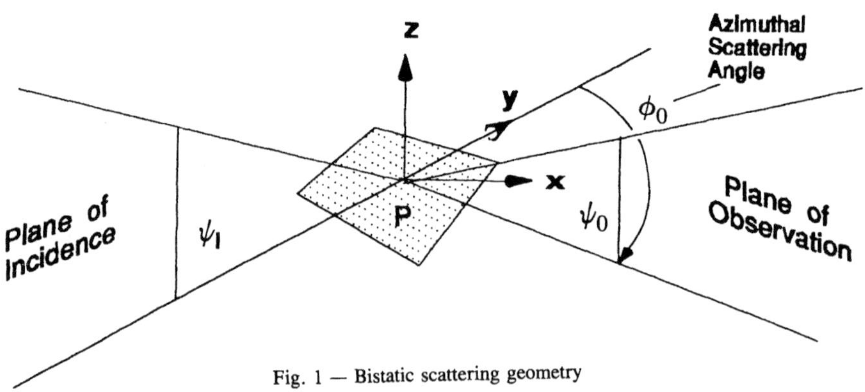

Fig. 1 — Bistatic scattering geometry

Monostatic backscatter is defined by setting $\phi_0 = \pi$; bistatic scattering takes place for any other values of ψ_0 and ϕ_0. The several probabilities associated with the shadowing process may be related symbolically as follows:

$$P(I,O) = P(I)P(I \mid O) \tag{1}$$

where $P(I,0)$ is the joint probability that the surface element is both Illuminated (I) and Observed (O), $P(I)$ is the probability that the element is illuminated, and $P(I \mid O)$ is the conditional probability that if illuminated, the element is also observed. In the monostatic case, illumination and observation take place over the same path, so if I, then O, and $P(I \mid O) = 1$. Thus we may write

$$\text{Monostatic Case:} \quad P_M(I,O) = P(I) \tag{2}$$

The bistatic case contains several possibilities. For example, if $\psi_0 > \psi_I$ with $\phi_0 = \pi$ (observation over the same surface as illumination, but at a greater angle), then once again $P(I \mid O) = 1$, since if the incident ray has escaped encounter with the surface peaks, then *certainly* a ray at a higher angle has escaped. In this case the bistatic $P(I,O)$ is the same as the monostatic. If the observation angle $\psi_0 < \psi_I$, then I and O are simply exchanged, and $P(I,O) = P(I \mid O)P(O)$ by the same reasoning; that is, the joint probability is always the single-path probability for the path with the smallest grazing angle.

However, when $\psi_0 \neq \pi$, the illumination and observation paths lie over different regions of the sea surface, and to the extent that the surface features in these regions are uncorrelated, the conditional probability $P(I \mid O)$ is independent of the path I and is simply equal to $P(O)$, the probability of observing the element of interest. Under most sea conditions correlation distances for the larger-scale surface features responsible for shadowing are of the order of the primary sea wavelength, so de-correlation will be the rule, and the probabilities can be written as follows:

Bistatic Case:

$$P_B(I,O) = P(I)P(O), \quad \phi_0 \neq \pi \tag{3a}$$

$$P_B(I,O) = P(I) \quad ,\phi_0 = \pi, \ \psi_0 > \psi_I \tag{3b}$$

$$P_B(I,O) = P(O) \quad ,\phi_0 = \pi, \ \psi_0 < \psi_I \tag{3c}$$

It should be noted that the first expression applies to the special case of forward scatter in the specular direction: $\phi_0 = 0, \ \psi_0 = \psi_I$.

The purpose of a shadowing theory is to give analytical expression to the basic path probability $P(\ )$ in terms of grazing angle and surface parameters. These probabilities are then used in the relations given above.

2.2. DIFFRACTION EFFECTS AT WAVE PEAKS

In this work, the problem of surface illumination will be approached from the basic concept of diffraction by a shadowing obstacle. The simplest surface model for incorporating diffraction effects into sea scatter is one containing two sharp peaks separated by a broad smooth trough. The cycloid provides such a curve, and is often related to the shape of ocean waves both formally [as in a Stokes wave: Kinsman, 1965], and informally [ask anyone to sketch an 'ocean wave']. It is given by the parametric equations:

$$x = x' - a \sin(Kx')$$

$$z = z' + a \cos(Kx') \tag{4}$$

where a controls the shape of the curve ($a = 1$ gives peaks with sharp points, while $a < 1$ gives rounded peaks). Figure 2 shows the case for $a = 1$ (trochoid), with the shadow line indicated for illumination at a grazing angle ψ. The axes are normalized so that the abscissa is given in fractions of a wavelength (distance between peaks), $s = x/\Lambda$, while the ordinate is the fraction of the peak-to-trough wave height H. It should be noted that Fig. 2 represents a two-dimensional model in which the wave crests are infinitely long in the direction perpendicular to the page. Such models can therefore be applied only in situations where the sea actually *is* long-crested, and the illumination is normal to the crests. Unfortunately, these conditions are seldom encountered. Real wind-driven seas tend to be relatively short-crested, with the water piled in heaps. Whether a real surface can be satisfactorily modeled in *any* practical way is a difficult question, which will be addressed in various contexts later. For now, however, the shadowing problem will be considered in terms of a simple two-dimensional model.

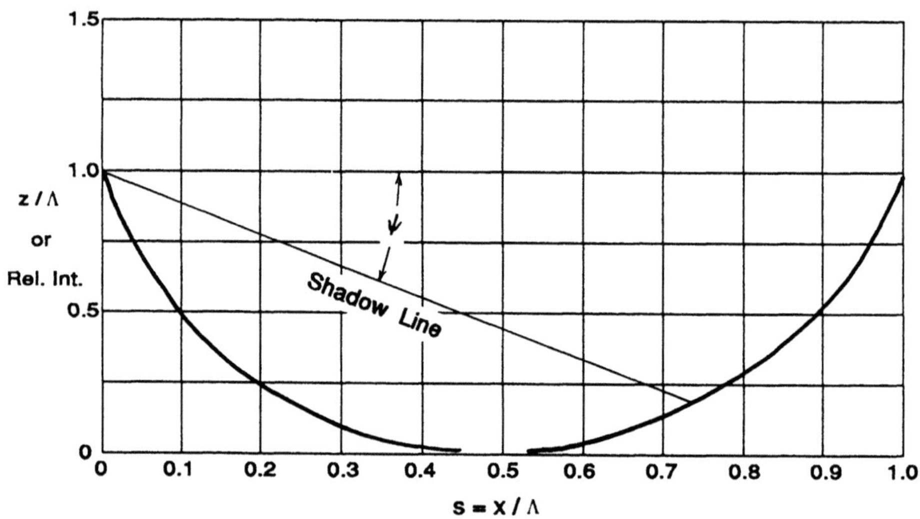

Fig. 2 — Trochoidal wave with shadow line for grazing angle ψ

Under the assumptions of geometrical optics, the wave surface in Fig. 2 would be fully illuminated to the right of the intersection of the shadow line with the surface, and totally dark to the left. The effect of diffraction in smoothing this transition may be approached as follows:

2.2.1. *Physical Optics.* Imagine a vertical plane erected perpendicular to the wave peak at the point of tangency with the shadow line. In the physical optics approximation, the field to the right of the plane is found from the Helmholtz integral over the plane using the field values appropriate to the unperturbed incident wave. The result is the familiar Fresnel diffraction pattern for a straightedge, and it is independent of the actual shape of the diffracting wave peak. This physical optics expression for the field in the region between the peaks in Fig. 2 is given in normalized form by

$$E_{PO}(R,\theta) = |F(w) - F(-\infty)| / \sqrt{2} \tag{5}$$

where $F(w)$ is the Fresnel integral

$$F(w) = \int_{o}^{w} e^{i\frac{\pi}{2}u^2} du ; \; w = \sqrt{kR\theta/\pi} \tag{6}$$

and θ is the angle relative to the shadow line (negative below, positive above).

This is the simplest approximation to diffraction effects, yet it quite dramatically illustrates the importance of including such effects in scattering calculations. In Fig. 3 the illumination field intensity ($|E|^2$) as derived from Eqs. (5) and (6) for a 3 cm wavelength (X-band), a 3° grazing angle, and a 15 knot wind speed, is plotted over the trochoidal surface of Fig. 2. (In this example, peak-to-trough wave height is taken as $H = .015 * W^2$ (m) and $L = .64 * W^2$ (m) is the surface "wavelength," where W is windspeed in m/sec.) Since the shadow line for a 3° grazing angle encounters the trochoidal surface at 0.47, the corresponding geometrical optics illumination profile is shown as a vertical dashed line at that point, separating light (intensity = 1, to the right) from dark (intensity = 0, to the left). For the Fresnel diffraction pattern, the diffracted field is always equal to 0.5 at the shadow line, so the relative intensity at that point is always 0.25. The gradual transition from light to shadow produced by diffraction can be characterized by a "transition fraction," F, defined by the fraction of the distance between wave peaks over which the field changes from 0 to 1 by the steepest tangent to the diffraction curve. It is seen that $F = 0.2$ for the conditions illustrated in Fig. 3. Note also that in this example F is measured between the intersection of the shadow line with the trough, at $s = 0.47$, and the point at which the diffraction profile crosses 1. This seems generally to be the case.

2.2.2. *Polarization Effects.* Scattering from the sea often shows strong polarization effects, so it is of interest to know whether the shadowing and diffraction processes being discussed display any polarization dependence. In the trochoidal model used to investigate shadowing and diffraction, the wave peak itself is essentially a knife-edge (examine the implications of

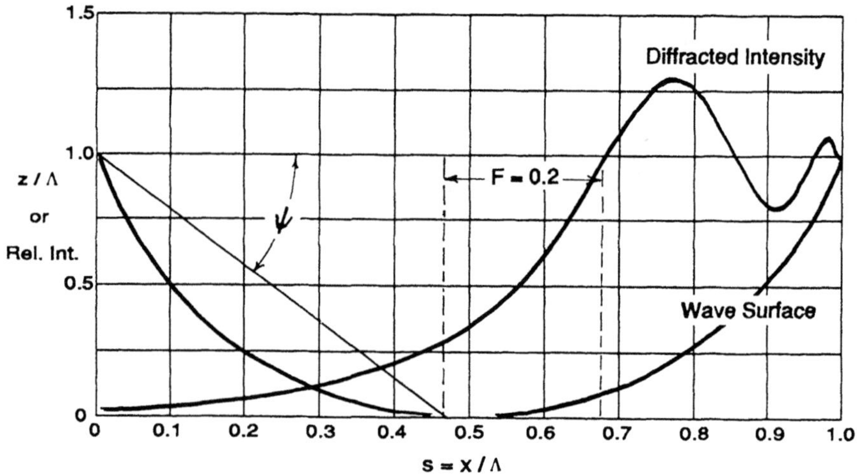

Fig. 3 — Diffraction pattern and "transition fraction" F for a trochoidal wave: X-band, 3° grazing, 15 kt wind

physical optics solution has already been invoked to find the diffracted field, it is an easy matter to append to this solution the so-called "edge waves" of Ufimtsev's Physical Theory of Diffraction [1957]. In this theory, the Physical Optics field given by (5) and (6) is "corrected" by adding an "edge wave" field:

$$E_{EW} = \frac{e^{i(kR+\pi/4)}}{\sqrt{2\pi kR}} D_{V,H}(\theta,\theta_0)$$

$$D_{V,H}(\theta,\theta_o) = -0.5\{[1 - \sin((\theta - \theta_0)/2)]/\cos((\theta - \theta_0)/2)$$

$$+ P\ [1 - \sin((\theta + \theta_0)/2]/\cos((\theta + \theta_0)/2)\} \qquad (7)$$

where θ_0 and θ are respectively the incident and scattering angles relative to the front face of the wave peak, and $P = +1$ for vertical, $P = -1$ for horizontal polarization. It turns out that in the low grazing angle, forward scatter case, the coefficient D is of order unity for both polarizations (although of opposite sign). Thus the importance of polarization effects is determined entirely by the magnitude of the cylindrical spreading factor in the first equation above. In virtually all cases of any practical interest, this factor is sufficiently small over the entire wave trough that polarization effects may be ignored in diffraction by wave peaks. In fact, it can be shown that polarization effects emerge from the theory only after the conditions for the application of the theory have already broken down.

If a different surface model had been chosen, say the cycloid of Eq. (1) for $a < 1$, which causes the peak to be rounded instead of sharp, then the polarization correction would take the form given by Wait and Conda [1959] for the field behind an obstacle with a rounded edge having a radius of curvature r:

$$E_{WC} = \frac{(kr/2)^{1/3}}{\sqrt{kR/2}} G_{V.H}((kr/2)^{1/3}\theta) \qquad (8)$$

where the function $G_{V,H}$ is an integral over Airy functions and is given in graphical form. Its amplitude ranges between 0 and 1. For broad smooth peaks, this correction could be considerably larger then the sharp-peak correction given in Eq. (7), but for curvatures of the order of a few wavelengths of the incident field these corrections may be ignored as well.

In summary, then, polarization will give at most small corrections to the diffracted field, and thus may safely be ignored when estimating the effects of diffraction on shadowing.

2.2.3. *When Can Geometrical Optics Be Used?* Figure 3 shows how diffraction spreads the illumination field over the trough region behind the wave peaks. If the "transition fraction" F, defined above, were to be zero, then geometrical optics would apply exactly. It can never be zero, of course, but if it were sufficiently small, geometrical optics might be expected to provide a reasonable model for estimating shadowing effects at low grazing angles. The criterion for "sufficiently small" must be arbitrary, but most would agree that values of $F < 0.1$ would provide a transition approaching optics-like sharpness, while for values of $F > 0.3$ the illumination profile is clearly controlled by diffraction. The basic data to which these criteria are to be applied may be found in a set of figures of the type illustrated in Fig. 4, where the diffracted field intensity profiles from (5) have been plotted for five grazing angles for a wavelength of 3 cm and wind speed of 15 kts.

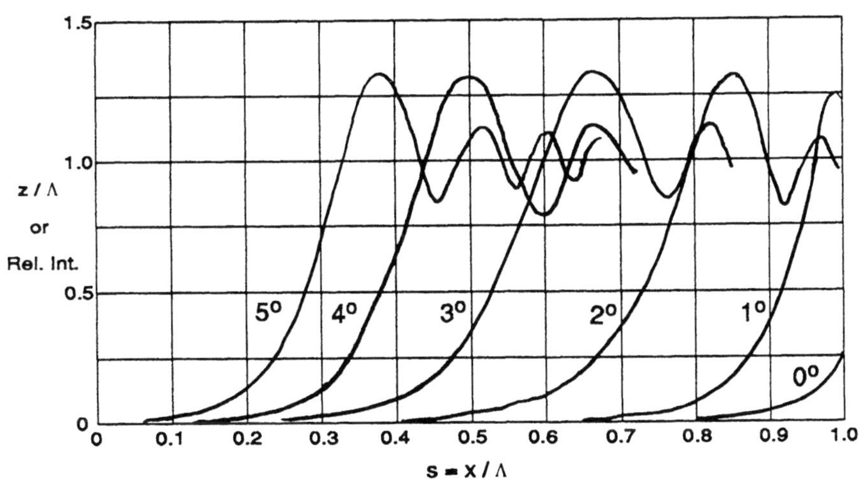

Fig. 4 — Diffraction patterns at X-band for 0-5° grazing angles, 15 kt winds

It is clear in Fig. 4 that for the given wavelength and wind speed, the shapes of the diffraction profiles, hence the fractions F, are relatively insensitive to the grazing angle. Thus lines of equal F may be plotted against wavelength/windspeed axes alone thereby roughly separating the regions defined by the criteria discussed above. Values of F were obtained for a set of wavelengths between 1 cm and 1 m and windspeeds between 5 and 25 kts [Wetzel, 1987a]. The results are shown in Fig. 5, where the labels "YES," "MAYBE," and "NO" are used to indicate the confidence to be placed in the use of the geometrical optics assumption in various domains of the wavelength/windspeed plane. As might be expected, the most popular radar wavelength (3 cm), at the mean global windspeed (15 kts), falls squarely in the middle of the ambiguous "MAYBE" domain. Nevertheless, this figure gives some useful perspective when trying to decide how seriously to take the predictions of shadowing theories.

The "YES" domain in Fig. 5 is disappointingly small, indicating that for the most popular radar frequencies, geometrical optics provides a really good approximation to the illumination profile only in the presence of gale force winds. While geometrical optics might be expected to provide at least some useful trends for points in the "MAYBE" domain, many microwave radar operations take place in the "NO" domain, where diffraction dominates and no satisfactory theory exists.

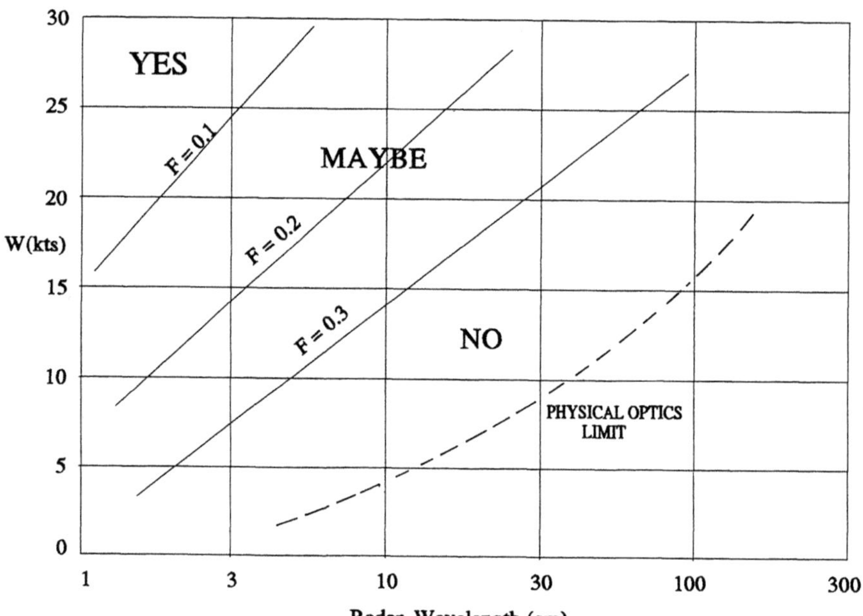

Fig. 5 — Confidence predictors for using geometrical optics

2.2.4. *Limitations and Extensions of Diffraction Theory.* The diffraction curves for the longer wavelengths and lower windspeeds and grazing angles are not to be taken seriously. Consider the dimensions of the surface features associated with a windspeed of 5 kts. The peak-to-trough height for the trochoidal model would be about 10 cm, so the distance of the diffracting peak above the trough is one wavelength at 10 cm (S-band), 1/3 wavelength at 30 cm (L-band), and 1/10 wavelength at 100 cm (UHF). The Physical Optics approximation underlying the diffraction theory expressed in Eqs. (5) and (6) cannot be expected to be valid under these conditions — in fact, for the extreme UHF case, the Rayleigh perturbation theory would seem more appropriate. This validity limit is expressed in Fig. 5 by the dashed curve labeled "Physical Optics Limit," which corresponds to wind speeds for which the radar wavelength is just equal to the peak-to-trough wave height.

Although it is the range of extremely small grazing angles, between 0 and 1 degree, that is of greatest interest, it is in this range that diffraction theory seems least informative. About all that can be said from examining this angular regime for the legitimate combinations of wind speed and radar wavelength, is that at very low grazing angles the illumination over most of the trough will always be low, and will rise fairly sharply close to the opposite peak. In a real sea, the lengths of the troughs and the heights of the peaks, which together determine the surface illumination, are governed by unknown statistics, so there can be at present no statistical theory of diffraction for a rough sea. Thus, if we wish to treat the real sea in any sense at all, we will have to rely on the implications of conventional, geometrical optical, shadowing theory. However, before leaving diffraction theory, let us take a quick look at a particular statistical extension of the standard diffraction problem that can add a bit of three-dimensional realism.

When the trochoidal wave model was introduced at the beginning of this section, it was noted that as a two-dimensional model, it is appropriate only to infinitely long-crested waves. However, in connection with the problem of radio propagation over hills and ridges, Furutsu [1965] considered the case in which the ridges contained statistical height variations along their crests. A sketch of such a ridge is shown in Fig. 6, where R is the distance behind the ridge, H is its mean height, and ψ is the grazing angle of the incident wave. The irregular crest height has a Gaussian distribution with variance $<|h^2|>$, and an arbitrary correlation function, while the basic theoretical assumption is that the physical optics integration plane can be divided into narrow strips of width dy above the crest height $h(y)$, and the integral done as usual over the $y-z$ plane above the crest. Certainly the irregular structure in Fig. 6 is a better model of a shadowing wave crest than the infinite two-dimensional model used earlier.

Here, as in most propagation problems involving irregularities, the "Fresnel length," $l_F = (\lambda R)^{1/2}$, can be expected to play a role. In Fig. 6, the first Fresnel Zone at the diffracting obstacle as seen looking back from the observation point has been added. The effects of fluctuations in crest height at the tangent point, and statistical variations in crest height along the crest, can both be discussed in terms of their relation to the Fresnel length (or radius of the first Fresnel zone). If h_0 is the rms height fluctuation about the mean height H, and l_c the correlation length of the height variations along the crest, then the diffraction patterns would look much like those for a fixed, infinite crest *provided*

$$h_0 << l_F \text{ and } l_c >> l_F . \tag{9}$$

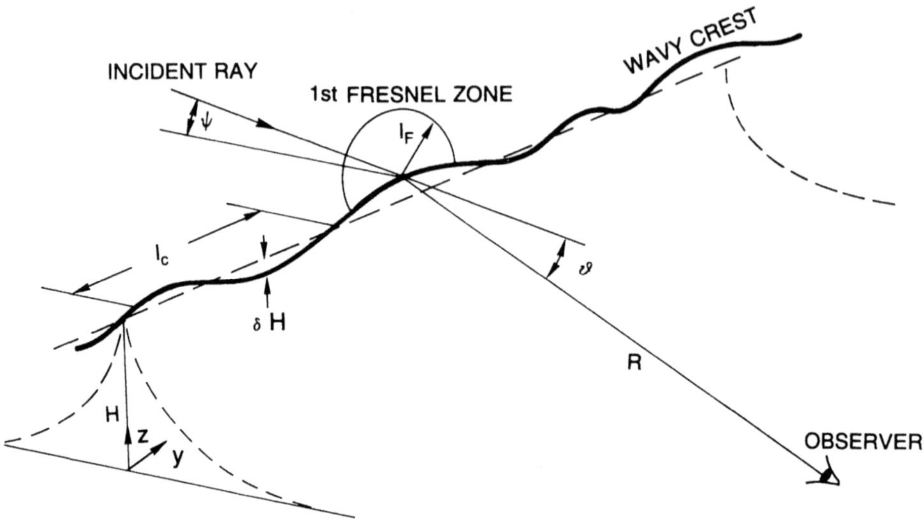

Fig. 6 — Diffraction with crest-height variations

Otherwise, surface statistics can have a significant effect on the diffraction pattern. Consider, for example, the illumination of the forward face of the right-hand crest in Fig. 2. In this case R will be the wavelength of the longest waves on a wind-driven sea, namely $R = 0.64 \cdot W^2$. The Fresnel lengths l_F (in meters) as seen looking back to the left hand, diffracting, crest for our five radar wavelengths (in cm) are given in the Table below:

Table 1 — Fresnel Lengths between Peaks (in meters)

Wavelength	1 cm	3 cm	10 cm	30 cm	100 cm	H/2 (m)
5 kts	.2	.4	.6	1.1	2	.05
15 kts	6	1.0	1.9	3.3	6	.45
25 kts	1	1.7	3.2	5.5	10	1.25

The last column is 1/2 the peak-to-trough wave height for the indicated wind speed, which is a measure of the maximum rms crest height fluctuation h_0 that could occur. It is seen that the first condition given in Eq. (9) is pretty well satisfied for all frequencies below K-band (1 cm wavelength). On the other hand, if seas are usually short crested and "piled in heaps," we would expect the second condition to be satisfied primarily at the shorter radar wavelengths, and for intermediate seas. From these considerations, then, it is possible to conclude that there is at least a small range of frequencies and wind speeds, centered about the X and S bands for about 15 knots of wind, in which deterministic diffraction theory based on the average wave heights should provide a reasonable estimate of surface illumination.

For readers curious about the effect of crest fluctuations on diffraction patterns, Fig. 7 contains a plot from the paper by Furutsu [*loc. cit.*]. Clearly, simple Fresnel-length arguments describe the more complicated diffraction geometry quite well.

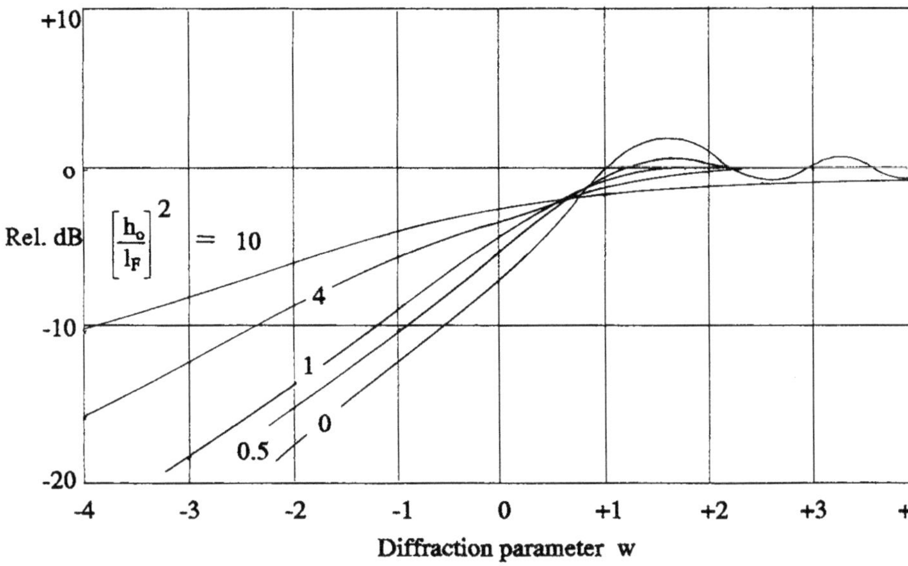

Fig. 7 — Effect of crest irregularity on diffraction pattern

2.3. SHADOWING THEORY BASED ON GEOMETRICAL OPTICS

Although Fig. 5 shows the domain of confidence in geometrical optics to be rather small, all existing theories of shadowing are based on this assumption. The assumption appears in slightly different forms in the various treatments. If the theory is developed from a Physical Optics scattering integral, as in Bass and Fuks or Sancer [*loc cit*], then the integrand will contain an "illumination factor," $I(R)$, which is taken to be unity if r is illuminated, zero if it is not. Smith and Wagner [*loc cit*] use purely statistical arguments to calculate $P(z,s)$ "the probability that an element of surface at height z with slope s is not shadowed by the intervening surface along the incident ray," and then integrate it in various ways to find the "shadowing function" mentioned earlier. Although multivariate statistics and correlation functions are sometimes mentioned in these papers, the problem is always simplified to one in which the surface heights and slopes are uncorrelated, with independent Gaussian distributions. Under these conditions, the Probability of Illumination $P(I)$ (refer to (2)) of a surface element at height z and slope s, on a Gaussian surface of rms displacement σ and mean slope s_0, when viewed at a mean grazing angle ψ (relative to the horizontal), is given by [see Smith, *loc cit*]:

$$P(I) = P(\eta;\, \xi, q) = (1 - .5 erfc\ (\xi))^{\Lambda(\eta)}\ U\ (q + \psi) \tag{10}$$

where

$$\Lambda(\eta) = \frac{1}{2}\left[\left(\frac{2}{\pi}\right)^{1/2}\frac{e^{-\eta^2/2}}{\eta} - erfc(\eta)\right] \tag{11}$$

with the definitions $\eta = \psi/s$, $\xi = z/\sqrt{2}\sigma$, $q =$ arctan (s). $U(x)$ is the unit step function, which acts as a *local* shadowing function by ensuring that an element on the back side of a wave peak will not be illuminated. These expressions can be further simplified by taking into account the special conditions encountered over the sea: namely, the mean slope, s_0, of the sea surface is usually 0.1-0.2, and the low grazing angles of major interest here will generally be less than 1°. Under these conditions, $\eta < 0.17$, so the complicated function (11) can be approximated in the simple form:

$$\Lambda(\eta) \approx \frac{1}{2}\left[\frac{0.8}{\eta} - 1\right]. \tag{11'}$$

The illumination probability given in (10) is based on a one-dimensional Gaussian process, but *bistatic* scattering generally takes place in a two-dimensional geometry, so that in this scattering configuration the two-dimensional aspects of the surface must be taken into account. Illumination and observation take place in two separate vertical planes intersecting the surface at the common point P, thereby defining two local wave slopes, q_I and q_0. The joint probability in (3a) is thus written as the product of two expressions of the form of (10), each containing the angles appropriate to its plane of definition.

$$P_B(I,0) = (1 - .5\,erfc(\xi))^{\Lambda(\eta_I)+\Lambda(\eta_0)}U(q_1 + \psi_I)U(q_0 + \psi_0) \tag{12}$$

2.3.1. *The Threshold Model.* In order to get a clearer idea of the implications of these expressions, consider first the behavior of the factor in (10) that depends only on surface height and mean grazing angle:

$$P^0(I) = (1 - .5\,erfc\ (\xi))^{\Lambda(\eta_I)} \tag{13}$$

This expression is easily calculated, and is shown in Fig. 8 for values of η between 0.003 and 0.3, corresponding to mean grazing angles between about 0.02° and 2° for a sea having a mean slope $s_0 = 0.1$. We see that the probability of illumination changes rapidly over a rather narrow range of surface heights, leading to the idea of ''illumination thresholding'' mentioned earlier. The ''threshold'' is taken as the value of ξ that makes $P^0 = 0.5$. Points on the surface for which $\xi > \xi_0$ are assumed to be totally illuminated, while those for which $\xi < \xi_0$ are totally shadowed. For bistatic scattering, the threshold is determined by the product of two such factors: $P^0(\xi_0,\eta_I) * P^0(\xi_0,\eta_0) = 0.5$. If the grazing angles over the I and 0 paths are the same, as the evaporative layer often forces them to be at extended ranges [Wetzel, *loc cit*], then for $\psi < 1°$ the approximation (11') for $\Lambda(\eta)$ indicates that

$$P^0(\xi,\eta)P^0(\xi,\eta) \sim P^0(\xi,\eta/2) \tag{14}$$

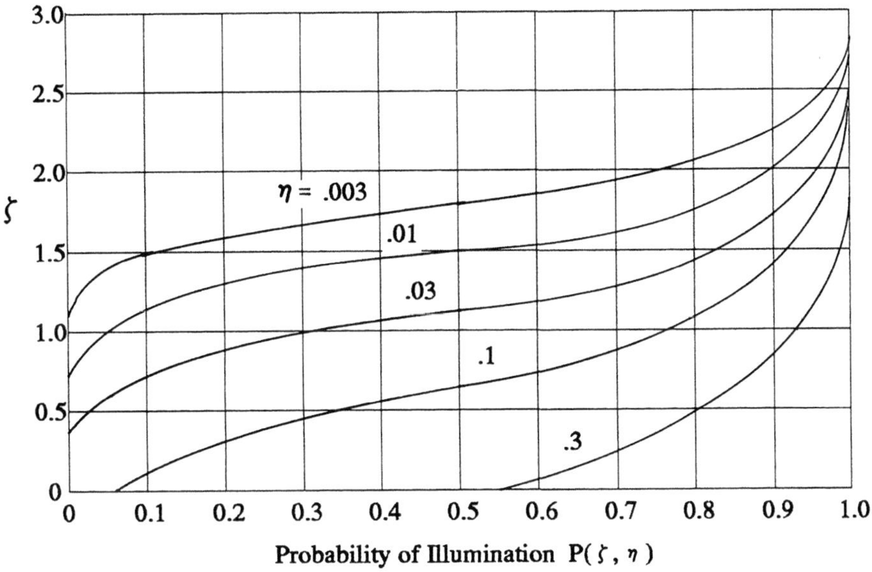

Fig. 8 — Illumination probability functions

or that the bistatic height factor is closely approximated by the monostatic height factor at half the grazing angle.

Figure 9 illustrates illumination thresholding for one realization of a random sea surface. The surface was obtained from a random phase expansion whose coefficients were defined by a Phillips spectrum for a wind speed of 15 knots; the length of the record is three dominant wavelengths (110 m). The threshold heights were obtained from Fig. 8 by setting $P^0(\xi^0, \eta) = .5$, and using the relation $z = \sqrt{2}\xi_0(\eta)\sigma$, where σ is the rms wave height of the surface ($\sigma = 0.28$ m for a 15 kt wind speed). Although one should not draw general conclusions from a single example, even this simple exercise displays some of the basic features of illumination thresholding. For the given grazing angles, only the largest peaks on the surface have a chance of being seen: for angles in the range 1° - 2°, there will be one peak per dominant wavelength (36m); for angles 0.6° - 0.3°, two peaks are visible over 3 wavelengths; for 0.3° - 0.1°, only one peak is visible, and for smaller angles no peaks are visible, which means, simply, that the probability per wavelength is less than 1/3.

This behavior was put on a semi-quantitative basis by Wetzel [*loc cit*] who showed that in this regimen of grazing angles, the definition of thresholding given above leads to a surface on which isolated scattering "islands" of mean length L_I are separated in a dark plane by a mean distance L_{SH}, where

$$L_I = Le^{\xi_0^2}[1 - erf\ (\xi_0)]\,, \tag{15}$$

$$L_{SH} = Le^{\xi_0^2}[1 + erf\ (\xi_0)]\,, \tag{16}$$

with the characteristic length L defined by

$$L = \pi(\sigma/s_0). \tag{17}$$

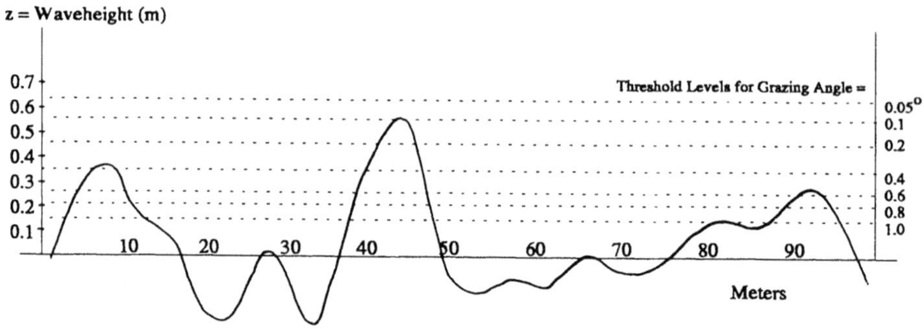

Fig. 9 — Threshold levels vs. grazing angles for simulated sea

It is interesting to compare the predictions of these analytical expressions with the results obtained pictorially in Fig. 9. The behavior of L_I and L_{SH} with normalized grazing angle is plotted with solid lines in Fig. 10 for $W = 15$ kts, $\sigma = 0.28$ m, $s_0 = 0.1$. As the grazing angle decreases from 1° to 0.1°, for example, the spacing L_{SH} between islands increases rapidly (from 14 m to 190 m), while the size of the exposed islands L_I decreases very slowly (from 6m to 3m). The reader will note that the lengths scaled from Fig. 9 for a single realization of the sea surface are, give or take a few inches in wave height, quite consistent with the corresponding numerical values in Fig. 10; e.g., for $\psi = 0.4°$, $L_I \sim 3$ m and $L_{SH} \sim 30$ m measured at the vertical dashed line.

These statistical results can further be used to explore the effects of mean sea slope s_0. The curves in Fig. 8, thus the threshold parameter ξ_0 which they select, depend only on $\eta = \psi/s_0$, the ratio of grazing angle to mean slope. Thus if the grazing angle and slope are changed by the same factor, then ξ_0 and all of the results depending on it, will remain the same. The only other slope-dependent parameter is the scale length L defined in (17). If we were to make the sea more "choppy" by doubling s_0 from 0.1 to 0.2, the effect on the lengths plotted in Fig. 10 would be to halve all of them, and displace them by twice the grazing angle (via redefinition of η), thereby producing the dashed curves. It is seen that the shadow length L_{SH} remains much the same, while the lengths of the exposed peaks are reduced by about half; e.g., for $\psi = 0.4°$, L_{SH} goes from 58m to 53m, while L_I is reduced from 3.7m to 1.6m. Thus for a fixed grazing angle, as the sea becomes choppier this model predicts that the separation between scattering islands remains much the same, while the areas of the exposed peaks get smaller.

2.3.2. *Clutter Intermittency.* The clutter scene presented to a radar in the deep shadowing regime will be strongly structured by the size and spacing of the scattering islands relative to the size of the radar footprint. This topic was discussed at length by Wetzel in the cited reference, where details may be found. As a rough approximation, it may be assumed that

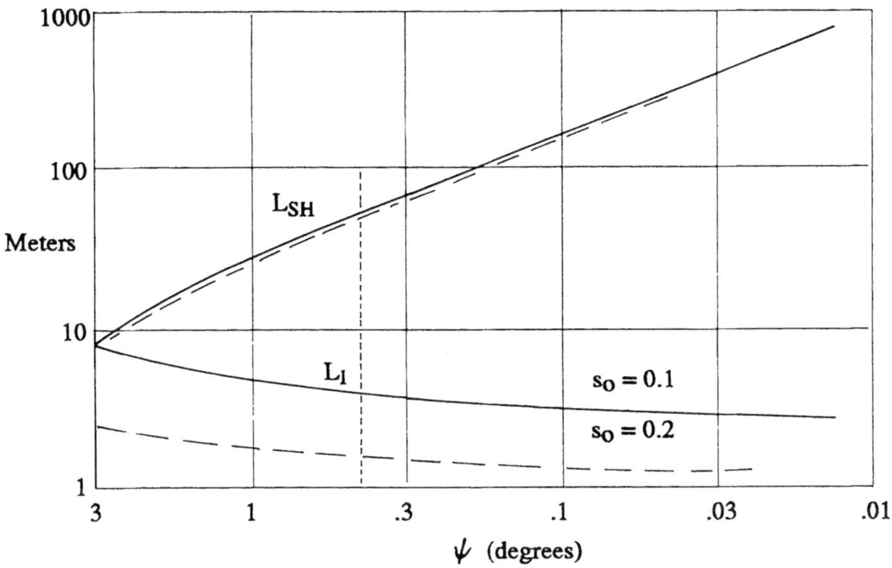

Fig. 10 — "Island" and "shadow" lengths on thresholded surface

the one-dimensional length parameter L_{SH} represents an average spacing between scattering islands on a two-dimensional sea surface, in which case the average number of peaks per unit area becomes

$$n = L_{SH}^{-2} \quad \text{peaks/m}^2 \tag{18}$$

For a monostatic pulse radar, the area A_f of the illumination footprint is given by $\beta Rc\tau/2$, where β is the beamwidth, R is the range, c is the speed of light, and τ is the pulse length. For bistatic radar, A_F is determined by the intersection of the transmitter and receiver beams, and the pulse length (under pulse limited conditions). In either case, the average number of surface peaks appearing in the radar footprint is

$$N = nA_F = A_F/L_{SH}^2 \text{ peaks in } A_F \tag{19}$$

Because of its effect on the statistics of the scattered signal, this number may be called an "intermittency index." It was originally introduced to describe the behavior of the clutter return as seen by a monostatic radar on an A-scope, or a *PPI*. For $N >> 1$, the instantaneous clutter signal is made up of returns from many peaks, so would display fluctuations about some mean value. For $N << 1$, the radar receiver will be looking at receiver noise most of the time, interrupted occasionally by a sharp spike from an isolated island. Similarly, a sequence of samples collected from a fixed range cell (or beam-intersection cell) will display different temporal statistics accordingly as N is greater or less than unity.

2.4. SUMMARY COMMENTS ON SHADOWING

Little more can be said at present about shadowing of the sea surface at low grazing angles. The effects of diffraction have been examined in a deterministic two-dimensional model, and some of the implications of statistical fluctuations in a truer, three-dimensional geometry have been explored. The domain of frequency vs windspeed in which geometrical optics provides a reasonable representation of surface illumination was found to be small, yet radar operations do occur in this domain, so the conventional geometrical optics shadowing theories are not totally without value. Besides, they are the only shadowing theories that presently exist.

While the isolation of scattering "islands" by illumination thresholding at low grazing angles is intuitively obvious, it is important to have quantified this process by introducing the lengths L_I and L_{SH}. These lengths define the size and distribution of the scattering regions available to a radar at low grazing angles, and therefore provide a basis for predicting how low-angle backscatter might depend on operational and environmental parameters.

3. Scattering

The second basic question raised in the Introduction concerns the kinds of scattering processes that can be expected to occur at the exposed wave peaks. A real wave peak could have almost any form. In a long greasy swell with no wind, it might even resemble the sinusoidal surface for which exact scattering solutions exist [DeSanto, 1975]. But usually the surface is complex, covered by small-scale structure in the form of cusps, wedges, microbreakers, hydraulic shocks, patches of turbulence and gravity/capillary waves, both wind-driven and parasitic. A particular region might contain a sharp crest on the verge of breaking, or, having broken, a cascade of plumes falling down its face, and there could be a cloud of spray droplets above it. Clearly, in the face of all of this real-world complexity, the selection of a convincing, and manageable, scattering model poses some difficulties.

In discussing scattering, however, it is necessary to be somewhat specific about the nature of the scattering obstacle. While the topography of the "scattering island" could take a variety of forms, it would not be unreasonable—at least for an energetic wind-driven sea—to take the basic underlying structure to be wedge-like, with a surface slope of about 30° suggested by the surface stability criterion for the onset of wave breaking. In order to avoid having to deal with a full three-dimensional sea, the problem will initially be simplified by assuming this wave-wedge to be sufficiently long crested to extend across the width of the island. Without such restrictive assumptions, it would be discouragingly difficult to obtain any useful information about the deterministic scattering processes that ultimately lie behind the statistical scattering behavior.

We will first examine the neighborhood of the peak, concentrating on possible sources of distributed scatter produced by small features amenable to modeling by perturbation theories. Next we evaluate the role of small wedges as scattering elements on the surface of the exposed crest, and finally consider some of the more powerful plume and corner-reflector scatterers that could produce scattering returns consistent with the observed "sea spikes."

Throughout this discussion, the importance of shadowing in structuring the scattered field at very low grazing angles will be continually emphasized.

3.1. DISTRIBUTED SCATTER: THE BRAGG HYPOTHESIS

Any small-amplitude approximation for scattering from a statistically rough surface leads to a scattering cross section proportional to the Fourier transform of the surface correlation function evaluated at twice the projection of the incident wave vector on the surface; i.e., the surface spectrum evaluated at the "Bragg" wavenumber. A similar result is obtained in other weak scattering theories, as in scattering from weak turbulence. With this approximate solution in hand, two questions remain: (1) Does the surface actually satisfy the conditions of the approximation? (2) What "spectrum" should be used in the formalism? Considering real seas at microwave frequencies, the answer to the first question is most certainly No, since the conditions of the approximation require that the maximum surface displacement be very much smaller than the illuminating wavelength (less than one-tenth). A detailed discussion of some of the limitations on the application of the first-order Bragg approximation is discussed by Plant (this book). Nevertheless, as further discussed in Plant's Chapter, the Bragg hypothesis continues to be used by arguing that the sea is "locally flat" when taken in small enough sections, and these small sections, now pushed about by the gross underlying surface, are responsible for the scattered field. This "composite surface" model requires that the surface spectrum be somehow "partitioned" between the region of small waves (Bragg scatterers) and the large waves on which they ride. There are serious logical and conceptual difficulties with this theory when applied to a real sea, but in the idealized model to be used here, these objections will not apply. In regard to the second question, it is difficult to decide how to choose the spectrum to be used with the Bragg hypothesis, because the sea spectrum has never been satisfactorily measured at wavelengths corresponding to Bragg resonance at microwave frequencies. Nevertheless, it is commonly taken to be the Phillips spectrum [Phillips, 1966], which is a k-to-the-minus-four spectrum roughly satisfied by real seas at much longer wavelengths. [See Donelan and Hui's chapter for the latest thinking on the subject of sea spectra.] However, if a particular scattering structure is being studied, then the spectrum chosen will be representative of that structure, as, for example, in the case of the "solitons" proposed by Middleton and Mellen [1985], of which more will be said shortly.

For all the criticism that can be leveled at the Bragg model, it has the virtue of delivering predictions that are often consistent with experience and are difficult to obtain from any other model. For this reason the Bragg model, though conceptually flawed, continues to be used. In what follows it is viewed as a paradigm for distributed scatter generally. It is intrinsically "distributed," in that it derives from either an integral or a wide-sense spatial transform over a flat surface, and it incorporates the surface boundary conditions, thus disclosing a polarization dependence. And finally, we must respect, if not accept uncritically, its apparent ability to agree with measurement in the range of *local* angles of incidence of interest to us here.

3.1.1. *Bistatic Scattering From Peaks in the Bragg Approximation.* In the simplest possible model, the exposed faces of the wave-wedge are uniformly covered with a carpet of Bragg-scattering wavelets, often called "ripples." The problem of Bragg scatter from a tilted surface element has been treated by many authors in a variety of different forms [for references see Valenzuela, 1978; Bass and Fuks, 1980; Plant, this book] with similar results. To provide a specific form for our discussion of low angle sea scatter, a bistatic formulation is included here in the Appendix, Section 1, where the bistatic cross section for scattering through angle ϕ_s at very small (close to zero) grazing angles is found for an element with arbitrary slope and azimuthal orientation (see Fig. 11). The effects of local shadowing are straightforward in this model, since a scattering path will exist only when the planes of illumination and observation intersect on the front face of the wedge, as expressed by the U-functions in (12). Note that the implications of the threshold shadowing theory of the previous sections have already been taken into account through the model of scattering "islands" we are using. Referring again to Fig. 11, it is clear that if the azimuth angle ϕ of the surface element is less than $\pi/2$, the illumination ($\hat{k}_i$) enters the back side of the element and will not be scattered into direction $\hat{k}_s$, while if $\phi > \phi_s + \pi/2$, the observer will be looking at the back side of the element and will not see the scattered field. This means that only those surface azimuths ϕ that lie in the range $\pi/2 < \phi < \pi/2 + \phi_s$ can contribute to bistatic scattering through an angle ϕ_s. (Note that this is a condition on both ϕ and ϕ_s, since it also resricts the range of ϕ_s for given ϕ.) The mean cross section per unit area is found by averaging ϕ over this range of angles for a particular probability density function $P(\phi)$. That is,

$$\overline{\sigma^0}(\phi_s) = \int_{\pi/2}^{\pi/2+\phi_s} \sigma^0(\phi,\, \phi_s) p(\phi) d\phi \tag{20}$$

where $\sigma^0(\phi,\phi_s)$ is given in (A17). The distribution of wave orientations ϕ could be anything from a delta-function, as when the crests line up in a breaking zone, to a uniform angular distribution, as in a short-crested confused sea. Since these represent two opposite extremes, both cases will be considered.

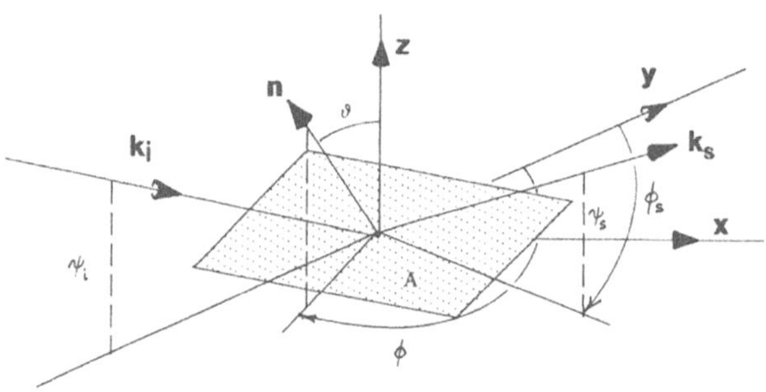

Fig. 11 — Bistatic scattering geometry for a flat crest face

Consider first *Case* 1, in which all the wedge faces are oriented with fixed angle ϕ_o:

$$p_1(\phi) = \delta(\phi - \phi_0) \tag{21}$$

In this case, the average in (20) is obtained by replacing ϕ by ϕ_0 in (A17). The cross sections for the two polarizations are plotted in Fig. 12 for a slope of 30° and for grazing angles close to zero, as explained in the Appendix following Eq. (A17). "Monostatic" means that the scattering angle $\phi_s = 180°$, while the wedge face is rotated from broadside (180°) to glancing incidence (90°); "bistatic" means that the wedge is fixed in broadside position,

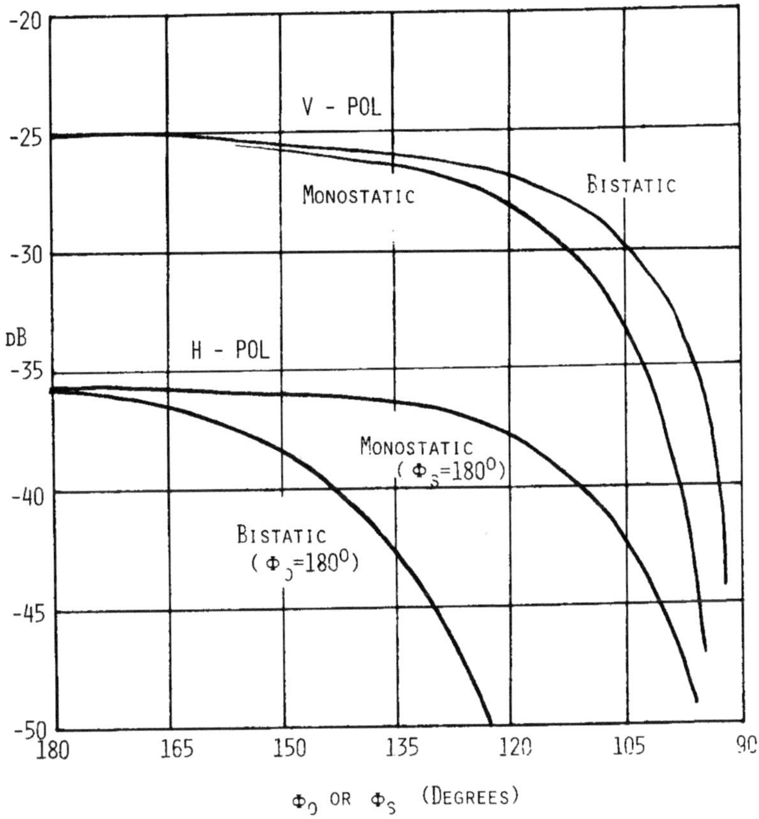

Fig. 12 — Angular dependence of wedge-face scatter—the Bragg model

while the receiver is moved from a monostatic (180°) to a glancing (90°) observation position.

For vertical polarization ($V-POL$), the scattering is controlled almost entirely by the first term in (A14) along with the outboard $\sin^2$ terms in (A17), which force the fall-off as the angles approach 90°. For horizontal polarization ($H-POL$), the most interesting result is the relative independence of the monostatic return on wedge orientation. This is due to the $H-V-H$ conversion that takes place as the surface normal of the tilted wedge rotates out of the plane of incidence, bringing the first term of (A14) into play for $p = H$. Thus the $H-POL$ return behaves with wedge orientation angle much like the $V-POL$ return, only lying about 10 dB lower. It should be noted that this is the same mechanism that produces the cross-polarized return in the two-scale Bragg model. It is derived in a simplified form for the monostatic case in Section 7 in Plant (this book).

Consider next *Case* 2, in which the wedge orientations are uniformly distributed between $\pi/2$ and $3\pi/2$, the allowed range of angles again containing only those for which there is an unshadowed path between source and observer. Here the probability density is given by

$$p_2(\phi) = 1/\pi \qquad (22)$$

and the trigonometric functions in (A17) have to be integrated over the restructed range of angles given in (20). The integrals were performed numerically in a simple computer program and the results are shown graphically in Fig. 13.

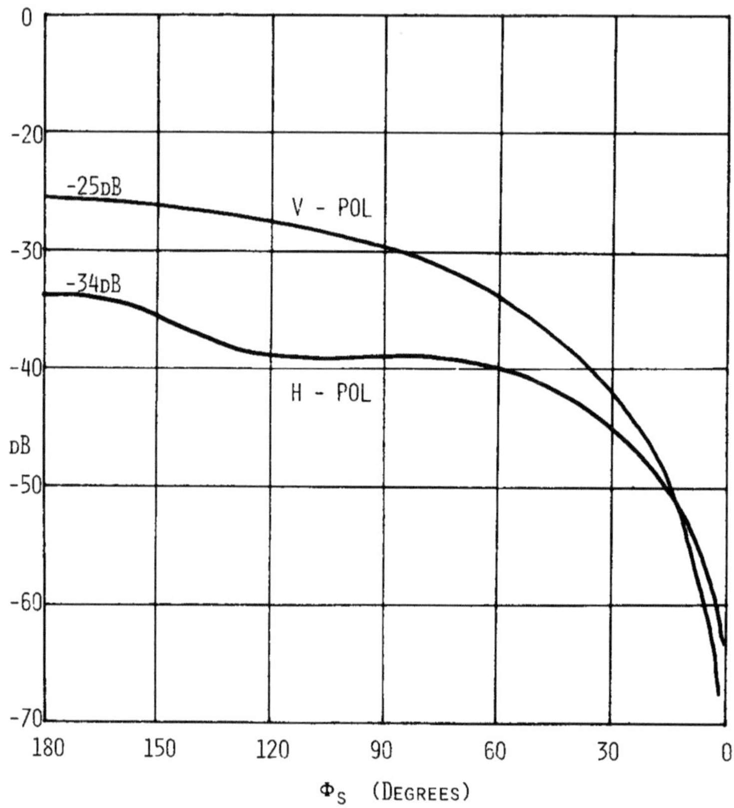

Fig. 13 — Bistatic cross sections for uniform crest angle distribution

The vertically polarized returns show a smooth 40 dB falloff over the range of bistatic angles, while the horizontal returns have a smaller variation, but seem to display a kind of "resonant" behavior. Note, however, that these are not true resonances, being artifacts of the changing dominance of the two terms in (A14) as the scattering angle moves through its range. Reference to (A8) and (A9) reveals that α_{HH} tends to peak at the extremes and have a hole in the middle, while α_{W} tends to be large in the middle. Thus the $H-POL$ curve starts off much like its bistatic counterpart in Fig. 12, but gets picked up by α_{VV} in (A14) over the intermediate angles, looking like a replica of the $V-POL$ curve until experiencing the effect of α_{HH} again as ϕ_s approaches zero. Of greater interest is the other end of these curves, which denote the values of radar *backscatter* cross section predicted by this model. It will be found shortly that they are in good agreement with measurement.

3.1.2. *The NRCS and a Shadowing Function For Extreme Grazing Angles.* The results obtained thus far still do not provide a prediction of the cross section per unit area (sometimes called the Normalized Radar Cross Section, or NRCS) that would be measured operationally. To do this we recall the basic picture of the scattering "island" as a prism with an interior angle of 120° and base width and crest length of L_I. Assuming the grazing angle to be smaller than 30°, so that only one face of the prism contributes in a scattering interaction (the other is in shadow), the average cross section of a single "island" is roughly the product of one of the cross sections per-unit-area found above and one-half the area of the island, $L_I \times (L_I/2)$. If the "intermittency index" defined by Eq. (19) is considerably greater than 1, then the cross section per unit area measured by a radar is just the product of the average cross section per peak, or "island," and the average number of peaks per unit area, given in Eq. (18). That is,

$$(NRCS)_p(\phi_s) = \overline{\sigma^0}(\phi_s)(1/2)(L_I/L_{SH})^2 \tag{23}$$

where the two lengths are given in Eqs. (15)-(17), and illustrated graphically in Fig. 10.

As noted at the beginning of Section 2, conventional shadowing theory produces a "shadowing function" that smoothly extinguishes the backscatter intensity linearly as the grazing angle approaches zero. In the threshold theory being used here, "shadowing" means something quite different. Here, the cross sections of the individual peaks vary little (L_I is weakly dependent on grazing angle), and the reduction in the NRCS results primarily from collecting fewer of these peaks under the pulse footprint as their separation increases. This behavior for extreme grazing angles is expressed by defining a new shadowing function, $S_T(\psi)$, as the length ratio in (23) with the lengths given in (15) - (17):

$$S_T = \frac{1}{2}(L_I/L_{SH})^2 = \frac{1}{2}\left[\frac{1 - erf\ (\xi_0)}{1 + erf\ (\xi_0)}\right]^2 \tag{24}$$

This function is plotted in Fig. 14, where a smooth 3 dB transition is made to the grazing angle (30°) corresponding to the slope of the prism face (to account for the factor (1/2) in (23)). The asymptotic form of the conventional shadowing function,

$$S_c \rightarrow \sqrt{(\pi/2)(\phi/s_0)} \tag{25}$$

is included for comparison. Keep in mind that these averages all make sense only if the intermittence index N in (19) is large. Otherwise the backscatter appears as a random occurrence of isolated spikes. However, when the conditions are right, it is clear that below about 2° the threshold-derived shadowing function will display a much steeper characteristic than does the conventional function (25).

3.1.3. *Comparison with Measurement.* Experimental support for the theory discussed in this section is hard to find, because surprisingly few reliable measurements have been made at very low grazing angles. Moreover, while we have expressed the NCRS in terms of statistical averages, some workers give the NCRS as the *median* cross section. In fact, the problem

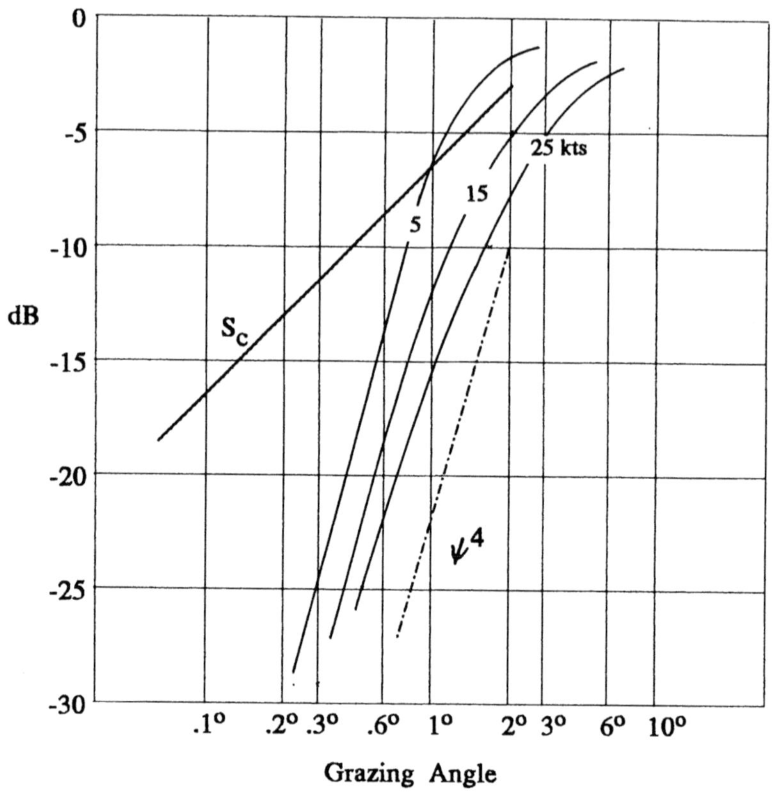

Fig. 14 — "Shadowing functions" at very low grazing angles

of finding a suitable statistical characterization for low angle clutter is not a simple one [see, for example, Trunk, 1976]. Of the few measurements that exist, a few of the more pertinent ones are reviewed below.

a. NRL 4 FR Data: The Four-Frequency-Radar (4FR) data taken by NRL around 1968 [Daley, *et al.*, 1968] were collected with reasonable care and are often used to "confirm" the Bragg scattering hypothesis [see e.g., Valenzuela, 1978]. The smallest grazing angle measured was 4°, which is somewhat below the mean sea slope angle and should correspond to the NRCS of (23) with $L_I = L_{SH}$. Figure 15 reproduces the figures shown in Valenzuela for a frequency of 4.455 GHz (about 7 cm wavelength), upwind aspect in strong winds (>22 kts). The results at 4° will therefore lie in the upper part of the "Maybe" region in Fig. 5, so the geometrical optics assumptions should be valid. The grazing angle can still be considered small, so under these conditions the appropriate values for the NRCS are taken as 3 dB (to account for the 1/2 in (23)) less than those at the left edge of Fig. 13: that is, -28 dB for V-polarization and -37 dB for H-polarization. These values are indicated by the big circles in Fig. 15, and the agreement with the measured values is seen to be quite good for both

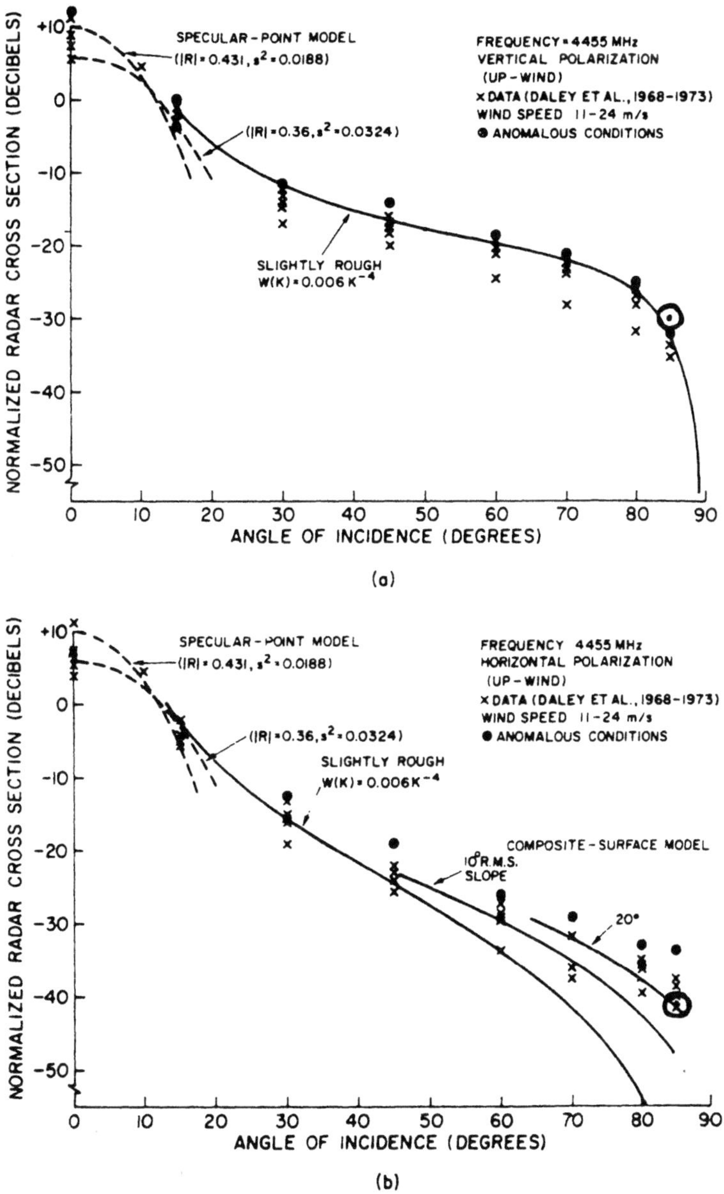

Fig. 15 — Comparison with NRL—4FR data

polarizations. It should be emphasized that these results are based on a much more sophisticated argument than the simple Bragg scatter curves shown on the figures, which assume a uniform carpet of scattering wavelets for $V-pol$, and a normally distributed slope distribution for the composite-surface values in the $H-pol$ plot. However, since the Bragg theory rests on such shaky assumptions, it is quite possible that the agreement with such measurements as these is purely accidental in both cases. Much more needs to be done before the Bragg theory can achieve intellectual respectability, and about all that can be said here is that these results share, with other applications of the Bragg hypothesis, this remarkable tendency to agree with these 4FR data.

b. The Critical Angle: Some 30 years ago, it was observed that when radar sea backscatter was plotted against range, there appeared to be a change in the rate of decrease of signal level at ranges corresponding to a grazing angle of 2° or so [Katzin, 1957]. This angle was called the "Critical Angle," and was ascribed to an interference phenomenon in which direct and (phase-reversed) reflected waves combined above the surface to produce a field at the scattering elements that decreased as the fourth power of the grazing angle. Both the concept and its implications are absurd, and were viewed with suspicion even at the time [Katz and Spetner, 1960]. However, very often *something* seems to occur in the data at these angles to give the idea of a "critical angle" some objective validity [see, e.g., Long, 1981]. Most recently, a 2° anomaly has been found in low angle sea backscatter data collected in the North Atlantic [Trizna, 1987].

The "interference" argument has persisted because the observed decay law does seem roughly to follow the fourth power of the grazing angle, although once a line has been drawn through a mass of spread data points, the eye can be often mislead into accepting an "agreement" that is ambiguous at best (see Fig. 16a). The Threshold Shadowing argument described here gives an alternative explanation for the observed behavior. A fourth power decay law is plotted as a dashed line in Fig. 14, and quite clearly it looks very much like the shadowing functions S_T based on Eq. (24). Figure 16b reproduces a figure from the Katz and Spetner paper cited above, showing some X-band data from the RRE in England at very low grazing angles. Superimposed is the "5 kt" shadowing function from Fig. 14, which is in very good agreement with the measured behavior.

c. Other Measurements of Low Angle Sea Backscatter: The lack of good low angle radar data is really rather surprising, considering how easily it could be collected from comfortable sites on the shore overlooking the sea. Two sources of such data are the measurements by Hunter and Senior overlooking the Atlantic from a site in the south of England [1964], and by Sittrop from the west coast of Norway [1975]. These investigators took the trouble to record wind speed and direction while measuring X-band radar backscatter at grazing angles between about 0.5 to 5 degrees. In Figs. 17a and 17b, these data have been separated into crosswind and upwind/downwind categories and compared, respectively, with the conventional (S_c) and threshold (S_T) shadowing functions. One expects that upwind/downwind aspects would be controlled by the deep shadowing responsible for island formation in the threshold model, while crosswind aspects would show the radar the troughs most of the time, and thus

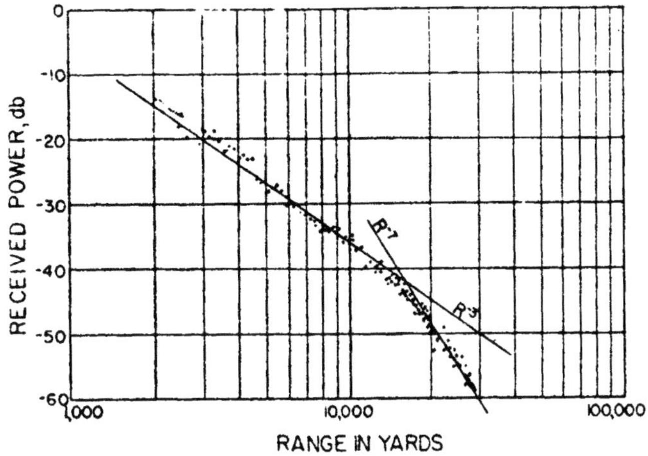

Fig. 16a — Evidence of a "critical angle"

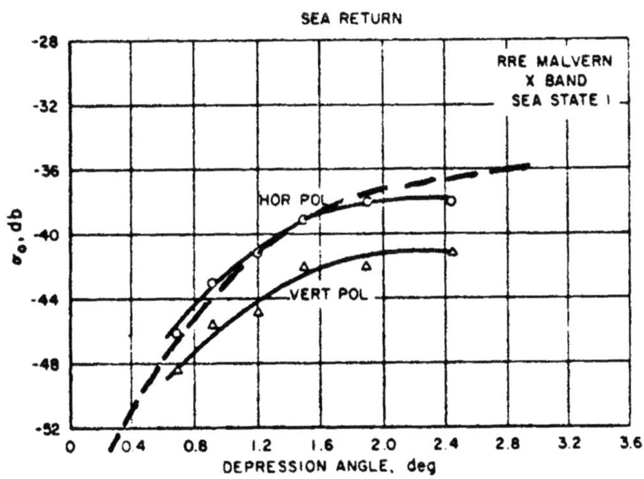

Fig. 16b — Polarization-independent low-angle behavior

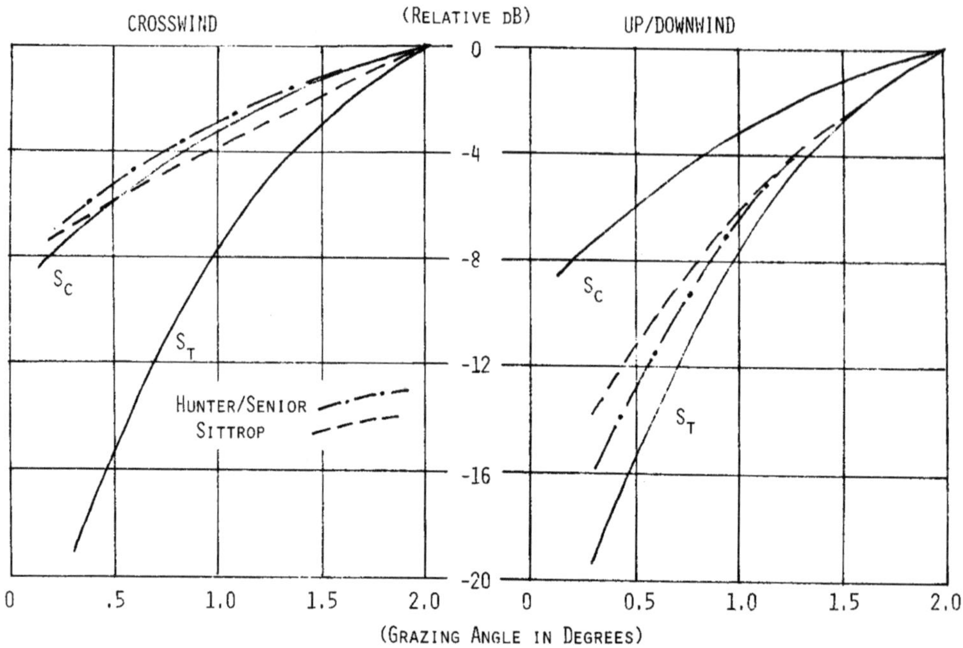

Fig. 17 — Comparison of low angle backscatter with shadowing theories

would display the more gradual backscatter decay described by the conventional shadowing function. There is certainly a strong qualitative difference between the backscatter behaviors for the two aspects, and they corroborate this hypothesis extremely well.

In a brief review of low-angle backscatter, Chan and Helmken [1983] adapt some measurements by others to show the effect of ducting. Their Fig. 7 is reproduced here as Fig. 18, where the 15 kt shadowing function from Fig. 14 is added as a dashed line. Again, the agreement is very good for the non-ducting conditions. It is of interest to note that the points plotted for the ducted case are not necessarily following a different shadowing function, but instead belong to larger angles than those obtained from simple free-space scaling. The situation expected in the similar case of the evaporation duct is discussed in Wetzel [1977].

d. Measurement of Sea Spikes: A dominant feature of low angle radar backscatter is the presence of Sea Spikes, which are most obvious with horizontal polarization under conditions of low intermittency index N (low grazing angles and small radar footprint), but must also make an important contribution to the NCRS under conditions of large N. The reason, of course, is that Sea Spikes are high-amplitude events, showing cross sections generally above 1 m^2, and at times reaching 100 m^2 or higher [Lewis and Olin, 1980; Hansen and Cavaleri, 1982; Trizna, 1987]. In the threshold model, Sea Spikes would have to be associated with the exposed islands of scattering

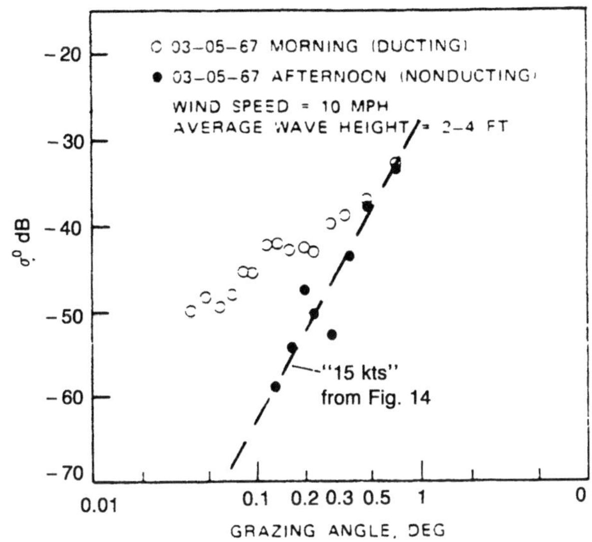

Fig. 18 — Example of low-angle backscatter

area $(L_l)^2/2$ which is of the order of 100 m^2. Considering the values of σ_o for distributed clutter shown in Fig. 12, it is obvious that Bragg scatter falls far short of being able to account for Sea Spikes. The source of this important scattering phenomenon must be sought elsewhere.

The agreement between theory and experiment demonstrated thus far is encouraging, particularly the tendency of the threshold shadowing function to predict the observed backscatter behavior in the limit of extremely small grazing angles - at least in those cases for which Fig. 3 gives geometrical shadowing theory a fighting chance. In regard to the backscatter cross sections themselves, it is of interest to note that the several independent measurements referenced here (NRL/4FR, Katzin, RRE, NWC, Sittrop, Trizna) permit the following "ballpark" estimate of low angle sea backscatter behavior at X-band frequencies: "For grazing angles of a degree or so, moderate wind speeds (about 15 kts), and for *both* H and V polarizations, the NRCS is -40 dB (m^2/m^2) give or take a few dB." This is also consistent with the value given in the oft-quoted radar reference book by Nathanson [1981]. Considering the wide variety in experimental procedures and cross section derivations, it is quite surprising to find this much consistency in data spanning 30 years.

3.1.4. *The "Soliton" Hypothesis of Middleton and Mellen.* In the previous applications in this section, the spectrum used was the general ocean wave spectrum of Phillips. As noted earlier, Middleton and Mellen [1985] have recently proposed a sea surface model in which the dominant scattering element at low grazing angles is taken to be a "Soliton" that develops out of a hydraulic shock that is produced, initially, by a puff of wind. Actually, close observation of a dynamic, wind-driven surface lends some credence to the wind-puff/hydraulic shock connection, although the production of solitons by this mechanism has met with skepticism [W. Pierson, private communication]. The hypothesis is highly speculative, and Middleton and Mellen have no strongly convincing arguments in support of it. Nevertheless, it has a certain appeal - particularly in a field where there have been few new ideas in over 20 years.

The profile of the soliton is taken to be a simple Gaussian shape with a certain height H, width L, and speed v:

$$z(r,t) = H \exp\left(-4(r - vt)^2/L^2\right). \tag{26}$$

From some reasonable statistical assumptions, a correlation function was calculated, whose Fourier transform provided a spectrum having the following form:

$$w(k) \sim (h^2 w^2)/[1 + (kw/2)^2]^2 \tag{27}$$

where h and w are the rms values of H and L in (26). It is of interest of note that when $kw > 2$, the spectrum assumes the k-minus-4 dependence familiar from the Phillips sea wave spectrum [Phillips, 1966], but also characterizing a distribution of wedges [Wetzel, 1986].

Middleton's interests lie primarily in acoustic scattering, so, assuming that the dimensions of these objects were small enough that the spectrum in (27) could be used in the Bragg formalism, he calculated the normalized backscatter cross section for an ensemble of such objects on an acoustically "soft" (pressure-release) surface like the sea surface. (In the duality between acoustic and electromagnetic scattering, this would correspond to horizontal polarization.) For low grazing angles (less than about 10°), the resulting NCS can be written in the form:

$$\sigma^0 \sim \left(\frac{h}{w}\right)^2 s_0^4 \left[\frac{(kw)^4}{[1 + (kw)^2]^2}\right] \tag{28}$$

where s_0 is the rms sea wave slope (which appears in the place of the grazing angle in the previous Bragg theory), and k is the sonar/radar wavenumber. Since the "soliton" object has the form of a Gaussian perturbation, it might be viewed as representative of isolated scattering features generally. That is, (26) is a scattering *something* that has a height and a length which are statistically characterized by rms parameters. Thus the cross section in (28) might be regarded as generic for such scattering objects. At least the asymptotic behavior is reasonable: when $kw > 1$, the cross section is independent of frequency, as one would expect in Bragg scattering using the k-minus-four spectrum derived above for these objects; when $kw < 1$, the cross section assumes the wavelength-minus-4 dependence of Rayleigh scattering, and very closely resembles Booker-Gordon scattering by weak atmospheric irregularities [1960]. But these results are only what one would expect *a priori* under the conditions describing this type of scattering; that is, there are no new scattering insights to be found in this approach.

3.2. IMPLICATIONS OF WEDGE MODELS

Close observation of almost any open water surface discloses that the basic surface features are wedge-like structures, rather than the sine waves or patches of gravity/capillary waves that permit either exact or small-perturbation scattering solutions. Hydrodynamically, these ''wedges'' might to be wave groups of the Stokes type, but with limited wave-train length and duration. Scattering from those wedges that are large compared to the incident wavelength may be treated by one of the short-wavelength approximations. The interior angles are limited by surface stability conditions to about 120°, so at low grazing angles, the wedge faces will generally be viewed at oblique incidence. If, in addition, the plane of incidence is perpendicular to the wedge edge, then the Physical Optics approximation tends to underestimate the scattering at such shallow scattering angles and asymptotic scattering theories emphasizing edge effects will begin to provide significant corrections. All such theories date from Sommerfeld's rigorous treatment of diffraction at edges in 1896, and while Sommerfeld himself reduced his solutions to more manageable Fresnel integrals and their asymptotic forms, the present theories of wedge scattering derive from the reinterpretation of these results independently by Keller (GTD) and by Ufimtsev (PTD) in 1957. But before trying to make sense out of some of these wedge scattering theories, it will be instructive to take a brief look at the few existing applications of wedges to sea scatter.

3.2.1. *Previous Applications of Wedge Models.* Although wedges are an obvious feature of the sea surface, scattering models based on wedges have made their appearance only recently. It should be noted that *all* such models rest on the implicit, and unrealistic, assumption that the wedge edge is normal to the plane of incidence.

a.) Kalmykov and Pustovoytenko [1976]: These authors describe X-band backscatter measurements at low grazing angles ($<3°$) in which ''bursts'' occur for both horizontal and vertical polarizations. They found that the horizontal returns were often greater than the vertical, contrary to the usual backscatter experience at higher grazing angles. In an attempt to provide some explanation for this behavior, they assumed that the ''bursts'' were due to breaking crests pointing toward the radar, as sketched in Fig. 19. They calculated the polarization ratios from the classical asymptotic forms, while we have used the GTD to obtain the wedge angle dependence plotted in Fig. 19. The results are the same, and may, or may not, have any validity in describing the physical situation. The authors wisely did not press the issue.

b.) Kwoh and Lake [1983]: Small two-dimensional wedge-like waveforms were generated mechanically in a wave tank, and profiled with a scanning laser slope gauge. A method-of-moments calculation of scattering by the resulting shapes showed angle and polarization dependence quite similar to the GTD results for a wedge, and were in reasonable agreement with measurements of the scattered field, at least for angles above the minimum grazing angle of 22°. Note that this work applies to angles in the intermediate range, which is commonly claimed as the province of the composite-surface Bragg model. It therefore has important general implications in the modeling of sea scatter, and should be read by anyone with an interest in the field.

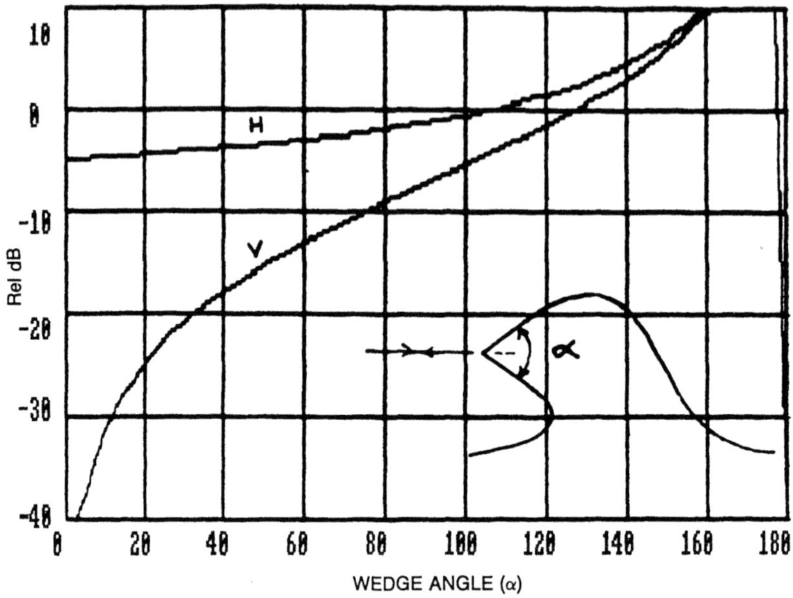

Fig. 19 — Wedge-angle dependence of backscatter: edge-on incidence

c.) Lyzenga, Maffett and Shuchman [1983]: Using what is essentially the PTD formalism for a 120° wedge, these authors found the *edge* contribution to wedge scattering, and showed that by adding an arbitrary portion of this component to the conventional Bragg component, an improved fit could be obtained with the NRL 4FR X-band data for horizontal polarization and grazing angles from 5 to 30 degrees.

d.) Wetzel [1986a]: Noting that the sea surface is a "wedgy" rather than a "wavy" surface, this author found such a surface to have the expected k-minus-four spectrum, but with a spectral scale factor that is proportional to wind speed. This wind speed dependence seems to track the NRL 4FR data quite well (if massively averaged), and a GTD calculation of wedge scattering provides both grazing angle and polarization agreement superior to that provided by the Bragg theory in the angular range in which Bragg theory is usually thought to apply. But as in all applications of discrete-target scattering models to the sea, there is no independent *a priori* basis for establishing target size and distributions, hence either mean or specific scattering levels.

3.2.2. *Problems in Applying Wedge Theory to Sea Scatter.* All of the above calculations used a simple two-dimensional wedge model, while for application to a real sea surface it will be important to consider scattering in *three* dimensions. In GTD, an extension to three dimensions occurs in allowing oblique incidence, in which case the scattered field appears in the form of a cone of rays—the so-called "Keller Cone"—whose half-angle is equal to the angle of incidence at the edge [Keller, 1957]. (See Fig. 20.) There *are no other scattered waves*; this is still a strict optical theory. The next extension is a generalization in which it is assumed that the incident field excites at each point of the edge a *fictitious physical current* of just such a form as to produce the scattered field on the optically defined Keller Cone emerging at that point. The edge may now be made arbitrary, and the current thus defined is integrated along the edge, as if it were a real current, to yield a value for a scattered field at points in space lying outside the Keller Cone. Several formalisms based on this idea have emerged in recent years: the Method of Equivalent Currents (MEC) as described by Knott and Senior [1974], the method of Incremental Length Diffraction Coefficients (ILDC) by Mitzner [1974], Michaeli's MEC (MMEC) by Michaeli [1984], to name the most important. Ufimtsev's PTD, which has already been used in section 2.2.2, is also concerned with edge currents, but indirectly, as a conceptual device for the correction of the Physical Optics approximation rather than in concrete form for the calculation of scattered fields. It should be noted that none of these methods is more "fundamental" or "rigorous" than the others. They are all, i.e., GTD, PTD, MEC, ILDC, MMEC, heuristic procedures for finding approximate solutions to scattering problems involving objects with edges. And there are still other methods (known by their acronyms as UAT, UTD, STD, HTD, etc.) which are designed primarily to deal with the singularities characteristic of the asymptotic approximations; these need not concern us here.

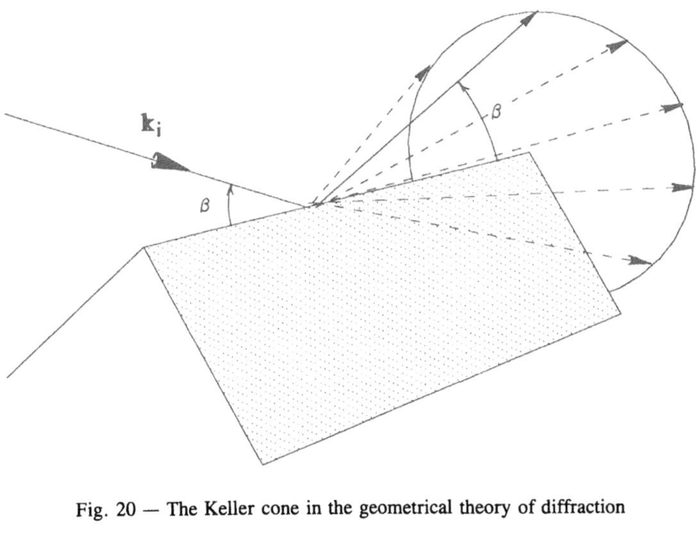

Fig. 20 — The Keller cone in the geometrical theory of diffraction

In order to disentangle ourselves from this mass of scattering theories and methodologies, we consider first the very basic problem of two-dimensional scattering from a wedge as described by the GTD. In Fig. 21, the incident ray, denoted by k_i, encounters an infinite 2-D wedge having an interior angle α and lying normal to the plane of incidence. $\hat{e}_V$ and $\hat{e}_H$ indicate respectively the orthogonal Vertical and Horizontal polarizations. In the GTD, the scatterered field consists of two parts: a field E_{GO} reflected from the illuminated face of the wedge according to the laws of geometrical optics, and a "diffracted" field, E_D, originating

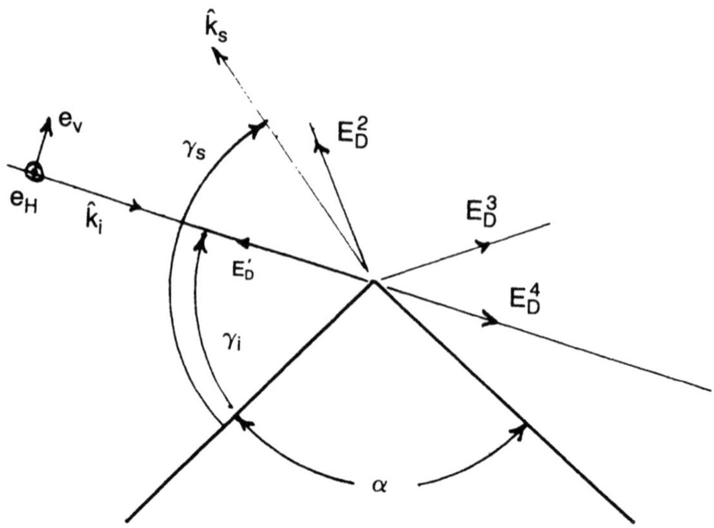

Fig. 21 — Scattering geometry in the GTD (2-D)

at the edge, consisting of a fan of rays which, for incidence normal to the edge, uniformly cover the plane of incidence. The field on these rays is defined by

$$E_{H,V}(\gamma_i,\gamma_s) = \frac{e^{i(kr+\pi/4)}}{\sqrt{2\pi kR}} D_{H,V}(\gamma_i,\gamma_s) \tag{29}$$

where

$$D_{H,V}(\gamma_i,\gamma_s) = \frac{\sin(\pi/n)}{n} \cdot \left[\frac{1}{\cos(\pi/n) - \cos(\gamma_s,\gamma_i)/n} - \frac{1}{\cos(\pi/n) - \cos(\gamma_s+\gamma_i)/N} \right] \tag{30}$$

with $n = 2 - \alpha/\pi$ (the *exterior* wedge angle is $n\pi$), and the incident and scattering angles measured from the front face of the wedge. The $-$ and $+$ signs are used for the H and V polarizations respectively. The four rays labeled in Fig. 21 have special significance: E_D^1 returns to the source and contributes the backscatter cross section of the wedge; E_D^2 defines the "reflection boundary" for the front face, and since its angle is $\gamma_s = \pi - \gamma_i$, it causes the second term in (30) to blow up; E_D^3 is a ray which would correspond to "specular reflection" if the wedge were sitting on a plane surface; E_D^4 defines the "shadow boundary" and since its angle is $\gamma_s = \pi + \gamma_i$, it causes the first term in (30) to blow up. These singularities are in no way related to actual field behavior, of course. They are artifacts of what is essentially an asymptotic expansion of the Fresnel Integral in (6), where the leading term is proportional to $1/w$, hence to $1/(\sqrt{kR/\pi D})$. This means that the GTD can be used only for angles well outside an angular zone of width $1/(\sqrt{kR/\pi})$ surrounding the singular rays. But except for these singular domains, the GTD has often been found to give reasonable approximations even under conditions fairly well removed from the "optical limit."

If end-defects are ignored, two-dimensional fields can be converted into three-dimensional fields "per unit length" by the factor $(k/2\pi R)^{1/2}$, from which a cross section defined in the usual way takes the form

$$\sigma_{H,V}(\gamma_i, \gamma_s) = L^2 D_{H,V}^2(\gamma_i, \gamma_s)/\pi \tag{31}$$

where L is the length of the wedge. This is the form used by Lyzenga, et al., [*loc. cit*] for their PTD corrections to Bragg scatter, and by Plonus, et al. [1978] in an interesting experimental comparison of various approximation methods. The backscatter cross sections versus interior wedge angle shown in Fig. 19 were obtained by substituting (30) into (31) with $\gamma_s = \gamma_i$, so that the incident ray lies along the wedge-angle bisector, and letting $L = 1$. Figure 22 shows the wedge angle dependence of scattering at zero grazing angle incidence across the top of the wedge. This corresponds to the normal wedgy sea of our low-angle model. The polarization dependence may be seen to be reversed for these two cases ($H > V$ in Fig. 19, $V > H$ in Fig. 22). In Fig. 23a, we show two examples of the grazing angle dependence for a stability-limited wedge angle of 120°. In Fig. 23a, the backscatter cross section versus grazing angle from 0 to 90 degrees, and in Fig. 23b, the *bistatic* cross section for scattering angles between 0 and 180° *in the plane of incidence*, is shown for a zero degree incident grazing angle. The plots show similar behavior, having an infinity in the direction of specular reflection from the front face. In the backscatter case, this direction corresponds to broadside scattering, which is generally an intractable case for GTD-like approximations (see, e.g., Sikta, *et al.*, 1983). Figure 23a was used by Wetzel [1986] to compare the implications of wedge scattering with some NRL 4FR backscatter data. The circles on this Figure show the *relative* behavior of the 4FR data with polarization and grazing angle.

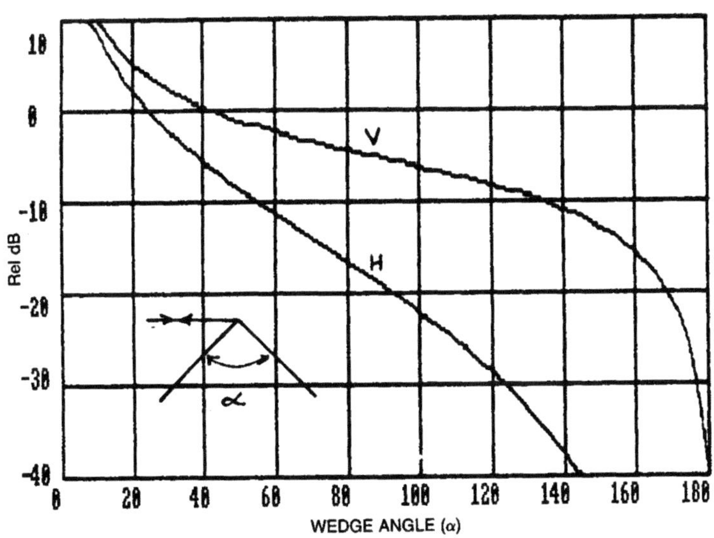

Fig. 22 — Wedge-angle dependence of backscatter: grazing incidence

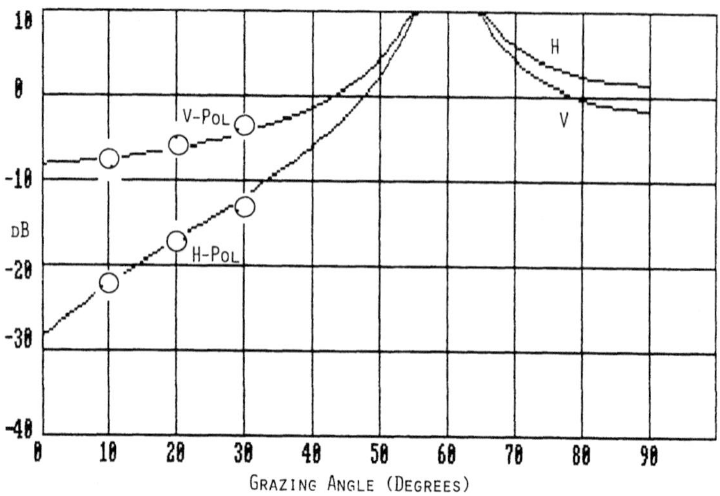

Fig. 23a — Backscatter cross section by the GTD (wedge angle =120°)

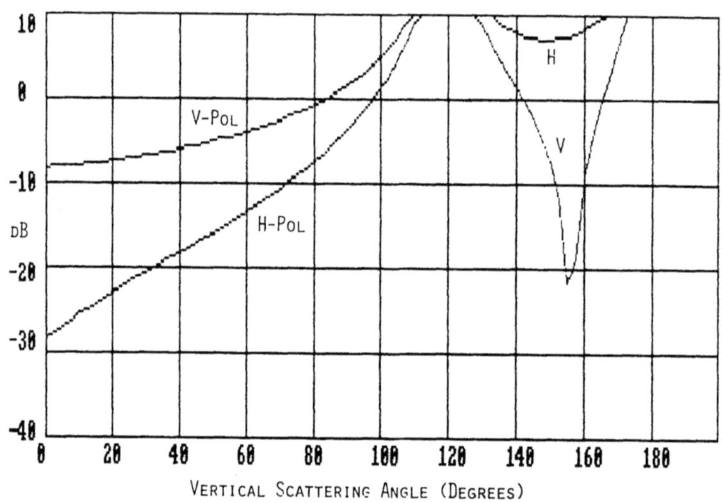

Fig. 23b — In-plane bistatic cross section by the GTD (120°)

But such curves, based on such theories, cannot tell us what we really want to know. They apply to a perfect wedge, aligned with perfect precision normal to the plane of incidence. As soon as the wedge deviates from the perpendicular, by however small an angle, the backscattered field vanishes. The reason for this is seen in Fig. 20, where the incident ray makes an angle $\beta \neq \pi/2$ with the wedge, and the diffracted rays now all lie on the so-called Keller Cone, whose interior half-angle is β. As mentioned earlier, *there are no other diffracted rays*—certainly none back into the direction of incidence to produce "backscatter." The bistatic case is no better, because the observer must lie on one of the elements of the Keller Cone, and again a scattering path is possible only for a single, precise orientation of the wedge edge. These unforgiving geometrical constraints can be relieved only by a return, in one form or another, to the basic idea of scattering from current distributions; that is, by using the fictitious ''equivalent currents'' in geometrical theories, or the approximate ''optical currents'' in Physical Optics.

(a) Method of Equivalent Currents: The bistatic cross section for a 120° wedge of length L in the low grazing-angle limit is calculated using the MEC in Section A.2 of the Appendix:

$$\sigma_{H,V}(L;\phi_w,\phi_s) = C_{H,V} L^2 \, sinc^2 \left[\frac{1}{2} kL \, (\cos\phi_w - \cos(\phi_s - \phi_w)) \right] \tag{A21}$$

where $C_{H,V} = -29\text{dB}\ (H-pol)$, -8 dB $(V-pol)$, and ϕ_w, ϕ_s are azimuthal wedge and scattering angles respectively (see Fig. A3). (To avoid confusion, it should be noted that in (31) and Fig. 23b, the bistatic angle ψ_s lies in the plane of incidence and sweeps up and over the top of the wedge, while in (A21) and Fig. A2, the bistatic angle is azimuthal and sweeps around the wedge in the horizontal plane.)

At this point it is important to recall one of the central problems with all models, of this sort, that introduce specific scattering structures that have no natural origin in the physics of the underlying surface. That is, while we *see* wedge-shaped structures on the surface of the sea, no one has ever bothered to measure or count ''wedges,'' so there is no knowledge of the distributions of their sizes and orientations. Nor is there a body of hydrodynamic theory from which such distributions might be inferred. For this reason it becomes necessary to come up with ''reasonable'' assumptions for these distributions, which usually translates as ''not completely idiotic,'' and leaves one with a corresponding skepticism about any conclusions drawn from them. Nevertheless, we cannot proceed without taking this necessary step.

It would be absurd to assume the scattering island to consist of a single large, coherently scattering wedge of length L_I, so we will replace this overly simplified model with one in which the faces of the illuminated islands are populated by ensembles of wedges of various sizes and orientations,having individual cross sections given by (A21). For backscatter, the scattering angle $\phi_s = \pi$, and since the sinc-function gives a very narrow beam around the wedge angle $\phi_w = \pi/2$, we can write (A21) as

$$\sigma_{H,V}(L,\delta) \approx C_{H,V} L^2 \, sinc^2(kL\delta) \tag{32}$$

where δ is the departure of ϕ_w from $\pi/2$. The ''3-dB'' points of the sinc^2 function occur at $\pm\, kL\delta \approx \pi/2$, so the range of wedge angles that could contribute to the backscattered signal is just

$$\Delta = 2\delta = \pi/kL\,. \tag{33}$$

Assuming the wedge angles ϕ_w to be uniformly distributed, the probability that a wedge of length L will provide a backscattered return becomes

$$P_{BS}(L) = \Delta/\pi = 1/kL \tag{34}$$

If each such wedge is further assumed to scatter independently, then the average cross section may be written as the average of (32) times (34) over the distribution $p(L)$ of lengths L:

$$<\sigma_{H,V}> \sim C_{H,V} \int_0^\infty L^2 \left[\frac{1}{kL}\right] p(L)dL = \frac{C_{H,V}}{k} \int_0^\infty L\, p(L)dL\,. \tag{35}$$

The wedge lengths L will be distributed between some minimum length (which seems by observation to be about 10 cm), and some much larger length of the order of the island dimension L_I; Not knowing what form this distribution might take, we choose a Rayleigh distribution

$$p(L) = (L/L_0^2)e^{-L^2/2L_0^2} \tag{36}$$

in which the mean length L_o lies somewhere between the extremes stated above. With this choice, the average cross section per wedge in (35) becomes

$$<\sigma_{H,V}> = \sqrt{\pi/2}\,(L_0/k)C_{H,V} \tag{37}$$

The "wedge" in this model is really a kind of flat prism, as sketched in Fig. A2, and since neither long, thin nor narrow, tall wedge faces are physically reasonable; or visually verified, we take a canonical aspect ratio of 2/1 for the length/height ratio of all wedge faces, so that the mean "area" of a wedge against the plane of the sea is about L^2. Assuming, now, that the wedges are distributed densely over the scattering islands, and that the grazing angle is sufficiently low so that only half are visible (well below the crest slope of about 30°), yet sufficiently high to be above the threshold shadowing knee in Fig. 14 (a degree or two), the NRCS from (23) takes the form

$$\text{NRCS}\,|_{H,V} = <\sigma_{H,V}>/2L_0^2 = C_{H,V}\sqrt{\pi/8}/(kL_0)\,. \tag{38}$$

Finally, in order to make some kind of reasonable estimate of L_0, we let L_0 be the geometric mean of L_{MIN} (about 10 cm) and L_I:

$$L_0 = \sqrt{(0.1 \times L_I)} \tag{39}$$

For moderate wind speeds L_I is about 10 m, so the mean wedge size is about 1 m. Using this value and the values of $C_{H,V}$ from (A21) in (38) for X-band frequencies ($k = 200/\text{m}$), the NRCS's become

$$\text{NRCS}\,|\,V = -33\ \text{dB}(m^2/m^2)$$

$$\text{NRCS}\,|\,H = -54\ \text{dB}(m^2/m^2) \tag{40}$$

(b) Physical Optics: it was noted earlier that when the plane of incidence is normal to the edge of the wedge, Physical Optics (P.O.) underestimates the scattering at the relatively large off-broadside angles for the flatter surface aspects seen in the low grazing-angle regime. This occurs because of the inaccuracy of the assumed current distribution at the edges for the particular scattering geometry. But as discussed in the previous section, any realistic wedge model of the sea surface will have its wedges oriented over a wide range of angles, so the population of wedges for which edge effects are important will be virtually zero. It is therefore of interest to look at the scattering implications of a straightforward Physical Optics approach.

The P.O. cross section of a rectangular plate with sides a and b, viewed at angles θ' and ϕ', where θ' is the angle between the plate normal and the incident direction and ϕ' is the angle of the plane of incidence relative to the a side, is given by Kerr [1952] in the form

$$\sigma(k,a,b,\delta',\phi') = \frac{k^2(ab)^2}{\pi} \cdot \left[\frac{\sin^2(ka \sin\theta' \cos\phi'}{(ka \sin\theta' \cos\phi')^2} \cdot \frac{\sin^2(kb \sin\theta' \sin\phi')}{(kb \sin\theta' \sin\phi')^2} \right] \cos^2\theta \quad (41)$$

Notice that there is no polarization dependence in the P.O. approximation as written here. However, the Fresnel reflection coefficients of the underlying surface on which the wedges sit will introduce polarization dependences into both the incident and scattered waves. This is a rather complicated problem in multiple scattering that will not be discussed here.

The scattering geometry may be viewed in terms of Fig. 11, where θ', ϕ' in (41) are replaced by $(\phi - \pi/2)$, $(\pi/2 - \psi)$, with ψ given in Eq. (A4). It will again be assumed that the exposed scattering islands are covered by wedges whose orientations, ϕ, are uniformly distributed, whose lengths, L_l, are Rayleigh distributed, and whose aspect ratios b/a are all equal to 1/2 (i.e., $a = L$, $b = L/2$, for all values of L.)

Performing the averages as in the last section, we write:

$$<\sigma> = \frac{k^2}{4\pi} \int_0^\infty L^4 F(kL) p(L) dL \,, \quad (42)$$

where $p(L)$ is given in (36), and $F(kL)$ is the angle integral,

$$F(kL) = \frac{1}{\pi} \int_{-\pi/2}^{+\pi 2} \text{sinc}^2(kL \cos\psi \sin\phi) \, \text{sinc}^2 \left[\frac{1}{2} kL \cos\psi \cos\phi \right] \sin^2\psi \cdot d\phi \quad (43)$$

with the definition (from (A4), with $\theta = 30°$, $\psi_i^o = 0°$)

$$\psi = \arcsin \left[\frac{1}{2} \cos \phi \right] \quad (44)$$

The solution of this set of relations is less formidable than it might appear. Numerical evaluation of the integral in (43) discloses that it can be quite accurately represented by

$$F(kL) = 2/(kL)^3 \,. \quad (44')$$

Substitution into (42) yields the very simple expression

$$<\sigma> = \sqrt{\pi/2} \, (L_o/K) \times (1/2\pi) \quad (45)$$

3.3 SCATTERING BY OTHER SURFACE FEATURES

Thus far we have looked at scattering from small amplitude surface perturbations amenable to a Bragg scatter formulation, and at wedge-like structures of indeterminate size, but large enough to be treated by optical approximations, both geometrical and physical. Other surface features might include:

(i) Pyramidal *cusp-like structures,* that probably scatter much like a short-crested wedge, or two wedges meeting at an angle.

(ii) *Parasitic capillaries*, that are visible on the front face of most forced waveforms, are of low amplitude, and appear to occupy a rather narrow spectrum with wavenumbers in the neighborhood of the transition between gravity and capillary waves.

(iii) The pools of *surface turbulence* associated with breaking waves. If the surface displacement spectrum of the turbulence were known, its scattering behavior could most likely be estimated from the Bragg formulation described earlier.

(iv) The *spray* associated with vigorous breaking in strong winds. Looking at waves breaking over a submerged breakwater that produced alot of spray, Kalmykov, *et al.*, [1976], used a rain model to explain their measurements. This model required an equivalent rain rate of 1 meter per hour (!) to match the observed levels of backscatter. Since it is difficult even to *conceive* of what a rain rate of such intensity might actually be like, it is possible that the use of this model was inappropriate.

There are two other kinds of structures that will occupy our attention for the remainder of this section: ''shocky''-looking structures seemingly associated with localized wind-puffs or the passage of a steep wave, and the ''plumes'' sliding down the front faces of breaking waves, as suggested by Longuet-Higgins and Turner [1974], and developed into a scattering model by Wetzel [1986]. The feature that characterizes both of these structures is the sharp entry into the underlying surface at their toe'' as they move along.

3.3.1. *Hydraulic Shocks or ''Sloshes.''* Consider first the subject of ''hydraulic shocks.'' When associated with shallow water effects such as a bore on a beach, breaking in the surf zone, or constricted flows in a channel (see Lighthill, 1978), they are usually referred to as ''hydraulic jumps,'' But as noted earlier, Middleton and Mellen [*loc cit*] have proposed that wind puffs on an open sea surface may induce small hydraulic shocks that decay into Gaussian solitons which, in turn, become a source of scattering. A bit of ''naked eye oceanography'' convinces one that small shocky-looking structures occur commonly on open water surfaces, sometimes in response to small-scale turbulent components in the surface wind field, and sometimes in association with steep waveforms. In fact, they are seen quite frequently at wind speeds above a few knots, taking the form of what might be called ''micro-breakers.'' Such an event may be seen at the arrow in the tank wave shown in Fig. 24. (Parasitic capillaries of the type also seen in the figure are found everywhere on almost any disturbed water surface.) Small shocky structures can be created on the surface of a basin of water by blowing at it obliquely with a household fan. In a laboratory setup of just this kind, a high-resolution radar looking into the wind recorded sharp spikes of backscatter which appeared to

which should be compared with the MEC result in (37). Since $1/2\pi$ can be expressed as -8 dB, we come to the remarkable conclusion that for this low grazing angle regime, the P.O. cross sections for H and V polarization are exactly the same as the MEC cross section for V polarization. Perhaps this is not so remarkable as it is reassuring, since GTD and P.O. quite often give similar results in scattering calculations. But the paths leading to (37) and (45) were enough different that one cannot help being struck by the coincidence of the two results. Aside from numerical constants, these expressions have the form (L_o^2) times $(1/kL)$, the first being a "size" factor proportional to the area occupied by the wedge, and the second being a "beam" factor, expressing the decreasing number of wedges seen as their average scattering pattern narrows. The uniform angular distribution used in these calculations are appropriate to an isotropic, or "confused" sea. But if a strongly directional sea were viewed up/down wind or crosswind, the form and frequency dependence of these cross sections might change significantly.

3.2.3. *Recapitulation.* Thus far, three scattering models for low grazing angle sea backscatter have been examined, along with some of the results of experimental measurements in this scattering regime. It would be useful at this point to pause for a moment and intercompare the values of the NRCS obtained from these various sources for X-band backscatter in moderate winds.

SOURCE	$V-Pol$	$H-Pol$
NRL 4FR (about 5°)	− 33 dB	−37 dB
Bragg Model (Sec. III.A.3)	−28 dB	−37 dB
MEC Wedge Model (Sec. III.B.2a)	−33 dB	−54 dB
P.O. Wedge Model (Sec. III.B.2b)	−33 dB	−33 dB
"Consensus" Measurements about 1°	−40 dB	−40 dB

For comparison against the NRL 4FR measurements at about 5 degrees, the cross sections from the scattering models developed here are given for grazing angles above the "knee" in Fig. 14. The "Consensus" measurements, mentioned at the end of Sec. 3.1.3, were all made at grazing angles of about 1°, so should be compared against the same model values lowered by the additional 7 or 8 dB required by the S_T curves in Fig. 14. One is tempted to conclude from this that the Bragg model is high for Vertical polarization, while for Horizontal polarization it is better at the higher angles (about 5°) while the P.O. model is much better at the lower angles (1°). The MEC model, based on the geometrical theories of diffraction, appears to underestimate the level of Horizontally polarized backscatter by 15 to 20 dB in this regime of grazing angles. However, none of these models can account for the powerful "Sea Spike" returns that characterize this regime, and which could themselves make a contribution to the averaging process defining the NRCS. These returns will be the subject of the next section.

Fig. 24 — Evidence of a "microbreaker" in a wave tank (Crapper, *"Capillary waves"*)

be associated with the formation of such structures [Hansen and Wetzel, *unpublished observation,* 1986].

Figure 25a shows what a puff-induced event might look like, based on both observation and physical plausibility. The vertical force of the localized puff causes a dimple in the water surface, while the horizontal component urges the edge of the dimple on the downwind side to move out from the center at a speed greater than the normal wave velocity for an impulsively excited surface, causing the forward face of the disturbance to steepen up and show a sharp entry into the quiet water ahead. The mysterious "micro-breakers" mentioned above look very much the same. While it is tempting to view a surface disturbance that steepens into a sharp front as a "shock wave," we must be careful not to confuse a "shock wave" with an asymmetrical "slosh." A shock wave tends to retain its form over some reasonable distance, while the event we are talking about here is short-lived, and its subsequent development is uncertain. If it relaxes into a smoother form of disturbance propagating from the puff site with unchanging shape, it becomes the presumed source of Middleton and Mellen's Solitons. But another alternative would have the initial "slosh" decay into a dispersal of wave components running at different speeds.

The details of these events appear to be completely unknown, although one can imagine that the *scale* of the phenomenon will play a role. It has been observed, for example, that the "ring wave" produced by the splash of a raindrop on a water surface propagates away from the splash site as a single, well-defined waveform [Wetzel, 1987]. The speed of this waveform is quite close to the minimum in the velocity vs. wavelength characteristic for water waves. Now, the neighborhood of this minimum contains a relatively broad range of

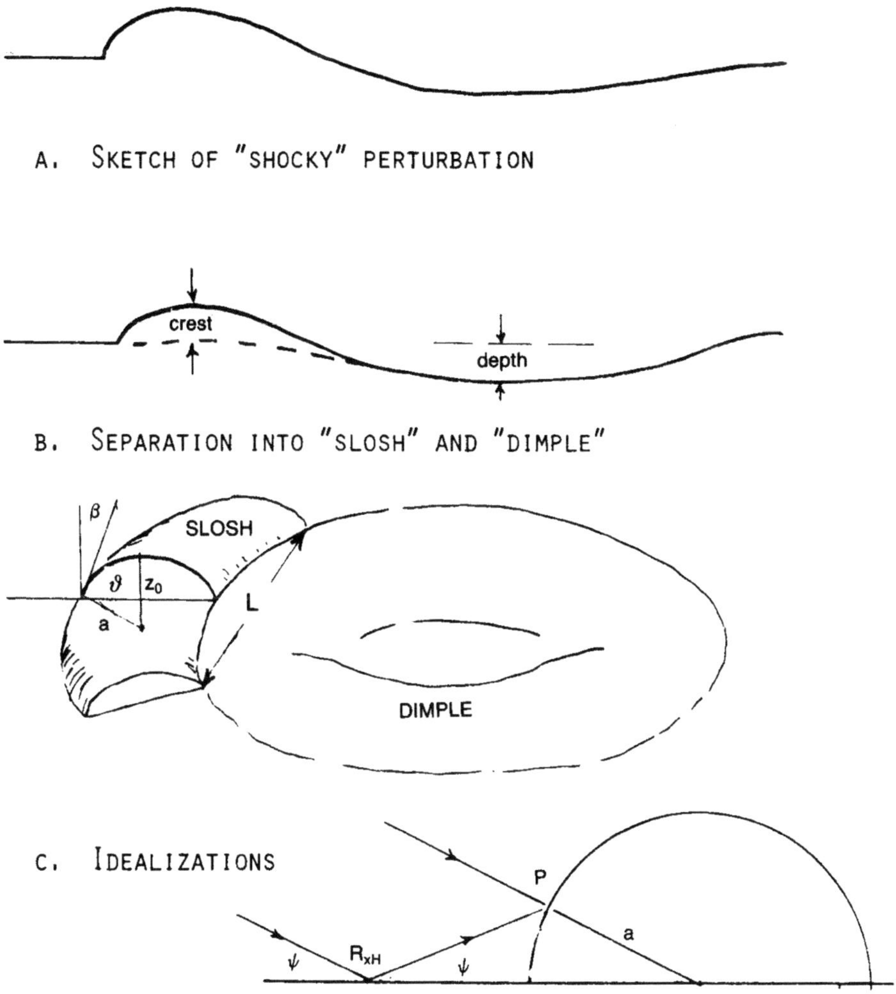

Fig. 25 — Modeling "shocky" surface disturbances

wavelengths with similar velocities, so it is reasonable to imagine that a *group* of waves having close to the same velocity might assemble itself from the chaos of the initial splash because the scale is right for such a result. Raindrop splashes are measured in centimeters, as are the critical wavelengths at the velocity/wavelength minimum. The scale lengths in the atmospheric turbulence responsible for surface "puffs," or in the perturbing waveforms responsible for "micro-breakers," should be considerably larger, however, so the wavelengths involved will lie along the rising branch of the gravity-wave dispersion curve, and the transient "slosh" will quickly disintegrate. We will try to make some sense of this topic by considering several scattering models that might apply to surface events which start out looking like the disturbance sketched in Fig. 25a.

In Fig. 25b the essential scattering feature of such a disturbance is idealized as a ridge of transverse length L, having the cross-section of a circular segment of radius a and entry angle β, and lying at the edge of a "dimple" of radius R. The scattering behavior of this disturbance will, or course, depend on the dimensions a, L and R, the angles β and ψ (local grazing angle), and the wavelength of the incident signal. Since this is so speculative a scattering mechanism and, as in the case of the wedge, there have been no measurements of size, speed or duration, it is difficult to decide what range of dimensions might actually be of interest. To be quite arbitrary, yet guided by observation of these events, we will take the "dimple" radii R to lie between 10 cm and 1 m, with "slosh" radii a some small fraction of R. For microwave wavelengths between 1 and 100 cm, then, we will have to consider both long-wavelength (perturbation) and short-wavelength (physical optics) approximations. The depth of the "dimple" will, by the conditions of its formation, always be a very small fraction of the radius, and, based on a "ring wave" scattering calculation by Wetzel [1987], we will assert without proof that a wide, shallow depression of the type sketched in Fig. 25a will have an extremely small scattering cross section whatever theory is used. This leaves us with the more sharply curved, elevated "slosh" feature as the major contributor.

(a) Long-Wavelength (Perturbation) Approximation: If the height of the circular segment in Fig. 25b is much smaller than a wavelength of the incident signal, a simple surface-perturbation theory of the type used by Wright [1966] can be used. His expression for the backscatter cross section due to a surface perturbation $\zeta(x,y)$ is given in the Appendix, Section A.3, Eqs. (A23), (A24), (A25). It should be noted that Wright's expression is a *deterministic* cross section for a specific perturbation, *not* a statistical result based on a surface correlation function, as in Rayleigh/Rice perturbation theory and the Kirchhoff theory of Beckmann and Spizzichino [1963]. The literature often seems confused on this point.

With reference to Fig. 25b, the integral in (A24) can be written

$$I(\psi) = \int_L dy \int_{-a\sin\theta_0}^{a\sin\theta_1} (\sqrt{a^2 - x^2} - b)e^{iKx}\, dx\,, \tag{46}$$

where $b = s_0/a$ controls the shape of the segment, as well as the "toe" angle β, and $K = 2k\cos(\psi)$. It is obvious that under the conditions for which the perturbation approximation is valid ($ka << 1$) the exponential may be put equal to 1, and the x-integral becomes simply the volume, $V(a,L)$, of the segment. The cross section of the "slosh" feature then takes the simple form:

$$\sigma_{PERT} = \frac{4}{\pi} k^4 |g_{V,H}|^2 V^2(a,L) \tag{47}$$

with the angle factors $|g_{V,H}|$ given by (A24) and (A25). (Note the interesting similarity to volume scattering in the Born Approximation for a volume of unity dielectric constant.) Taking the circular segment to be a semicircle, the volume is $V = a^2 \times L/2$, so the cross section in (47) becomes

$$\sigma_{PERT} = \pi(ka)^4 L^2 |g_{V,H}(\psi)|^2, ka << 1\,, \tag{48}$$

where L is the transverse length of the structure. (It is of further interest to note that this is exactly the form initially obtained by Wright [1966] for the cross section of a patch of area L^2 containing Bragg-resonant capillaries of rms height a.) The angular g-factors are plotted as the solid curves in Fig. 26 for a dielectric constant $\epsilon = 50$, corresponding to sea water at X-band frequencies. It should be noted that these angular factors are the same ones that appear in *all* Bragg theories of electromagnetic backscatter, as, for example, in the "slightly rough" Bragg curves in Fig. 15.

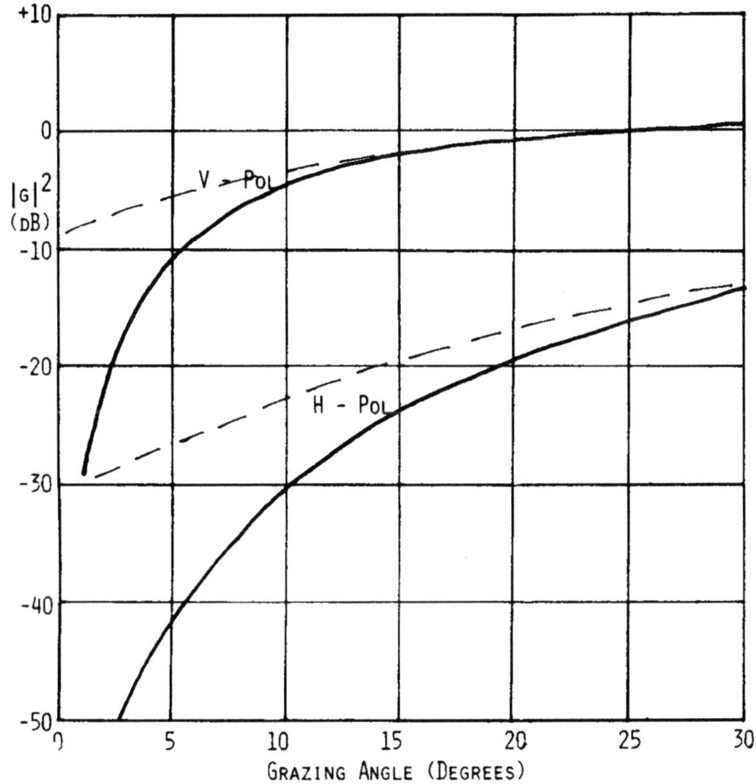

Fig. 26 — Angle factors $|g_{V,H}|^2$ in perturbation theory

(b) Short-Wavelength (Physical Optics) Approximation: When the circumference of the segment in Fig. 25b is long compared to a wavelength ($ka >> 1$), it is appropriate to use the physical optics approximation. Here the scattering takes place at the point of specular reflection, shown as the point P in Fig. 25c. For a conducting cylinder of radius a and length L, the cross section in this approximation is given by $.5 \times (ka) \times L^2$ [Kerr, 1951; we have added a factor of .5 to account for the reflection coefficient of water at microwave frequencies]. However, the water surface ahead of the scatterer produces a reflected wave which must be added to the incident wave in the neighborhood of P. Strictly speaking, for very large ka, the surface-reflected wave will itself be specularly reflected at P, away from the incident direction. However, for relatively small values of grazing angle ψ, and modest values of ka, the scattering patterns of the direct and surface-reflected waves at P should overlap back at the source. The effect may be simulated by multiplying the specular-point cross section by an angle- and polarization-dependent factor based on the geometry of Fig. 25c:

$$F^2 = |E_{DIRECT} + E_{REFLECTED}|^2$$

$$F^2_{V,H}(a,f,\psi) = |e^{ika\cos 2\psi} + R_{V,H}(f,\psi)e^{ika}|^2 \tag{49}$$

where $R_{V,H}(f,\psi)$ is the (complex) reflection coefficient for an incident wave of frequency f and grazing angle ψ. We call F^2 the "Surface Proximity Function" and have plotted it for both polarizations, in Fig. 27a for $ka = 2,4,6$, and in Fig. 27b for the larger values $ka = 10,20$. The physical optics approximation for scattering from a large "slosh" may thus be written

$$\sigma_{PO} = \frac{1}{2}(ka)L^2F^2_{V,H}(a,f,\psi). \tag{50}$$

(c) Effects of Surface Slopes and Wind Speed: The angle in the factors $|g|^2$ and F^2 is the *local* grazing angle. These scattering structures are excited on the sea surface, so this local angle will be $\psi = \psi_0 + \alpha$, where ψ_0 is the grazing angle relative to the mean surface (horizontal), and α is the local slope angle of the underlying surface. When the scattering structure is tilted toward the incident ray, the local grazing angle increases, while when tilted away, the angle decreases, reaching its minimum of zero when $\alpha = -\psi_0$. Let us assume that the sea slope distribution is Gaussian [Kinsman, 1965], characterized by an rms slope s_o which, for the relatively small angles involved ($<30°$) can be replaced by the slope angle. The averages of the angle factors in (48) and (49) therefore become, respectively,

$$<|g_{V,H}(\psi_0)|^2> = \frac{1}{\sqrt{2\pi}s_0}\int_{-\psi_0}^{\infty} |g_{V,H}(\psi)|^2 e^{-\alpha^2/2s_0^2}\, d\alpha, \tag{51}$$

$$<F^2_{V,H}(a,f,\psi_0)> = \frac{1}{\sqrt{2\pi}s_0}\int_{-\psi_0}^{\infty} F^2_{V,H}(a,f,\psi)e^{-\alpha^2/2s_0^2}\, d\alpha. \tag{52}$$

The dependence of mean square slope on wind speed may be found in the celebrated sun-glitter measurements by Cox and Munk [1954]. They measured the glitter from both *clean* surfaces, which included the flashes from all the small-scale structure, and "oiled" surfaces, on which the capillaries were suppressed. Since we are primarily interested in the larger slopes of the underlying surface, we will use the latter, for which the mean-square slope s_o^2 is given by

$$s_0^2 = 0.008 + 0.00156W, (W \text{ in m/sec}). \tag{53}$$

The use of (53) in (51) and (52) gives an estimate of the wind-speed dependence of the "slosh" cross sections (viewed normal to L) for given dimensions a and L, which will themselves be functions of wind speed. For a moderate wind speed of 15 kts, the rms slope is

0.14 (about 8°). We have shown the effect of such a wind on the angle factors $|g|^2$, as calculated from (51), by the dashed curves in Fig. 26. The result is akin to the Two-Scale Bragg model illustrated on the $H-Pol$ curves in Fig. 15. The effect on the Proximity Factors for Physical Optics is more complex, since the averages depend on the scatterer size and the illuminating frequency, as well as on the rms slope. In Fig. 28, the effect of a 15 kt. wind is shown for the same values of ka used in Fig. 27. There is little effect on the $V-Pol$ returns, other than to bring up the very low angle segment of the curve and to smooth the angular resonances at the higher ka's. The effect on the $H-Pol$ curves is much more pronounced, and the $H-Pol$ returns actually match or exceed the $V-Pol$ returns over the entire angular region at the higher values of ka. The reason for this lies, of course, in the Brewster angle "bite-out" that removes the reflected $V-Pol$ field at angles commensurate with the rms slope angles (about 8°). Contributions to the total scattered field from this scattering mechanism would therefore tend to show, even at the higher grazing angles, a considerably smaller spread between the $V-Pol$ and $H-Pol$ returns than that predicted by the Bragg theory. This is just what is observed (see, e.g., Fig. 15).

Unfortunately, there is *no a priori* basis for estimating the "slosh" dimensions required to obtain numerical values for the cross sections defined in (48) and (50); as mentioned above, they have never been measured (and those who have not observed natural water surfaces might question whether they even exist). However, a small-scale "slosh" would probably, like most small-scale perturbations forced into being on a water surface, have the dimensions of waves at the minimum of the velocity/wavelength curve; that is, a couple of centimeters. Thus we will take a minimum value of the radius a to be 1 cm. Taking a minimum "dimple" diameter to be about 1 foot, we assign the least value for R to be 10 cm, placing the division between the perturbation and physical optics approximations ($ka = 1$) at the middle of the microwave region—C-Band (5 GHz, 6 cm). Thus for "sloshes" of this size, we will use (48) for S-Band and below, and (50) for X-Band and above. The result for low grazing angles (<5°) is sketched in Fig. 29, where a wind speed of 15 kts is assumed, and the transition between the two approximations is drawn in as a smooth dashed line. For larger values of a the curves would move up, the transition point would occur at lower frequencies, and the polarization difference at a given frequency would decrease.

Although not yet established as a legitimate scattering obstacle, the existence of these "shocks" or "sloshes" would help explain certain puzzling aspects of low angle sea clutter, particularly small amplitude "sea spikes." As noted earlier, the turbulent wind field of a 20 inch fan blowing obliquely over a water filled basin in the laboratory and viewed with a high resolution horizontally polarized radar, produced a random sequence of sharp transient returns of significant amplitude [Hansen and Wetzel]. These returns could be visually correlated with the appearance of the type of micro-sloshes we have been discussing. In their behavior they were quite reminiscent of the "Sea Spikes" measured by Lewis and Olin [1980] in open water.

Figure 30 reproduces some low grazing angle (1.4°) X-band results obtained by Lewis and Olin [*loc. cit.*] under two quite different sea surface conditions. The records on the top were obtained from a "wind-blown sea with many whitecaps," while those below were characteristic of "calm" water. The general structure and temporal behavior of the returns for the two conditions are virtually identical, the only difference being a scale difference of 40 dB. Moreover, (for a given surface condition) for each of the surface conditions ("rough", "calm") the $V-Pol$ and $H-Pol$ returns are seen to have the same amplitudes. The only two models that can account for comparable $H-Pol$ and $V-Pol$ returns are: (1), the nose-on

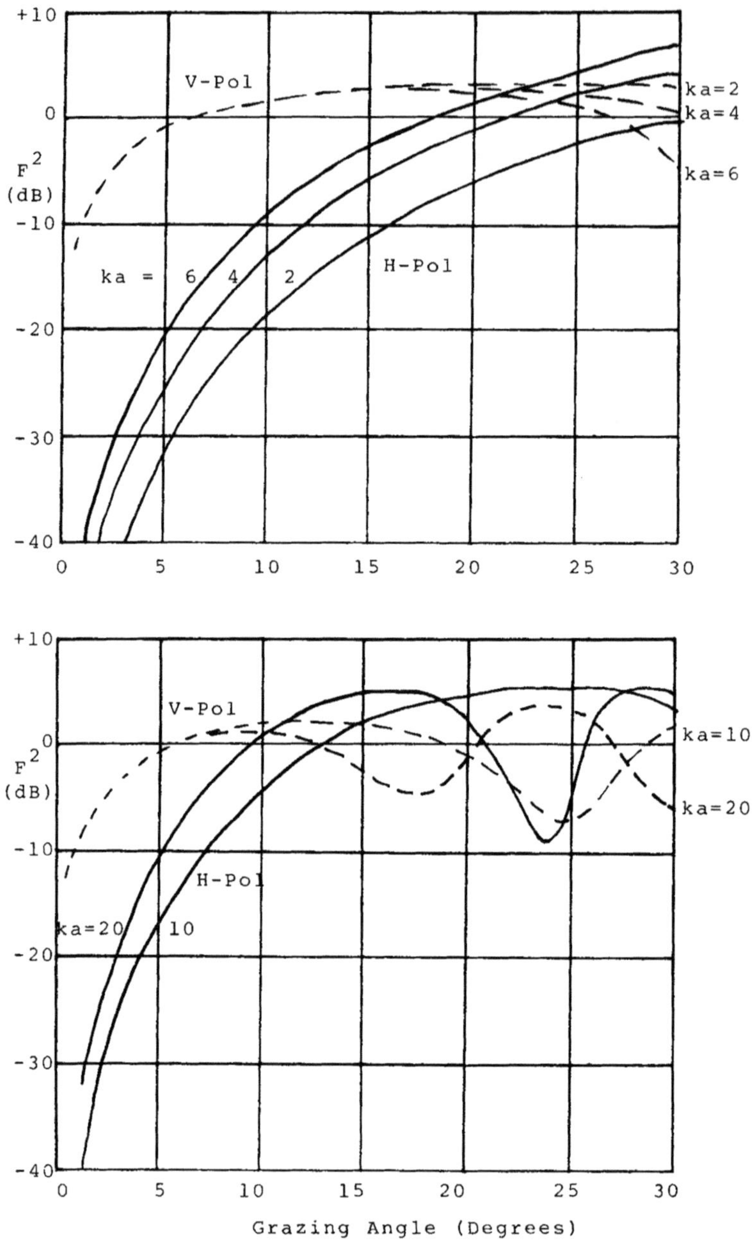

Fig. 27 — The surface proximity factor $F^2_{V,H}$.

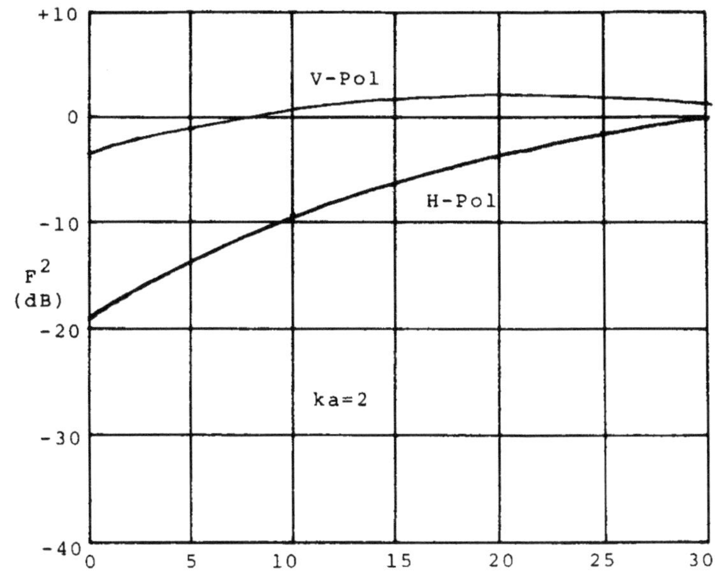

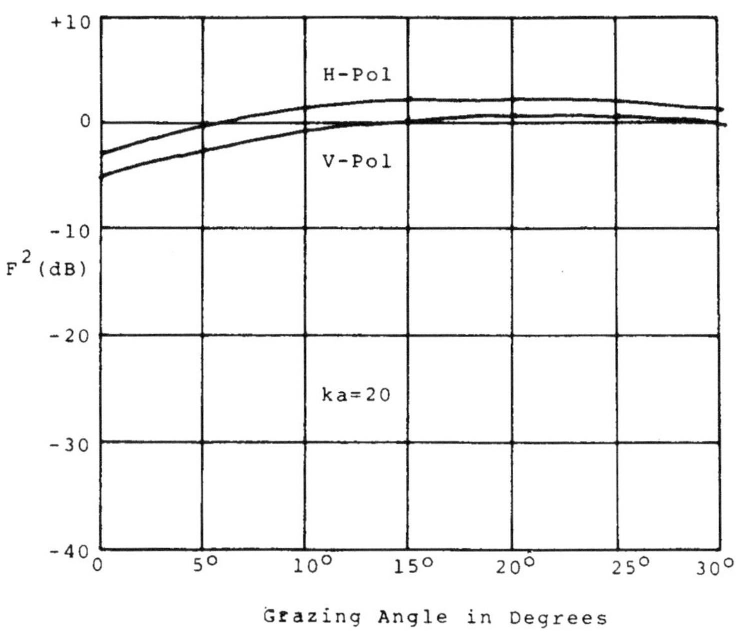

Fig. 28 — The surface proximity factor with 15 kt Wind (rms slope = 0.15)

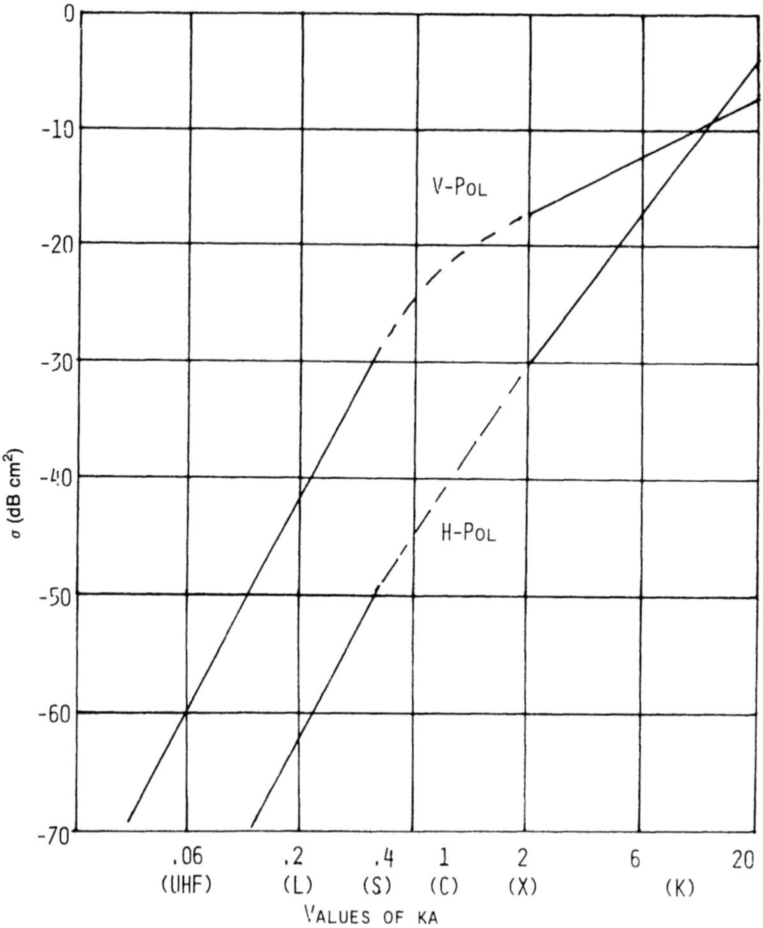

Fig. 29 — Low-angle ''slosh'' cross sections (1 by 20 cm, 15 kt wind)

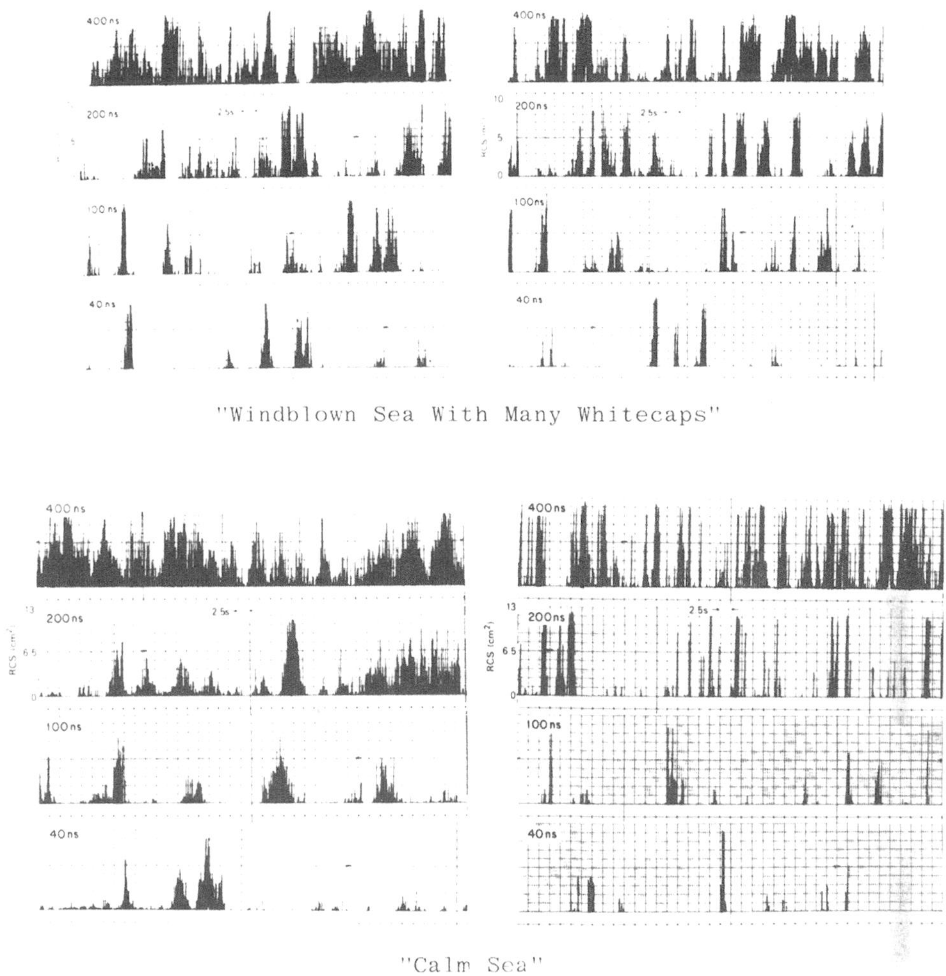

Fig. 30 — Low angle backscatter displaying "Sea Spikes" (Lewis and Olin, 1980)

geometrical optics wedge model of Kalmykov and Pustovoytenko [*loc, cit.*] (see Fig. 19a); and (2), a Physical Optics model of some kind; either the glancing-incidence wedge model of section 3.2.2b), or the "slosh" model being discussed here. Since model (1) is highly contrived and unrealistic, we are left with the latter choices as the most likely sources of low-angle transient returns. As Sherlock Holmes has said:

> "My dear Watson! When all possible explanations save one have been eliminated, that one, however unlikely it might seem, must contain the true solution."

Of course, we might not have exhausted "all possible explanations," but we have considered quite a few, and while the "slosh" hypothesis cannot fully account for the "calm water" returns observed by Lewis and Olin, it is the only model whose trends are all in the right direction.

3.3.2. *Sea Spikes and the "Plume" Model for Breaking Waves.* Certainly the most obvious place to look for scattering features in an active sea at low angles is at the peaks of breaking waves. But what are these scattering features? The peaks display different shapes according to which of the several types of breaker is involved (plunging, surging, spilling), and then there is turbulence, and spray. The well-ordered models we have been discussing thus far apply to structures and surfaces that are difficult to identify in the chaos of a heavily breaking sea. Moreover, scattering from breakers is characterized by the appearance of "Sea Spikes" having cross sections several orders of magnitude greater than those predicted by any of the mechanisms discussed thus far.

We have already seen "Sea Spikes" in the records collected by Lewis and Olin at the top of Fig. 30. Other investigators have observed them as well [Kalmykov and Pustovoytenko, *loc. cit.*; Long, 1975], and radar operators encounter them on a regular basis (hence their colloquial nomenclature). Peak cross sections are of the order of 10 m^2 (or greater) for both polarizations, but these peaks occur sporadically, so the average cross section will be much smaller. Actually, when considering low angle cross sections, it is important to know how the quoted values of NRCS were obtained. Were they based on raw averages (including sea spikes), amplitude-limited data (chopping off the sea spikes), or selected data (removing data sections containing sea spikes)? Or were they averages at all? Some authors present their data in terms of *median* cross sections.

It is easy to infer from the breaking-wave records in Fig. 30 that, for this set of measurements at least, the mean "flux" of sea-spike events is roughly $0.001/m^2$-sec. Since the duration of the H-Pol sea spikes was roughly one second, and their mean amplitude about 5 m^2, this flux translates into a *mean* NRCS of about -23 dB, which is considerable greater than the low-angle results quoted earlier in Section 3.2.3. However, when cross sections are presented as *median* values (as in the early reports of the NRL-4FR data, and the more recent measurements by Trizna [1988]), the quoted NRCS levels would fail to reflect the contributions from the powerful sea spikes.

Cumulative distributions are much more informative, and Trizna has found them to reveal some interesting aspects of low angle backscatter. Figure 31 gives an example of how such distributions change with wind speed (X-band, H-Pol, upwind, about 5° grazing, for three wind speeds increasing from left to right). Clearly the statistical behavior is complex, and no single distribution tells the whole story. Trizna views the data in terms of multiple distributions, each characterizing the contributions from a particular class of scattering features. The straight lines on Fig. 30 represent Weibull "fits" to data presumably generated by the different "scattering classes." The left hand (low wind) curve follows a standard Rayleigh distribution over its entire length. The high amplitude segments labeled "Sea Spikes" appear to have tighter distributions, as if sea spikes were produced by a special scattering mechanism quite different from that responsible for the main body of backscatter returns. Such a mechanism, based on a particular hydrodynamic model of a breaking wave, has been explored in some detail by Wetzel [1986]. Some of its features are summarized below.

Although the word "breaker" usually summons up a mental image of a curled mass of crashing water (the "plunging" type), the breaking waves occurring in the open sea are generally of the "spilling" type, in which the interior crest angle at a wave peak sharpens to the point of instability (120°), and a series of water masses emitted at the crest slide down the front face of the wave under the force of gravity. Longuet-Higgins and Turner [1974] describe this process in their "plume" model of a spilling breaker, idealized in Fig. 32a. The "plume" has a characteristic shape (sketched in Fig. 32b) and entrains air to form the

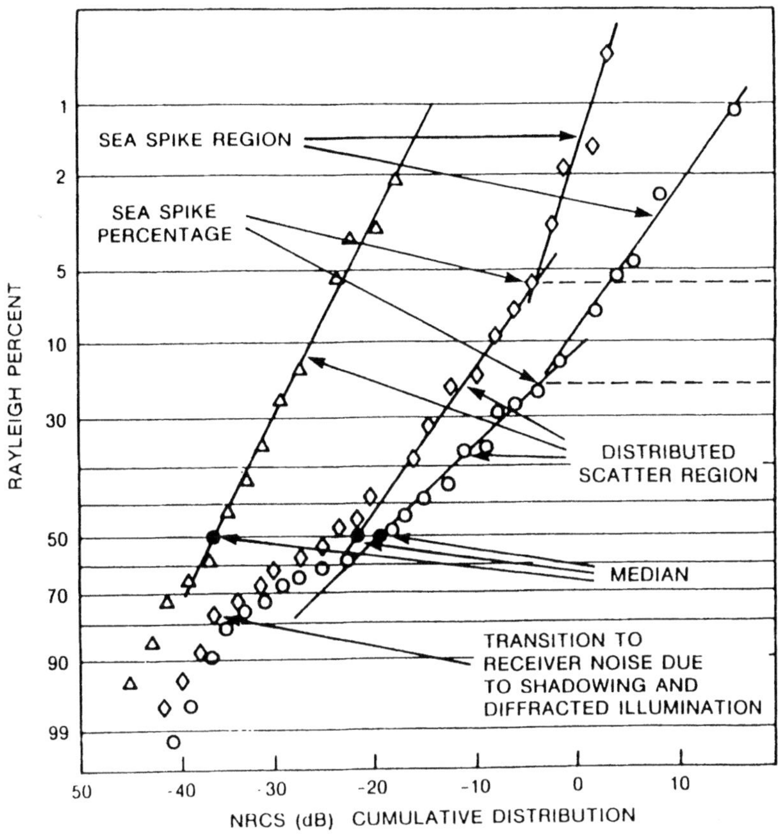

Fig. 31 — Cumulative NRCS distributions vs wind strength (Trizna, 1987)

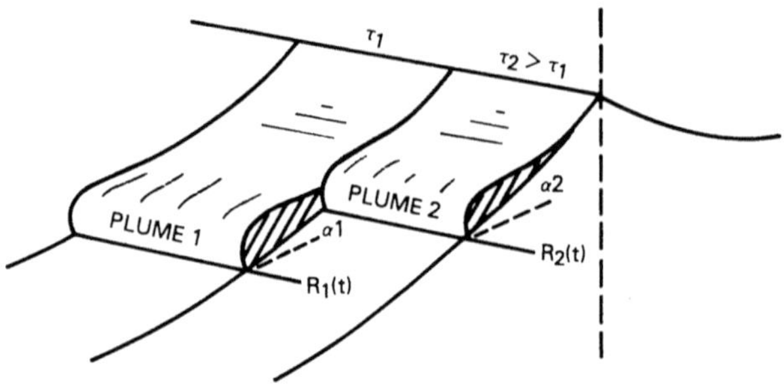

Fig. 32a — Idealized plumes on a breaking wave face (Wetzel, 1986)

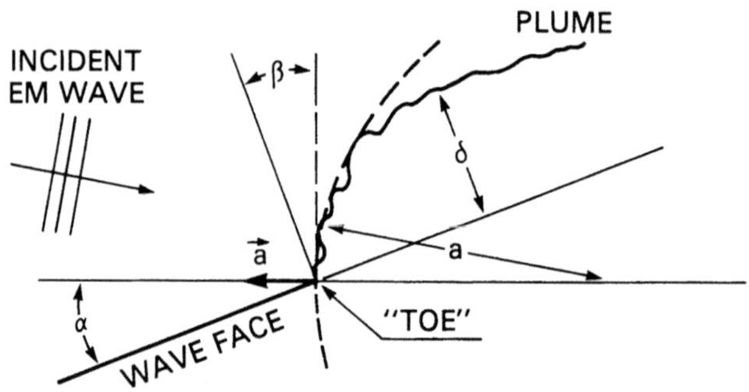

Fig. 32b — Sketch of "plume" geometry (Wetzel, 1986)

"whitecap." The scattering model for these features resembles very closely the Physical Optics model for "sloshes" described above, the major differences being in the scale of the event and the range of expected local grazing angles.

The plumes emerge from the crest of the breaking wave, so initially the underlying wave face angle will be steep, a value close to the value of 30° which we have been using in other models for the crest peaks. The radius of the plume will be considerably larger than that of the "sloshes" discussed above, lying between 3 and 15 cm, just to take a wild guess, although one would expect the larger plumes to have rougher surfaces, thus reducing their resemblance to smooth cylinders, particularly at the higher frequencies. The toe length L, over which the plume can be expected to resemble a cylindrical scatterer is another unknown, but we will assume it is of the order of a few plume diameters. Using (50) for the plume cross section, and the implications of Fig. 27 for the factor F^2, we find that these assumptions lead to individual plume cross sections lying between 0.25-25 m^2 for the popular X-band frequencies ($k = 2/\text{cm}$). A breaking wave will generally emit several plumes, so if the cross sections are assumed to combine incoherently, which is the worst case, the "sea spikes" associated with a breaking wave could quite easily have cross sections lying in the range between 1-100 m^2. Interestingly, the range of cross sections measured by Lewis and Olin at the top of Fig. 30 was 1-10 m^2, while the "sea spike" regions in Trizna's cumulative distributions (Fig. 31) contained individual returns from the open ocean in excess of 100 m^2. Clearly, using dimensional parameters well within the realm of credibility, the "plume" model is capable of delivering "sea spike" cross sections in the amplitude range actually observed.

The polarization dependence of the plume cross sections can be inferred from the surface-proximity factors shown in Fig. 27. For the smaller plumes (e.g., $ka = 6$) the H-Pol returns could exceed the V-Pol by almost 10 dB close to the peak, but would decay rapidly as the plume slid down the wave face (toward decreasing local grazing angles), while the V-Pol return would increase. Since the acceleration of the plume down the wave face makes the local grazing angle a function of time, the temporal behavior of small sea spikes due to plumes should resemble that sketched in Fig. 33a. The H-Pol returns are seen to be much "spikier," and associated with the wave peaks, while the V-Pol returns are smoothly distributed over the length of the wave. This is just the kind of behavior often remarked upon by observers of low angle sea backscatter, particularly at grazing angles of several degrees (above the knee of the shadowing curve). Corresponding behavior for the larger plumes ($ka = 20$) is shown in Fig. 33b. Here the H-Pol and V-Pol returns oscillate out of phase, but with similar mean amplitudes, as the plume moves down the wave face.

Unfortunately, there is little reliable experimental evidence available which could either confirm or deny this scattering hypothesis. What evidence does exist is largely circumstantial: reasonable assumptions about plume dimensions yield observed scattering levels; the polarization dependence of plume scattering is consistent with observations; the range of oscillations in large-plume returns (Fig. 33b) is consistent with the width of the sea spike region at the higher windspeeds in Trizna's distribution functions (Fig. 31). Other evidence in support of the plume model may be found in the reference cited above. We conclude our discussion of the plume model with the following paragraph from that reference: "We have found that one implication of the plume model—the existence of a curved scattering surface rising sharply out of the underlying wave surface—leads to radar scattering cross sections which, with a reasonable assumption of plume thickness, can be as large as those observed experimentally. We note, however, that the scattering results are based on two idealizing assumptions: (1) that the radar wavelength is of the order of [or less than] the plume thickness; (2) that the front surface of the plume is "smooth." As the radar wavelength increases, the scattering moves into the Rayleigh regime and falls off sharply with the fourth power of frequency. At lower frequencies, therefore, the plumes become invisible and our attention must shift to larger features, such as the wedgelike crest of the underlying wave. On the other hand, as the radar wavelength decreases, the roughness (see Fig. [32b]) over the scattering face of the plume can no longer be ignored, and the scattering amplitude must be multiplied by a roughness factor, $\exp[-2(kh_p)^2]$, where h_p is the rms roughness of the plume surface. Moreover, at short wavelengths, the likelihood of finding strong-scattering coherent segments along the plume front diminishes. It can be seen, therefore, that plume scattering of the type we have been discussing can occur only within a rather narrow window in the microwave spectrum, the location and width of this window being determined by unknown properties of the plumes, such as their thickness, shape, and roughness. Nevertheless it is interesting and instructive to sketch the scattering behavior of a (reasonable?) plume having a thickness of 3 cm (about an inch) and an rms roughness of 10% of its thickness. Figure 34 places this behavior in the context of other scattering mechanisms that might be expected to play a role in microwave scattering from breaking waves. Although this breakdown into scattering domains is speculative, and dependent on as yet poorly understood breaker morphology, it has the virtue of differentiating between the kind of scattering one might expect at different frequencies, and guiding attention to the most appropriate mode."

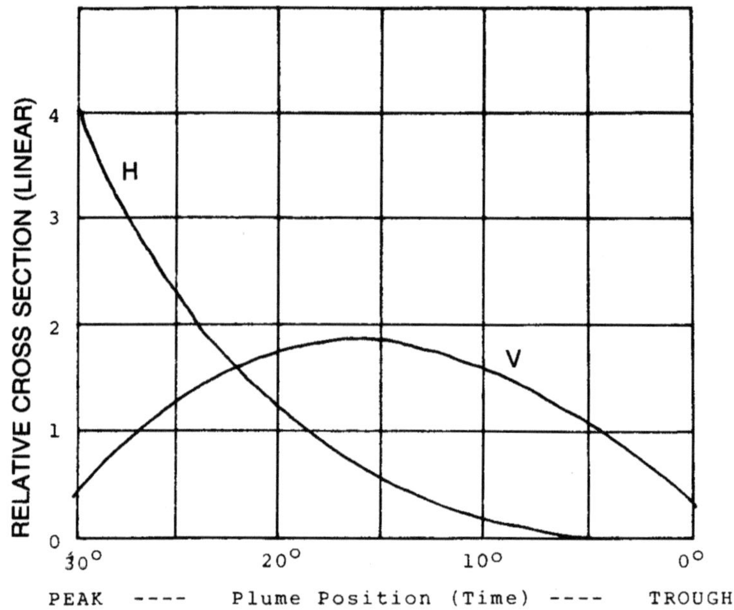

Fig. 33a — Temporal behavior of V and H-Pol cross sections (ka = 6)

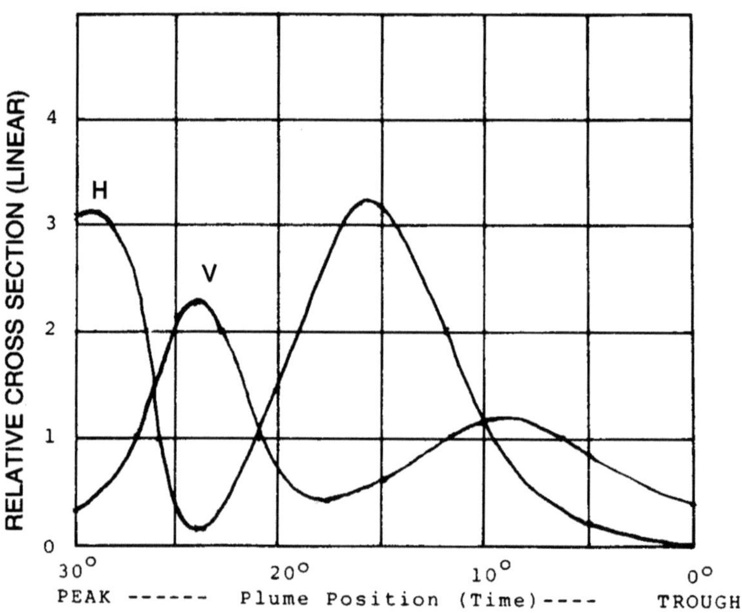

Fig. 33b — Temporal behavior of V and H-Pol cross sections (ka = 20)

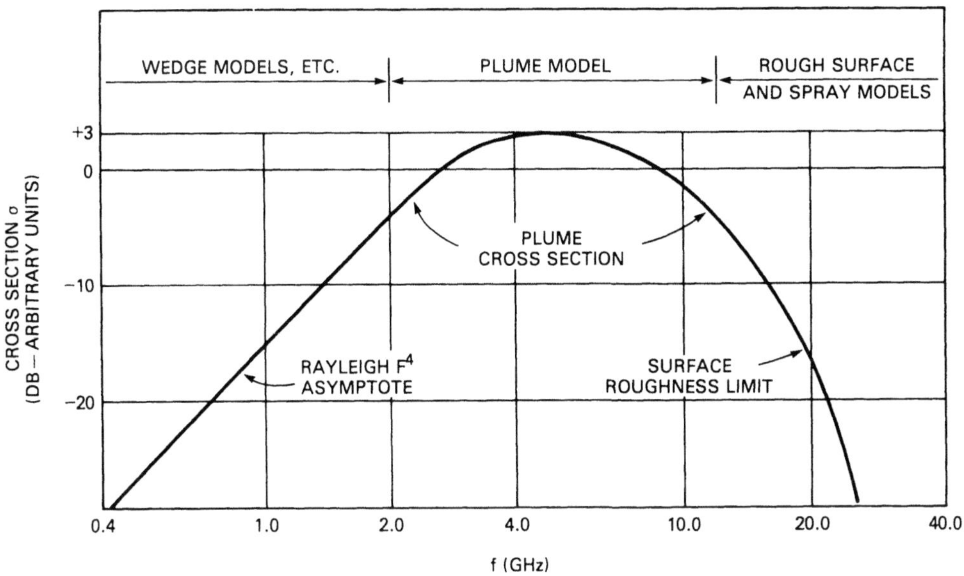

Fig. 34 — Possible breaker scattering mechanisms over the microwave spectrum (Wetzel, 1986)

The first question to be asked in any discussion of low angle sea backscatter is how the radar sees the surface; that is, what is the nature of the shadowing process? All existing rough-surface shadowing theories are based on geometrical optics, but approaching the illumination problem through diffraction theory made it possible to establish limits in windspeed/wavelength space for the application of both Geometrical and Physical Optics (Fig. 5). In addition, the estimated sensitivity of the diffraction model to radar polarization and wave crest irregularities (Figs. 3 and 7, respectively) was found to be sufficiently weak that simple two-dimensional physical optics should apply over a useful range of operating parameters.

For those operating parameters which allow use of the geometrical optics approximation, a "threshold" model for sea surface illumination applies at extreme grazing angles, providing a scattering field of isolated "islands"emerging from a dark background. Since the area and spacing of these scattering islands are functions of grazing angle and wind speed, a modified shadowing function for very low grazing angles could be derived (Fig. 14). The implications of this shadowing function were verified in a variety of experimental situations (Figs. 15-18).

Having identified, and given some quantitative measure of the preferentially illuminated parts of the sea surface, it was possible to test the implications of various models for scattering from these regions. Bistatic scattering models were developed for composite-surface Bragg scattering, and for wedge scattering by both the Method of Equivalent Currents and Physical Optics (Appendix). Values of the low-angle NRCS obtained from these models were in reasonable agreement with experiment (Section 3.2.3), but also raised the curious suspicion that low angle scattering might be model-independent, which is especially curious in view of the almost diametrically opposite assumptions underlying the Bragg (small perturbation) and optical approximations.

A new scattering feature—the "slosh"—may be described by a simple heuristic scattering model. Scattering by small "sloshes" would be expected to behave with time and polarization much like observed low angle, high resolution returns from calm water, and the "plume" model of scattering from breaking waves seems able to account for many of the observed characteristics of the Sea Spikes that populate the low grazing angle clutter scene.

It would appear that the modeling of low angle sea scatter has come about as far as possible without more detailed knowledge of the surface features ultimately responsible for the scattered field. Although a "spectrum" has never been known to scatter an electromagnetic wave, models (like the Bragg model) that require only the specification of a spectrum may at least appeal to the oceanographer's description of the surface. These models need never ask what the surface actually looks like. A spectrum is a highly averaged, fictitious entity in which all of the phases required to construct a real scattering surface are lost. In order to quantify "feature" scattering—from wedges, spray, sloshes, plumes, etc., i.e., we must know their dimensions, lifetimes, distributions, etc. And since spectral theories cannot explain low angle sea scatter, the accumulation of this information must be the next order of business if scattering in this angular regime is to be understood. The existing assortment of scattering theories and approximations should certainly suffice to produce acceptable levels of accuracy once the actual scattering obstacles are defined.

Appendix

SCATTERING FORMULARY

A.1. A Composite-Surface Bragg Model for Bistatic Scattering

Consider a surface element A with orientation defined by its normal $\hat{n}$, illuminated from direction $\hat{k}_i$ and scattering into direction $\hat{k}_s$, as in Fig. 11. The incident vector $\hat{k}_i$ is assumed to lie in the $y-z$ plane with grazing angle ψ_i^0, while the scattering vector $\hat{k}_s$ is defined by grazing angle ψ_s^0 and azimuth (or *scattering*) angle ϕ_s. The normal $\hat{n}$ is defined by its inclination θ and azimuth angle ϕ. In terms of the fixed $x-y-z$ coordinate system,

$$\hat{n} = \sin\theta\,\sin\phi\hat{x} + \sin\theta\cos\phi\hat{y} + \cos\theta\hat{z} \tag{A1}$$

$$\hat{k}_i = 0\hat{x} + \cos\psi_i^0\hat{y} - \sin\psi_i^0\hat{z} \tag{A2}$$

$$\hat{k}_s = \cos\psi_s^0\,\sin\phi_s\hat{x} + \cos\psi_s^0\cos\phi_s\hat{y} + \sin\psi_s^0\hat{z} \tag{A3}$$

Scattering from A depends on the *local* grazing angles ψ_i and ψ_s relative to the surface A. These angles are given by

$$\sin\psi_i = -\hat{n}\cdot\hat{k}_i = -\sin\theta\cos\psi_i^0\cos\phi - \sin\psi_i^0\cos\phi \tag{A4}$$

$$\sin\psi_s = +\hat{n}\cdot\hat{k}_s = \sin\theta\cos\psi_s^0\cos(\phi-\phi_s)+\sin\psi_s^0\cos\theta \tag{A5}$$

For a coordinate system *on the scattering patch A*, the bistatic cross section per unit area is given by Barrick and Peake [1967] in the form

$$\sigma_{pp}^0(\psi_s,\phi_s) = 4\pi k_0^4 \sin^2\psi_i \sin^2\psi_s \,|\alpha_{pp}|^2 W(K_1,K_2) \tag{A6}$$

where k_0 is the radar wavenumber, and pp denotes the polarization (i.e., $pp = HH$ or VV). W is the wave spectrum, and the surface wave vector components are given by

$$K_1 = k_0(\cos\psi_s \cos\phi_s - \cos\psi_i) \tag{A7.1}$$

$$K_2 = k_0 \cos\psi_s \sin\phi_s \tag{A7.2}$$

and the coefficients α_{pp} are complicated functions of the angles above and the electrical properties of the scattering surface. They are given approximately by

$$\alpha_{HH} = -\cos\phi_s \tag{A8}$$

$$\alpha_{VV} = (\cos\psi_i \cos\psi_s - \cos\phi_s)/(\sin\psi_i + 1/\sqrt{\epsilon})(\sin\psi_s + 1/\sqrt{\epsilon}) \tag{A9}$$

where ϵ is the dielectric constant of the surface (about 50 for sea water at microwave frequencies).

However, the *local* components of the incident and scattered field vectors on A are generally different from those defined in the radar reference system. That is, a rotation of $\hat{n}$ out of the plane of incidence will convert an initial horizontal or vertical polarization into a mixture of horizontally and vertically polarized components on A. In Fig. A1 we represent the incident and scattered field vectors in the radar reference system (x,y,z) as $\vec{E}_V$ and $\vec{E}_H$, and the tilted, rotated surface element A as described by its normal $\hat{n}$.

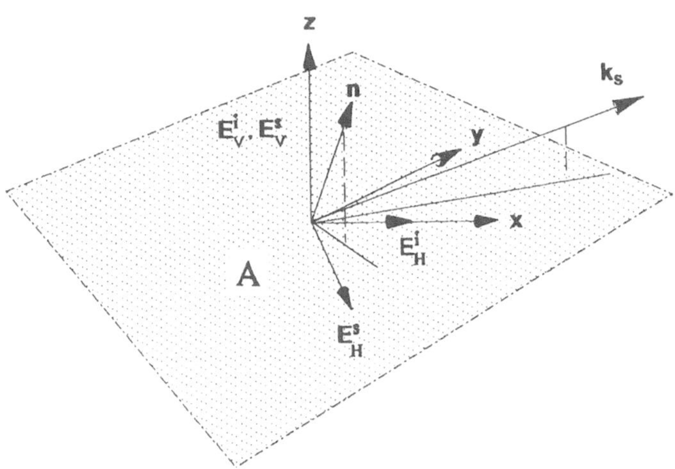

Fig. A1 — Field Transformations on a Tilted Plane

The transformation of the incident fields to the surface A where the scattering actually takes place is provided by the following vector relationships (p denotes the polarization, H or V):

"Vertical" component on A due to the incident fields:

$$\vec{E}^{Ai}_{VP} = (\hat{n} \cdot \vec{E}^{i}_{p})\hat{n} \tag{A10}$$

"Horizontal" component on A due to the incident fields:

$$\vec{E}^{Ai}_{HP} = (\hat{n} \times \vec{E}^{i}_{p}) \times \hat{n} \tag{A11}$$

The *local* scattered fields for these two polarizations are proportional to the angular coefficients in (A8) and (A9), so we write:

$$\vec{E}^{As}_{HP} \sim \vec{E}^{Ai}_{HP} \cdot \alpha_{HH}\,;\ \vec{E}^{As}_{VP} \sim \vec{E}^{Ai}_{Vp} \cdot \alpha_{VV} \tag{A12}$$

Finally, the scattered field in the *radar* reference frame is found from the component of (A12) in the scattered field directions (denoted by a carat) defined in the Figure:

$$E^{s}_{p} \sim |E^{i}_{p}|\ \{(\hat{n} \cdot \vec{E}^{i}_{p})\ \hat{n}\,\alpha_{VV} + (\hat{\eta} \times \vec{E}^{i}_{p}) \times \hat{n}\,\alpha_{HH}\} \cdot \hat{E}^{s}_{p}. \tag{A13}$$

Note that cross-polarized terms, which we ignore here, could also be obtained from this formalism.

The angular factors $|\alpha_{pp}|^2$ in (A6) were originally obtained from the square of the ratio of scattered to incident field strengths, so we will replace them in our composite-surface bistatic model with the corresponding expressions formed from (A13):

$$|\alpha_{pp}|^2 \rightarrow |\{(\hat{n} \cdot \vec{E}^{i}_{p})\hat{n}\,\alpha_{VV} + (\hat{n} \times \vec{E}^{i}_{p}) \times \hat{n}\,\alpha_{HH}\} \cdot \hat{E}^{s}_{p}|^2. \tag{A14}$$

A similar expression was obtained by Valenzuela [1968] for the monostatic (backscatter) case, which is often used as the current "Two-scale" or "Composite-Surface" Bragg model for radar backscatter from the sea (for a recent application see Donelan and Pierson, 1987).

Finally, we assume that the wave spectrum W in (A6) is isotropic at the Bragg resonant wavenumber, and becomes a function only of

$$K = (K_1^2 + K_2^2)^{1/2} = 2k_0 \cos \psi_i \sin(\phi_s/2) \tag{A15}$$

Using the Phillips [1966] spectrum for W, i.e.,

$$W(K) = 0.005K^{-4}, \tag{A16}$$

Equation (A6) reduces to

$$\sigma^0_p(\phi,\phi_s) = 1.6 \times 10^{-5}\,\frac{\sin^2\psi_i \sin^2\psi_s}{\sin^4(\phi_s/2)}\ |\alpha_{pp}|^2. \tag{A17}$$

The grazing angles are defined in (A4) and (A5), and the angular factor $|\alpha_{pp}|^2$ defined in (A14) is assembled with the help of (A1), (A8), and (A9).

Keeping all of the angles straight in a completely general formulation of bistatic scattering in the small-amplitude Bragg approximation is a real nightmare, so when using these expressions for examples in Section 2.1.1 we will introduce some rational simplifications. For example, the surface slope angle will be taken as 30°, which is the limiting angle for surface stability and thus quite likely to be encountered close to the wave peaks. The radar-referenced grazing angles ψ_i^0 and ψ_s^0 will be very much smaller than 30°, so we will take them both to be vanishingly small. The *local* grazing angles given in (A4) and (A5) will be approximated by the first terms in those expressions, noting, however, that these terms contain the cosine of the azimuthal angles, so they will vanish for values of these angles at the extremes of the allowed range of ϕ_s. Even with these simplifications, the expressions are sufficiently complex that little would be gained by displaying them here. The reader is invited to reproduce them, if he wishes, from the recipe given above.

A.2 Bistatic Wedge Scattering by the MEC

Consider a wedge of length L and interior angle α with its edge lying in direction $\hat{t}$ in the $x-y$ plane. The incident and scattering directions are indicated by $\hat{k}_i$ and $\hat{k}_s$ respectively in Fig. A2.

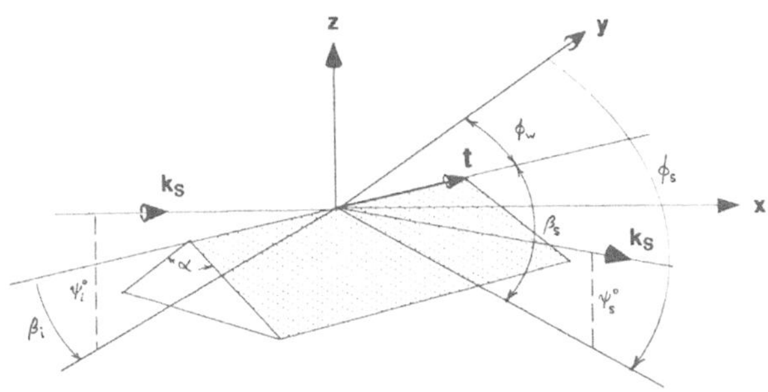

Fig. A2 — Geometry of wedge Scattering

As in the previous case, we assume that the grazing angles ψ_i^0 and ψ_s^0 are so much smaller than the wedge face angle $(\phi - \alpha)/2$ that they can be ignored. All of the angles then lie in the $x-y$ plane.

In the Method of Equivalent Currents (MEC) as described by Knott and Senior [1974], the scattered field is expressed as an integral over a fictitious filamentary current laid along the edge C of the wedge, where this "equivalent current" is of just such a form as to reproduce the results of the Geometrical Theory of Diffraction (GTD) in those special directions allowed by the GTD, and to produce a fictitious scattered field in all other directions. It is taken as an article of faith that the fields so defined represent a generalization of the GTD fields, and have some relation to reality. The fictitious currents are themselves non-physical and non-causal, since they depend upon the direction from which the edge is observed—i.e., on the location of the observer. (This is a marvelous example of *teleology*, where the unknowable end configures the means to that end.) The scattered field defined by this method is given by the expression:

$$E_s(L\,;\,\phi,\phi_s) \;=\; \frac{E_0 e^{ikR}}{2\pi R}\cdot\int_C \frac{e_{it}\,e_{st}(X\;-\;Y)\;+\;h_{it}hi_{st}(X\;+\;Y)}{\sin\beta_i\;\sin\beta_s}\;e^{ikw\hat{t}\cdot(\hat{k}_i-\hat{k}_s)}dw \qquad \text{(A18)}$$

where X and Y are the GTD diffraction coefficients,

$$(X,Y) = \frac{\sin(\pi/n)}{n}\cdot\left[\frac{1}{\cos(\pi/n)-\cos((\gamma_s-\gamma_i)/n}+\frac{1}{\cos(\pi/n)-\cos((\gamma_s+{}_i)/n}\right] \qquad \text{(A19)}$$

In (A18), the coefficients $e_{it,st}$ and $h_{it,st}$ are the components of the incident (i) and scattered (s) electric and magnetic fields along the direction of the edge ($\hat{t}$); the rest of the angles are defined in Fig. A2. In (A19), $n\;=2\;-\;\alpha/\pi$, and the angles $\gamma_{i,s}$ are the incident and scattered angles measured from the common wedge face in a plane *perpendicular* to the edge, so at very low grazing angles they are both approximately equal to the wedge face slope angle $(\pi\;-\;\alpha)/2$. For Vertical polarization, the e's vanish, and the remaining h's cancel the sines in the denominator of (A18); for Horizontal polarization, the h's vanish and the e's cancel the sines. This leaves an elementary $\sin x/x$ integral in (A18), and the scattered field reduces to the simple form:

$$E_{H,V}(L\,;\,\phi_s,\,\phi_w) = E_0\frac{e^{ikR}}{2\pi R}\;(X\;\mp\;Y)\cdot L\cdot\;\mathrm{sinc}\;\left[\frac{1}{2}\;kL\;(\cos\phi_w\;-\cos(\phi_s\;-\phi_w)).\right] \qquad \text{(A20)}$$

where the $-$ and $+$ signs are used for $H-$ and $V-$polarization, respectively. Using the low grazing angle simplification indicated above for (X,Y), and defining cross section in the usual way, $[\sigma\;=\;(4\pi R^2)\,|\,E_s/E_0\,|^2]$, the bistatic cross section for a wedge of length L and edge orientation angle ϕ_w becomes

$$\sigma_{H,V}(L\,;\,\phi_w,\phi_s)\;=\;C_{H,V}L^2\,\mathrm{sinc}^2\cdot\;\left\{\frac{1}{2}kL\;(\cos\phi_w\;-\cos(\phi_s\;-\phi_w))\right]. \qquad \text{(A21)}$$

For our standard stability-limited wedge angle of 120°, the polarization coefficients have the numerical values

$$C_H\;=\;-29\;\mathrm{dB},\;C_V\;=\;-8\;\mathrm{dB}. \qquad \text{(A22)}$$

A.3 Perturbation theory for Discrete Scattering Features

There is an interesting, and often overlooked, difference between the Rayleigh/Rice perturbation approximation [Rice, 1951], on which the cross section in (A6) is based, and the scattering integral approach used by Wright [1966]. In the latter, the cross section is expressed in terms of an integral over the surface perturbation, much as in a small amplitude physical optics approximation, but with the boundary conditions treated somewhat differently in order to provide the polarization dependence often missing in P.O. calculations (however, see Leader, 1971). Wright's result takes the form of the cross section of a deterministic perturbation $\zeta(x,y)$ over a flat surface:

$$\sigma_{V,H}(\psi) = \frac{4k^4}{\pi} \,|\, g_{V,H}(\psi) \,|^2 \,|\, \int_S \int \zeta(x,y) e^{i2h\cos\psi \cdot x} dxdy \,|^2 \tag{A23}$$

where the angle coefficients may be written in the form [Valenzuela, 1978],

$$g_V = \frac{(\epsilon - 1)[\epsilon(1 + \cos^2\psi) - \cos^2\psi]\sin^2\psi}{[\epsilon\sin\psi + (\epsilon - \cos^2\psi)^{1/2}]^2} \tag{A24}$$

$$g_H = \frac{(\epsilon - 1)\sin^2\psi}{[\sin\psi + (\epsilon - \cos^2\psi)^{1/2}]^2} \tag{A25}$$

References

Barrick, D.E. and W.H. Peake (1967), "Scattering from surfaces with different roughness scales: analysis and interpretation," Rept. No. BAT-197A-10-3, Battelle Memorial Institute, Columbus, Ohio.

Bass, F.G. and I.M. Fuks (1964), "Calculation of shadowing for wave scattering from a statistical rough surface," Izv. vuzov, Radiofizika **7**, 101-112.

Beckmann, P. (1965), "Shadowing of random rough surfaces," IEEE Trans. Antennas Propagat., **AP-13**, 384-388.

Beckmann, P. and A. Spizzichino (1963), *The Scattering of Electromagnetic Waves from Rough Surfaces,* (Macmillan, New York.)

Booker, H.G. and W.E. Gordon (1950), "A theory of radio scattering in the troposphere," Proc. IRE, **38**, 401.

Chan, W.H.L. and H.F. Helmken (1983), "Low grazing angle radar backscatter from the ocean surface," Rept. NWC TP 6450, Naval Weapons Center, China Lake, California.

Cox, C.S. and W.H. Munk (1954), "Statistics of the sea surface derived from sun glitter," J. Mar. Res., **13**, 198-227.

Daley, J.C., Ransone, J.T., Burkett, J.A. and J.R. Duncan (1968), "Seaclutter measurements on four frequencies," Rept. No. 6806, Naval Research Laboratory, Washington, D.C.

DeSanto, J.A. (1975), "Scattering from a sinusoid: derivation of linear equations for the field amplitudes," J. Acoust. Soc. Am. *57*, 1195-1197.

Deschamps, G.A., Boersma, J. and S.W. Lee (1984), "Three-dimensional half-plane diffraction: Exact solution and testing of uniform theories," IEEE Trans. Antennas Propagat., **AP-32**, 264-271.

Furutsu, K. (1966), "A statistical theory of ridge diffraction," Radio Sci., 1,79-98.

Hunter, I.M. and T.B.A. Senior (1966), "Experimental studies of sea surface effects on low-angle radars," Proc. IEE, **113**, 1731-1740.

Kalmykov, A.I., Kurekin, A.S., Lementa, Y.A., Ostrovskiy, I.Y. and V.V. Pustovoytenko (1976), "Scattering of microwave radiation by breaking sea waves," Gor'kiy Radiofiz., **19**, 1315-1321 (translation).

Kalmykov, A.I., and V.V. Pustovoytenko (1976), "On polarization features of radio signals scattered from the sea surface at small grazing angles," J. Geophys. Res., **81**, 1960-1964.

Katz, I. and L.M. Spetner (1960), "Polarization and depression angle dependence of radar terrain return," J. Res. NBS-D., **64-D**, 483-486.

Katzin, M. (1957), On the mechanisms of radar sea clutter," Proc. IRE, **45**, 44-54.

Keller, J.B. (1957), "Diffraction by an aperture," J. Appl. Phys., **28**, 426-444.

Kerr, D.E. (ed.) (1951), *Propagation of Short Radio Waves,* Chap. 6, (McGraw-Hill, New York).

Kinsman, B. (1965), *Wind Waves,* (Prentice-Hall, Englewood Cliffs, N.J.).

Knott, E.F. and T.B.A. Senior (1974), "Comparison of three high-frequency diffraction techniques," Proc. IEEE, **62**, 1468-1474.

Kwoh, D.S. and B.M. Lake (1984), "A deterministic, coherent, and dual-polarized laboratory study of microwave backscattering from water waves, Part I: Short gravity waves without wind," IEEE J. Ocean. Engin., **OE-9**, 291-308.

Leader, J.C. (1971), "The relationship between the Kirchoff approach and small perturbation analysis in rough surface scattering theory," IEEE Trans. Antennas Propagat., **AP-19**, 786-788.

Lewis, B.L. and I.D. Olin (1980), "Experimental study and theoretical model of high resolution backscatter from the sea," Radio Sci,. **15**, 815-826.

Lighthill, M.M. (1978), *Waves in Fluids,* (Bantam Press, New York.)

Long, M.W. (1975), *Radar Reflectivity of Land and Sea*, Heath, Boston.

Longuet-Higgins, M.S. and J.S. Turner (1974), "An entraining plume model of a spilling breaker," J. Fluid Mech. **63**, 1-20.

Lynch, P. and R.J. Wagher (1970), "Rough surface scattering: shadowing, multiple scatter, and energy conservation," J. Math. Phys. **11**, 3032-3042.

Lyzenga, D.R., A.L. Maffett and R.A. Shuchman (1983), "The contribution of wedge scattering to the radar cross section of the ocean surface," IEEE Trans. Geosci. Romote Sens., **GE-31**, 502-505.

Michaeli, A. (1984), Equivalent edge currents for arbitrary aspects of observation," IEEE Trans. Antennas Propagat., **AP-32**, 252-258.

Middleton, D. and R.,H. Mellen (1985), "Wind-generated solitons: A potentially significant mechanism in ocean surface wave generation and surface scattering," IEEE J. Ocean Eng., **OE-10**, 471-476.

Middleton, D. (1986), "A proposed soliton mechanism in wind-wave surface generation and scattering," NUSC Tech. Rept. 7775, 19 Dec. 1986, Naval Underwater Systens Center, Newport/New London.

Phillips, O.M. (1966), *The Dynamics of the Upper Ocean,* (Cambridge University Press, Cambridge.)

Phillips, O.M. (1985), "Spectral characteristics of breaking waves" *The Ocean Surface, Wave Breaking, Turbulent Mixing and Radio Probing*, Y. Toba, H. Mitsuyasu, eds., (D. Reidel Publishing Co., Boston).

Pierson, W.J. Jr. and M.A. Donelan (1987), "Radar-scattering and equilibrium ranges in wind generated waves - with application to scatterometry," J. Geophys., Rev. Vol. 91 (C5), 4971-5029.

Plonus, M.A., Williams, R. and S.C.H. Wang (1978), "Radar cross section of curved plates using geometrical and physical diffraction techniques," IEEE Trans. Antennas Propagat., **AP-26,** 488-493.

Rice, S.O. (1951), "Reflection of electromagnetic waves from slightly rough surfaces," Commun. Pure Appl. Math, **4**, 361-378.

Sancer, M.I. (1969), "Shadow-corrected electromagnetic scattering from a randomly rough surface," IEEE Trans. Antennas Propagat., **AP-17**, 577-585.

Saxton, J.A. and L.A. Lane (1952), "Electrical properties of sea water," Wireless Engineer, **29**, 269-275.

Sikta, F.A., Burnside, W.D., Chu, T>T. and L. Peters, Jr. (1983), "First - order equivalent current and corner diffraction scattering from flat plate structures; IEEE Trans. Antennas Propagat., **AD-31**, 584.

Sittrop, H. (1974), "X- and Ku-Band radar backscatter characteristics of sea clutter," Proc. of the URSI Comm. II Specialist Metting on Microwave Scattering and Emission from the Earth, (E. Schanda, Ed.), Berne, Switzerland.

Smith, B.G. (1967), "Geometrical shadowing of a random rough surface," IEEE Trans. Antennas Propagat., **AP-15**, 668-671.

Sommerfeld, A. (1896), "Mathematische theorie der diffraktion," Math. Ann., **47**, 317-374.

Trizna, D. (1988), "Measurement and interpretation of North Atlantic ocean marine radar sea scatter," NRL Report 9099.

Trunk, G.V. (1976), "Non-Rayleigh Sea Clutter: Properties and Detection of Targets," NRL Report No. 7986, June 25, 1976.

Ufimtsev, P.I. (1957), "Approximate computation of the diffraction of plane electromagnetic waves at certain metal boundaries: I. Diffraction Patterns at a wedge and a ribbon," Soviet Physics; Tech. Phys., **3** 1708-1718.

Valenzuela, G.R. (1978), "Theories for the interaction of electromagnetic and ocean waves-a review." Boundary-Layer Meteor., **13**, 61-85.

Wagner, R.J. (1967), "Shadowing of randomly rough surfaces," J. Acoust.Soc. Amer., **41**, 138-147.

Wait, J.R. and A.M. Conda (1959), "Diffraction of electromagnetic waves by smooth obstacles for grazing angles," J. Res. NBS-D. Rad. Prop., **63D**, 181-197.

Wetzel, L.B. (1977), "A model for sea backscatter intermittency at extreme grazing angles," Radio Sci., **12**. 749-756.

Wetzel, L.B. (1986a), "A minimalist approach to sea backscatter - the wedge model," URSI Open Symposium on Wave Propagation: Remote Sensing and Communication, July 28-Aug 1, 1986, Durham, N.H., Pre-Print Vol. pp. 3.1.1-3.1.4.

Wetzel, L.B. (1986b), "On microwave scattering by breaking waves," *Wave Dynamics and Radio Probing of the Ocean Surface,* O.M. Phillips and K. Hasselmann, eds., pp. 273-284, (Plenum Press, New York).

Wetzel, L.B. (1987), "On the theory of electromagnetic scattering by a raindrop splash," NRL Memorandum Report 6103, Dec. 31, 1987.

Wright, J.W. (1966), "Backscatter from capillary waves with application to sea clutter," IEEE Trans. Antennas Propagat., **AP-14**, 749-754.

Chapter 13

DEPENDENCE OF RADAR BACKSCATTER ON ENVIRONMENTAL PARAMETERS

W. J. PIERSON, JR.
CUNY Institute of Marine and Atmospheric Sciences
The City College of the City University of New York

1. Retrospective Introduction

1.1. BACKSCATTER AND ENVIRONMENTAL PARAMETERS

Radar backscatter has been measured from instruments on SKYLAB and SEASAT with the goal of determining the winds from the backscatter data. New scatterometers such as NSCAT, proposed for ADEOS, have the same goal. Radar backscatter comes from the wavy ocean surface. It is thus assumed that there is some form of direct connection between the local wind speed and direction and the waves in the same local area. It is necessary to understand both the waves and the characteristics of radar backscatter so as to be able to obtain the winds. Any environmental feature of the waves that causes backscatter to change other than those due to some well defined property of the local wind can cause the backscatter to vary in a way not directly related to the local wind.

Before describing as many as possible of the environmental properties that may affect radar backscatter, it is important to be able to describe how well the present status of remote sensing is able to describe the waves and winds when various quantities that describe them are compared with conventional data.

1.2. SKYLAB

The SKYLAB Earth Resources Experiment Package (NASA SP-399), orbited in 1974, was a suite of instruments on a manned spacecraft designed to study the planet Earth. Some of the instruments were forerunners of those used on GEOS-3, SEASAT and GEOSAT and those to be used on unmanned spacecraft in the planning and design stages. The instrument called S193 was a combination scanning pencil beam radar-radiometer and altimeter, which lead to the GEOS-3 altimeter, the SEASAT SASS (SEASAT A Scatterometer), SMMR (Scanning Multichannel Microwave Radiometer), the SEASAT altimeter and the GEOSAT altimeter. The study of the data from S193 provided encouraging results for the possibility that scatterometry could provide data on the winds over the ocean. Even then attempts were made to try to determine the uncertainties in the determination of the winds obtained by conventional means with those obtained by S193 (Pierson, et al. 1978).

G. L. Geernaert and W. J. Plant (eds.), Surface Waves and Fluxes, Volume II, 173–220.

1.3. GEOS-3

The primary objective of GEOS-3 was to determine a better geoid over the oceans. It was also possible to obtain estimates of the significant wave height from changes in the shape of the return pulse.

1.4. SEASAT

The SEASAT A User Agency Working Group was involved in the preliminary plans for the instrument suite to be carried on SEASAT. Many hours of discussions were involved in describing how the data from the various instruments on SEASAT could be used for intercomparison purposes and to form a mutually supportive interrelated data set such that, for example, wave conditions and vector winds could be determined more accurately. The SASS could, in principle, recover something about wind speed and direction. The SMMR could, in principle, recover something about wind speed and account for the attenuation of the radar beam for the SASS. The altimeter could, in principle, yield the significant wave height, and, from the normalized radar backscattering cross section at nadir, recover something about wind speed at nadir. The Synthetic Aperture Radar (SAR) could, in principle, provide information on the waves, possibly in the form of wave spectra.

Although there were three different ways, in principle, to obtain wind speeds from remotely sensed quantities measured by SEASAT instruments (the SASS, SMMR and Altimeter), point by point possible comparisons were not done for a large number of orbit segments. Mean fields for the magnitude of the wind vector were prepared on a global basis that showed fairly large discrepancies such that the SMMR winds were higher than the SASS winds and the SASS winds were higher than the altimeter winds in a systematic way. There has been no reconciliation of these discrepancies. The mutually supportive role of the different instruments was not exploited to the fullest extent possible.

1.5. DATA INTERCOMPARISON

All of these remotely sensed quantities, plus others not mentioned, must, by tradition, be compared with conventionally measured values of comparable quantities. This tradition began for data obtained over the land, such as by LANDSAT, where, for example, geological features, such as faults, agricultural crops and types of forest stands could be identified. It was then a simple matter to verify the presence of a fault, the kind of crop growing in a field or the type of forest. The concept of ''ground truth'' and an absolute verification arose in this context.

For the remote sensing of the winds over the ocean and the waves on the surface of the ocean, the available conventional data do not provide a simple yes or no, right or wrong, verification. The conventional data also have sources of scatter that result from random variations in the measurements and calibration errors.

Moreover, the rather oversimplified concepts developed during the evolutionary process of studying data from SKYLAB, SEASAT and GEOSAT and from aircraft, fixed platforms and other conventional systems are now being superseded by more complicated interpretations of the data that suggest that there need not necessarily be a unique relationship between a single remotely sensed quantity such as radar backscatter and a desired geophysical quantity such as the properly defined and measured wind speed and direction.

In one sense, the comparison of wind speeds and, less often, directions determined from remotely sensed data with wind speeds and directions determined from conventional sources has not improved very much since SKYLAB. One can look up the references and compare Fig. 5-55 from Pierson, et al. (1978), Figs. 9, 10, 12 and 13 from Jones, et al. (1982) and Fig. 8 from Dobson, et al. (1987) and find that the scatter in these plots of the wind speed obtained by remote sensing versus the wind speed obtained by conventional means is comparable.

Similarly, as two examples, the remotely sensed significant wave heights as in Figs. 1 and 2 of Queffeulou (1983) and in Fig. 9 of Dobson, et al. (1987), when compared with significant wave heights from buoys, still scatter by about the same amounts. Queffeulou (1983) also illustrated (his Figs. 12 to 15) synoptic conditions such that the significant wave heights from the SEASAT altimeter varied by very large amounts over small distances along the track. Changes from 4 to 9 m and from 8 to 3 m over 108 and 130 nautical miles are difficult to model in wave forecasting.

2. Scatter Plots

2.1. GENERAL PROPERTIES

Scientific papers in geophysics in general and in air-sea interaction and wave measurement and wave forecasting studies in particular are replete with scatter plots similar to those mentioned above, as a small set of examples. Some of these scatter plots strain credulity that there is a relationship between the two quantities that are plotted.

When remotely sensed quantities are compared to conventionally measured quantities an important aspect of the comparison is the uncertainty of the conventionally measured quantity in the sense that the conventionally measured quantity may not be comparable to the remotely sensed quantity.

In any scatter plot, it is possible to identify at least four, and often more, sources of variability for each of the plotted points. For one of the two coordinates, there are both the possibility of systematic errors in the determination of the quantity involved and the virtual certainty, at present, that the quantity does not represent an average value that is truly representative of the quantity it it were to have been averaged over either a longer, more representative, time interval or a larger, more representative area, or distance. The term "more representative" is still to be defined. The same statement holds for the second of the two coordinates.

2.2. AN EXAMPLE

An example from Halpern (1987) reveals a part of the problem. The abstract is given below:

> "Wind measurements at 3.8 m from a surface-following toroid and at 4.0 m from a vertically stable spar were compared to estimate the influence of mooring motion. Buoys were separated by 9 km. Data were obtained at 15-min intervals for 41 days in the equatorial Pacific, where the maximum 15-min averaged wind speed was 9.0 ms^{-1} and wind speeds averaged 4.5 ms^{-1}. Toroid wind speeds were 3.5% greater than the spar data, and the correlation

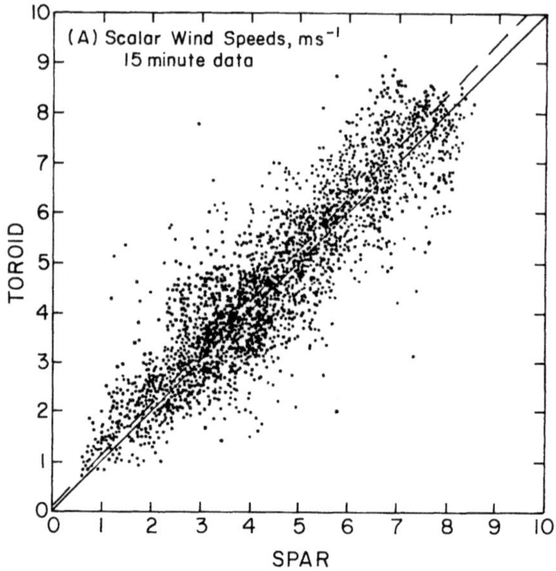

Fig. 1 — Concurrent Fifteen Minute Scalar Wind Speed Averages From Two Buoys 9 km Apart in the Equatorial Pacific (Redrafted From Halpern, 1987).

coefficient between 15-min toroid and spar data was 0.92. The frequency of the 50% noise level was 0.125 cph, and the correlation coefficient between 8-hour averaged toroid and spar data was 0.98. Toroid 8-hour averaged wind speeds referenced to 10-m height were 2% larger than corresponding spar data."

Figure 1, redrafted from Halpern (1987), compares 15 minute averages of the wind speed measured by one buoy with 15 minute averages of the wind speed measured at the same time by the other. The scatter from point to point far exceeds the possible errors introduced by the different types of buoy platform.

If either of the two axes were relabled, as say, SASS, or GEOSAT, or at some future data, NSCAT, the scatter would be quite similar to those that were obtained for the SASS based on JASIN data and for GEOSAT as referenced above. Yet, in this example, there are over 3900 points, and the scatter represents the differences in the magnitude of the vector wind as averaged for 15 minutes at two places on the ocean a mere 9 km apart.

Apart from the qualifications in the interpretation of the data given by Halpern, there is the question of whether or not winds from a surface following buoy as compared to the winds from a spar buoy are more accurate for use in models of the wind field and whether or not the winds from the National Data Buoy network need to be corrected theoretically for possible bias, especially when the waves are high. There is some theoretical guidance in the analysis by Benjamin (1959).

2.3. SIGNIFICANT WAVE HEIGHT

Similarly, the scatter in significant wave heights contains a contribution that results from the sampling variability in the estimation of the wave spectrum. This variability can be expressed in terms of the estimated wave spectrum and can vary from 10 to 25%, or so, of the estimated significant wave heights. Examples can be found in Neumann and Pierson (1966) (pages 348 and 349) where for one sea plus swell condition the probability was 0.90 that the random interval from 13.3 feet (4.1 m) to 16.9 feet (5.2 m) contained the "true" significant wave height with the best estimate being 15 ft (4.6 m). For a wind sea the corresponding interval extended from 34.3 ft (10.5 m) to 40.7 ft (12.4 m) with a best estimate of 37.4 ft (11.4 m). A recent example of the use of the sampling variability of significant wave height estimates is Komen (1987, his Fig. 5). Donelan and Pierson (1983) have shown that the problem of sampling variability for significant wave height estimates and spectral estimates is unavoidable even for laboratory conditions with an unchanging wind speed and fetch.

2.4. MODELS VERSUS CONVENTIONAL DATA

The wind vectors that are found in the specification of a computer based planetary boundary layer model are smoothly varying from one grid point of the model to the next over oceanic scales. They represent the well defined synoptic scale winds for the planetary boundary layer. Presently, these winds are determined by means of conventional measurements made by transient ships and anchored buoys. That transient ships measure the winds poorly has been shown by Dischel and Pierson (1986).

Zambresky (1987) has made a study that compares the winds and the waves specified by the operational wave forecasting methods used at the Fleet Numerical Oceanographic Center as determined earlier by the SOWM (Spectral Ocean Wave Model) for the Northern Hemisphere and more recently by the GSOWM (Global Spectral Ocean Wave Model). When the change was made from the SOWM to the GSOWM, a change was also made in the way the synoptic scale wind fields, which are used to drive the models, were obtained from the available conventional data. Many of the features of SOWM were retained in the GSOWM as also described by Clancy, et al. (1986).

A figure in Zambresky (1987, Fig. 2) shows four scatter plots; two for winds and two for significant wave heights. Two scatter plots are for the SOWM, one where the winds used to compute the waves are plotted against the magnitude of the 8.5 minutes average winds measured by buoys deployed by the National Data Buoy Center, and the other where the significant wave heights from the model are plotted against the significant wave heights from the data buoys The other two are for the GSOWM.

The winds used for GSOWM are clearly an improvement of the winds used for the SOWM. The points scatter by a larger amount for the SOWM than for the GSOWM when each different planetary boundary layer model is compared to the data buoy 8.5 min average winds. The scatter in the data suggests that the SOWM winds tended to be too high when the actual winds were low and too low when the actual winds were high as does the regression line. The GSOWM winds have a regression line that is closer to the 45° line of best fit but these data still retained a residual bias from high to low winds.

The improvement in the specification of the GSOWM significant wave heights compared with the SOWM significant wave heights is also evident and follows the same trends as the improvement of the winds. Since both the wind field specification and the wave model were improved at the same time, Zambresky notes that the improvement in the wave height specifications for GSOWM is the result of both improved winds and improvements in GSOWM. The contribution from each is difficult to determine from the scatter plot data.

The scatter plots in this example have proven to be useful, but the comparison with the conventionally obtained data puts a limitation on reaching the goal of even better models and specifications. Both the magnitudes of the measured winds and the estimates of the significant wave height have an element of variability, which introduces randomness in both the averages of the wind, which is the result of turbulence, and the value of the significant wave heights. Given the best possible wind field specifications and the best possible wave spectral specifications the conventionally obtained data will still cause scatter in such plots. Better ways than scatter plots can be developed to compare either models or remotely sensed quantities and conventional data. There may also be better ways to measure the conventional data. Zambresky also notes that at times over a portion of the model, the winds are poor and that then the waves heights are poor. When the winds are good, as in Zambresky (1987, Fig. 3) the waves are good.

2.5. DIMENSIONLESS QUANTITIES FOR THE MARINE BOUNDARY LAYER

The simplest of all models of the variation of the wind with height above the wavy ocean surface for the lower 50 m, or so, is Eq. (1) for a neutrally stratified atmosphere.

$$\overline{U}(z) = \frac{u_*}{\varkappa} \ln (z/z_0) \tag{1}$$

If z_0 and u_* are known then the time averaged wind at a given height can be computed for conditions such that $\overline{U}(z)$ does not change "rapidly" as a function of time. This equation is a simplified result from the Monin-Obukhov (1954) theory, which was derived for over-land conditions. The reference cited refers to evaluating the unknown parameters, z_0 and u_*, over grass, and it has yet to be shown that the theory holds over a wavy ocean surface.

The quantity, u_*, is related to the downward flux of eddy momentum as in Eq. (2).

$$u_*^2 = - < u'w' > \tag{2}$$

The original theory considered only two components of the wind given by Eqs. (3) and (4) with an implied function of time for u' and w'.

$$\vec{\overline{U}}(z,t) = \vec{\overline{U}}(z) + u'(z) \tag{3}$$

$$w = w'(z) \tag{4}$$

so that a mean wind direction, say, $\overline{\chi}$ needs to be found. Only the fluctuations of the wind speed about the mean wind vector, $\overline{U}(z)$, in the mean wind direction are considered to contribute to the eddy momentum flux. There assumed to be no lateral component in the form of a $<u'w'>$.

One way to be able to compute the variation of the mean wind with height from Eq. (1) is to try to relate u_* to the magnitude of the time averaged wind in the $\bar{\chi}$ direction, $\bar{U}(10)$, by means of Eq. (5), where C_{DN} is the neutral drag coefficient.

$$u_*^2/(\bar{U}(10))^2 = - < u'w' > /(\bar{U}(10))^2 = C_{DN} \tag{5}$$

From Eqs. (1) and (5), one can obtain Eqs. (6) and (7), with z_0 in meters and $\bar{U}(z)$ in ms^{-1}.

$$\bar{U}(z) = (C_{DN}(\bar{U}(10)))^{1/2}\bar{U}(10) \ln (z/z_0)/\varkappa \tag{6}$$

$$z_0 = 10 \exp (- \varkappa/(C_{DN}(\bar{U}(10))^{1/2})) \tag{7}$$

Numerous experiments over the ocean have attempted to determine C_{DN} which appears to be a function of $\bar{U}(10)$. Five of the expressions (there are others) for C_{DN} found in this way are those of Donelan, as in Donelan and Pierson (1987), Large and Pond (1981), Smith and Banke (1975), Geernaert, et al. (1987) and the WAMDI group (Hasselmann, et al. 1988) as in Table 1. The one used by Donelan and Pierson (1987) was chosen to be representative of fully developed seas since Donelan (1982) has shown that a much larger drag coefficient is needed to describe the wind stress for storm surge prediction over the Great Lakes because the waves are steeper and not fully developed. The WAMDI drag coefficient was empirically determined to produce the desired spectral wave specifications.

One procedure is to obtain time histories of U and w for some duration, say T_1 seconds, average the wind somehow to obtain $\bar{U}$ for this time interval, find u' and w', correct the results to a height of 10 m and compute C_{DN} from Eq. (5). A plot of sample values of C_{DN} versus $\bar{U}(10)$ then reveals considerable scatter, which is removed by regressing C_{DN} against $\bar{U}(10)$. As shown by Fig. 2 and Table 1 the results obtained by different investigators can be quite different.

Apart from the various sources of variability identified by T. V. Blanc in a series of papers (Blanc, 1983a, 1983b, 1983c, 1985, 1986a, 1986b, 1987), even if the instrumentation were perfect there would still be sources of variability associated with the time averaged wind and with the value of the area under the cospectrum of u' and w' as in $<u'w'>$. The time averaged wind over T_1 seconds may be either too low or too high compared to a more representative time average over T_2 seconds where $T_1 < T_2$. The downward flux of momentum may be either too high or too low because the turbulent wind at a somewhat higher elevation may be either higher or lower than the wind at the measurement height and produce either a larger, or smaller, flux at the height of the measurements.

These effects are illustrated in a simple way by Fig. 2. The values of $10^3\ C_{DN}$ are graphed as a function of $\bar{U}(10)$ for the five models in Table 1 and coded as D, G, LP, SB and W, for WAMDI. The average value from the five models for C_{DN} for 5, 10, 15 and 20 ms^{-1} is shown by the circled center dot.

As an illustration of the possible effect of sampling variability, suppose that the estimates of $\bar{U}(10)$ differ from the more stable value for the center circled dot by plus or minus one meter per second, or plus or minus 10%, whichever is worse, and that the estimate of $<u'\ w'>$ varies by plus or minus 10% about a more stable value represented by the center dot. One could Monte Carlo the effect of sampling variability by a further assumption that the sample values scatter about their respective expected value according to a normal distribution with standard deviations as assumed above. The dashed polygons on Fig. 2 outline the

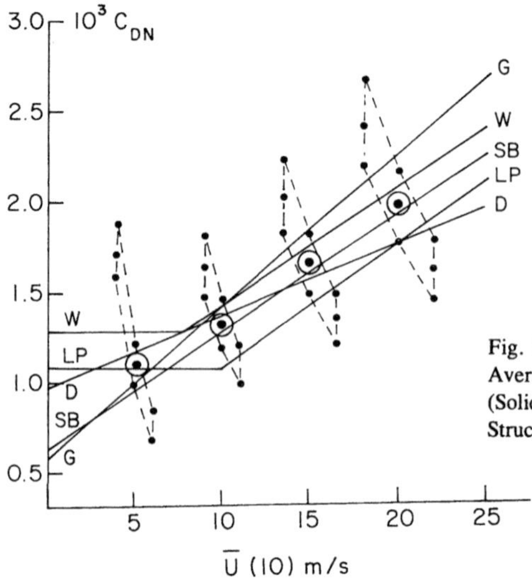

Fig. 2 — Neutral Drag Coefficient, C_{DN}, Versus the Time Averaged Wind at 10 m According to Various Authors (Solid Lines). Scatter Region for Data Points for Error Structure Assumed in Text (Dashed Lines and Circles).

Table 1 — Values for the Neutral Drag Coefficient, $10^3\ C_{DN}$, According to Various Authors

Donelan	$0.96 + 0.041\ \overline{U}(10)$	
Geernaert, et al.	$0.577 + 0.0847\ \overline{U}(10)$	
Large and Pond	1.14 $0.49 + 0.065\ \overline{U}(10)$	 if $10 < \overline{U}(10) < 50$
Smith and Banke	$0.63 + 0.066\ \overline{U}(10)$	
WAMDI	1.2875 $0.8 + 0.065\ \overline{U}(10)$	if $0 < \overline{U}(10) < 7.5$ if $7.5 < \overline{U}(10) < 50$

effects of the variability of $\overline{U}(10)$ and $<u'\,w'>$. For example, the highest point for a mean wind of 10 ms^{-1} represents a sample value for the wind of 9 ms^{-1} and a sample value of 1.10 times the more representative value of $<u'\,w'>$. For this point the value at the center circled dot plots at 9 ms^{-1} and $1.1/(0.9)^2$ equals 1.36 times the assumed true value of C_{DN}. All of the other points that form the polygons are based on similar calculations.

The assumed $\pm 10\%$ variability may be too high. Values of $\pm 5\%$ could yield a value of C_{DN} that was as much as 16% too high or as much as 14% too low. The effect of two standard deviations about the expected value would then be similar to the plotted polygons.

The assumed scatter in the sample values of C_{DN} about the assumed averaged values for C_{DN} more than covers the five models. For a rather small sample of calculated values of C_{DN} versus $\overline{U}(10)$, the scatter of the points could quite possibly yield any one of the model curves. Figure 2 shows a tendency for the plotted points to be systematically too high and to produce values of C_{DN} that are too high at a wind speed that is too low. Also the scatter of the sample values is probably inversely related in some way to duration of the average that was used.

Part of the difficulty in finding a form for C_{DN} that would be acceptable for everyone who wants to use values of the time averaged wind to calculate values of the wind stress and u_* lies in forming the expression for C_{DN} as a ratio of two randomly varying quantities. Even if the probability density functions that describe the sampling variability of $\overline{U}(10)$ and $<u' w'>$ were known, it would probably not be possible to derive the probability density function for estimates of C_{DN}. (See Mood, et al., (1974) on ratios of random variables.)

Perhaps one way to obtain more convincing results would be not to regress C_{DN} against $\overline{U}(10)$. One really wants to relate u_*^2 or $<u' w'>$ to the average wind at 10 m. This can be done by finding the values of A_1, A_2 and A_3 in Eq. (8) by minimizing the quantity Q in Eq. (9).

$$-<u'w'> = u_*^2 = A_1\overline{U}(10) + A_2(\overline{U}(10))^2 + A_3(\overline{U}(10))^3 \tag{8}$$

$$Q = \sum_{1}^{N} (u_{*n}^2 - A_1\overline{U}(10)_n - A_2(\overline{U}(10)_n)^2 - A_3(\overline{U}(10)_n)^3)^2/N \tag{9}$$

The derivatives of Q as in $\partial Q/\partial A_1 = 0$, $\partial Q/\partial A_2 = 0$ and $\partial Q/\partial A_3 = 0$ yield a set of simultaneous linear equations in three unknowns, A_1, A_2, A_3, where the 3×3 matrix is determined from the values of $\overline{U}(10)_n$. The values of u_* computed from $\overline{U}(10)$ are the best fit to the data in a mean square sense. The values of $C_{DN} = u_*^2/\overline{U}(10)^2$ can be computed and compared with values found by the usual method. They could be quite different.

2.6. ATMOSPHERIC STABILITY

Even if u_* and z_0 could be defined by an appropriate choice among the many different proposed values for C_{DN} so that Eq. (1) could be used for a neutrally stratified atmosphere, the effects of atmospheric stability will still complicate the analysis. The air temperature at a known height, the specific humidity at a known height and the sea surface temperature are needed to determine whether the atmosphere is unstably, neutrally or stably stratified. The effect of stability is accounted for by the Monin-Obukhov length, L, which modifies the shear of the wind with height in terms of a function written as $\phi_u(z/L)$. The function, $\phi_u(z/L)$ is essentially unity close enough to the sea surface for all values of L.

Since L can approach plus and minus infinity, z/L evaluated at 10 m is usually tabulated. If u_* is known, $U(z)$ will vary in different ways depending on whether $z/L < 0$ (unstable), $z/L = \pm 0$, or $z/L > 0$ (stable). At, say, a height of one meter, for the same u_* (and implied z_0), $\overline{U}(1)$ for $z/L < 0$, $z/L = 0$ and $z/L > 0$ will be imperceptively different. For a height of 10 m, for the same u_*, if $\overline{U}_N(10)$ represents the wind at 10 m for neutral conditions, then $\overline{U}_u(10) < \overline{U}_N(10)$ for $z/L < 0$ and $\overline{U}_s(10) > \overline{U}_N(10)$ for $z/L > 0$, where subscripts u and s represent unstable and stable conditions, respectively.

At heights above the sea surface of 10 and 19.5 m, the differences in the theoretical values of the time averaged winds can be substantial for the same friction velocity. That only C_{DN} needs to be defined in order to treat the variation of wind with height for all conditions according to present theories is shown in the programs published by Liu and Blanc (1984) where Lines 1580 and 1600 give three different choices for the neutral drag coefficient in the FORTRAN program. The programs are based on the paper by Liu, et al. (1979).

For the SOWM and the GSOWM, the winds used to drive the model are the winds that would be measured at 19.5 m if the atmosphere were neutrally stratified. This is the effective neutral wind, and it was also used to verify the winds from the SEASAT SASS. The "properly defined and measured wind" of Section 1.5 could be this wind for a long enough, but not too long, average.

It is possible to take any set of data for a wind, a temperature and a humidity measured at three different heights (or the same height), plus the sea surface temperature, to use any one of a number of different models of C_{DN}, one after the other, and compute u_*. The effective neutral winds obtained at either 10 or 19.5 m for each of the different u_* will differ by only a few cm/sec just as long as the same value of C_{DN} is used throughout. The stress on the ocean surface will differ considerably, however.

Geernaert (1988a, 1988b) has provided some new interpretations for the influence of long and short waves on C_{DN} in coastal regions and provided data and an interpretation of it to show that the shearing stress at various heights above the waves is not always in the mean wind direction for that height.

2.7. DIMENSIONLESS QUANTITIES FOR WAVE FORECASTING MODELS

In contrast to wave forecasting models that do not use wave interaction theory such as the SOWM and GSOWM as described by Clancy, et al. (1986) and as reviewed by Zambresky (1987), the various parametric wave prediction models as for example, Hasselmann (1976), Hasselmann, et al. (1977), depend on the relationships between various dimensionless quantities so as to be able to describe the self similar properties of a duration dependent, or fetch limited, wave spectrum. The concept of self similar spectra complicates the prediction of swell because it is difficult to determine when a sea changes into a swell.

Parametric wave prediction models depend on finding functional relationships between the following dimensionless variables as in Eqs. (10) to (12)

$$\epsilon = g^2 E / (\overline{U}(10))^4 \tag{10}$$

$$\nu = f_m \overline{U}(10)/g \tag{11}$$

$$\xi = gF / (\overline{U}(10))^2 \tag{12}$$

and perhaps other quantities such as the equilibrium range constant, α, the JONSWAP peak enhancement factor, γ, as in Hasselmann, et al. (1973), the JONSWAP shape factor, Eq. (13),

$$\lambda = \epsilon \nu^4 \alpha = \alpha E (f_m)^4 / g^2 \tag{13}$$

and the Wallops fetch factor, Eq. (14).

$$E \ \nu/\xi = E f_m / F\overline{U}(10) \tag{14}$$

In the above equations,

$$E = \int_0^\infty \hat{\Phi}(f)\,df \tag{15}$$

where $\hat{\Phi}(f)$ is the estimated spectrum, f_m is the frequency of the spectral peak and F is the fetch, or distance upwind to a coast line, in meters.

All of the quantities, E, $\bar{U}(10)$, f_m,F, α and γ have an element of randomness associated with them. Since the spectra are estimates, the quantity, E, is a random variable. The frequency of the spectral peak, as discussed by Pierson (1977) and Donelan and Pierson (1983), is a random variable. Fitting the spectral shape by means of the value of the estimated spectrum at the spectral peak results in an over estimation of the peak enhancement factor, γ, and produces strange results for the study of the sampling variability of spectral estimates as in Günther (1981).

The concept of a fetch such that a constant wind has blown long enough over the entire fetch to generate a unique wave system at a point of measurement has been questioned by both the results of Walsh (1985) and Donelan, et al. (1987). Wave spectra estimated by means of the surface contour radar, Walsh (1985), have shown that the wavenumber spectra increases downwind from Chesapeake Bay at an angle to the wind direction such that there seems to be a part of the spectrum coming from upwind and another part from the bay. Donelan, et al. (1985) have shown that winds that blow at an angle to the long axis of Lake Ontario produce spectra such that the dominant waves travel along the long axis of the lake.

The upwind distance to the nearest shore is not a good measure of F. Also the larger the value of F for oceanic scales, the less representative the value of $\bar{U}(10)$ over the entire fetch.

The same general analysis of scatter plots of, say, ν versus ξ and ϵ versus ξ, which consider the effect of sampling variability, would show that the points so obtained would scatter about their expected values by rather large amounts. The scatter is so large, in fact, that double logarithmic scales have to be used. The JONSWAP fit (Hasselmann, et al. 1973) gave $\nu = 3.5\ \xi^{-0.33}$ and $\epsilon = 1.6 \times 10^{-7}\epsilon$.

These dimensionless quantities in terms of $\bar{U}(10)$ have been computed from data obtained by the National Data Buoys in the Great Lakes by Liu (1985) and Liu and Schwab (1987). In Liu (1985) for ξ near $4 - 10^{-4}$ the points for ϵ scatter by more than three orders of magnitude. The thick cloud of points though very wide does not fit the JONSWAP model. Liu and Schwab replot the dimensionless wave data in terms of u_* computed from various drag coefficients. There is little, or no, improvement in the scatter.

There are a number of wave spectral forecasting models that have been intercompared by the Sea Wave Modeling Project (SWAMP 1985). An intercomparison of the dimensionless energy versus the dimensionless duration (not defined above) for ten of these models along with GSOWM and ODGP (Cardone, et al. 1976) is given in Cardone, (1987, Fig. 2). The differences among the various curves that define the relationship between these two quantities are considerable and may be partially explained by the kinds of scatter described above and by the particular data bases used by the various modelers.

The Wave Modeling Development and Implementation (WAMDI Group, Hasselmann, et al. 1988) have developed a third generation spectral wave specification and forecasting model. Komen (1987) showed some preliminary results. This third generation model produces some results that are closer to the first generation models than they are to the parametric second generation models. The second generation has been skipped for many aspects of this third generation model, and the paper lists many faults of the parametric model.

Among the changes are the replacement of ϵ by E^* and $\bar{U}(10)$ by u_* in Eq. (10), the use of f^*_p instead of ν, and u_* instead of $\bar{U}(10)$ in Eq. (11), and the use of F^* instead of ξ, and u_* instead of $\bar{U}(10)$ in Eq. (12). Also the peak enhancement factor, γ, is no longer a part of the prediction scheme. The parametric equations for E^* versus X^* and $f_p{}^*$ versus X^* are quite different. The JONSWAP method of fitting an estimated spectrum is not used, perhaps as a result of the comments of Pierson (1977) and the results of Gunther(1981) and Donelan and Pierson (1983).

This third generation model has many empirical features such as the choice of a particular drag coefficient so as to be able to model u_*, an approximation to a complete wave interaction theory, the empirical changes in the growth term (their Eq. (2.95)), compared to the experiments on which it is based, and the empirical form for the dissipation term.

The dimensionless spectrum as a function of dimensionless frequency for various dimensionless fetches no longer has the saw-tooth appearance of the parametric model. It would be very difficult to demonstrate a real difference between a WAMDI spectrum for a given wind and fetch and a GSOWM spectrum for that same wind and fetch, especially near dimensionless frequencies of 0.011 and 0.012.

A major change from previous models is the use of a friction velocity to parameterize a model and the choice of a particular neutral drag coefficient. This change will be discussed below.

3. Improved Data Records and Data Analysis

3.1. RANDOM FLUCTUATIONS

Both the winds measured at any height above the sea by any method and the waves measured by any method at either a point as a function of time or over an area by either aircraft or spacecraft have elements of randomness. The elements of randomness for the winds are the various scales of turbulence. In turbulence the water affects the waves also as a result of wave breaking, but wave models treat waves as if they are more or less irrotational and consequently do not treat wave breaking and turbulence directly. Both turbulence models and wave specification and forecasting models are validated as if the time averaged winds and the wave spectra were not changing as a function of time.

Wind field analyses for the marine boundary layer attempt to specify the synoptic scale winds at some given height above the mean sea surface. These winds are used for the synoptic scale forecasting of the weather. They are also the winds that drive the various forecasting models for the waves. Those environmental parameters that describe the turbulent properties of the winds for time scales less than and space scales smaller than the synoptic scale are consequently an important part of the problem of being able to relate the backscatter to the waves and then to the winds. Those environmental parameters that determine the significant wave height, wave breaking, turbulence in the water, and wavenumber spectra at the synoptic scale are important so that the waves that the winds generated can be described.

It is not possible to predict what a two minute average of the wind speeds and directions obtained by a data buoy would be twenty minutes into the future nor is it possible to predict how high a given wave crest will be above mean water level even two to three minutes into the future.

It is possible to describe the temporal variations of the winds and the waves about longer records in terms of probability models and spectra and to validate these models in terms of the statistical analysis of data. With these probability models that part of the differences between the measurements of the winds and the waves can be accounted for and not attributed to errors in the remotely sensed winds and waves. Probability models and statistical analyses are also needed for the remotely sensed quantities. If the same quantity is measured by two different systems and if the random effects for the two systems are accounted for correctly by a probabilistic model, then any residual differences would be the result of some real errors in either, or both, of the systems used to obtain estimates of the variable that was obtained.

For the winds, the element of randomness in measurements by an anemometer on a data buoy can be suppressed with reference to the time averaged wind by obtaining longer records. For the waves, longer time histories for wave measuring buoys could help. Sensing systems on aircraft for both the winds and the waves show promise of obtaining more useful, more stable averages for the winds and less variable estimates of the wavenumber spectra and significant wave height for the waves. Some examples of measurement methods that reduce the effects of randomness, or sampling variability, follow.

3.2. FLUCTUATIONS IN THE WINDS

Figure 3 is from Ezraty (1986). The data were obtained over the North Atlantic west of France. The large ticks are 30 min apart and the small ones 5 min apart. One minute averages are plotted. The record, except for three gaps, is almost 17 hours long. There is a dramatic change in wind direction at about 7 hours into the record. Prior to this, the wind is almost from due south and the wind direction fluctuates about some average direction by only a slight amount. The wind speeds fluctuate by only a few ms^{-1} about what could be either a one hour running average or a smooth curve drawn by eye. The smooth curve would show an increase in speed from about 10 ms^{-1} to about 16 ms^{-1}. The fluctuations in the wind speed are rapid during a given hour, and an average over one hour would be representative of the wind speed and direction to be used in a synoptic scale analysis.

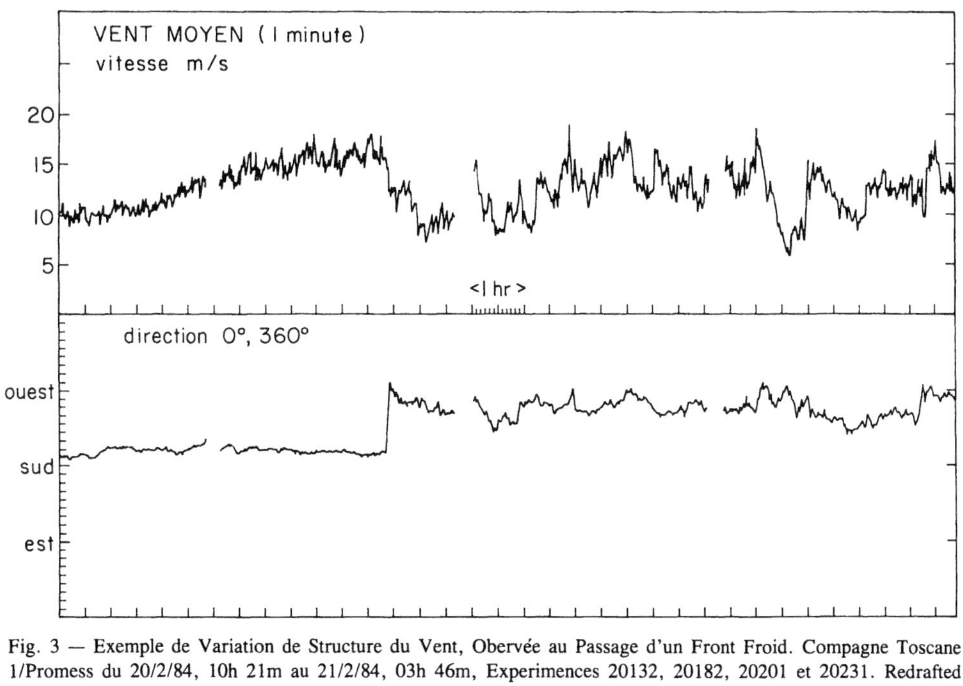

Fig. 3 — Exemple de Variation de Structure du Vent, Obervée au Passage d'un Front Froid. Compagne Toscane 1/Promess du 20/2/84, 10h 21m au 21/2/84, 03h 46m, Experimences 20132, 20182, 20201 et 20231. Redrafted for Reproduction Purposes. The Large Time Ticks are 30 min Apart and the Illustrated Small Ticks are 5 min Apart. Direction Ticks are 10° Apart. One Minute Averages Were Obtained and Plotted in the Original Figure.

The sharp wind shift determines the time of a frontal passage to within just a few minutes at the location of the measurements. As shown by Brown (1983, Fig. 5), the SEASAT-SASS wind fields also located fronts to considerable accuracy. In Fig. 3, after the frontal passage within the colder more turbulent air from the west, both the wind speeds and directions fluctuate by large amounts even over running averages of a half hour to an hour. It is doubtful that any of the presently available synoptic scale analysis methods and forecasting models could properly reproduce even the time smoothed fluctuations in the wind after the frontal passage. A remotely sensed value for the wind for an area near, but not exactly at, the site of the anemometer measurement could differ from any wind calculated from any time average of these data. Yet both could be correct and free from systematic errors.

The synoptic scale time averaged vector wind at a given height over the sea might be defined in a somewhat circular way as that vector wind as a function of space and time that best defines the winds for the scales being used by the numerical model. The time scales (frequencies) and wavelengths (wavenumbers) are seldom stated explicitly. A time step by step listing of the wind speeds and directions at a set of grid points of a model could quickly define the frequencies kept by a model by a spectral analysis. The wavenumber spectra must correspond to wavelengths more than twice the grid spacing: otherwise aliasing will occur. The duration of a time average to best determine the synoptic scale wind is probably a function of the location of the measurements and the current synoptic pattern.

With the assumption that a one hour average of the wind better represented the synoptic scale than a briefer average, Pierson (1983) modeled the mesoscale spectrum as f^{-1} and the microscale spectrum, preceded by a bump, as $f^{-5/3}$. It was then possible to compute this standard deviations of the fluctuations of winds averaged for T seconds relative to a one hour average.

Ezraty, et al. (1985) have checked this theory against data obtained from anemometers along the coast of France. The comparisons are shown in Table 2.

Table 2 — Values of the Standard Deviation of 2, 10, 20 and 30 Minute Averages of the Wind Speed in ms^{-1} Relative to a One Hour Average, According to a Theory (Pierson, 1983), and as Measured (Ezraty, et al. 1985). The Range for the Theory for Different Stabilities is Given

T(min)	Theory for 10 ms^{-1}	Measured for 9.2 ms^{-1}	Ratio (M/T)
2	0.75 — 0.59	1.20	1.48 — 1.90
10	0.56 — 0.40	0.93	1.66 — 2.33
20	0.45 — 0.32	0.75	1.67 — 2.34
30	0.30 — 0.27	0.61	2.03 — 2.26

For these data, the theory is too low by a factor of about 0.67 to 0.43. Two and ten minute averages fluctuate about a longer sixty minute average by more than 10%. A thirty minute average halves the variability of a two minute average.

Data for winds measured at 10 m for a row of anemometers about 10 km apart have been extensively analyzed by M. Champagne-Philippe (1985). Spectra of the magnitude of the wind, cross spectra, coherence and phase have been computed for pairs of recorded values measured simultaneously at spacings of multiples of about 10 km. For periods shorter than about 5 min, the coherence is negligible and the phase is more or less random at all spacings including 10 km.

For frequencies corresponding to periods greater than 10 to 15 min, the coherence is high and there is a realistic phase shift for a spacing of 10 km. Lower coherencies and larger phase shifts occur for larger spacings. The higher frequency fluctuations in the wind are unrelated from one anemometer to the next. The longer the averaging time, within reason, the more these fluctuations are filtered out and the more representative the wind for the synoptic scale.

Aircraft in low level horizontal flight can measure the three components of the wind vector, temperature and pressure at a given height, z, above the waves. The high velocity of the aircraft would provide data more nearly related to wavenumber spectra than to frequency spectra. The "Results of the Royal Society Joint Air-Sea Interaction Project" are based on data obtained at various times by 14 ships, 3 aircraft and 35 moorings during July 17 to September 5, 1978. Data from the airborne instruments were combined with data from instruments supported by tethered balloons and other data by Nicholls, et al. (1983) to estimate the values of all but one of the six terms in the turbulent kinetic energy budget. The values in the budget were based on data from the sides of a square 70 km on a side and from a ship near the center. The aircraft flew as low as 30 m above the sea surface.

3.3. THE VARIABILITY OF WAVE DATA

The variability of wave spectral estimates in terms of confidence intervals and of the significant wave height as in Eq. (16) in terms of the total degrees of freedom (Neumann and Pierson, 1966) is well documented.

$$\bar{H}_{1/3} = 4\,E^{1/2} \tag{16}$$

The only way to increase the degrees of freedom for a given spectral resolution and the total degrees of freedom is to increase the duration of a time history if only one wave measuring system is available. The presently typical 17 to 20 min long time histories might be twice as long, but this does not narrow either the confidence intervals or the interval with a probability of, say, 90% of containing the significant wave height, by very much.

Instruments for use on aircraft have been developed so as to be able to compare wave spectra estimated by means of data from synthetic aperture radars such as SIR B on past shuttle flights with other data. In addition to airborne real and synthetic aperture radars and the radar ocean wave spectrometer, the instrument of interest here is the surface contour radar (SCR) as described by Walsh, et al. (1985, 1987).

Airborne instruments trade time for area. The SCR comes close to providing values of the sea surface, $\eta(x,y)$, on a grid of equally spaced points in a checkerboard pattern. The slight time dependence is an advantage. The Fourier transform of $\eta(x,y)$ yields $\Phi(k_x,k_y) = \Phi(-k_x,-k_y)$ with symmetry through the origin. The time element permits the removal of the false part of the wavenumber spectrum.

Wave data values about one second apart for about 17 to 20 min yield spectral estimates that are uncertain to within a factor of two for typical frequency resolutions and total degrees of freedom from about 90 to 150 based on spectral shape. The correlations between successive data points in the time history cause a series of 1024 points, or so, to be the equivalent of an independent random sample with the above total degrees of freedom.

If the wavy surface is represented by $\eta(p\Delta x, q\Delta y)$ for $P = 0$ to 511 and $q = 0$ to 511, the data would consist of a square of 512 by 512 points for a total of 262,144 values. This is equivalent to 265 × 1024 values or what would appear to be the equivalent of 256 time histories. This is not exactly true because adjacent rows of data are not independent. The Fourier transform of $\eta(p\Delta x, q\Delta x)$, yields spectral values at the wavenumber $(k_x)_i \equiv 2\pi i/512\Delta x$, $(k_y)_j = 2\pi j/512\Delta y$ in terms of sines and cosines such that $C_{ij}^2 = a_{ij}^2 + b_{ij}^2$ is Chi Square with two degrees of freedom. In wavenumber space a 9 by 9 sum of the C_{ij}^2 would produce a smoothed spectral estimate with 162 degrees of freedom. The total degrees of freedom would be found from Eq. (17)

$$\mathrm{TDF} = \frac{2(\sum_j \sum_i C_{ij}^2)^2}{(\sum_j \sum_i C_{ij}^4)} \tag{17}$$

where the sum over i is from zero to 255 and the sum over j is from -255 to $+256$.

The value for the TDF could easily be a factor of 10 larger than the TDF for a 1024 sec time history. For TDF's of 90 to 150 the estimates of the significant wave heights, $\bar{H}_{1/3}$ from (16), can be multiplied by 0.89 and 1.13 for 90 degrees of freedom and by 0.91 and 1.10 for 160 degrees of freedom to define an interval with a probability of 90% of enclosing the "true" significant wave height. For 900 and 1600 degrees of freedom, the corresponding values to use are 0.96 and 1.04 and 0.97 and 1.03 respectively. A properly calibrated set of values representing the sea surface for a square grid of points could therefore provide values for the significant wave height within about 4% or 3% of the correct value 90% of the time. The scatter in the various plots of wave height and dimensionless values that contain E described above would be greatly reduced.

The SCR provides an estimate of the vector wavenumber spectrum $\hat{\Phi}((k_x)_i, (k_y)_j)$ after smoothing. It is thus not necessary to postulate anything about the angular spreading function and the properties of the cos χ and cos 2χ terms as in the analysis of the data from pitch roll buoys. Steele, et al. (1985) and Pierson (1987) describe the way pitch and roll buoy data are used to try to determine some of the properties of the angular spreading function.

It is, however, useful to compute the first two Fourier coefficients of the angular spread from the SCR data. These can be compared with the parametric plot of the values of C_1/A_0 and C_2/A_0 as the value of S is varied in Eq. (18) where $\bar{\chi}$ is the dominant wave direction and $A(s)$ normalizes $F(\chi)$ to one over the range of χ from $-\pi$ to π.

$$F(\chi) = A(S)\,(\cos((\chi - \bar{\chi})/2))^{2S} \tag{18}$$

The points for C_2/A_0 versus C_1/A_0 from the Cartwright (1963) data are shown in Fig. 4 as well as the points from SCR data during ARSLOE. The buoy data (not shown) from ARSLOE scatter by about the same amount as the 1963 data.

Equation (16) and the dimensionless relation $\epsilon = 1.6 \times 10^{-7}\xi$ can be used to derive Eq. (19).

$$\bar{H}_{1/3} = 16 \times 10^{-4}\,\bar{U}(10)\,F^{1/2}\,g^{-1/2} \tag{19}$$

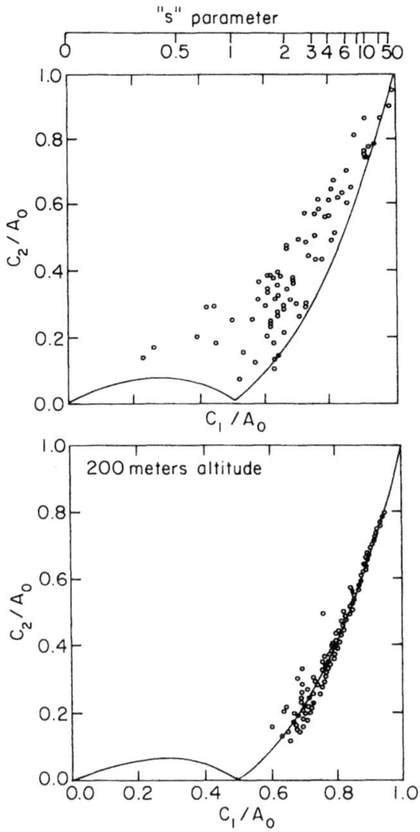

Fig. 4 — Parametric Plots of the Normalized Amplitudes of the First and Second Harmonics of the Angular Spread of Directional Wave Spectra for Pitch-Roll Ship Data (Cartwright 1963) and SCR Data. (Redrafted From Walsh, et al. 1987).

For non-fully developed seas, for a fixed F the wave height varies linearly with $\overline{U}$; for a fixed $\overline{U}$ the wave height varies as $F^{1/2}$. If the significant height for a fully developed sea is proportional to the square of the wind speed, then the fetch for a fully developed sea is also proportional to the square of the wind speed as can be derived from (19) and $\overline{H}_{1/3} \sim \overline{U}^2$.

Graphs of $\overline{H}_{1/3}$ versus F from Eq. (19) are shown for $\overline{U}$ equal to 10, 15 and 20 ms^{-1} in Fig. 5 from Walsh, et al. (1987). Also shown are data from two SCR flights for nominal winds of 20 ms^{-1} and 11 ms^{-1}. The constant 16×10^{-4} may be a bit small. Compared to the laborious gathering of data for sets of single points, a single offshore flight provides many data points as a function of fetch for a presumed constant offshore wind with the sampling variability of wave height and E greatly reduced. The waves for the nominal 20 ms^{-1} wind were duration limited.

The variations of the significant height as a function of fetch is erratic for both of the nominal wind speeds in the figure. Whether or not the random interval with a 90% chance of

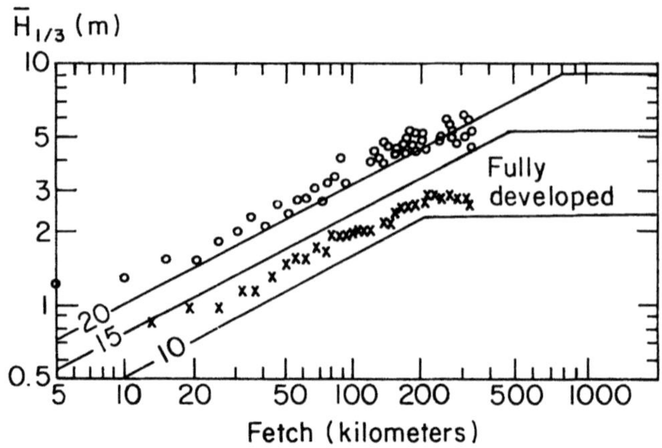

Fig. 5 — Significant Wave Height as a Function of Wind Speeds of 10, 15 and 20 ms^{-1} and Fetch (Continuous Lines) and Data From the SCR for Nominal Winds of 11 and 20 ms^{-1}. (Redrafted From Walsh, et al. 1987).

enclosing a model significant height, if computed, would permit a straight line according to (19), requires further analysis.

Neither the wave data from data buoys nor the wave data from the SCR, or any other airborne and spaceborne instrument, provide any information on the very high wavenumber very high frequency waves on the ocean that correspond to the near capillary and far capillary waves with wavelengths from 30 cm, or so, to 2 cm and on to 3 mm.

The significant wave height can also be determined from the shape of the altimeter return pulse. It has been recovered from GEOSAT data since its launch. The algorithm for recovering the significant wave height has had a long history beginning with GEOS-3. The waveform analysis of the GEOSAT data by Hayne and Hancock (1987) is an improvement over previous waveform analyses and will provide more accurate values for the significant wave height. The reduction in the scatter of various waveform derived quantities is impressive. A reanalysis of all of the data that are based on GEOSAT significant wave heights published prior to the cited report may well be merited.

3.4. BUOY, WAVE AND WIND DATA

In wave forecasting models, there is also the concept of a fully developed sea such that ϵ in (19) would become a constant for a sufficiently long fetch and a sufficiently long time. The significant wave height, $\overline{H}_{1/3}$, given by $4E^{1/2}$, would then be related to the square of the wind speed at 10 m as in Eq. (20) which is dimensionless or Eq. (21) which gives the height of a fully developed wind sea as a function of $\overline{U}(10)$.

$$g\,\overline{H}_{1/3}/(\overline{U}(10))^2 = C_1 \qquad (20)$$

$$\overline{H}_{1/3} = C_1\,(\overline{U}(10))^2/g \qquad (21)$$

Forming new dimensionless quantities in terms of u_* leads to some interesting inconsistencies. If, for example,

$$g\, \overline{H}_{1/3}/u_*^2 = C_2 \tag{22}$$

holds for fully developed seas, it is possible to find from Fig. 3 of Hasselmann, et al. (1988) that E^* for a long fetch equals 905 approximately. After simplification and with the WAMDI drag coefficient, the result is Eq. (23).

$$\overline{H}_{1/3} = 10.24 \cdot 10^{-2}(\overline{U}(10))^2 \quad \text{if} \quad \overline{U}(10) < 7.5 \text{ m/s}$$

$$\overline{H}_{1/3} = 0.01223(0.8 + 0.065\, \overline{U}(10))(\overline{U}(10))^2 \quad \text{if} \quad 7.5 < \overline{U}(10) < 50 \tag{23}$$

To confound the interpretation further, Moskowitz (1964) found that the wave height for a fully developed sea was proportional to the square of the mean wind at 19.5 m and Pierson and Moskowitz (1964) gave the constant as in Eq. (25). It was required that the wind be for a neutrally stratified atmosphere. The height of 19.5 m was used to avoid choosing among the various drag coefficients of that time as in Pierson (1964).

$$\overline{H}_{1/3} = C_3(\overline{U}(19.5))^2/g \tag{24}$$

Ewing and Laing (1987) used wave and wind data from DB1 moored in water 170 m deep off Great Britain to study properties of fully developed seas defined along lines similar to those of Moskowitz (1964). The data buoy, DB1, had been in service from June 1978 to March 1982 and during this time meteorological conditions and wave spectral forms corresponding to the concept of a fully developed sea occurred only 33 times. From their Table 1, the significant wave heights for 33 spectra are plotted against $\overline{U}(10)$ for neutral stability in Fig. 6. This scatter plot of wave height versus wind speed can be interpreted in terms of Section 2.1, 2.6 and 3.2 in particular. Pierson and Moskowitz (1964) gave Eq. (25) for the fully developed significant wave height.

$$\overline{H}_{1/3} = 2.12 \cdot 10^{-2}\, (U(19.5))^2 \tag{25}$$

If as in Ewing and Laing, the relation, $\overline{U}(10) = 0.93\, \overline{U}(19.5)$ is used to relate the 19.5 m wind and 10 m wind, Eq. (25) becomes Eq. (26).

$$\overline{H}_{1/3} = 2.45 \cdot 10^{-2}(\overline{U}(10))^2 \tag{26}$$

The 19.5 m wind can be related to the 10 m wind for the Smith and Banke values for C_{DN} by

$$\overline{U}(19.5) = \overline{U}(10)[1 + 0.515(0.63 + 0.066\, \overline{U}(10))^{1/2}] \tag{27}$$

Changing the values in parentheses would give $\overline{U}(19.5)$ for any C_{DN} in Table 1.

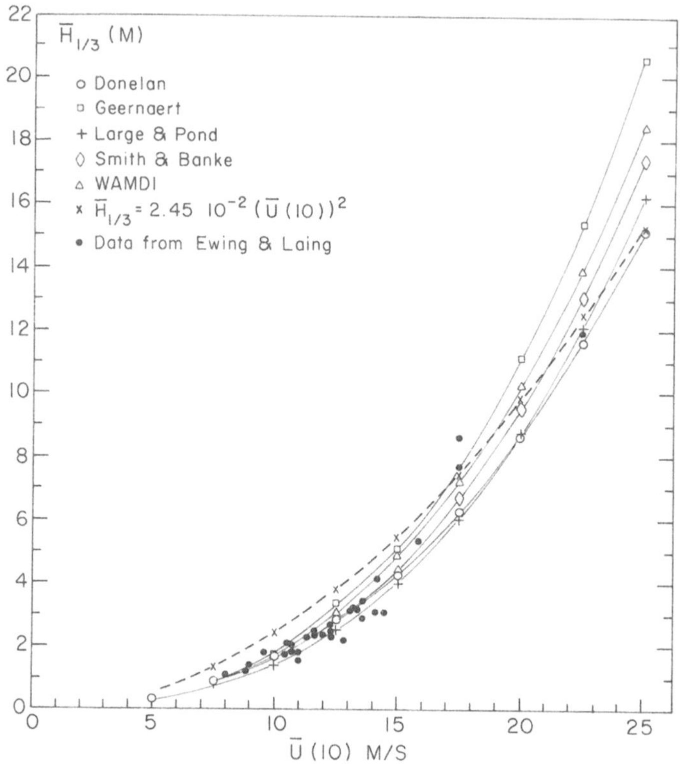

Fig. 6 — Significant Wave Height Versus the Mean Wind Speed at 10 m from Ewing and Laing (1987). Significant Wave Height Versus the Mean Wind Speed at 10 m Modified from Pierson and Moskowitz (1964) (x). Fits to Eq. (22) Based on the Asymptotic Long Fetch Value of E^* from the WAMDI Group and the Five Drag Coefficients of Table 1 as Coded.

The dashed line connected by X's is the graph of Eq. (26). Substituting (27) into (25) and graphing the result produces only minor differences as do also the other drag coefficients. The most liberal interpretation of possible sources of scatter in $\bar{H}_{1/3}$ and $\bar{U}(10)$ for $\bar{U}(10) < 16$ ms^{-1} cannot explain all data points lying below the curve. If the point cloud for $\bar{U}(10) < 16$ ms^{-1} were fitted to a square law the result might be that $\bar{H}_{1/3} = 4$ m when $\bar{U}(10) = 15$ ms^{-1} (as good as any) so that the constant in (26) would be $1.7778 \cdot 10^{-2}$. This yields a model significant wave height of 9 m for the 11.84 m $H_{1/3}$ data value for the 22.5 ms^{-1} wind. The sampling variability of $\bar{U}(10)$ and $\bar{H}_{1/3}$ would have had to have been quite high to have produced such a difference. Could it be that the significant wave height for a fully developed sea is not proportional to the square of the time averaged wind speed at 10 m? Could the spectra for winds over 17 m/s have been affected by refraction because the water was only 170 m deep? Reducing the constant in Eq. (25) to about 73% of its value so as to fit for low winds in Fig. 6 would have a substantial effect on wave specifications and forecasts for GSOWM.

The various drag coefficients in Table 1 can each be used to calculate the significant wave height as a function of $\bar{U}(10)$ with the assumption that the large fetch value of F^* is about 905.

The wave height for the Donelan value for C_{DN} is

$$\bar{H}_{1/3} = 1.022 \cdot 10^{-2} \, (\bar{U}(10))^2 \, (0.96 + 0.041 \, \bar{U}(10)) \tag{28}$$

for Geernaert, et al. the result is

$$\bar{H}_{1/3} = 1.022 \cdot 10^{-2} \, (\bar{U}(10))^2 \, (0.0577 + 0.0847 \, \bar{U}(10)) \tag{29}$$

and for Smith and Banke

$$\bar{H}_{1/3} = 1.022 \cdot 10^{-2} \, (\bar{U}(10))^2 \, (0.63 + 0.066 \, \bar{U}(10)) \tag{30}$$

The various curves that result when the drag coefficients given in Table 1 are each used to calculate wave height versus $\bar{U}(10)$ are all shown in Fig. 6. Changing the large fetch value of F^* could raise or lower each of the curves. In contrast to the dependence of the wave height for a fully developed sea on the square of the wind speed, a dependence on friction velocity implies a square plus cubic dependence on the wind.

For the Ewing and Laing data set, finding B_1 and B_2 for Eq. (31) could provide a new relationship for $\bar{H}_{1/3}$ versus $\bar{U}(10)$ and perhaps another candidate for a drag coefficient.

$$\bar{H}_{1/3} = B_1(\bar{U}(10))^2 + B_2(\bar{U}(10))^3 \tag{31}$$

The above result based on 33 spectra from Ewing and Laing is suggestive of an alternative for fully developed seas. More data based on similar constraints from data buoys would be useful.

The need to choose an appropriate relation for u_* as a function of $\bar{U}(10)$ along with the parameterization of F^* versus X^* provides many different choices for $u_*(\bar{U}(10))$ and $E^*(X^*)$. For Fig. 6, the result of these choices is that for low wind speeds fully developed seas will be lower and for high wind speeds fully developed seas will be higher than they would be if the wave model were parameterized in term of $\bar{U}(10)$.

Any perceived improvement due to the use of u_* instead of $\bar{U}(10)$, may be in part fortuitous. For example, the winds used to drive the GSOWM model were shown by Zambresky (1987) to be too high when they ought to be low and too low when they ought to be high when compared to data buoy winds. If the winds were more nearly correct so that they would agree better with the data buoy winds, then the GSOWM waves would be lower for low winds and higher for high winds.

If the winds used to drive the WAMDI wave model are too high for low winds and too low for high winds, the use of u_* according to Fig. 6 would counteract the effect of an incorrect wind field and give the appearance of an improved model.

For the spectra proposed by Pierson and Moskowitz, the length of time the winds were averaged by the weather ships is not known. The waves were recorded by the Tucker Shipborne wave recorder. The DB1 averaged the winds for 10 minutes and recorded and processed the wave data in a way similar to the one described by Pierson (1987). Differences in anemometer exposure, averaging times, and wave recorder design and calibration are all possible explanations for this lack of agreement in addition to those explanations of Ewing and Laing. The calibration of some of the more recent accelerometers used on buoys has been known to drift toward lower recorded values (personal communication L. Baer, 1987).

In passing, it should be noted that neither the SOWM not the GSOWM uses the spectrum proposed by Pierson and Moskowitz (1964). Nor is the significant wave height exactly proportional to the square of the 19.5 m wind for fully developed seas. For details, see Pierson (1982).

3.5. IMPROVING WINDS FROM DATA BUOYS

Data buoys, as of early 1988, average the wind for 8.5 to 10 minutes, typically, once an hour and report this and other data via a geostationary satellite. Three deep water buoys are now reporting ten minute averages of the wind for fifty minutes of each hour and an 8.5 minute average to allow time for data transmission. Averages of these averages for any increment of ten minutes is then possible. From Fig. 3, the various scales of averaging can be estimated visually. Plans are under way to change the deep water National Data Buoys of the United States to this mode of operation. From the above material, this increased data rate and the ability to average the winds for times longer than 8.5 minutes will be essential for the validation of winds obtained from scatterometer data.

3.6. MODERN SYSTEMS

Since SKYLAB, from 1974 to 1989, there have been fabulous advances in the ability to obtain, process and analyze meteorological and oceanographic data. Digitized data and high speed computers have made it possible to carry out in a year or so what would have required a much longer time twenty or thirty years ago. SKYLAB backscatter data totaled about 1.5 orbits around the Earth. SEASAT was continuous for almost 100 days with over thirteen orbits per day. GEOSAT was orbited on March 12, 1985 and as of March 1989 was still functioning.

The National Data Buoys of the United States and similar buoys for other nations have replaced the visually read and manually recorded reports, prior to transmission, from the weather ships by completely automated data recording and transmission for nearly all of the quantities reported by a ship. The ability to record data digitally and process it by a computer inexpensively is increasing rapidly such that data rates, especially for remote sensing, at present would have been unimaginable for the recent past.

With the increased volume of data from moored buoys and polar orbiting spacecraft and the ability to analyze this large volume of data effectively, so as to make progress in defining the winds and the waves by remote sensing methods it will be necessary to define and understand the effects of randomness in all scales of measurement and to treat these effects of randomness in the analysis and interpretation of both data and models. Only by doing this can be presently inconsistent results such as those illustrated in previous sections be reconciled. It may even be necessary to modify many of the present theories on waves and winds.

It is, however, possible at present to obtain improved measurements of the winds and the waves by conventional means, by means of aircraft and by means of remote sensing. These data can be used to find an improved model for recovering winds from backscatter data for those oceanic areas where there are no buoys, where aircraft will not fly and that are not in an altimeter foot print.

4. Backscatter Data

4.1. THE RANDOMNESS OF BACKSCATTER DATA

The randomness of the winds and the waves and of the quantities and functions needed to define the winds and the waves is a feature that must be better understood and better controlled for the intercomparison of winds, waves and backscatter data. The normalized radar backscattering cross section σ^0 is itself a randomly varying quantity.

A scatterometer, which is a term coined by R. K. Moore, is a unique kind of radar that makes possible the estimation of backscatter values that are very low. Nevertheless, backscatter from the wavy sea surface is incoherent and noise-like. The recorded backscatter values vary randomly about an expected value, σ_M^0, according to Eq. (32).

$$f(\sigma^0) = \frac{\exp\left(-\frac{1}{2}(\sigma^0 - \sigma_M^0)^2/(\alpha(\sigma_M^0)^2 + \beta\sigma_M^0 + \gamma)\right)}{(2\pi)^{1/2}((\alpha(\sigma_M^0)^2 + \beta\,\sigma_M^0 + \gamma)^{1/2}} \tag{32}$$

In (32), α,β and γ are design features of the particular scatterometer. The quantity, σ_M^0 is the model that for a given polarization relates the winds to the waves and then to the backscatter. The probability density function for σ^0 is an interesting distribution because there is only one unknown population parameter. If the expected value, σ_M^0 is known, the variance is known and does not have to be determined independently as for most samples for a normal distribution. If σ_M^0 could be expressed as a function of the time averaged wind speed and direction for each polarization, then the effect of sampling variability can be simulated by Monte Carlo methods. The transformation given by Eq. (33) changes Eq. (32) to Eq. (33).

$$t = (\sigma^0 - \sigma_M^0)/(\alpha(\sigma_M^0)^2 + \beta\,\sigma_M^0 + \gamma)^{1/2} \tag{33}$$

$$f(t) = (2\pi)^{-1/2}\exp(-t^2/2) \tag{34}$$

In (34), t is a random variable with a zero expected value and a unit standard deviation.

A Monte Carlo simulation of backscatter values can be generated by picking values of t from (34) and generating σ^0 from Eq. (35).

$$\sigma^0 = \sigma_M^0 + t(\alpha(\sigma_M^0)^2 + \beta\,\sigma_M^0 + \gamma)^{1/2} \tag{35}$$

Pierson (1989) has treated the probabilistic and statistical properties of Eq. (32) especially in terms of the design features of a new scatterometer.

4.2. SASS WINDS

The SEASAT A scatterometer (SASS) data on backscatter were converted directly to vector winds without any consideration of refinements that could be introduced by considering those properties of the waves that could affect backscatter. It was assumed that the backscatter from the sea surface could be related to the wind speed and direction by Eqs. (36a) and (36b) where $\overline{U}(19.5)$ is the effective neutral wind at 19.5 meters above the ocean surface

(Schroeder, et al. 1982). The functions of χ in these equations are even functions over $-180° < \chi < 180°$ where $\chi = 0$ corresponds to upwind and θ is the incidence angle.

$$\sigma^0_{VV\ \mathrm{dB}} = A_{VV}(\chi,\theta) + B_{VV}(\chi,\theta) \log_{10} \overline{U}(19.5) \tag{36a}$$

$$\sigma^0_{HH\ \mathrm{dB}} = A_{HH}(\chi,\theta) + B_{HH}(\chi,\theta) \log_{10} (\overline{U}(19.5) \tag{36b}$$

With ambiguities removed, the resulting wind fields proved to be of great value. The essentially global, over water, coverage during 12 hours is illustrated by Woiceshyn, et al. (1985). The referenced figure shows all of the major features of the winds over the oceans, including the equatorial doldrums, the trade winds, two typhoons, the horse latitudes and extratropical cyclones.

For the model given by Eqs. (36) and the wind fields that resulted therefrom, it was difficult to demonstrate the accuracy of the winds over the full range of over ocean conditions in terms of comparisons with conventionally measured winds from either ships or data buoys, in part because of the too brief averages for the conventional winds. The JASIN program provided the best comparison data. Two special issues of the Journal of Geophysical Research (Vol. 87, C5, 1982 and Vol. 88, C3, 1983) show these results and Schroeder, et al. (1982) define the functions in (36a) and (36b).

Many, many studies have been published based on the winds recovered from the SASS data. For those parts of the world ocean where other kinds of data are sparse, the SASS data seemed so realistic that many useful results were obtained. Comparisons with the few and far between conventional reports, even if made, would be inconclusive because of the poor quality of the conventional data.

For these reasons, a number of years passed before Woiceshyn, et al. (1986) were able to use the winds recovered from the SASS to show that the wind speeds based on the vertically polarized backscatter values and the horizontally polarized backscatter values were inconsistent one with the other so that at least part of the model had to be incorrect. Woiceshyn, et al. (1987) then showed that there were also systematic errors in wind direction recoveries.

4.3. RECENT MODELS

It is thus necessary to find a better relationship between the vector wind and the backscatter. Perhaps it could be of the form given by Eqs. (37a) and (37b) where any implied power law has been removed. Straight lines on a double logarithmic plot are a rather limited class of possible functions for the description of backscatter dependence on the winds.

$$\sigma^0_{VV} = \sigma^0_{VV}(\overline{U}(19.5),\ \chi,\ \theta) \tag{37a}$$

$$\sigma^0_{HH} = \sigma^0_{HH}(\overline{U}(19.5),\ \chi,\ \theta) \tag{37b}$$

For relatively simple forms such as Eqs. (36a and b), and with the aspect angle dependence represented by the first few terms of a Fourier series, with the second harmonic dominant, it is possible to use cut and try methods that by-pass an understanding of the dependence of the backscatter on the waves that were generated by the winds. Subsequent to SEASAT, attempts have been made to define the vector wave number spectrum of the waves for fully developed seas as a function of wind speed and to use the two scale Bragg scattering theory, plus specular reflection, as reviewed by Valenzuela (1978), to calculate the backscatter. Models along these lines are those of Durden and Vesecky (1985), Iwata (1985), Plant (1986) and Donelan and Pierson (1987). These various models differ from one another in the level of approximation used to evaluate the theory and in the wavenumber spectrum that is used. Unpublished results for still other models are being circulated. The models are for fully developed seas.

Let $\Phi(k, \chi; \vec{U}(z_1))$ represent the wavenumber spectrum in polar coordinates, with a differential element $dk\ k\ d\chi$ where k is the wavenumber, χ is the direction toward which the spectral component is traveling and $\vec{U}(z_1)$ is the effective mean vector wind at the height, z_1.

For a two scale theory, the specular slopes need to be found as in Eq. (38) for the x (or upwind-downwind) component.

$$S_x(k_1, \overline{U}(z_1)) = \int_0^{k_1} \int_{-\pi}^{\pi} k^2 \cos^2 \chi\ \phi(k,\chi; \overline{U}(z_1))\ dk\ k\ d\chi \tag{38}$$

and, with cos χ replaced by sin χ for a similar equation for S_y. These two slopes then are used to calculate specular reflection which may or may not be important for incidence angles as high as 30°.

That portion of the wavy surface with the high wavenumber Bragg waves on it is also tilted relative to a level surface so that two other slopes with k_1 above, changed to $k_2(\theta)$ such as $S'_x(k_2,\overline{U}(z_1))$ and $S'_y(k_2,\overline{U}(z_2))$ are needed.

With the effect of tilt accounted for by a probability density function involving S'_x and S'_y the proper wavenumber is selected from the spectrum to contribute to the Bragg scattering according to

$$k = 2k_0 \sin \theta \tag{39}$$

where θ is the incidence angle and k_0 is the radar wavenumber. For modeling the wavenumber spectrum, data in the frequency domain are at times transformed to wavenumbers by means of the linear dispersion relation as in Eq. (40).

$$\omega^2 = gk + \tau k^3/\rho \tag{40}$$

where τ is surface tension and ρ is density. The currents at the water surface for the longer waves make the apparent frequencies of the shorter waves quite different from the true values and give false wavenumbers.

If the waves for high wavenumbers have been started by some initiation process, the further growth to saturation according to Donelan and Pierson (1987) would be at first exponential, subject to the constraint that the right hand side of Eq. (41) is greater than zero and that $\overline{U}(\pi/k)$ be greater than $C(k)$ as defined by Eq. (42). Also $\alpha = \alpha(\tau k^2/\rho_w g)$ and $n = n(\tau k^2/\rho_w g)$.

$$\frac{\partial \phi(k,\overline{\chi})}{\omega \phi(k\overline{\chi})\partial t} = 0.194 \frac{\rho_a}{\rho_w} \left[\frac{\overline{U}(\pi/k)}{C(k)} - 1 \right]^2 - 4\nu k/C(k)$$

$$-\alpha(k^4 \Phi(k,\overline{\chi}))^n \tag{41}$$

$$C(k) = (gk^{-1} + \tau k/\rho)^{1/2} \tag{42}$$

The first term on the right hand side represents the growth of the spectrum at the various wavenumbers in the upwind direction. The wind that causes a given spectral component to grow is the wind at a height of π/k above the wavy surface, and since $k = 2\pi/\lambda$ this is the same as the height, $\lambda/2$.

The second term represents the effect of molecular viscosity. Monochromatic waves with wavenumbers corresponding to wavelengths less than 5 cm, or so, would die down to nothing in just a few minutes (Kinsman; 1965, 1984, pg. 501).

The third term is important if the wave spectrum, $\Phi(k,\bar{\chi})$, is high enough to result in micro breaking effects dominantly affecting the same wavenumber as the other terms. Both α and n have to be determined from backscatter data. They are related to the behavior of the group velocity of waves with varying wavenumbers.

For very low winds, the wind at π/k may not be strong enough to generate waves since for some wind speeds Eq. (43) may apply in that the effect of viscosity may overcome the generation term.

$$0.194 \frac{\rho_a}{\rho_w} \left(\frac{\bar{U}(\pi/k)}{C(k)} - 1 \right)^2 - 4\nu k / C(k) < 0 \tag{43}$$

There is some threshold wind speed, $\bar{U}(\pi/k)$, for a given k that changes the inequality to an equality in (43). This wind is given by Eq. (44). The viscosity, ν, of sea water is mostly a function of the temperature of the water. As the temperature varies from 30° to 0° C, the viscosity almost doubles.

$$\bar{U}(\pi/k) = C(k) + 2(\nu k C(k) \rho_w / 0.194 \rho_a)^{1/2} \tag{44}$$

The wind speed at π/k can also be expressed in terms of the wind at 10 m if any one of the drag coefficients in Table 1 is used with A and B representing the appropriate constants.

$$\bar{U}(\pi/k) = (\bar{U}(10) \left[1 - \frac{(A + B\,\bar{U}(10))^{1/2}}{\varkappa} (\ln k - \ln(\pi/10)) \right] \tag{45}$$

The application of (45) to the real ocean environment is cautioned since the log-profile does not necessarily apply immediately above a wavy surface with periodic surface currents induced by orbital velocities.

After a brief time, the last term on the right hand side of Eq. (41) becomes so large that the entire right hand side balances to zero. The value of a slice through the wavenumber spectrum at upwind is then given by Eq. (46),

$$\Phi(k,\bar{\chi}) = k^{-4} \left[\frac{0.194}{\alpha} \frac{\rho_a}{\rho_w} \left(\frac{\bar{U}(\pi/k)}{C(k)} - 1 \right)^2 - \frac{4\nu k}{\alpha C(k)} \right]^{1/n} \tag{46}$$

and according to Donelan and Pierson (1987) the full polar wavenumber spectrum would then be Eq. (47) for high wavenumbers.

$$\Phi(k,\chi) = \Phi(k,\bar{\chi})\,\mathrm{sech}^2(h_1(\chi - \bar{\chi})) \tag{47}$$

Two important features of the high-wavenumber spectrum of capillary waves are illustrated by Eqs. (44), (45) and (46). Unless $\bar{U}(\pi/k)$ is greater than the right hand side of Eq. (44), the spectrum for that wavenumber and viscosity is zero. For a given wavenumber, the wind has to be stronger for cold water than for warm water for the spectrum to be non zero. For wind speeds below a certain water temperature dependent value there will be no Bragg waves generated according to the model, and since the wave slopes for a fully developed sea for these light winds will also be small, there may be no detectable backscatter. The omitted curves for light winds in Fig. 6 suggest that very light winds may be incapable of generating any waves at all.

From Eq. (45) as $\bar{U}(10)$ is increased, $\bar{U}(\pi/k)$ will increase, but for still higher winds at 10 m, the value of $\bar{U}(\pi/k)$ can decrease. The shear of the wind above the viscous boundary layer can be so strong for high winds that the wind decreases at π/k while increasing at 10 m. Thus Bragg backscatter may saturate and then decrease at higher wind speed.

The slopes of the tilting gravity waves are found from the spectrum of Donelan, et al. (1985) and the waves defined by (46) are tilted back and forth by these slopes. Also the wind is varied about its mean value by means of a normal distribution with an assumed standard deviation. A triple integration by computer is required over two slope components and the wind variability to obtain, for a given incidence angle, polarization, average wind speed and direction, a value for the two scale Bragg backscatter. To this must, at least, be added to contribution from specular reflection for the lower incidence angles.

It follows that when compared to some best fit power law, as in Eqs. (36), the actual backscatter will be higher for some intermediate range of wind speeds, $\bar{U}_1(10) \le \bar{U}(10) \le \bar{U}_2(10)$, and lower for winds speeds both below, $\bar{U}(10) < \bar{U}_1(10)$, and above, $\bar{U}_2(10) < \bar{U}(10)$, this range.

This particular model was fitted to backscatter data at K_u band and which have documented by Schroeder, et al. (1984). Additional data obtained during FASINEX (Geernaert, et al. 1988) support the concept of backscatter dependence on the effective neutral wind, which defines both $\bar{U}(\pi/k)$ and u_*, given a correct drag coefficient. Laboratory data (Keller and Plant, 1988) partially support the result that for low wind speeds the backscatter could be hardly detectable.

Should it ever be shown that the low wavenumber part of the gravity wave spectrum is related to winds at increasing elevations with decreasing gravity wavenumber, then still another dimensionless form for a fully developed sea could be Eq. (48). The sheltering hypothesis of Jeffreys (1924, 1925) is quite different from the present models for the generation of waves by wind.

$$g\bar{H}_{1/3}/(\bar{U}(z(k)))^2 = C_4 \tag{48}$$

In the model of Donelan and Pierson just described, $\bar{U}(\pi/k)$ enters the theory through the wave growth rate. Another common formulation of the growth rate which gives nearly the same numerical values relates wave growth to the wind stress, ρu_*^2, on the surface. When this formulation is used, backscattering cross-sections turn out to depend on u_* instead of $\bar{U}(\pi/k)$ (Plant, 1986). Since the latter measure of wind is difficult to measure and must be inferred from measurements well above the surface, a dependence on the measurable u_* is somewhat more satisfying. Which of these formulations, if either, more accurately reflects reality is a matter of ongoing research at the present time.

4.4. SLOPE EFFECTS

Conditions that represent pure fully developed seas seem to be rather rare. Equation (47), or something like it, may still represent the high wavenumber part of the spectrum. The slope components for the gravity wave part of the wave spectrum would be different because the spectrum in equations like Eq. (38) would not be for a fully developed sea. Pitch-roll buoys as in Steele, et al. (1985) and Ewing and Laing (1987) and the ROWS described by Jackson, et al. (1985a, 1985b) obtained some information on wave slope. Buoys do not do well for those wavenumbers corresponding to waves with lengths less than about twice the buoy diameter. Unfortunately, this part of the spectrum is not well covered by measurements.

Weissman (1987) use the Large and Pond values for C_{DN} and data on total wave slope for waves longer than 4 m for stable conditions. His Fig. 4 shows backscatter versus friction velocity with slope as a parameter. The effect of slope is pronounced. For unstable conditions, the data show the role of slope to be relatively small.

If it is assumed that the high wavenumber part of the spectrum is in equilibrium with the local wind, then the changes in slope of the gravity waves could cause the backscatter to vary. The gravity waves change as a function of increasing fetch for a constant wind over the fetch as in Eq. (19) and according to Fig. 5. Ross and Jones (1978) reported both the winds and the backscatter for one experiment as a function of fetch. There was no realistic change in either the backscatter or the winds as the fetch varied.

5. Air-Sea Temperatures, Humidity and Atmospheric Soundings

5.1. MEASUREMENTS OF AIR-SEA TEMPERATURES AND WATER VAPOR CONTENT

The preceding material has argued that the "properly defined and measured wind" for relating backscatter to the vector wind is the effective neutral wind at some selected height based for comparison purposes on a measured vector wind at some height that is from a long enough, but not too long, average and that has been corrected by a model for the effects of atmospheric stability and the variation of wind with height to the chosen height. The water temperature is needed to determine the viscosity as in (47), but the wind at π/k cannot be found from the wind at some typical height, z_1, unless the air temperature and the water vapor content (or relative humidity) are known.

The sea surface temperatures are well enough known globally by means of VHRR from remote sensing and other data to account for viscosity effects, but need to be more accurate for air-sea temperature differences. The presently distributed analysis field produced by the Marine Products Branch of the National Meteorological Center of NOAA is a blended and averaged sea surface temperature field over a period of about two weeks based on infrared data, reports from ships of opportunity and data buoys. Even for this scale of averaging, the gradients in temperature are very strong, for example, near the Gulf Stream-Labrador Current and Kiro Shio-Oya Shio. The day to day location of the cold wall boundary, and the meanders of the Gulf Stream are lost as well as cold and warm core rings. A 2° to 4° error for viscosity effects is unimportant compared to the other sources of error.

Data buoys presently provide water and air temperature, but not humidity, so that validation is possible only if ship reports of dew point are available. To use the recovered effective neutral winds to obtain the actual winds that would be measured at some height relatively close to the water requires data that are not presently available over the ocean at a sufficient density and accuracy. The development of marine boundary layer models that would combine prediction and specification techniques to define the air temperature and the water vapor content for the swaths to be covered by the next generation of scatterometers would provide a means to go back and forth from effective neutral winds to actual winds. A recent National Data Buoy Technical Bulletin contains articles that describe programs for the measurement of precipitation (E. Michelena) and humidity (B. Hundermark).

5.2. TURBULENCE EFFECTS

For low and moderate winds, turbulence is quite different in stable air compared to unstable air and still more different during convective activity. Figure 1, for example, was used for its shock value: that part of the Pacific is likely to be dominated by convective activity. The spectra of Pierson (1983) did not give correct results for the more usual situations. Whether or not the different degrees of turbulence in stable versus unstable air produce different spectra for the turbulence and whether or not the turbulent fluctuations for a fixed effective neutral wind and value of u_* can produce changes in the Bragg wavenumber spectrum has yet to be investigated by measurements over water systematically. Tchen and Pierson (1988) outline some models along these lines that show the differences in the turbulent spectra for stable and unstable stratifications. Tchen, et al. (1985) have demonstrated a spectral gap in the shear turbulence of a stably stratified atmosphere.

5.3. ATMOSPHERIC SOUNDINGS

It may be necessary to have some idea of the correct wind direction if ambiguities in direction remain after the development of the next generation of scatterometers. Remote soundings combined with surface winds can permit the definition of the sea level atmospheric field, the effects of the thermal wind on the variation of wind with height and the wind at the gradient wind level. The thermal wind can influence the angle at which the wind crosses the sea level isobars.

6. Recapitulation

6.1. WIND MEASUREMENTS AND MODELS

Comparisons of even the magnitude of the vector wind by means of all sorts of scatter plots for either of the different methods of measurement or measurements by different methods and boundary layer models all lead to substantial scatter in the plotted points. The situation has not improved by much since 1974.

There are many published models for the relationship between the friction velocity and the time averaged wind speed at 10 m, four of which have been illustrated. It is quite possible that there is no unique relationship between the friction velocity and the mean wind at 10 m because of different wave conditions as suggested by Donelan (1982), Geernaert, et al. (1987) and many others. There is still the possibility of a unique neutral drag coefficient for nearly fully developed seas that would apply most of the time over most of the ocean.

The environmental parameters that describe the time averaged vector wind for various atmospheric stabilities, the turbulent scales and spectra for these winds, the shear of the wind with height, air temperature, water temperature and water vapor content are all needed to define the winds over the ocean that generate the waves that cause the backscatter. These quantities are not determined well enough by most present conventional methods to eliminate a major source of the discrepancies between remotely sensed and conventionally measured values. Airborne measurements of mean and turbulent winds and essentially continuous measurements from buoys may provide sufficient data so that present-day systems can clarify the situation.

6.2. WAVE MEASUREMENTS AND MODELS

The proposed spectral form for fully developed wind seas based on the similarity theory of S. A. Kitaigorodskii (Pierson and Moskowitz, 1964), now often referred to as the PM spectrum, yields an equation for the significant wave height as a function of the time averaged wind speed that is not sensitive to the choice of the height at which the wind speed is measured. An appropriate equation can be given for a wind at either 10 or 19.5 m. The spectra obtained by Donelan, et al. (1985), the DHH spectra, also results in a significant wave height that varies as the square of the 10 m wind. This wave height is considerably larger than that of the PM spectrum (see Donelan and Pierson, 1987), so that the DHH elevation spectrum lies above the PM spectrum. The DHH spectrum may be too high (personal communication, M. Donelan 1987), but insofar as radar backscatter is concerned, slope spectra are more relevant. Modifications of the area under the elevation spectrum while increasing the intermediate and higher wavenumber part of the spectrum could produce all sorts of different slope spectra for a large range of elevation spectra.

It is important to be able to obtain reliable bounds on estimates of wave height and wave slopes and on the wavenumber and frequency spectra of wave elevation and wave slopes. Part of the present uncertainty may be the result of the scatter in the data caused by the combined uncertainties in the estimates of elevation and slope, the vector wind, and the other required atmospheric parameters.

Wave measurement systems on aircraft provide the potential for obtaining more reliable and more stable wave data. As one example, the data points of Walsh from Fig. 5 (if correct) for a fully developed sea at a nominal wind of 11 ms^{-1} fits the curve defined by the X's in Fig. 6. To fit the other curve, the wind would have had to have been about 25 ms^{-1}. Intercomparisons of wave data from buoys and from aircraft could improve the calibration of the various systems.

6.3. BACKSCATTER DATA AND BRAGG SCATTERING THEORY

Only one of a number of models for the backscatter at various wavenumbers according to Bragg scattering theory has been described. There may well be several more as time passes, and the one described may need revision. Attempts to check these various models and to intercompare them by means of backscatter, wind and wave data are in progress. One purpose of such models could be to provide guidance toward an algorithm to be used to relate winds to waves and to backscatter. There are numerous laboratory studies and theoretical analyses that suggest that specular reflection and two scale Bragg scattering theory are not adequate.

6.4. EXPERIMENTAL REPEATABILITY

A foundation of the scientific method is that when an experiment results in a new model, or description, of some physical fact and is reported by an investigator, another investigator can repeat the experiment and obtain the same result, within experimental error. So far in this chapter, those experiments and data analyses that have been described have not demonstrated this basic, and essential, feature. Some of the possible reasons for this unfortunate state of affairs have been given.

The rapid evolution of instrumentation, data logging and data analysis systems accounts in part for the differences that have resulted. Some way must be found to obtain definitive results when the random waves, turbulent winds, and the randomly varying backscatter are studied by conventional means; this process is an important precursor to obtaining the best possible results from remote sensing.

7. Other Wave Effects that Affect Backscatter

7.1. INTRODUCTION

To begin this part of the chapter, the ''Proceedings of the Symposium, Measuring Ocean Waves from Space,'' cited a number of times above, contained a panel discussion which questions were asked by the attendees and answered by panel members. A part of a comment by Merrill Skolnick is quoted below:

''My second point, perhaps more controversial, has to do with Bragg scatterers [SIC]. Bragg scattering is the theory of choice at the moment. It has two very nice things going for it. First, it explains very well the sea-scatter effects one sees with high-frequency radar, that is, long-wave radar looking at water waves on the order of tens of meters wavelength. There is no question that Bragg scattering is a nice theory. People can work with it. But when one looks at the higher microwave frequencies (an L-band radar might be high enough), there is nothing that resembles Bragg scattering.

''One wonders why people are not more concerned about the sea spikes that Phillips mentioned on the first day. Little waves do not ride on big waves. Pictures of capillary waves or small waves riding on large waves always show a lot of discontinuities comparable to a wavelength. A discontinuity can be expanded into a Fourier series, and a component of that discontinuity will have a Bragg-scattering component. But observations of short, high-resolution microwaves show that the sea spikes last for a few seconds. They are nonstationary in time and they are discontinuous in space. Again, Bragg-scattering theory might give satisfactory answers, but it does not reflect the real world.''

A reply to this comment by me follows:

''I concur with everything you have said, but I do not think it is a particularly difficult problem to solve. There is some good work by Lyzenga, by Kwoh and Lake, and also by Mike Banner in Australia on all those effects. They can be combined with Bragg scattering and specular backscatter in a logical way to help explain radar images of the ocean. So things will probably soon become clear. For example, many scientists are working to perform the kind of experiments that will help resolve some of the questions that you have raised. I am sure everybody is aware of them and has some pretty good ideas of what to do about them."

The comment may be a bit too pessimistic and the reply somewhat overoptimistic (based on just having had the paper by Donelan and Pierson, 1987, accepted).

The warning by Skolnick is clear, however. There is enough uncertainty in the form of the wavenumber spectrum for L, C, X, K_u and K_a band backscatter so that a given specular reflection plus two scale Bragg scattering model has to be tuned (tuned: adjusted to fit observations of backscatter) so as to provide agreement between the wind and the backscatter. If effects other than Bragg scattering for a two scale model are present, the tuned wavenumber spectra (in the example of Donelan and Pierson (1987) the α and the n) may be too high because effects other than Bragg effects are interpreted as Bragg effects.

There are wave properties other than specular and two-scale Bragg scattering that affect the total backscatter. These are even less well understood than the Bragg wavenumber spectrum plus gravity wave slope spectra, but there are both over ocean and laboratory measurements that show these effects.

7.2. LABORATORY EXPERIMENTS

In a laboratory, specific wave conditions can be created and either backscattered power or the normalized backscattering cross section can be measured. Experiments by Wright (1966), Kwoh and Lake (1984) and Banner and Fooks (1985) have respectively isolated backscatter from Bragg waves, wedges and breaking waves. This is difficult to do with oceanic measurements, but there may be ways to isolate possible contributions to backscatter from different wave features that produce backscatter.

The experiment by Wright (1966) studied backscatter from Bragg waves and showed the predicted relationship between the wave amplitude and the backscattered power. The waves appear to have been low enough so that wedges and breaking waves did not happen.

For the experiment by Kwoh and Lake (1984), the waves were produced by a mechanical wave generator. The wave generator was adjusted in such a way that wedge-like wave forms of various degrees of sharpness occurred at different times in the radar test section. The waves were essentially long crested across the test section. A complication arose because of backscatter from parasitic capillary waves that formed the advancing front of the wedges as predicted theoretically by Longuet Higgins (1963). This was eliminated by using data from wedges where this did not happen. Radar scattering into the entire upper half plane, including backscatter to the radar transmitter, was measured for radar waves incident at 40°, 55° and 67.7° for both polarizations and compared to wedge scattering theory (Lyzenga, et al. 1983, Kwoh and Lake, 1984, and Wetzel, this volume). The theory and the data agreed well over the upper half plane. The variation of the backscatter versus polarization and incidence angle was also obtained and shown as a scatter plot. Although there was considerable variability, possibly due to the variable sharpness of different wedges, there were systematic differences in polarization and the values of σ^0 for the different incidence angles. The wavy surface as the waves propagated down the tank had no small scale roughness elements that could produce backscatter so that breaking waves and Bragg scattering waves were not present.

Banner and Fooks (1985) found a way to produce a wave-like form on a moving current that stood still in the test section and that had many of the features of a spilling breaker. The simulated breaking wave had two major features. One was that the down spilling water formed something like a hydraulic jump with a very sharp tilted angle at the toe of the jump.

The other was that the top part of the spilling breaker from the crest to the breaker jump was covered by undulations and oscillations with lengths comparable to wind generated Bragg scattering waves. These appear to be somewhat isotropic, at least in the photograph of the breaking wave. If these undulations are perhaps to be described as waves, the hydrodynamics would be completely different from those of the hydrodynamics of wind generated waves as in Longuet-Higgins and Turner (1974).

Backscatter came from all parts of this wave form, and it was difficult to separate that part from the jump from that part from the wave-like undulations. The data did however provide values similar to those obtained theoretically by Wetzel (1986) for backscatter from the jump in a breaker. There were no Bragg waves and no wedges during this investigation.

7.3. OCEAN EXPERIMENTS

Many experiments from oceanic towers and by means of aircraft have been carried out in the recent past. Results from these experiments have not yet in general appeared in the literature, but they can be expected to appear in the near future.

R. K. Moore (personal communication) has done an experiment concerned with the sea spikes mentioned above. The time histories of the backscattered power for various look directions relative to upwind, incidence angle near 45°, wind speeds from 3 to 15 ms^{-1} and non-fully developed seas from 1 to 3 m were available. Large, but brief, spikes were seen intermittently in the time history. The difficulty is in the "spike-no spike" threshold decision, which is arbitrary to a certain extent. For the threshold definition of a spike that was used, the contribution from the spikes increased the backscatter by anywhere from 0.5 to 1.0 dB.

Thus if "non sea spike" backscatter is two scale Bragg, then the contribution from the spikes, which could be from wedges and spilling breakers is from 12% to 25% of the Bragg part for these experiments. The dependence on non-fully developed effects of the number and dimensions of wedges and of the number of and dimensions of spilling breakers per unit area is not known.

Donelan and Pierson (1987) derived a definite form for the wavenumber spectrum as in Eq. (47), and Kahma and Donelan (1988) provided experimental support for the dependence on viscosity. Research to determine aspects of this part of the full wavenumber spectrum by means of stereo photographs of small areas of a large wave surface so as to resolve very short wavelengths has been described by Shemdin and Huang (1988) and Banner, et al. (1988). Data reduction is tedious and a considerable number of stereo pairs needed processing before some results were obtained. Research along these lines should provide data on the actual form of the high wavenumber spectrum so that "tuning" will not be needed. The results of these two different investigations do not agree with each other nor with the spectrum described by Donelan and Pierson (1987).

Oil slick experiments began with Cox and Munk (1954) long before the era of remote sensing. An experiment of creating artificial oil slicks of oleyl alcohol and imaging them with an X-band real aperture radar for swaths with the radar looking upwind, crosswind and downwind has been done by Kozu, et al. (1987). The slicks formed three perfect rectangles of different sizes at different ranges. The slick-no slick difference in the received signal power was about 7 dB for upwind, downwind for an X-band radar operating with vertical polarizations and 6.6° incidence angle. A dominant effect of an oil slick would be to eliminate the X-band Bragg scattering waves. Short wave attenuation has been studied by Bock (1987).

7.4. WHITECAPS, FOAM, AIR BUBBLES, RAIN AND SPRAY

Scientists who look at the sea, as Kinsman (1965) requested, tend to see the sea through mental filters. These filters at times seem to be rather narrow-band so that the full spectrum of properties of the seaway is not applied to the study of a particular problem such as the one under analysis. Although (rearranging the above comment by Skolnick) "Bragg-scattering theory might give satisfactory answers," unless whatever corrections are needed for other effects are incorporated in a model, such a theory may "not reflect the real world."

Two recent books reflect how different scientists see the sea. One is "The Ocean Surface; Wave Breaking, Turbulent Mixing and Radio Probing," edited by Y. Toba and H. Mitsuyasu. The other is "Oceanic Whitecaps and their Role in Air-Sea Exchange Processes," edited by E. C. Monahan and Gearoid MacNiocaill. Very little of either book is concerned with microwave radar backscatter from other than wedges and Bragg waves, compared with air-sea interaction effects, waves forecasting and the subject matter of this section. It is not the purpose of this chapter to discuss all of the articles in these two books. Some of these articles are briefer versions of papers published elsewhere.

The chapter by Fairall in this text discusses whitecaps in terms of air-sea interaction, but the question remains as to the effect of all of the consequences of breaking waves and falling drops on radar backscatter. Those who study waves tend to think in terms of a moving surface, $\eta(x, y, t)$, such that above this surface there is air and below it there is water. Further favorite assumptions are that the water is a perfect fluid without vorticity (turbulence) and (usually) viscosity.

Most present models for the wavy surface are, at most, third or fourth order perturbation expansions of a linear model as a starting point with free surface boundary conditions that are often improperly stated (my opinion) for the higher order corrections. A notable exception is the work of Hui and Tenti (1982, 1985) and Lamb, et al. (1987). The latter reference is the first theory for how a wavemaker generates waves and describes the propagation of these waves once generated.

Drennan (1989) has used the results of Hui and Tenti (1982, 1985) to solve the Stokes water wave problem. This dissertation is a major contribution to classical hydrodynamics to the extent that this one hundred and nine year old problem can be considered to have been completely solved. It is noteworthy that the method proposed by Stokes does not converge for the steepest waves and that a completely new method was needed.

Actual wind generated waves are as nonlinear as they can get. The equations that are used to describe waves cannot describe the free surface for a breaking event. The whitecaps that form, even for a spilling breaker, are a mixture of sea water and air bubbles. The foam left behind by a transient breaking wave lasts for a considerable time as a white patch on the water as the bubbles rise to the surface. Drops of water can break away from the whitecap and travel through the air blown by the wind. As the rising bubble breaks through the sea surface a jet of water forms from which drops separate that are blown by the wind. These drops can fall back to the water, or they can evaporate into the air leaving behind a sea salt nucleus. (See also the chapter by Fairall; this book.)

Those aspects of whitecaps, foam, bubbles and spray of interest in this section are mostly found in the book edited by Monahan and MacNiocaill. Of particular interest (to me) are the works of Thorpe (1986), Resch (1986) and O'Muircheartaigh and Monahan (1966).

Thorpe (1986) illustrated the detection of backscattered sound from undersea bubbles that result from waves breaking in water 34 m deep. His main conclusions are quoted below:

"The main conclusions of this study using sidescan sonar which bear on the properties of bubbles are described below (see Thorpe and Hall, 1983).

"(a) Separate bubble clouds are created by individual waves which break, often several in succession, in groups of waves which propagate downwind.

"(b) The breaking crests are seldom more than a few meters long (e.g. in the 205 deg direction in Fig. 3) Waves generally continue to break as they advance over a greater distance so that, at formation, the ratio of the downwind length to the crosswind width of the clouds is about 1.4. It is likely that this ratio will increase during the lifetime of a bubble cloud if, as is likely, there is significant vertical shear in the near-surface currents; it may also change if the turbulence acting to spread the cloud is non-isotropic in the horizontal. There is however no evidence in the sonographs to show these processes. The streaks fade with time.

"(c) The horizontal size of bubble clouds increases with wind speed, the largest being about 4 m in width for winds of 8 ms^{-1}.

"(d) The bubble clouds produced by breaking waves appear initially to have some motion in the wave direction, but their motion becomes indistinguishable from the general drift of the surface within about 10s of their formation.

"(e) The life-times of bubble clouds detectable by sonar are typically 1-5 mins (see the length of streaks in Fig. 3).

"(f) In winds exceeding about 7 ms^{-1} (when a stratus layer of bubbles exists) or when heavy rain reduces the number of breaking waves but produces bubbles in a uniform manner over the whole water surface, bands of bubbles are formed. These have separations of about 5 m, and usually continue for a least 5-10 mins. They are parallel to the wind direction, although they may be carried across wind by tidal currents. They appear to be caused by the convergence induced by Langmuir circulation.

"(g) Individual bubble clouds become difficult to distinguish from the background bubbles in winds of 14 ms^{-1} (when you can't tell the clouds from the storm!).

"(h) The horizontal advection of bubble clouds, and hence the near-surface current, are influenced by internal waves and fronts. Bubble clouds appear most dense in regions of horizontal convergence (see the front in Fig. 4). The distribution of clouds of course reflects the distribution of breaking waves, and this may be affected by the currents induced by fronts or internal waves."

O'Muircheartaigh and Monahan (1986) have obtained data on oceanic whitecap coverage W, in %, as a function of the time averaged wind speed at 10 m, and other environmental parameters. The data were originally fitted to a power law of the form given by Eq. (49), which is reminiscent of a simplified version of the power law formulation for backscatter as in Eqs. (36).

$$W = \alpha U^{\lambda} \tag{49}$$

This formulation was found to be inadequate. One of the data sets was fitted by nonlinear least squares methods to the data to obtain Eq. (50).

$$W = 1.36 \cdot 10^{-2}(\bar{U}(10)/10)^{1.73} - 6.42 \cdot 10^{-4} \tag{50}$$

This lead to the definition of a quantity called $\bar{U}_B(10)$, the 10 meter Beaufort wind, which is the minimum wind speed to produce breakers. For Eq. (50), it is 1.71 ms^{-1}.

The minimum wind speeds required to generate various spectral wavenumbers as in Eq. (44) is in the range of 1.71 ms^{-1}. For the first appearance of whitecaps, this work has been extended to include the effects of air-sea temperature differences in more recent publications. It is intriguing that certain effects of the wind on waves are predicted in two different ways not to occur until the wind exceeds various threshold values (Monahan and O'Muircheartaigh (1986)).

As described by Wetzel (1987), A. M. Worthington succeeded in making photographs of water drops falling into still water and what happened thereafter. The work was reported to the Royal Society (London) in 1982. Wetzel references this work as Worthington (1963). As a parallel, but later, study, Blanchard and Woodcock (1957) showed what happens when a bubble of air rises to the sea surface and breaks through. Resch (1966) has schematized the sequence of events from whitecaps production, bubble entrainment and mixing, rising bubbles and their breaking through the sea surface.

As shown in Fig. 7 a falling drop of water first produces a crown shaped splash with essentially circular symmetry and little droplets irregularly spaced around the crown. The crown then changes to a jet of rising water characterized at its peak by Wetzel as a stalk and button. This falls back into the water and generates an expanding ring of circular ripples that radiate away. A rising bubble, (Blanchard and Woodcock, 1957, Resch, 1986) collapses when it breaks through the water surface and the latter part of a sequence similar to a falling drop then follows.

The last two stages on the left of Fig. 7 are for the radiating ripples similar to those when a stone is cast into water. They would be repeated on the right. The superposition of viscous damped Bessel functions with appropriate phase and group velocities for capillary and gravity waves has been used by LeMehaute (1988) to model the expanding ring ripples.

Wetzel (1987) has modeled the backscatter at X-band from the various stages shown on the left of Fig. 7 for a single drop and compared his theoretical results with the backscatter measurements of Hansen (1984, 1986, 1987). The bare drop, the crown, the stalk and the ripple ring were all treated. The difference in the measurements between vertical polarization and horizontal polarization is striking. As the cylindrical form of the stalk grows to a maximum and then falls back, resonant modes for vertical polarization were excited as organ pipe type standing waves with perhaps as many as eleven resonant peaks as a function of time. There is, in contrast, only one peak in the response for horizontal polarization.

Further measurements and analyses along the lines of this fundamental study by Wetzel with drop size distributions appropriate to different rainfall rates and with bursting bubbles of various sizes can model a backscatter effect that is probably nearly isotropic as a function of wind direction and dependent only on incidence angle. Research on backscatter from splashing raindrops in windwave tank experiments has been described by Bliven and Norcross (1988) and Bliven, et al. (1988). These field and laboratory results should be intercompared.

Mestayer and Lefauconnier (1988) have produced flying spray for various humidities. Drops can evaporate rapidly into dry air. This raises an interesting area of research as to the

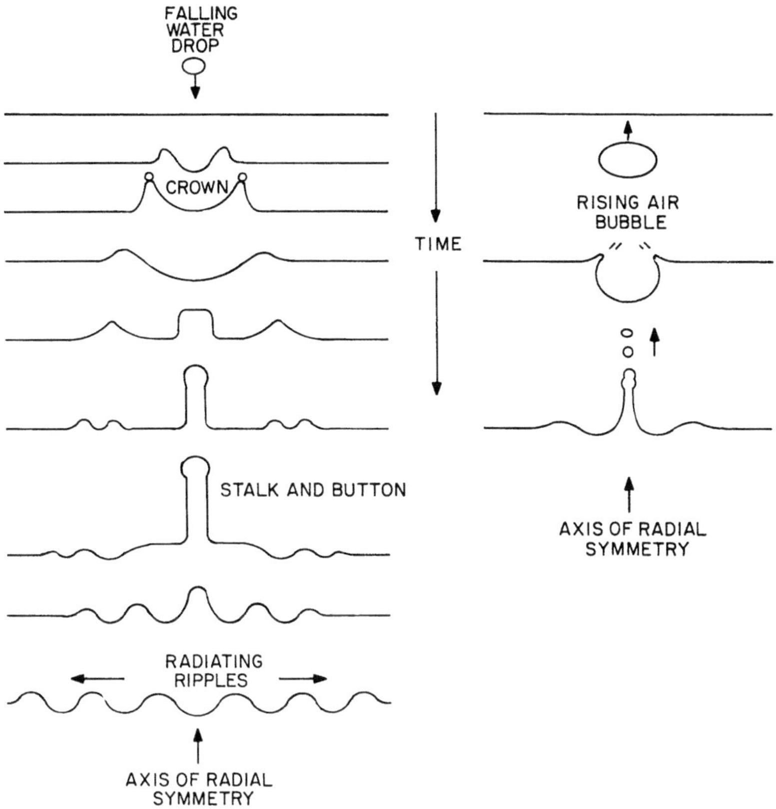

Fig. 7 — Schematic Drawings With Radial Symmetry of the Sequence of Events as a Drop of Water Falls Into the Water (Left) and as a Bubble of Air Rises Through the Water and Bursts Into the Air, Ejecting Water Drops Into the Air (Right). The Last Part of the Bubble Sequence (Not Shown) is Similar to the Water Drop Sequence. -

appropriateness of the bulk aerodynamic method for calculating the evaporation from the sea surface. Wu (1988) has analyzed data on bubble distributions and the acoustic scattering cross section as a function of $\overline{U}(10)$ from Thorpe (1982) and other data. From these plots, which are power laws, the acoustic scattering cross section becomes very, very low for winds under about 3 ms^{-1}. An equation similar to Eq. (50) might fit the rather meager data just as well.

Foam might be defined as bubbles of air in the water that are so close together that many of them touch so that the internal structure could be like a sponge with the solid part of the sponge replaced by sea water. The percentage of the air in the water, or of the water in the air, would be some function of time and position for each foam patch.

Foam and whitecaps are known to emit strongly at passive microwave frequencies as in Droppleman (1970) and Ross and Cardone (1974). Electromagnetic theory then suggests that microwave radiation might be strongly absorbed and not backscattered at the same frequency. Microwave backscatter from foam patches, to my knowledge, has not been measured.

Spray, drops, bursting bubbles and foam may have important effects on backscatter at K_a, K_u and X-band but less important effects at C and L-band. An area of investigation having to do with roughness scales smaller than the radar wavelength in use has been hinted at in the work of Wetzel (1986), who treated these scales as a detuning effect. Is a three scale backscattering model needed?

After a linear model for the motion of ships in linear waves was developed a number of years ago, a rather skeptical naval architect (Ogilvie, 1964) wrote: "From these stochastic models, one can derive all kinds of interesting conclusions about the sea, some of which will be true." The study of breaking waves by means of linear models can probably be described in the same way.

Two series of photographs are of interest in this context. One is the photographs from aircraft, found in Neumann and Pierson (1966), of waves for various wind speeds in a hurricane. The other is the series of black and white photographs entitled "State of the Sea Photographs for the Beaufort Wind Scale" published by the Meteorological Branch of the Department of Transport, Canada, with a foreward by J. R. H. Noble dated 1969 and reprinted in 1971. The C.C.G.S. Stonetown and C.C.G.S. St. Catherines from which the photographs were taken have been decommissioned. Their data have been of great value, especially during the GOASEX program for the SEASAT SASS.

The Canadian reference also contains a table, which is repeated here as Table 3. To convert from feet to meters multiply by 0.305 and to convert from knots to ms^{-1} multiply by 0.514. The definitions of the Beaufort scale in terms of wave properties, which were later on fitted to ranges of wind speed in knots, are rather hard to come by in the more recent literature. The definition of the waves fits the concept of a speed defined by $\overline{U}(10)$ and the threshold speed for the generation of Bragg scattering waves of L-band and higher frequencies.

The photograph for calm, apart from radiating short waves caused by the slight motions of the ship, shows a glass-like smooth sea surface with low short crested waves perhaps 10 to 30 m long. The calm was preceded by light winds. Could these waves have been generated locally by the mechanism of Donelan and Pierson (1987, see page 4980)? Swell would be considerably longer. The fetch was surely long enough.

There must be larger areas of the sea surface for gravity waves, even for high winds, that are close to the mean level, else the waves would not be quasi-Gaussian. For these same areas the slopes also cannot be large. Breakers and, to scale, wedges would be rare over these areas, but these areas could be covered by Bragg scattering waves. The relative size of the areas covered by Bragg waves, wedges, spilling breakers and so forth give the edge to Bragg scattering.

The photographs under discussion show increasingly higher and more abundant short waves within the resolution of the photographs as the winds increase up to about 44 knots. At higher wind speeds, the Bragg-like scatterers seem to decrease. There could be other reasons for this, in addition to the model of Donelan and Pierson (1987), such as increased turbulence in the water and a damping effect on the short waves.

Large cloud drops and falling rain that has not yet reached the wavy surface could merit another subsection, but, instead, will be described only briefly here. Large enough drops, either falling rain or cloud, scatter the radar waves both on the way to the water waves below and on the way back from the water waves below. The amount of the radar waves scattered back to the radar by the cloud drops and rain seems to be less that the amount lost to and from the waves. The net effect is an attenuation of the backscattered power which reduces the backscattered power. This attenuation can be estimated from multichannel passive microwave data as described by Moore, et al. (1982). Should the backscatter increase less

Table 3 — Beaufort Scale of Wind Force

Beaufort Number	Mean Wind Speed in knots	Limits of Wind Speed in knots	Descriptive Terms	Appearance of the Sea if the fetch and duration of the blow have been suff- icient to develop of the sea full	Probable Wave Height* in feet	Probable Maximum Wave Height* in Feet
0	0	1	Calm	Sea Like a mirror	0	0
1	2	1-3	Light air	Ripples with the appearance of scales are formed, but without foam crests.	1/4	1/4
2	5	4-6	Light Breeze	Small wavelets, still short but more pronounced: crests have a glassy appearance and do not break.	1/2	1/2
3	9	7-10	Gentle Breeze	Large wavelets; crests begin to break. Foam of glassy appearance. Perhaps scattered white horses.	2	3
4	13	11-16	Moderate	Small waves, becoming longer. Fairly frequent white horses.	3 1/2	5
5	19	17-21	Fresh Breeze	Moderate waves, taking a more pronounced long form. Many white horses are formed. Chance of some spray.	6	8 1/2
6	24	22-27	Strong Breeze	Large waves begin to form. The white foam crests are more extensive everywhere. (Probably some spray).	9 1/2	13
7	30	28-33	Near Gala	Sea heaps up and white foam from breaking waves begins to be be blown in streaks along the direction of the wind.	13 1/2	19
8	37	37-40	Gale	Moderately high waves of greater length. Edges of crests being to break into the spindrift. The foam is blown in well marked streaks along the direction of the wind	18	25
9	44	41-47	Strong Gale	High waves; dense streaks of foam along the direction wind. Crests begin to topple, tumble, and roll over. Spray may affect visibility.	23	32
10	52	48-55	Storm	Very high waves with long overhanging crests. The resulting foam, in great patches, is blown in dense white streaks along the direction of the wind. On the whole, the surface of the sea takes on a white appearance. The tumble of the sea becomes heavy and shock-like. Visibility affected.	29	41
11	60	56-63	Violet Storm	Exceptionally high waves. (Small and medium sized ships might be a time lost to view behind the waves). The sea is completely covered with long white patches of foam lying along the direction of the wind. Everywhere the edges of the crests are blown into froth. Visibility affected.	37	52
12	68	64-71	Hurricane	The air is filled with foam and spray. Sea completely white with driving spray. Visibility very seriously affected.	45	—

*The last two columns are added as a guide to show roughly what may be expected in the open sea, remote from land. In enclosed waters, or when near land with an off-shore wind, wave heights will be smaller and the waves steeper. The probable maximum wave height is reached by about one wave in ten.

rapidly with increasing wind than a power law, or saturate and then decrease, attenuation by clouds and falling rain could be a needed important correction so as to be able to recover high winds correctly.

A difficulty with accounting for the effects of rain and large cloud drops is that they generally have scales smaller than the resolution cells of scatterometers and microwave radiometers. It is difficult to weight these effects in proportion to the total area of a cell.

8. An Experiment

8.1. CUMULATIVE EFFECTS

Donelan and Pierson (1987) tuned their model to K_u-band vertically polarized transmitted and received backscatter data with fairly good results. The model then did not agree with data for horizontally polarized transmitted and received backscatter. A possible explanation was postulated to be the combined effects of backscatter from the toes of spilling breakers, wedges and the turbulent undulations on the breaker. To these could be added possible effects of spray, foam and drops produced by bursting bubbles. The needed additions for horizontal polarization were found to be reasonable. Also additions to the backscatter for vertical polarization, though much larger in antilog form, made the model agree more closely with the data.

8.2. σ^0_{VV} COMPARED WITH σ^0_{HH}

The AAFF Langley Circle Flight maneuvers obtained data on the variation of horizontally and vertically polarized K_u-band backscatter as a function of wind direction for a fixed incidence angle and wind speed (Schroeder, et al., 1984). The same maneuver done by Masuko, et al. (1986) obtained data for X and K_a band, with a possible problem of interpretation associated with the use of logarithmic amplifiers. Nevertheless, the data for X, K_u and K_a band all show major differences between vertically and horizontally polarized backscatter for a given high wind speed and incidence angle. One difference is that horizontally polarized backscatter is higher than would be predicted for a two scale model if the model is first fitted to vertical polarization and then evaluated for horizontal polarization. The other is that upwind and downwind backscatter values are very nearly, but not quite, equal for vertical polarization to the extent that random effects can conceal upwind-downwind differences. For horizontal polarization upwind backscatter is always observed to be stronger than downwind backscatter for the higher winds and higher incidence angles. Although the latter observation may be explained by the higher tilt modulation expected at H-pol for a Bragg-type theory (Plant, 1986), there does not seem to be any theoretical explanation for these former observations with this type of model.

Accounting for all of the different effects of the waves that can affect backscatter is a bookkeeping problem. For a given large enough illuminated area, incidence angle, wind direction and wind speed, as a hypothetical example, 5% of the received power could be from specular reflection, 81% from two scale Bragg, 8% from wedges and 6% from other effects. The various contributions come from different portions of the wavy surface with some effects concentrated near the crests of the highest waves in groups of waves. Appropriately weighted models for vertical and horizontal polarization for these effects might then be able to explain the present differences between the simpler theories and the measurements. The weights would differ for L-band, C-band, X-band, K_u-band and K_a-band.

8.3. AN AIRCRAFT SYSTEM

Ulaby, et al. (1986) obtained useful information from land passes by means of two-color images that combined vertically polarized and horizontally polarized data. Narrow beam, short pulse, real aperture radars that alternate between line scans of vertical and horizontal polarization could produce images that would emphasize the relative importance of different effects for each polarization. With sufficient resolution, portions of each wave can be isolated. Sea spikes should be very bright returns in the image which would show where they come from. A photographic slit camera for times with good lighting could document the visual scene. Simultaneously operated synthetic aperture radars might help settle some of the theoretical differences in the interpretation of these radars. Such an aircraft radar system can still keep these various effects apart (as in laboratory experiments and theory) and yet permit these effects to be put back together to get the total effect.

9. Concluding Remarks

A backward step seems to be necessary to understand the dependence of radar backscatter on environmental parameters. Better ways to measure the winds and to determine the turbulent structure of the winds so as to suppress the randomness are needed. The variation of the time averaged wind as a function of height for the marine boundary layer needs to be better understood. If $C_{DN}(\overline{U}(10))$ is a definable concept in turbulence theory as a boundary layer closure (why 10 m?) $\overline{U}(\pi/k)$, u_* and $\overline{U}(10)$ are each definable in terms of either of the other two quantities. The physical concepts for these quantities differ considerably, however. The wavenumber spectra of wind generated waves need to be estimated for greatly decreased sampling variability. These long standing areas of research have lead to contradictory results that need to be resolved.

Many features of a wind driven sea have been identified and studied either experimentally or theoretically (or both) in either the laboratory or over the ocean as contributors to backscatter, especially at X-band, K_u-band and K_a-band. These different environmental effects need to be combined so as to be able to model vertically and horizontally polarized backscatter as a function of wind speed, wind direction, wave spectra and non-linear wave properties. An experiment has been suggested to help sort out some of these effects.

10. Acknowledgements

The research on which this material is based and the preparation of this chapter has been supported by the National Aeronautics and Space Administration. The opportunity to revise and extend this chapter, originally completed in January 1988, at the page proof stage is appreciated.

11. References

Baer, L. and C.L. Vincent, Special Issue on Atlantic Remote Sensing Land Ocean Experiment, ARSLOE, IEE J. Oceanics Eng. OE-8 201-271, 1983.

Banner, M.L., and E.H. Fooks, On the microwave reflectivity of small-scale breaking water waves, Proc. R. Soc. London, Ser. A, **399**, 93-109, 1985.

Banner, M.L., I.S.F. Jones and J.C. Trender, Wavenumber spectra of short gravity waves, J. Fluid Mech. (in press), 1988.

Benjamin, T.B., Shearing flow over a wavy boundary, J. Fluid Mech., **6**, 161-205, 1959.

Blanc, T.V., A practical approach to flux measurements of long duration in the marine atmospheric surface layer, J. Clim. Appl. Meteorol., **22**, 1093-1110, 1983a.

Blanc, T.V., An error analysis of profile flux, stability, and roughness length measurements made in the marine atmospheric surface layer, Boundary Layer Meteorol., **26**, 243-267, 1983b.

Blanc, T.V., Typical influences of moisture on profile measurements in the marine atmospheric surface layer, Boundary Layer Meteorol., **25**, 411-415, 1983c.

Blanc, T.V., Variation of bulk-derived surface flux, stability and roughness results due to the use of different transfer coefficient schemes, J. Phys. Oceanogr., **15**, 650-669, 1985.

Blanc, T.V., Superstructure flow distortion corrections for wind speed and direction measurements made from Tarawa class (LHA1-LHA5) Ships, NRL Rep. 9005, 20 pp., Nav. Res. Lab., Washington, D.C., 1986a, (Available as NTIS ADA175223 from Natl. Tech. Inf. Serv., Springfield, Va.).

Blanc, T.V., The effect of inaccuracies in weather-ship data on bulk-derived etimates of flux, stability and sea-surface roughness, J. Atmos. Ocean. Technol., **3**, 12-16, 1986b.

Blanc, T.V., Accuracy of bulk-method-determined flux, stability, and sea surface roughness, J. Geophys. Res., **92**, C4, 3867-3876, 1987.

Blanchard, D.C. and A.H. Woodcock, Bubble formation and modification in the sea and its meteorological significance, Tellus **9**, 2, 145-158, 1957.

Bock, E.J., On ripple dynamics, J. Colliod and Interface Sci. **119**, 2, 326-334, 1987.

Bliven, L.F., J.P. Giovanangeli and G. Norcross (1988) A study of rain effects on radar scattering from water waves. Preprint volume. Seventh conference on Ocean-Atmosphere Interaction Jan 31 to Feb 5 1988 Anaheim, Calif. Amer. Meteor. Soc.

Bliven, L.F. and G. Norcross (1988) Effects of rainfall on scatterometer derived wind speeds. Proc IGARSS'88 Edinburg, Scotland 13 to 16 Sept 1988 ESA SP 284 (IEEE 88 CH 2497.6).

Brown, R.A., On a satellite scatterometer as an anemometer, SEASAT Special Issue II. Scientific Results, J. Geophys. Res. **88**, C3 1663-1673, 1983.

Cardone, V.J., W.J. Pierson and E.G. Ward, Hindcasting the directional spectrum of hurricane generated waves, J. Petrol. Tech., **28**, 385-394, 1976.

Cardone, V.J., The present status of operational wave forecasting, Proceedings of the Symposium: Measuring Ocean Waves from Space, John Hopkins APL Tech. Digest, **8**, 1, 24-32, 1987.

Cartwright, D.E., The use of Directional spectra in studying the output of a wave recorder on a moving ship, *in* Ocean Wave Spectra, Prentice Hall, Englewood Cliffs, N.J., 203-207, 1963.

Champagne-Philippe M., Structure spatial et temporelle du vent horizontal dans la region de transition entre la turbulence et institut francais de recherche pour l'Expoitation de Le mer, Centre de Brest, B.P., 29263 Plouzane, France, 103 pp., 1985.

Clancy, R.M., J.E. Kaitala and L.V. Zambresky, The Fleet Numerical Oceanography Center Global Spectral Ocean Wave Model, Bull., Am., Meteorol., Soc., **67**, 498-512, 1986.

Cox, C.S., and W. Munk, Statistics of the sea surface derived from sun glitter, J. Mar. Res., **13**, 198-227, 1954.

Dischel, R.S. and W.J. Pierson, Comparisons of wind reports by ships and data buoys, Proc. MDS '86, Marine Data Systems International Symposium, Marine Tech. Soc., Gulf Coast Sect., 447-466, 1986.

Dobson, E., F. Maldonado, J. Goldhirsch and J. Wilkerson, Validation of GEOSAT altimeter-derived wind speeds and significant wave heights using buoy data, The Navy GEOSAT Mission, Johns Hopkins APL Technical Digest, **8**, 2, 222-233, 1987.

Donelan, M.A., The dependence of the aerodynamic drag coefficient on wave parameters, *in* Proceedings of the First International Conference on Meteorology and Air-Sea Interaction of the Coastal Zone, 381-387, American Meteorological Society, Boston, Mass., 1982.

Donelan, M.A. and W.J. Pierson, The sampling variability of estimates of spectra of wind-generated gravity waves, J. Geophys. Res. **88**, C7, 4381-4392, 1983.

Donelan, M.A., J. Hamilton, and W.H. Hui, Directional spectra of wind-generated Waves, Phil. Trans. R. Soc., London, **A315**, 509-562, 1985.

Donelan, M.A. and W.J. Pierson, Radar scattering and equilibrium ranges in wind generated waves with application to scatterometry, J. Geophys. Res., **92**, C5, 4971-5029, 1987.

Drennan, W.H., Accurate calculations of the Stokes water wave. Ph.D. Dissertation, University of Waterloo, Waterloo, Ontario, Canada, 1989.

Droppleman, J.D., Apparent microwave emission of sea foam, J. Geophys. Res. **75**, 696-698, 1970.

Durden, S.L. and J.E. Vesecky, A physical radar cross section model for a wind driven sea with swell, IEEE J. Oceanic. Eng., **0E-10**, 4, 445-451, 1985.

Ewing, J.A. and A.K. Laing, Directional spectra of seas near full development, J. Phys. Oceanogr., **17**, 10, 1696-1706, 1987.

Ezraty, R., Using buoys and ships to calibrate ERS-1 altimeter and scatterometer. Proceedings of a Workshop on ERS-1 Wind and Wave Calibration, Schliersee, FRG, 2-6 June 1986, (ESA SP-262, Spet. 1986).

Ezraty, R., P. Queffeulou and M. Champagne, Illustration of wind field time and space statistics during the Toscane-T Compaign, Proc. Third Internalt. Coll. Spectral Signatures of Objects in Remote Sensing, Les Arcs, France 16-20, Dec. (ESA SP 247), 1985.

Geernaert, G.L., S.E. Larsen and F. Hansen, Measurements of the wind stress, heat flux and turbulent intensity during storm conditions over the North Sea, J. Geophys. Res., **92**, C12, 134127-13140, 1987.

Geernaert, G.L., Variation of the angle between the surface layer wind vector and the wind stress vector and its dependence on thermal stratification and advection, J. Geophys. Res. **93**, 8215-8220, 1988a.

Geernaert, G.L., Influence of long and short waves on the drag coefficient in coastal regions *in* Proc. Seventh Conf. on Ocean Atmosphere Interaction, Jan. 31 - Feb. 5, 1988, Amer. Meteor. Soc. Anaheim Cal. (to be pulished), 1988b.

Geernaert, G.L., W.J. Plant and D.L. Schuler, K_u-band NRCS measurements and their dependence on wind azimuth, wave state, wind stress and wind speed during FASINEX, *in* Proc. Seventh Conf. on Ocean Atmosphere Interaction, Jan. 31 - Feb. 5, 1988, Amer. Meteor. Soc. Anaheim Cal. (to be published), 1988.

Günther, H., A parametric surface wave model and the statistics of the prediction parameters, Hamburger Geophys. Einzelschr., **55**, 1-90, 1981.

Halpern, D., Comparison of moored wind measurements from a spar and toroidal buoy in the eastern equatorial pacific during February - March 1981, J. Geophys, Res., **92**, C8, 8303-8306, 1987.

Hansen, J.P., High resolution radar backscatter from a rain disturbed sea surface, LSNR-84 Record, Oct. 22-24, Tokyo, Japan, 1984.

Hansen, J.P., A system for performing ultra high resolution backscatter measurements of splashes, Proc. of the Int. Microwave Theory and Tech. Symp. Baltimore M.D., June 2-6, 1986.

Hansen, J.P., Experimental characterization of high resolution radar backscatter from rain splash, NRL Memo Rept. (in preparation) 1989.

Hasselmann, K., T.P. Barnett, E. Bouws, H. Carlson, D.E. Cartwright, K. Enke, J.A. Meerburg, H. Gienapp, D.E. Olbers, K. Richter, W. Sell, and H. Walden, Measurements of wind-wave growth and swell decay during the joint North Sea wave Project (JONSWAP), Deut. Hydrog. Z., **A8**, (Suppl.), 1-95, 1973.

Hasselmann, K., D.B. Ross, P. Muller and W. Sell, A parametric wave prediction model, J. Phys. Oceanogr., **6**, 200-228, 1976.

Hasselmann, K., On the spectral energy balance and numerical prediction of Ocean waves, *in* A. Favre and K. Hasselmann, *Eds.*, Turbulence Fluxes Through the Sea Surface, Wave Dynamics and Prediction, 531-543. New York: Plenum Press, 1977.

Hasselmann, S., K. Hasselmann, E. Bauer, P.A.E.M. Janssen, G.J. Komen, L. Bertotti, P. Lionello, A. Guillaume, V.J. Cardone, J.A. Greenwood, M. Reistad, L. Zambresky, and J.A. Ewing, The WAM Model—A third generation ocean wave prediction model. The WAMDI group, J. Phys. Oceanog., **18**, 12, 1775-1810, 1988.

Hayne, G.S. and D.W. Hancok III, Wave form analysis for GEOSAT day 96, The Navy GEOSAT Mission, Johns Hopkins APL, Technical Digest, **8**, 2, 255-259, 1987.

Hui, W.H. and G. Tenti, A new approach to steady flows with free surfaces, Z. Angew. Math. Phys., **33**, 569-589, 1982.

Hui, W.H. and G. Tenti, Nonlinear water wave theory via pressure formulation *in* Toba, Y. and H. Mitsuyasu *eds* The Ocean Surface, Wave Breaking, Turbulence Mixing and Radio Probing, 17-24, D. Riedel Publishing Co., 1985.

Iwata, N., An Algoithm of microwave backscattering from a Perturbed Sea Surface, *in* Toba, Y. and H. Mitsuyasu eds, The Ocean Surface: Wave Breaking, Turbulent Mixing and Radio Probing, D. Riedel, 283-288, 1985.

Jackson, F.C., W.T. Walton and P.L. Baker, Aircraft and satellite measurements of ocean wave directional spectra using scanning-beam microwave radar, J. Geophys. Res., **90**, C1, 987-1004, 1985a.

Jackson, F.C., W.T. Walton and C.Y. Peng, A comparison of in situ and airborne radar observations of ocean wave directionality, J. Geophys. Res., **90**, C1, 1005-1018, 1985b.

Jeffreys, H., On the formation of waves by wind, Proc. R. Soc. London, Ser. A., **107**, 189-206, 1924.

Jeffreys, H., On the formation of waves by wind, II, Proc. R. Soc. London, Ser. A., **110**, 341-347, 1925.

Jones, W.L., L. C. Schroeder, D.H. Boggs, E.M. Bracalente, R. A. Brown, G.J. Dome, W.J. Pierson and F.J. Wentz, The SEASAT-A satellite scatteromenter: The geophysical evaluation of remotely sensed wind vectors over the ocean, J. Geophys, Res., **87**, C5, 3297-3317, 1982.

Kahma, K.K., and M.A. Donelan, A laboratory study of the minimum wind speed for wind wave generation, J. Fluid Mech., **192**, 339-364, 1988.

Keller, M.R., and W.J. Plant, X-band scatterometer measurements at low winds in a wave tank *in* Proc. Seventh Conf. on Ocean Atmosphere Interaction, Jan. 31 - Feb. 5,

1988, Amer. Meteor. Soc. Anaheim Cal. (to be published), 1988.

Kinsman, B., Wind Waves: Their generation and propagation on the ocean surface, 676 pp, Prentice Hall 1965, Dover Reprint 1984.

Komen, G.J., Recent results with a third-generation ocean-wave model, Proceedings of the Symposium: Measuring Ocean Waves from Space, Johns Hopkins APL Tech. Degest, **8**, 1, 37-41, 1987.

Kozu, T., T. Umchara, T. Ojima, T. Suitzu, H. Masunko and T. Inomata, Observations of oil slicks on the ocean by X-band SLAR. IGARSS 87CH 24349, 73x5-740, 1987.

Kwoh, D.S. W., and B. M. Lake, A Deterministic, coherent, and dual-polarized laboratory study of microwave backscattering from water waves, I, Short gravity waves without wind, IEEE J. Oceanic Eng., **0E-9**, 5, 291-308, 1984.

Lamb, K.G., G. Tenti and W.H. Hui, General theory of wavemaker: I Theoretical results, Z. Agnew, Math, Phys., **38**, 391-408, 1987.

Large, W.G. and S. Pond, Open ocean momentum flux measurements in moderate to strong winds, J. Phys. Oceanogr., **11**, 324-336, 1981.

LeMehaute, B., Gravity capillary rings generated by water drops, J. Fluid Mech., **197**, 415-427, 1988.

Liu, W.T., K.B. Katsaros and J.A. Businger, Bulk paramenterizations of air-sea exchanges of heat and water vapor including the molecular constraints at the interface, J. Atmos. Sci. **36**, 1722-1735, 1979.

Liu, W.T. and T.V. Blanc, The Lui, Katsaraos and Businger, 1979, bulk atmospheric flux computational interaction program in FORTRAN and BASIC. NRL Memorandum Report 5291, 1984.

Liu, P.C. Testing parametric correlations for wind waves in the Great Lakes, J. Great Lakes Res., **11**, 4, 478-491, 1985.

Liu, P.C. and D.J. Schwab, A comparison of methods for estimating u_* from given U_z and air-sea temperature differences, J. Geophys. Res., **92**, C6, 6488-6499, 1987.

Longuet-Higgins, M.S., The generation of capillary waves by steep gravity waves, J. Fluid Mech., **16**, 1, 138-159, 1963.

Longuet-Higgins, M.S. and J.S. Turner, An entraining plume model of a spilling breaker, 1, J. Fluid Mech., **63**, 1-20, 1974.

Lyzenga, D.R., A. L. Maffett, and R.A. Shuchman, The contribution of wedge scattering to the radar cross section of the ocean surface, IEEE Trans. Geosci. Remote Sens., **GE-21**, 4, 502-505, 1983.

Masuko, H., K. Okamoto, M. Shimada and S. Niwa, Measurements of microwave backscattering signatures of the ocean surface using X-band and Ka-band airborne scatterometers, J. Geophys. Res., **91**(C11), 13065-13083, 1986.

Mestayer, P. and C. Lefauconnier, Spray droplet generation, transport and evaporation in a wind wave tunnel during the humidity exchanges over the sea experiments in a simulation tunnel. J. Geophys. Res. **93**, C1, 572-586, 1988.

Monahan, E.C. and G. MacNiocaill, *eds* Oceanic whitecaps and their role in air-sea exchanges processes, D. Reidel Publishing Co., 294 pp, 1986.

Monahan, E.C. and I. G. O'Muircheartaigh, Whitecaps and the passive remote sensing of the ocean surface, Int. J. Remote Sensing **7**, 5, 627-642, 1986.

Monin, A.S. and A.M. Obukhov, Basic laws of turbulent mixing in the ground layer of the atmosphere, (in Russian), Tr. Geofiz. Inst. Akad. Nauk SSSR, **24**, 151, 163-187, 1954.

Mood, A.M., F.A. Graybeal and D.C. Boes, Introduction to the theory of statistics, Third Edition, McGraw Hill, New York, 564 pp. 1974.

Moore, R.K., I.J. Birrer, E.M. Bracalente, G.J. Dome and F.J. Wentz, Evaluation of atmospheric attenuation from SMMR brightness temperature for the SEASAT satellite scatterometer, J. Geophys. Res., **87**, C5, 1982.

Moskowitz, L., Estimates of the power spectrums for fully developed seas for wind speeds of 20 to 40 knots, J. Geophys. Res., **69**, 24, 5161-5179, 1964.

NASA SP-399, SKYLAB EREP Investigations Summary Scientific and Technical Information Office, Supt. of Docs. No. NAS 1.21:399 U.S. Govt. Printing Office Stock. No. 033-000-0741-8, 1978.

Neumann, G. and W.J. Pierson Jr., Principles of Physical Oceanography Prentice-Hall, Englewood Cliffs, N.J. 1966.

Nicholls, S., B. Brummer, F. Fielder, A. Grant, T. Hauf, G. Jenkins, G. Readings and W. Shaw, The structure of the turbulent atmospheric boundary layer, *in* Charnock, H. and R.T. Pollard Eds. Results of the Royal Society Joint Air-Sea Interaction Project (JASIN) 71-89, London, The Royal Society. Also Phil. Trans. R. Soc. Lond. A. **308**. 1983.

Ogilvie, F.T., Recent progress toward the understanding and prediction of ship motions, Fifth Symposium On Naval Hydrodynamics, Wageningen Pub. ACR-112, Office of Naval Research, Washington, D.C. 1964.

O'Muircheartaigh, I.G. and E.C. Monahan, Statistical aspects of the relationship between oceanic whitecap coverage, wind speed and other environmental factors *in* Monahan, E.C. and MacNiocaill *eds* Oceanic Whitecaps and Their Role in Air-Sea Exchanges Processes, 125-128, D. Reidel Publishing Co., 1986.

Pierson, W.J., The interpretation of wave spectrums in terms of the wind profile instead of the wind measured at a constant height, J. Geophys. Res., **69**, 24, 5191-5203, 1964.

Pierson, W.J. and L. Moskowitz, A proposed spectral form for fully developed wind seas based on the similarity theory of S.A. Kitaigorodskii, J. Geophys. Res., **69**, 24, 5181-5190, 1964.

Pierson, W.J., Comments on 'A parametric Wave prediction model,' J. Phys. Oceanogr., **7**, 127-134, 1977.

Pierson, W.J., The spectral ocean Wave model (SOWM), a northern hemispheric computer model for specifying and forecasting ocean wave spectra, DTNSRDC-821011, David W. Taylor Naval Ship Research and Development Center, Bethesda, Md. 1982.

Pierson, W.J., The measurement of the synoptic scale wind over the ocean, J. Geophys. Res., **88**, 1683-1708, 1983.

Pierson, W.J., Some approaches for comparing remote and in-situ estimates of directional wave spectra, Proceedings of the Symposium: Measuring Ocean Waves from Space, Johns Hopkins APL Technical Digest, **8**, 1, 48-54, 1987.

Pierson, W.J., Probabilities and statistics for backscatter estimates obtained by a scatterometer, J. Geophys. Res. (1989) in press.

Pierson, W.J., W.E. Marlatt, Z.H. Byrns and W.R. Johnson, Oceans and atmosphere in SKYLAB EREP investigations summary NASA SP-390, Scientific and Technical Information, Supt. of Docs. No. NAS 1.21:399 U.S. Govt. Printing Office Stock No. 033-000-0741-8, 1978.

Plant, W.J. A two-scale model of short wind-generated waves and scatterometry, J.Geophys. Res., **91**, C9, 10735-10749, 1986.

Queffeulou, P., SEASAT wave height measurement: A comparison with sea-truth data and a wave forecasting model-applications to the geographic distribution of strong sea states in storms, SEASAT Special Issue II Scientific Results, J. Geophys. Res., **88**, C3, 1779-1788, 1983.

Resch, F., Oceanic air bubbles as generators of marine aerosols, in Monahan E.C. and MacNiocaill, G. *eds* Oceanic Whitecaps and Their Role in Air-Sea Exchanges Processes, 101-112, D. Reidel Publishing Co., 1986.

Ross, D.B. and V. Cardone, Observations of oceanic whitecaps and their relation to remote measurements of surface wind speed, J. Geophys. Res., **79**, 444-452, 1974.

Ross, D.B., and W.L. Jones, On the relationship of radar backscatter to wind speed and fetch, Boundary Layer Meteorol., **13**, 133-149, 1978.

Schroeder, L.C., D. H. Boggs, G. Dome, I.M. Halberstam, W.L. Jones, W.J. Pierson and F.J. Wentz, The relationship between wind vector and normalized radar cross section used to derive SEASAT-A satellite scatterometer winds, SEASAT Special Issue I, J. Geophys. Res. **87**, C5, 3318-3336, 1982.

Schroeder, L.C., W.L. Jones, P.R. Schaffner and J.L. Mitchell, Flight measurements and analysis of AAFE RADSCAT wind speed signature of the ocean, NASA Tech Memo. TM 85646, 144 pp. 1984.

Shemdin, O.H. and P.A. Huang, Comparison of measured and predicted sea surface spectra of short waves, J. Geophys. Res., **93**, C11, 13883-13890, 1988.

Smith, S.D. and E.G. Banke, Variation of the sea surface drag coefficient with wind speed, Q.J.R. Meteorol. Soc., **101**, 665-673, 1975.

Steele, K.E., J.C.K. Lau and U.H.L. Hsu, Theory and application of calibration techniques for a NDBC directional wave measurement buoy, IEEE J. Oceanic Eng. **0E-10**, 382-396, 1985.

SWAMP (Sea Wave Modeling Project) Group (24 Authors), Ocean Wave Modeling, Plenum Press, New York, 1985.

Tchen, C.M., S. E. Larsen, H. Pecsili and T. Mikkelsen, Large-scale spectral structure with a gap in the stably stratified atmosphere, Physica Scripta, **31**, 616-620, 1985.

Tchen, C.M. and W.J. Pierson, A group-kinetic theory with a closure by memory loss for modeling turbulence in the atmosphere and the oceans, Meteor, Atm. Physics, **38**, 95-105, 1988.

Thorpe, S.A., On the clouds of bubbles formed by breaking wind waves in deep water, and their role in air-sea gas transfer, Phil. Trans. Roy. Soc. (London) Ser. A, 304, 155-210, 1982.

Thorpe, S.A. and A.J. Hall, The characteristics of breaking waves; bubble clouds and near-surface currents observed using side-scan sonar, Continental Shelf Research 1, 353-384, 1983.

Thorpe, S.A., Bubble Clouds: A review of their detection by sonars, of related models and of how K_y may be determined. *in* Monahan, E.C. and MacNiocaill, G. Eds Oceanic Whitecaps and Theor Role in Air-Sea Exchanges Processes, 57-68, D. Reidel Publishing Co., 1986.

Toba, Y. and H. Mitsuyasu, *eds* The ocean surface; wave breaking, turbulent mixing and radio probing, D. Reidel Publishing Co., 585 pp, 1985.

Ulaby, F.T., F. Kouyate, B. Brisco and T.H. L. Williams, Textural information in SAR images, IEEE Trans. Geosci. Remote Sens., **GE-24**, 2, 235-246, 1986.

Valenzuela, G.R., Theory for the interaction of electromagnetic and oceanic waves- A review, Boundary Layer Meteorol., **13**, 61-85, 1978.

Walsh, E.J., D.W. Hancok III, D.E. Hines, R.N. Swift and J.F. Scott, Directional wave spectra measured with the surface contour radar, J. Phys. Oceanogr., **115**, 566-592, 1985.

Walsh, E.J., D.W. Hancock III, D.E. Hines, R.N. Swift, and J.F. Scott, Wave measurement capabilities of the surface contour radar and the airborne oceanographic LIDAR, Proceedings of the Symposium: Measuring Ocean Waves from Space, Johns Hopkins APL Technical Digest, **8**, 1, 74-81, 1987.

Weissman, D.E., Modeling the interactions between the ocean and the environment for microwave radar sensing, Radio Science, **22** - 1, 87-99, 1987.

Wetzel, L., On microwave scattering by breaking waves, *in*, O.M. Phillips and K. Hasselmann, Wave Dynamics and Radio Probing of the Ocean Surface, pp. 273-284, Plenum, N.Y., 1986.

Wetzel, L.B., On the theory of electromagnetic scattering from a raindrop splash, NRL Memorandum Report 6103, Naval Research Lab., Washington, DC., 42 pp, 1987.

Woiceshyn, P.M., S. Peteherych, M.G. Wurtele, D.H. Boggs, M.M. Borowski, G.F. Cunningham, A. Davies, and G. Mittitt, Cover Photo: SEASAT scatterometer marine wind analysis, IEEE J. Oceanic Eng., **OE-10**, 4, 342, 1985.

Woiceshyn, P.M., M.G. Wurtele, D.H. Boggs, L. F. McGoldrick, and S. Peteherych, The necessity for a new parameterization of and empirical model for wind ocean scatterometry, J. Geophys. Res., **91**, 2273-2288, 1986.

Woiceshyn, P., M. Wurtele, D. Anderson, D. Boggs, G. Cunningham and P. Jansse, Wind field inference from ocean microwave backscatter: the problem as a whole (abstract only) 1987 International Geo-science and Remote Sensing Symposim IGARSS '87 IEEE 87 CH 2434-9, **2**, 987-988, 1987.

Worthington, A.M., A study of splashes, The Macmillan Co., New York, N.Y., reprint of 1908 edition, original work reported in Proc. Roy. Soc. (London), **34**, 217 pp. (1882), 1963.

Wright, J.W., Backscattering from capillary waves with application to sea clutter, IEEE Trans. Antennas Propag., AP-14, 749-754, 1966.

Wu, J., Bubbles in the near-surface ocean: a general description, J. Geophys. Res., **93**, C1, 587-590, 1988.

Zambresky, L.F., The operational performance of the Fleet Numerical Oceanography Center Global Spectral Ocean-Wave Model, Proceedings of the Symposium: Measuring Ocean Waves from Space, Johns Hopkins APL Technical Digest, **8**, 1, 33-36, 1987.

Chapter 14

MICROWAVE TECHNIQUES FOR MEASURING DIRECTIONAL WAVE SPECTRA

F. C. JACKSON
NASA Goddard Space Flight Center
Greenbelt, MD

D. R. LYZENGA
Environmental Research Institute of Michigan
Ann Arbor, MI

1. Introduction

A variety of instruments have been developed during the past thirty years for measuring ocean wave spectra. The simplest of these devices is designed to measure, either directly or indirectly, the surface elevation at a given location in space as a function of time. The observed temporal fluctuations of the surface elevation are described statistically in terms of the frequency spectrum. Using the dispersion relation for surface gravity waves, each frequency component may also be associated with a unique wavelength, assuming that the apparent frequency of the waves is not influenced by currents or by the motion of the sensor through the water. However, no information on the direction of propagation of the waves is obtainable from a point measurement of this type.

For many scientific and engineering purposes, such as the study of air-sea interactions or the evaluation of wave forces on coastal structures and beaches, information on the directional distribution of the surface waves is also needed (e.g. Dean, 1975; Wiegel, 1982). Such information can be obtained from devices such as pitch-and-roll buoys or by deploying arrays of point measurement devices and appropriately processing the results of the measurements. However, to obtain adequate directional resolution for long waves it may be necessary to use large arrays which are difficult to deploy, particularly in deep water. Stereophotographic techniques (e.g. Cote et al, 1960; Holthuijsen, 1983) or other optical methods such as the Stilwell technique (Stilwell, 1969; Monaldo and Kasevich, 1981) may also be used for measuring directional spectra, but these are dependent on favorable environmental conditions and require intensive processing.

The perceived need for new and improved methods of measuring the directional wave spectrum has fortunately coincided with recent developments in radar technology, and as a result considerable research in the last two decades has been devoted to studies of the possible uses of radar for wave measurements. One area of research has involved the use of radar devices operating in the high-frequency (HF) region of the spectrum, with electromagnetic wavelengths on the order of 10 meters (Crombie and Watts, 1968). Another area of research

G. L. Geernaert and W. J. Plant (eds.), Surface Waves and Fluxes, Volume II, 221–264.

developed in the microwave region of the spectrum, with wavelengths in the centimeter range (Zamarayev and Kalmykov, 1969). Interest in microwave devices is largely due to the possibility of operating such devices from spacecraft platforms, which would allow large-scale synoptic observations of wave conditions over wide areas of the ocean surface.

An overview of the various types of radar systems and techniques which have been developed or proposed for measuring ocean wave spectra is presented in the following section, and a subset of these techniques is discussed in more detail in the remainder of this chapter. The overview is meant to be representative, rather than exhaustive. For a more complete picture of the research in the field, the interested reader should avail himself of other reviews, books, and compendia bearing on the subject, for example, by Barrick (1972), Beal (1981), Vesecky and Stewart (1982), Dean (1982), Allen (1983), Hasselmann et al. (1985), Phillips and Hasselmann (1986), Atlas et al. (1986), and Beal (1987). The main object of this contribution is to provide the reader with a basic understanding of what we consider to be the most potentially important techniques for aircraft and satellite application. Toward this end, we have attempted to present the material in a unified framework.

2. Summary of Available Radar Techniques

Several different types of radar systems and techniques have been developed for measuring ocean wave spectra. These systems may be categorized in terms of the radar frequency or waveform, the interaction mechanisms which form the basis for the measurement, or the platforms from which the systems operate. A brief description of each of these systems is given below, and more detailed analyses are presented in the following section for those systems which are capable of being operated from spacecraft platforms.

One possible measurement technique uses radar wavelengths which interact directly (via a Bragg scattering mechanism) with surface waves having wavelengths on the order of tens of meters. Several groups have reported interesting and useful results using such techniques (e.g. Trizna et al., 1977; 1980). However, because of the difficulty of generating directional beams at longer wavelengths, most of the energy-containing ocean waves with wavelengths of 50-500 m cannot be measured directly by this method. To get around this limitation, indirect high-frequency (HF) techniques have been developed that make use of the second-order temporal (or Doppler) spectrum of the backscattered signal to obtain the spectrum of the longer waves (Barrick, 1972; Lipa et al., 1981). Because of the difficulty in synthesizing the required antenna apertures and because of strong interference by the ionosphere, HF techniques are not well-suited to satellite application. Recent research has therefore tended to focus on the microwave portion of the electromagnetic spectrum, with wavelengths on the order of 1 cm to 1 m.

An extension of the HF radar technique to microwave frequencies is the dual-frequency or delta-k technique (Plant, 1977; Alpers and Hasselmann, 1978; Johnson and Weissman, 1984). In this type of system, two closely spaced microwave frequencies are used, such that the difference or "beat" frequency corresponds to a wavelength comparable to that of the surface waves of interest. Because of the shorter electromagnetic wavelengths involved in this technique, much smaller antennas and larger standoff distances can be used. Consequently operation from aircraft and even spacecraft platforms can be contemplated. This technique will therefore be analyzed in some detail in the following section of this chapter. As will be seen, the interaction mechanism in this case is not as direct as in the long-wave radar system, and involves the variation or modulation of the backscattered microwave power along the wave profile, which may be environmentally dependent.

A more direct microwave measurement of the surface elevation or wave height spectrum can be obtained by conventional beam-limited ranging techniques applied in the vertical or near-vertical direction. An example of such a system is the Surface Contour Radar (Walsh et al., 1985; 1987). This system is basically a scanning altimeter, with a footprint which is small enough to resolve the surface wavelengths of interest. The surface elevation is measured with a vertical resolution of about 0.15 meters across a swath width equal to about half the platform altitude, which is normally 400 to 800 meters. The results obtained by this device are probably equal or superior to those obtained by most *in-situ* measurements. However, as a beam-limited device it is not feasible for spacecraft application because of the large antenna size which would be required to maintain an acceptable spatial resolution.

Two types of airborne short pulse radar techniques have been developed that, like the delta-k radar technique, do not involve actual imaging of the sea surface, but rely on a wavefront matching condition to discriminate among wave directions. The FM-ROWS (Plant et al., 1987) measures the wave amplitude by observing the Doppler shift (or frequency modulation) of the returned signal due to the orbital motions of the surface waves, while the AM-ROWS (Jackson et al, 1985a,b; 1987b) uses the variation in the amplitude of the returned signal to infer the wave slope. Both techniques have been demonstrated from aircraft platforms, and could in principle be operated from spacecraft platforms as well. The AM-ROWS technique is more likely to be successful in space because it is not as seriously affected by the motion of the platform. However, this technique is less direct, since, like the dual-frequency or delta-k technique discussed above, it relies on variations in backscattered power across the wave profile. The measurement theory and data analysis techniques involved in this device are discussed in section 3.2.2, and some example results are shown in section 4.

The final type of microwave device to be considered in this review is the side-looking imaging radar. This category includes both real-aperture radar (RAR) and synthetic aperture radar (SAR) systems. Although originally developed for imaging land surfaces and targets, these systems have been used in essentially the same configuration for estimating ocean wave spectra by Fourier transforming the images obtained over the ocean (e.g. Gonzalez et al., 1979; Alpers et al., 1981; McLeish and Ross, 1983; Shuchman et al., 1983; Beal et al., 1983, 1986; Monaldo and Lyzenga, 1986). Both systems use short pulses to obtain resolution in the cross-track direction. However, RAR uses a narrow antenna beam to obtain resolution in the along-track direction, while SAR uses the Doppler shift of the received signal to obtain resolution in this direction. As a result, the RAR resolution degrades rapidly with altitude, while SAR is capable (in principle) of obtaining fine-resolution images from large distances, including spacecraft altitudes, with moderate antenna sizes. The principles of operation and the limitations of SAR for measuring wave spectra are discussed in section 3.2.3, and example results from a spaceborne SAR system are presented in section 4.

3. Theory

3.1. SURFACE SCATTERING

Microwave measurements of ocean wave spectra depend on the use of wave-induced variations in the amplitude and phase of the electromagnetic radiation which is backscattered from the ocean surface. Hence, a knowledge of the electromagnetic scattering properties of the surface is required in order to interpret the measurements. The theory describing this scattering has been discussed in detail in previous chapters of this book. We will therefore concentrate in this section on the statistical properties of the backscattered radiation, including both the random and "deterministic" variations which are related to the small-scale and large-scale structure of the surface.

3.1.1. *The Complex Surface Reflectivity.* The backscattering properties of the ocean surface, including both the amplitude and phase of the scattered field, are conveniently described in terms of a complex reflectivity density $r(x,t)$ (see e.g. Hasselmann et al., 1985). This reflectivity can be modelled as a zero-mean Gaussian noise process that is modulated by the cross section variation associated with the large, or dominant, wave structure. The base noise process derives from the coherency of the incident radiation (cf. Beckmann and Spizzichino, 1963) and has its counterpart at optical frequencies in coherent laser speckle. Even when the surface is only moderately wind roughened, the phase fluctuations are sufficiently rapid and deep that the surface fields (as projected onto the horizontal plane $z = 0$) become incoherent within a few electromagnetic wavelengths so that the fading noise process can be considered to be spatially white. For analysis (and exposition) purposes, it is convenient to consider the reflectivity statistics in two ways, first, in terms of a complete expectation or ensemble average, denoted by $<...>$ that includes all realizations of the variable cross section, and, second, in terms of a conditional or 'partial' expectation, denoted by $<...>_p$, that depends on a particular realization of the large wave structure. The second order statistics for the latter type of averaging can be expressed as

$$< r(\vec{x},t) r^*(\vec{x}\,',t') >_p \; = \delta(\vec{x} - \vec{x}\,')\rho(\vec{x}, t; t - t')\sigma_0(\vec{x}, t) \tag{1}$$

where $\delta(\vec{x} - \vec{x}\,')$ is the Dirac delta function, $\rho(\vec{x}, t; t - t')$ is the temporal autocorrelation function, which is normalized so that $\rho(\vec{x}, t; 0) = 1$, and $\sigma_o(\vec{x},t)$ is the radar cross section per unit area of the surface.

Fourth order moments of the reflectivity involving the complete expectation are required for analytical expressions of the detected reflectivity spectra from the outputs of the several remote sensing systems we shall be considering. These moments can be related to the second moment results using the Gaussian moment theorem (Reed, 1962) for the noise process and assuming the independence of base noise and modulation processes. In the simplest case where the detailed time evolution of the echo signal process can be ignored, the fourth moment statistics can be considered in terms of an ensemble of spatial realizations of the reflectivity density and the moment expressed as

$$\begin{aligned} < r(\vec{x}) r^*(\vec{x}\,') r^*(\vec{x}\,'') r(\vec{x}\,''') > \; &= \delta(\vec{x} - \vec{x}\,')\,\delta(\vec{x}\,'' - \vec{x}\,''') < \sigma_o(\vec{x})\sigma_o(\vec{x}\,'') > \\ &+ \delta(\vec{x} - \vec{x}\,'')\,\delta(\vec{x}\,' - \vec{x}\,''') < \sigma_o(\vec{x})\sigma_o(\vec{x}\,') > \end{aligned} \tag{2}$$

Since $\sigma_o(\vec{x}, t)$ is assumed to be a stationary and homogeneous process, it follows, for example, that

$$<\sigma_o(\vec{x})\sigma_o(\vec{x}')> = \Gamma_\sigma(\Delta\vec{x}) = \sigma_o^2[1 + \gamma_m(\Delta\vec{x})] \tag{3}$$

where $\Gamma_\sigma(\Delta\vec{x}) = \Gamma_\sigma(-\Delta\vec{x})$ is the autocovariance function, $\Delta\vec{x} = \vec{x}' - \vec{x}$ is the lag vector, $\sigma_o = <\sigma_o(\vec{x}, t)>$ is the mean cross section, and where γ_m is the normalized autocorrelation function of the cross section modulation. The two dimensional spectrum of the cross-section modulation may then be defined as the Fourier transform of γ_m,

$$S_m(\vec{K}) = (2\pi)^{-2} \int \gamma_m(\Delta\vec{x}) \exp(-j\vec{K}\cdot\Delta\vec{x})\, d(\Delta\vec{x}) \tag{4}$$

for any arbitrary wave vector $\vec{K} = (K_x, K_y)$. More general results involving temporal statistics could be given here, but as these are only required for a complete analysis of the SAR ocean response--the other techniques being relatively insensitive to scatterer motion effects--we will dispense with further details at present.

3.1.2. *Reflectivity Modulation Mechanisms.* Surface waves with relatively long wavelengths (compared to the electromagnetic wavelength) cause variations in the amplitude and phase of the backscattered field which are described by $\sigma_o(\vec{x}, t)$ and $\rho(\vec{x}, t; t - t')$. Variations in the radar cross section may be caused by variations in the small-scale roughness, referred to as hydrodynamic modulation, or by purely geometric effects, which are referred to as tilt modulation.

The hydrodynamic modulation of the small-scale roughness by longer surface waves is discussed elsewhere in this book. The interactions responsible for this modulation may include the straining of short waves by the orbital motions associated with the longer waves, variations in the wind stress across the profile of the longer waves, and possibly nonlinear interactions such as wave breaking. These interactions are not completely understood and may be environmentally dependent, making predictions difficult, although a fairly extensive data base of measurements exists (e.g. Plant et al., 1983; Keller et al., 1985; Schroter et al., 1986).

The tilt modulation term includes two effects (cf. Figure 1). The first effect is due to the change in the radar cross section per unit area with incidence angle. For small incidence angles, $\theta < 15°$, the dominant scattering mechanism is quasi-specular reflection (e.g. Valenzuela, 1978), and the fractional change in the radar cross section per unit area for a change $d\theta$ in the local incidence angle is given by

$$d(\ln\sigma_o) = -(\tan\theta / <\eta_x^2>)d\theta \tag{5}$$

assuming the surface is Gaussian-distributed with a slope variance $<\eta_x^2>$ in the plane of incidence. This slope variance should not be confused with the total surface slope variance as, for example, might be measured optically. On account of diffraction, the surface that is sensed by the radar is essentially a filtered version of the surface in which waves on a scale smaller than the electromagnetic wavelength are effectively removed (cf. G.S. Brown, this volume).

For larger incidence angles, $\theta > 20°$, the dominant scattering mechanism is Bragg scattering. Assuming a perfectly conducting surface and a k^{-4} short wave spectrum, the fractional

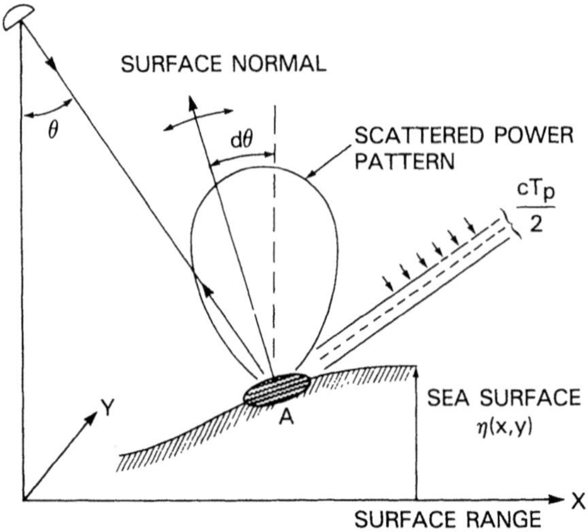

Fig. 1 — Illustration of the tilt-modulation mechanism for producing wave contrasts. The component of the large-scale surface slope $\eta_x(x, y)$ in the plane of incidence acts to modulate the radar cross section by (1) increasing or decreasing the intercepted area A within a slant range interval $dR = cT_p/2$, and (2) by rotating (through an angle $d\theta$) the scattered power pattern produced by the smaller wave structure (which is presumably responsible for most of the scattering).

change in the radar cross section per unit area for a change $d\theta$ in the incidence angle in this regime is

$$d(\ln\sigma_o) = -4 \cot \theta(1 \pm \sin^2\theta)^{-1} d\theta \tag{6}$$

where the plus sign refers to vertical polarization and the minus sign refers to horizontal polarization (e.g., Valenzuela, 1978). In both cases, the local σ_o decreases with increasing local incidence angle.

The second effect which contributes to tilt modulation is due to the change in the surface area within a resolution cell with the local incidence angle. For a pulse-limited system, the surface area contributing to a given range resolution cell is proportional to the cosecant of the local incidence angle. This effect produces an additional change in the effective radar cross section per unit (horizontal) area, given by

$$d(\ln\sigma_o) = -\cot \theta d\theta \; . \tag{7}$$

Note that this effect is greatest at small incidence angles, and contributes in the same sense as the first term, i.e. it causes a decrease in the received power with increasing local incidence angle. For a beam-limited system, the surface area within a resolution cell at a given range is proportional to the secant of the local incidence angle, and the corresponding change in the effective radar cross section is given by

$$d(\ln\sigma_o) = +\tan\theta d\theta . \tag{8}$$

In this case, the effect is in the opposite direction, i.e. it causes an increase in the received power with increasing local incidence angle. However, the effect is small at most incidence angles and may be expected to contribute significantly only at near-grazing incidence.

These results may be used to infer the changes in the received power due to variations in the surface slope. Since the deviation in the local incidence angle from its nominal value (in radians) is approximately equal to the component of the wave slope in the plane of incidence, the effective radar cross section per unit area may be written as

$$\sigma_o(\vec{x},t) = \sigma_o[1 + |m| \eta_x(\vec{x}, t)] \tag{9}$$

where $|m|$ is the value of $d(\ln\sigma_o)/d\theta$ including both effects described above, and η_x is the component of the surface slope* in the plane of incidence. If the waves are assumed to be small-amplitude, linear dispersive waves, then the surface elevation $\eta(\vec{x}, t)$ can be expressed in terms of the spatial Fourier transform $F_\eta(\vec{K})$ and the cross section accordingly expressed as

$$\sigma_o(\vec{x}, t) = \sigma_o[1 + \mathrm{Re}\int mK_x F_\eta(\vec{K}) e^{j(\vec{K}\cdot\vec{x} - \Omega_k t)} d\vec{K}] \tag{10}$$

where $m = j|m|$ and where for deep water waves the wave frequency $\Omega_k = \sqrt{gK}$, where $K = |\vec{K}|$ and g is the acceleration of gravity. We note that $F_\eta(\vec{K})$ contains information on the absolute direction of wave travel, and the directional energy spectrum, proportional to the average value of $|F_\eta|^2$, similarly contains this directional information. However, to obtain F_η from the reflectivity field, a temporal as well as spatial sample is required, and in many important applications (e.g., satellite measurements), this temporal information is lacking, since the sample area A is viewed for only a small fraction of a wave period. Thus, the field of σ_o is viewed essentially in an instant (say, at $t = 0$), and only a finite resolution estimate of the polar symmetric wavenumber height variance, or energy, spectrum

$$S_\eta(\vec{K}) = \lim_{A\to\infty} (\pi^2/A)[<|F_\eta(\vec{K})|^2> + <|F_\eta(-\vec{K})|^2>] \tag{11}$$

can be recovered from the reflectivity data (see, for example, Atanassov et al., 1985). That is, according to the linear tilt model, one obtains an estimate of S_η according to the relationship

$$S_m(\vec{K}) = |m|^2 K_x^2 S_\eta(\vec{K}) \tag{12}$$

*Here for convenience we are letting η_x represent the slope of the underlying long waves with wavelengths greater than the radar resolution, i.e. on the order of 10-20 m.

where S_m is the polar symmetric reflectivity modulation spectrum defined by Eq. (4). Although this equation was derived specifically for tilt modulation, hydrodynamic effects can also be included by appropriately modifying or generalizing the modulation transfer function m. In this more general case, m may be expressed as the sum of the tilt modulation transfer function (MTF) m_t described above and a hydrodynamic MTF m_h, i.e.,

$$m = m(\vec{K}) = m_t + m_h \tag{13}$$

The linear tilt MTF is independent of the wavenumber $\vec{K}$ whereas m_h is in general a function of $\vec{K}$. In the simplest case of hydrodynamic modulation of a Bragg wave with a relaxation time that is long compared to the period of the large wave m_h is given by (e. g., Alpers and Hasselmann, 1978)

$$m_h = -4.5(K_x/K) \tag{14}$$

Note that this hydrodynamic modulation of the small scale roughness is 90° out of phase with the slope signal, i.e., it is a maximum at the long wave crests. More general expressions for the hydrodynamic modulation are given by Alpers and Hasselmann (1978), for example, and by Smith (this volume). According to the above expressions, the hydrodynamic MTF dominates for large incidence angles, $\theta \gtrsim 45°$, while the tilt MTF dominates for small angles of incidence, $\theta \lesssim 45°$. In the case of near-vertical incidence, quasi-specular scatter, $\theta \lesssim 15°$, hydrodynamic modulation effects can be ignored and the MTF can be approximated by m_t alone; for pulse-limited illumination, we have from the above,

$$|m| \cong |m_t| = \cot\theta + \tan\theta/\langle n_x^2\rangle \tag{15}$$

It is again emphasized that while it is possible to calculate the hydrodynamic MTF from wave-current interaction theory, the physics of the modulation of the Bragg wave is, in reality, quite complex, and in the final analysis one must resort to empirical and strongly environmentally dependent MTFs as discussed above. Geometrical tilting on the other hand is a comparatively simple and well understood mechanism. Thus, there is reason to suppose that the near-nadir, quasi-specular scattering regime holds a greater promise for remote microwave measurements of directional wave spectra, and it behooves us therefore to investigate in somewhat more detail the precise mechanism of the reflectivity modulation in near-nadir backscatter.

3.1.3. *Quasi-Specular Scatter Solution for S_m.* As a first step toward a more detailed modelling of the reflectivity modulation in the near-nadir scatter regime, one may assume the scattering to be described entirely according to geometrical optics (cf. G. S Brown, this volume, for a discussion of quasi-specular scatter). One may start with the physical optics integral expression for the scattered field, form the fourth-order moments, and then proceed to the high frequency limit as done by Jackson (1981). Alternatively, one may start directly with a geometrical optics formulation for the scattered power as done by Jackson (1987a). Either way, one may derive the following result for the two dimensional reflectivity modulation spectrum (in the case of pulse-limited illumination):

$$S_m(\vec{K}) = (2\pi)^{-2} \int d(\Delta\vec{x}) \exp(-j\vec{K}\cdot\Delta\vec{x}) \cdot \iint d\eta d\eta' \exp[jK_x \cot\theta(n' - \eta)] \cdot$$

$$\left[\frac{p_6(\eta, \nabla\eta, \eta', \nabla\eta'; \Delta\vec{x}) - p_3(\eta, \nabla\eta) p_3(\eta', \nabla\eta')}{[p_2(\nabla\eta)]^2} \right]_{\nabla\eta = \nabla\eta' = \nabla\eta_{sp}} \quad (16)$$

where p_6 is the joint probability density function (pdf) of height and vector slope at two points in the horizontal plane $\vec{x}$ and $\vec{x}'$ separated by the lag vector $\Delta\vec{x} = \vec{x}' - \vec{x}$; p_3 is the pdf of height and slope at either $\vec{x}$ or $\vec{x}'$;p_2 is the pdf of slope alone (proportional to the average cross section σ_o); and where $\nabla\eta_{sp}$ is the specular slope,

$$\nabla\eta_{sp} = (\eta_x, \eta_y)_{sp} = (\tan\theta,\ 0) \quad (17)$$

where again the x, y coordinate system is defined with the x-axis in the plane of incidence and the y-axis orthogonal (Figure 2). The appearance of the phasor in the height is due to the fact that the cross section σ_o $(\vec{x}, t)$ at any point $\vec{x}$ in the horizontal plane actually consists of contributions from all surface specular points lying in a constant slant range interval $dR = dx\sin\theta - dz\cos\theta$ (cf. Fig. 1). This results in a confounding of the surface range x and surface elevation. The effect, which is well known in radar terrain mapping, is significant when the incidence angle is on the order of or smaller than the steepness of the dominant wave and results in a filtering of the shorter wave components. For a Gaussian surface, this filtering is described by the function $\exp(-K_x^2 cot^2\theta <\eta^2>)$. To some extent (the precise degree to which depending on the incidence angle), this filtering may be offset by overall spectral whitening from the nonlinear facet distribution over the long wave slopes (see Jackson, 1981, for some numerical results). In the 'optimal' case where the incidence angle is significantly larger than the dominant wave slope, but not significantly larger than the rms slope $<\eta_x^2>^{1/2}$ the scattering solution (16) may be linearized to yield a result identical to the above linear tilt model result for S_m, Eqs. (12) and (15), as shown by Jackson (1981) in the case of an approximately Gaussian-distributed sea surface. The optimal incidence angle, i.e., that angle for which nonlinearities in the response are minimized, is not precisely established at present, in part because spectral fidelity criteria have yet to be established for the wave-slope response. However, one may infer from the above that the optimal angle is in the neighborhood of $\theta = \theta_{opt} = \text{atan} <\eta_x^2>^{1/2}$ i.e., where m_t as a function of θ is a minimum.

Measurements of the near-nadir cross section at K_u-band frequency (~2 cm wavelength) show that the slope pdf is nearly Gaussian and isotropic with a total mean square slope parameter $<(\nabla\eta)^2> \sim 2<\eta_x^2>$ that is nearly a linear function of wind speed over the range of 5-15 m/s, with a magnitude of about 60% of the Cox and Munk (1954) optically-derived values (Jackson, 1988; Jackson et al., 1989). At a nominal wind speed of 10 m/s, $<(\nabla\eta)^2> \sim 0.04$, and so it would follow that the corresponding "optimal angle" is about 8 degrees. These direct measurements of the mean square slope parameter (via the cross section magnitude and shape) are in remarkably good agreement with mean square slopes inferred indirectly using the linear tilt model and measured modulation spectra (Jackson et al., 1985a);

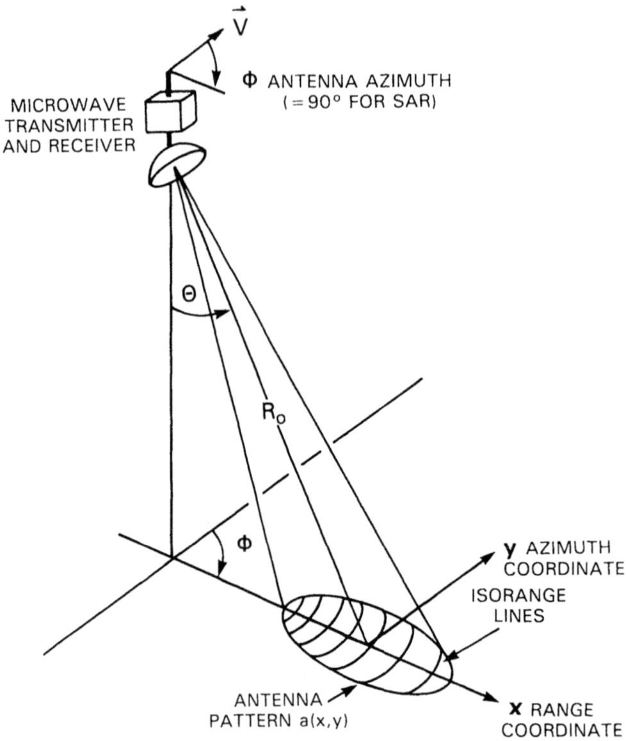

Fig. 2 — Basic geometry for air- or spaceborne radar ocean wave measurements. In the case of the nonimaging radar techniques discussed in the text (such as the delta-k and short pulse techniques) the antenna beam is rotated in azimuth through 360° to map all wave directions, while in the case of imaging radars (real and synthetic aperture), the antenna beam is held fixed, usually toward broadside ($\phi = 90°$). Except in the case of real aperture radar (RAR), the antenna beam 'footprint' on the surface is large compared to the scale of the waves in both the range (x) and azimuth (y) dimensions.

thus, there is strong empirical as well as theoretical evidence supporting the basic validity of the linear tilt model in quasi-specular scatter measurements of directional wave spectra.

In the transition region between quasi-specular and Bragg diffraction regimes, the scattering is, of course, more complex. However, motivated by the fact that the incidence angle of the Seasat SAR ($\theta = 23°$) was located within this transition region, efforts have been made to model the cross section modulation and the SAR response in this particular angular range (e.g., Holliday et al., 1986). The SAR response, we note, is particularly difficult to model in this range because of the dependence of the SAR response on the temporal statistics of the reflectivity density. These temporal statistics are considered briefly below.

3.A.4. *The Temporal Autocorrelation Function* $\rho(\bar{x}, t; \Delta t)$. The temporal statistics of the reflectivity density are described to second order by the temporal autocorrelation function $\rho(x, t; \Delta t)$ in Eq. (1). In order to model this dependence (short of a full solution of the scattering problem from first principles), the surface may be visualized as a collection of scatterers moving with random velocities. The complex reflectivity may then be expressed as

$$r(\bar{x}, t) = \sum_n a_n \exp(2jRV_n t) \tag{18}$$

where a_n represents the complex amplitude of a scatterer having a radial velocity V_n. If we assume that these amplitudes are not correlated with the velocities and that the scatterers are independent, then the reflectivity covariance function is given by

$$< r(\bar{x},t) r^*(\bar{x},t') >_p = \sum_n < |a_n|^2 >_p < \exp[2jkV_n(t-t')] >_p . \tag{19}$$

If we assume in addition that the scatterer velocities are identically normally distributed, then the correlation function becomes

$$\rho(\bar{x}, t; \Delta t) = \exp[-(\Delta t)^2/2T_c^2] \exp[j\Omega_c \Delta t)] \tag{20}$$

where $\Delta t = t - t'$, $T_c = (2k\sigma_v)^{-1}$ is the coherency time, $\Omega_c = 2k\bar{V}_r$ is the mean Doppler frequency, and $\bar{V}_r = \bar{V}_r(\bar{x}, t)$ is the mean radial (line of sight) velocity and $\sigma_v = \sigma_v(\bar{x}, t)$ is the standard deviation (width) of the velocity distribution. The scattering velocities are generally assumed to be related to the orbital motions of the surface waves, with sub-resolution-scale waves contributing to σ_v and longer waves contributing to the mean velocity. For Bragg scatter, this is a reasonable assumption since the lifetimes of the Bragg waves may be expected to be longer than the decorrelation time associated with the orbital velocities, and so the Bragg wave can be considered simply to be advected by the orbital motions of the longer waves. In the case of specular scattering, however, the lifetimes of individual facets can be quite small and so their detailed temporal evolution must be considered more carefully. Winebrenner and Hasselmann (1988), starting from the physical optics integral expression for the specularly scattered field, have calculated the temporal autocorrelation function for a Gaussian sea-surface. They find a form of $\rho(\bar{x}, t; \Delta t)$ identical to (20) with values of T_c not very different from those expected from orbital velocity effects alone. These results will be discussed further in section 3.2.3.

We note that higher-order surface motions, particularly surface accelerations, have been neglected in the above simplified model of the temporal autocorrelation, although to a certain extent the temporal dependence of the mean velocity field $\bar{V}_r(x, t)$ does reflect higher order surface temporal behavior. The effect of surface accelerations, particularly on the SAR ocean imaging mechanism, is still not completely understood at present, although some definite progress toward such an understanding has been made recently by Alpers and Bruening (1986). Their results will be discussed briefly in section 3.2.3.

3.2. SYSTEM DESCRIPTIONS

Three types of microwave systems for measuring ocean wave spectra are discussed in this section. Each system consists of a transmitter which generates a high-frequency electromagnetic field, an antenna (or set of antennas) which radiates this electromagnetic field over a solid angle Ω_o and receives the electromagnetic field scattered from the surface, and a receiving/recording system which converts the received electromagnetic field into a lower-frequency electrical signal which is appropriately sampled and recorded. Subsequent processing of the recorded signals to obtain and display the wave spectrum may be done immediately or at a later stage, depending on the complexity of the processing required.

The relationship between the transmitted and received electromagnetic field strengths, E_t and E_r, may be written as

$$E_r(t) = \int E_t(t - 2R/c)\, a(\bar{x}, t - R/c)\, r(\bar{x}, t - R/c)\, d\bar{x} \tag{21}$$

where $R(\bar{x}, t)$ is the instantaneous range distance from the transmit/receive antenna to the point $\bar{x}$ on the mean surface ($z = 0$), c is the speed of light, $a(\bar{x}, t)$ is a factor which includes the two-way antenna gain as well as propagation (R^{-2}) losses, and $r(\bar{x}, t)$ is the complex reflectivity density of the surface, as described in the previous section. Since $a(\bar{x}, t)$ and $r(\bar{x}, t)$ are slowly-varying functions of time, the time delay R/c may be neglected in their arguments. However, the time delay in E_t is important, since it is the basis for the range measurement which is implicit in each technique.

The differences among the various radar systems considered below lie in the form (i.e. the time dependence) of the transmitted signal, the illumination geometry, and the processing of the received signals. In the dual-frequency or delta-k technique, the transmitted waveform is essentially the sum of two c.w. signals with slightly different frequencies, whereas in the short pulse and synthetic aperture radar cases a pulsed waveform (usually with a linear frequency modulation) is used to obtain spatial information about the scene. Descriptions of these techniques are presented in the following paragraphs. In the discussions to follow, the reader may find it helpful to refer to Figure 2, which provides a general picture of the scattering and measurement geometry applicable to the several systems considered in detail herein, and to Figure 3, which provides elementary block diagrams of these systems.

3.2.1. *Dual-frequency and Three-frequency Systems.* In the dual-frequency (DF) or delta-k technique, the transmitted signal consists of two sinusoidal components with frequencies ω_1 and ω_2 separated by a difference frequency $\Delta\omega$. These components may be generated and transmitted simultaneously, for example, by modulating a carrier so as to produce two sidebands at the appropriate frequencies (Plant, 1977), or they may be generated and transmitted

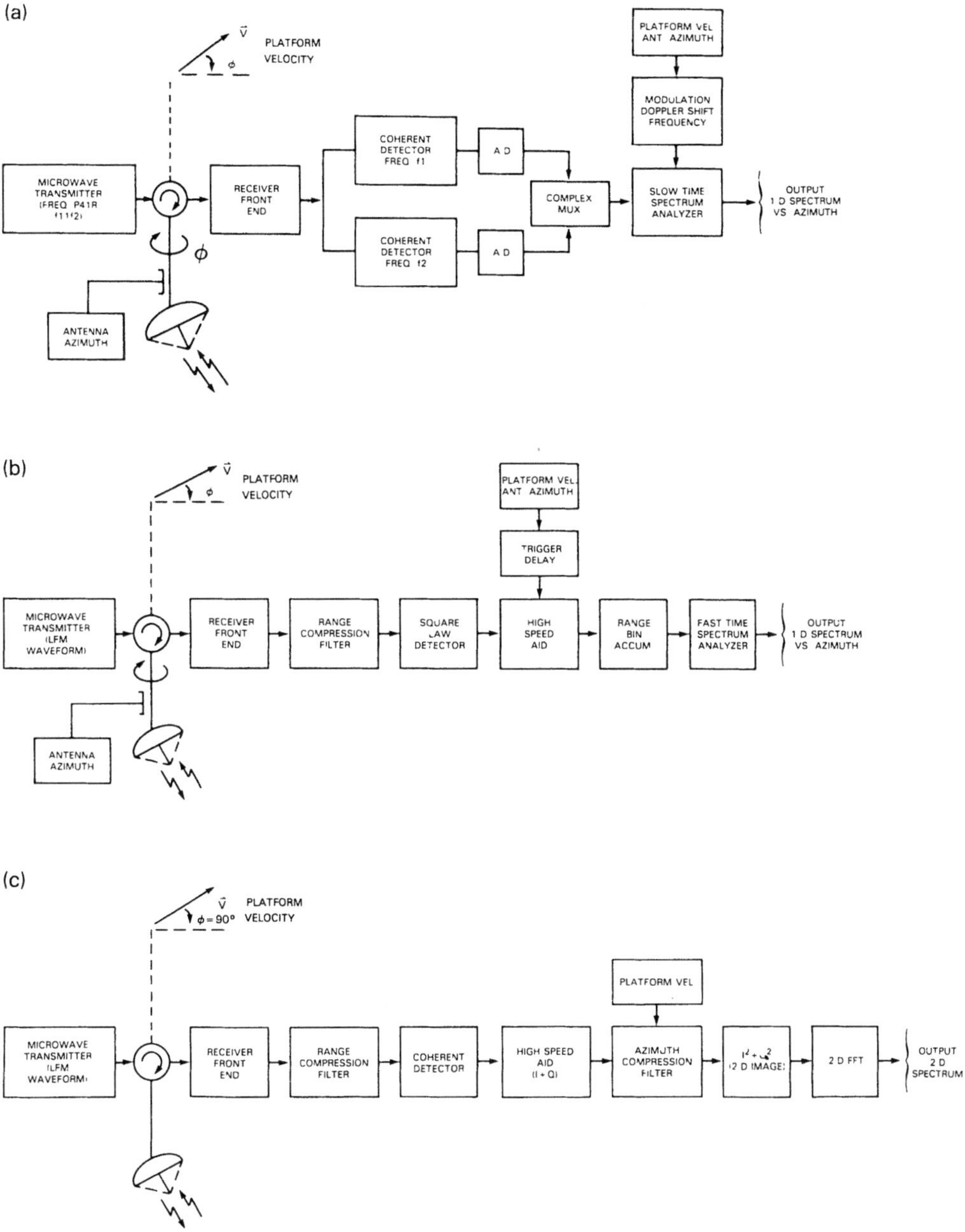

Fig. 3 — Elementary block diagrams of a dual-frequency (delta-k) system (a), a short-pulse (AM-ROWS) system (b), and a synthetic aperture radar (SAR) system (c). Presently, in all these systems, the processing segments following the sample and holds are not a part of the radar systems proper, and signal processing must be done off-line (on the ground).

sequentially on alternate pulses (Johnson and Weissman, 1984). The received signal from either of these components can be written as

$$E_i(t) = E_o \int a(\vec{x}, t) r(\vec{x}, t) e^{j(\omega_i t - 2k_i R)} d\vec{x} \tag{22}$$

where E_o is the amplitude and $k_i = \omega_i / c$ $(i = 1, 2)$ is the wavenumber of the transmitted signal. The received signals from the two transmitted frequencies are combined to produce an output signal which can be represented by

$$s_{12}(t) = E_1(t) E_2^*(t) e^{-j(\Delta\omega)t} \tag{23}$$

where $\Delta\omega = \omega_1 - \omega_2$. This signal can be obtained, for example, by mixing the RF signals received in two separate channels or by square-law detecting and bandpass filtering at $\Delta\omega$, and then downconverting by $\Delta\omega$ to baseband. Also, as depicted in Figure 3a, the signal at each frequency can be coherently detected and the (complex) amplitudes cross-multiplied to yield (23).

Using the delta-correlation property of the reflectivity, as discussed in the previous section, the expected value of this output signal can be shown to be given by

$$< s_{12}(t) >_p = E_o^2 \int a^2(\vec{x}, t) \sigma_o(\vec{x}, t) e^{-j2(k_1 - k_2)R} d\vec{x}. \tag{24}$$

The range distance R can be expressed to first order in the ground range coordinate x as

$$R = R_o + x \sin\theta \tag{25}$$

where θ is the incidence angle and R_o is the range distance to the center of the coordinate system $x = y = 0$, as shown in Figure 2. With this substitution, Eq. (24) can readily be seen to be the finite-resolution Fourier transform of the radar cross section $\sigma_o(\vec{x}, t)$ at the 'radial' wavenumber $\vec{K}_r = (K_x, 0)$ where

$$K_x = 2(k_1 - k_2)\sin\theta = 2(\Delta\omega / c)\sin\theta. \tag{26}$$

Assuming that the spatial variations in the radar cross section can be described by a linear modulation transfer function, as discussed in the previous section, this signal may then be related to the wave height or slope spectral density at the wavenumber $\vec{K}_r$. By changing the frequency separation and rotating the antenna, the two-dimensional wave spectrum can thus be inferred.

The accuracy with which the wave spectrum can be estimated using the dual frequency technique is limited by coherent interference or Rayleigh fading effects. These effects cause random fluctuations in $s_{12}(t)$ in addition to the periodic variations due to the motion of the waves being sensed. These two kinds of temporal variations can be differentiated by taking the power spectrum of $s_{12}(t)$. As shown by Plant (1977) and Alpers and Hasselmann (1978) this power spectrum consists of a broad background fading spectrum (Figure 4a), superimposed upon which is a sharp line at the frequency

$$\Omega_{res} = \vec{K}_r \cdot \vec{V} \pm \Omega_K \tag{27}$$

where $\vec{V}$ is the platform velocity vector, $\Omega_K = \sqrt{gK_x}$ is the intrinsic wave frequency, and where $\vec{K}_r \cdot \vec{V}$ is the platform motion induced Doppler shift of the modulation frequency $\Delta\omega$. In the case of radars mounted on fixed platforms the Doppler frequency shift is given by Ω_K alone. For satellite platforms Ω_K may be neglected; but it can be significant for aircraft platforms. The integral of the power spectrum of $s_{12}(t)$ over all frequencies yields the total DF signal variance

$$<P_{df}> = <s_{12}(t)s_{12}^*(t)> = <E_1(t)E_2^*(t)E_1^*(t)E_2(t)> \tag{28}$$

which, on making use of the fourth moment result (2), can be written as

$$<P_{df}> = E_o^4 \iint a^2(\vec{x})a^2(\vec{x}\,'')\Gamma_\sigma(\vec{x} - \vec{x}\,'')\cdot \exp[-j\vec{K}_r\cdot(\vec{x} - \vec{x}\,'')]d\vec{x}d\vec{x}\,''$$

$$+ E_o^4 \iint a^2(\vec{x})a^2(\vec{x}\,')\Gamma_\sigma(\vec{x} - \vec{x}\,')d\vec{x}d\vec{x}\,'$$

$$= <|<s_{12}(t)>_p|^2> + <|E_o^2 \int a^2(\vec{x}, t)\sigma_o(\vec{x}, t)d\vec{x}|^2>$$

$$\cong E_o^4\sigma_o^2 A(0)(2\pi)^2 S_m(\vec{K}_r) + (E_o^2\sigma_o A_o)^2 \tag{29}$$

where $A_o = \int a^2(\vec{x})d\vec{x}$ is the illuminated area and $A(0) = \int a^4(\vec{x})d\vec{x}$ is the convolved gain pattern, approximated by its value at zero lag, and where the approximation assumes large illuminated areas. The first term in Eq. 29 corresponds to the sharp line in the temporal spectrum of $s_{12}(t)$ as illustrated in Figure 4a, while the second term corresponds to the broad noise background, which is approximately equal to the square of the average echo power, $W_o = E_o^2\sigma_o A_o$, in accordance with the approximate Rayleigh distribution of the field amplitudes (Beckmann and Spizzichino, 1963). Since thermal noise is not as a rule a limiting factor in the measurement, the ratio of these two terms is the fundamental measure of detectability, and is therefore referred to as the signal-to-noise ratio, or SNR (Alpers and Hasselmann, 1978; Jackson, 1981). If the illumination pattern is assumed to be a Gaussian function with one-sigma widths L_x and L_y respectively in range and azimuth,

$$a(x, y) = \exp[-\frac{1}{2}(x^2/L_x^2 + y^2/L_y^2)] \tag{30}$$

then $A_o = \pi L_x L_y$ and $A(0) = A_o/2$, and the SNR becomes

$$SNR_o = (2\pi^2/A_o)S_m(\vec{K}_r) \tag{31}$$

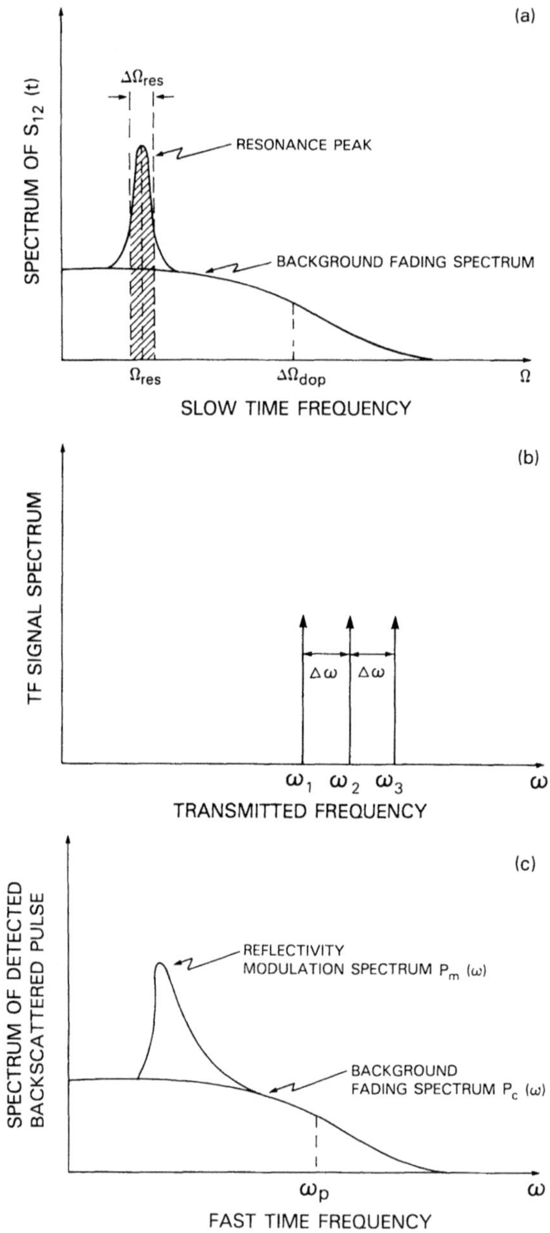

Fig. 4 — Signal spectra for various microwave techniques: (a) the temporal (slow time) spectrum of the backscattered dual-frequency signal $s_{12}(t)$, (b) the three-frequency transmitted signal spectrum, and (c) the mean power variance spectrum of a backscattered short pulse of bandwidth ω_p.

It is evident from Eq. (31) that for large illuminated areas (high platform altitudes) the SNR will become unacceptably small unless some means are taken to reduce the background clutter. Alpers and Hasselmann (1978) showed that the SNR in the DF technique can be dramatically improved if, rather than simply detecting the DF signal $s_{12}(t)$, one first filters the signal in the narrow band about the 'resonance' spike (Figure 4). If the transmitted DF signal is continuous-wave (CW), or if pulsed DF signals are transmitted at a sufficiently high rate (pulse repetition frequency (PRF) greater than twice the Doppler bandwidth), then the echo signal can, by the sampling theorem, be considered to be continuously sampled, and in this special case it can be shown that the SNR, after filtering at the resonance frequency and detecting, is

$$SNR = a_g (\Delta \Omega_{dop} / \Delta \Omega_{res}) \; SNR_o \tag{32}$$

where a_g is a factor on the order of 1 that accounts for the shape of the resonance peak and the details of the spectrum analysis (e.g., the bandwidth of the Doppler or analysis filter, or equivalently, the integration time); $\Delta \Omega_{res}$ is the width of the resonance peak, and $\Delta \Omega_{dop}$ is the bandwidth of the background noise spectrum. The improvement in SNR over SNR_o is typically about 30 dB since the ratio of carrier to modulation frequency Doppler spreads is about 10^3 (e.g., 10 GHz/10 MHz for X-band). The width of the resonance line can be expressed as $\Delta \Omega_{res} = \Delta \vec{K} \cdot \vec{V}$ where $\Delta \vec{K}$ is the dispersion or uncertainty in the modulation wavenumber determined by geometrical factors, namely, wavefront curvature and range extent of illumination (Alpers and Hasselmann, 1978; Jackson, 1981; Johnson and Weissman, 1984). With the Alpers-Hasselmann filtering method, the DF SNR is of the order of 0 dB for typical satellite geometries, and so the technique would appear to be feasible for satellite application. For typical aircraft geometries, the SNR is seen to be quite respectable, and aircraft tests (with a fixed-azimuth antenna) have indeed shown this to be the case (Johnson and Weissman, 1984).

Another method of reducing the background clutter noise is to vary the frequencies of the transmitted signals while holding the difference frequency constant. Averaging the resulting values of the output signal power P_{df} reduces the random variation of the signal, although it does not reduce the mean background signal. However, by appropriately combining the dual-frequency signals from two contiguous frequency pairs, the mean background signal can be reduced to zero (Plant and Reeves, 1984). The method involves forming the dual-frequency signal $s_{12}(t)$ from the frequencies ω_1 and ω_2, as shown in Eq. (23), and a second dual-frequency signal $s_{23}(t)$ from the frequencies ω_2 and ω_3, where $\omega_3 - \omega_2 = \omega_2 - \omega_1$ (cf. Fig. 4b). These two signals are then combined to form the output signal

$$P_{tf}(t) = s_{12}(t) s_{23}^{*}(t) \tag{33}$$

by means of another multiplication and low-pass filtering operation. Using the Gaussian moment result (2) again, the expected value of this three-frequency signal is readily seen to be

$$\begin{aligned} < P_{tf} > &= < | < s_{12}(t) >_p |^2 > + < | < s_{32}(t) >_p |^2 > \\ &= 2 < | < s_{12}(t) >_p |^2 > \end{aligned} \tag{34}$$

where the last follows immediately on noting that the frequency separation $\Delta\omega = \omega_3 - \omega_2 = \omega_2 - \omega_1$ is the same for the two frequency pairs (Fig. 4b). The three-frequency (TF) spectrum is thus twice the value of the DF signal spectrum and contains no noise background. Since the zero expected value of the background noise P_C should be approached asymptotically as $P_n \sim W_o^2/(\Delta\Omega_{dop} T_{in} \rightarrow 0$ where W_o is the average power and T_{in} is the integration time, then it follows that, for typical Doppler spreads of several Kilohertz, noise background levels 30 dB or so below W_o^2 can be achieved with integration times on the order of a second. Thus, the TF technique holds promise for a significant improvement in performance for space-based wave spectrum measurements, and is currently being tested from an aircraft platform (W. Plant, personal communication).

3.2.2. *Short-Pulse Systems.* An alternative to the dual-frequency technique for measuring ocean wave spectra is the conventional radar technique of using short pulses to probe the surface. Among the earliest reported measurements of wave spectra using a short pulse radar are those of Zamarayev and Kalmykov (1969). Their grazing incidence measurements made with an X-band (3 cm) shore mounted radar show surprisingly* good agreement between the spectra of the signal power from range scans (''A-scope'' displays) and buoy inferred slope spectra. More recently, scanning marine radars have been used to map the spatial reflectivity pattern (from ''PPI'' displays) from which three dimensional wavenumber-frequency spectra have been inferred from time sequences of azimuth scans (Young et al., 1985; Atanassov et al., 1985).

As a rule, when aircraft platforms are used, steeper incidence angles are required to ensure an adequate signal level, and, in the case of satellite measurements, angles of incidence (nadir angles) close to vertical incidence are generally mandated by both power and geometrical constraints. A block diagram illustrating an aircraft or satellite-based short-pulse system is shown in Figure 3(b). In principle, this technique could be implemented by generating very short pulses of electromagnetic radiation, with lengths selected to yield the desired spatial resolution. However, in practice various pulse compression methods are often used in which longer, modulated pulses are generated and subsequently converted into equivalent short pulses by appropriately processing the received signals.

In short pulse (SP) systems, it is convenient to distinguish between two time scales, a 'fast' time scale, in which time shall be denoted by τ, and a 'slow' time scale, in which time shall be denoted by t as before. The fast time scale corresponds to the pulse length of individual transmitted pulses T_p and to the duration of individual pulse returns $T_r = 2(R_2 - R_1)/c$ where R_1 and R_2 are the slant ranges to the surface at the near and far edges of the antenna footprint. Thus, fast time is measured typically in nanoseconds or microseconds. The slow time scale is determined by the interpulse period and by the time scale of the Doppler evolution process as determined primarily from aspect angle changes. In particular, the slow time scale is determined by the reciprocal of the Doppler spread, which for pulse-limited illumination, is given by

$$\Delta\Omega_{dop} = 2kV\beta_a \,|\sin\phi| \tag{35}$$

where β_a is the lateral (azimuthal) antenna beamwidth, k is the propagation constant, and ϕ is the look direction (azimuth) relative to the velocity vector $\vec{V}$. Since Doppler spreads and pulse

*In view of the complex nature of near-grazing incidence backscatter (Wetzel, 1987).

repetition frequencies (PRFs) are typically on the order of Kilohertz, slow time has a scale of milliseconds. We note that the separation of time scales implicitly assumes that the geometry and surface motion is frozen during an individual pulse return time. Such an assumption is by and large a legitimate one for the application at hand.

Consider a transmitted pulse of the form

$$E_t(\tau) = p(\tau)\exp(j\omega_o\tau) \tag{36}$$

where $p(\tau)$ is the pulse modulation and $\omega_o = k_o c$ is the carrier frequency. In simple short pulse systems, the pulse length is short compared to the scale of the waves to be probed. In pulse compression systems, the physical pulse length is usually quite long (so as to maximize the average echo power relative to the thermal noise); often, chirped waveforms of the form

$$p(\tau) = p_o(\tau)\exp(-j\dot{\omega}\tau^2/2) \tag{37}$$

are used where p_o is the envelope of length T_{po} and $\dot{\omega}$ is the frequency sweep rate.*

Substituting the pulse waveform into Eq. (21) yields for the received field

$$E_r(\tau,t) = \int a(\vec{x},t)r(\vec{x},t)p(\tau-2R/c)\exp[j\omega_o(\tau-2R/c)]d\vec{x} \tag{38}$$

If E_t is a pulse compression waveform, then the received echo signal may be subjected to a matched filter operation** before detection to produce the compressed signal,

$$E_c(\tau,t) = \int E_r(\tau', t)E_t^*(\tau'-\tau, t)d\tau' = E_r \otimes E_t^* \tag{39}$$

where $\otimes$ denotes convolution (here with respect to fast time). The compressed signal is then detected, usually with a square law detector, to produce a signal $W \alpha |E_c|^2$ that is proportional to the backscattered power versus fast time or, equivalently, versus surface range x. The Fourier transform of this signal, averaged over the clutter or speckle fluctuations is

$$<F_w(\omega, t)>_p = (2\pi)^{-1} \int <W(\tau, t)>_p e^{-j\omega\tau}d\tau \tag{40}$$

Two applications of the Fourier transform convolution theorem and the second moment result (1) may be used to show that

$$< F_w(\omega, t) >_p = G_p(\omega) \int a^2(\vec{x}, t)\sigma_o(\vec{x}, t)\, e^{(-j2\omega R/c)}d\vec{x} \tag{41}$$

where $G_p(\omega) = |F_p|^2 \otimes |F_p|^2$, where F_p is the Fourier transform of $p(\tau)$, and where $\otimes$ here denotes convolution with respect to frequency ω. In the simple short-pulse case, where

*Chirped pulses may, for example, be produced by surface acoustic wave (SAW) devices and compressed using reciprocal SAW filters.

**See, for example, Deley (1970).

no pulse compression is required, then E_r is effectively convolved with a delta function and $G_p = F_p \otimes F_p^*$. The result Eq. (41) is identical to the DF result Eq. (24) except that the spectral response is weighted by the point target frequency response function $G_p(\omega)$. Because of the random signal fading in both fast time and slow time, the actual detected echo power versus fast time or range will be corrupted by a broad fading spectrum. To simplify the analysis, we consider here, as in the case of the DF signal analysis, the ensemble average output of a short pulse system without regard to the details of the slow time evolution process. That is, for example, we consider the average output of the spectrum analyzer shown in Figure 3b in the case where no pulses are integrated prior to the spectrum analysis. In this case, the output power spectrum may be expressed as

$$<P_{sp}(\omega, t)> = <|(2\pi)^{-1} \int W(\tau, t)e^{-j\omega\tau}d\tau|^2>. \tag{42}$$

Assuming the backscatter is stationary in t, then on making use of the fourth moment result (2), we find after some manipulation, and on omitting the DC or antenna pattern term, that

$$<P_{sp}> = (2\pi)^2(\sigma_o)^2 A(0)[G_p^2(\omega)S_m(\vec{K}) + H_p(\omega)(d\omega/dK_r)(L_y/\sqrt{2\pi})] \tag{43}$$

where the approximation holds for large illuminated areas A_o and where we have again assumed a Gaussian antenna gain pattern, so that $A(0) = A_o/2 = (\pi/2)L_x L_y$, as previously defined. The first term on the right hand side of Eq. 43 is identically equal to the signal spectrum $<|<F_w(\omega, t)>_p|^2>$, while the second term represents the background fading noise spectrum. In the noise background term, $d\omega/dK = c' = c/(2 \sin\theta)$ is the Jacobian relating modulation frequency to surface wavenumber $K = K_x$ magnitude of $\vec{K}_r$. If we assume a Gaussian compressed pulse spectrum $|F_p|^2 \propto \exp(-\omega^2/\omega_p^2)$ the normalized fading spectrum $H_p(\omega)$ is given by

$$H_p(\omega) = |F_p|^4 \otimes |F_p|^4 = (\sqrt{2\pi}\omega_p)^{-1}G_p^2(\omega) \tag{44}$$

where, as previously defined, $G_p = |F_p|^2 \otimes |F_p|^2$ and the same result applies in the case of simple SP systems if F_p is assumed to have the same one sigma width and F_p^2 is replaced by F_p. For large time-bandwidth product (TBP) waveforms, $\dot{\omega}T_{po}^2 >> 1$, $\omega_p \cong \dot{\omega}T_{po}$, and so the one-sigma range resolution parameter $\rho_x = c'/\omega_p$. It follows from Eqs. (43) and (44) that the SP system signal spectrum to background noise spectrum ratio (SNR) can be expressed as (e.g., assuming $\omega << \omega_p$):

$$SNR_o = N_r(2\pi^2/A_o)S_m(\vec{K}_r) \tag{45}$$

where $N_r = L_x/\rho_x = \omega_p T_r$, $T_r = L_x/c'$ again being the echo return time. Eq. (45) is identical to the DF SNR result, Eq. (31) except for the factor N_r, which can be interpreted as the echo signal TBP or equivalently, the number of independent samples of the fading noise process across the range extent of the antenna footprint L_x. For a space system, one might have as typical values $L_x = 10$ km and $\rho_x = 10$ m, in which case $N_r = 10^3$, so that the SNR improvement over the comparable narrowband DF SNR_o is of the order of 30 dB. The SNR may also be put in the form

$$SNR_o = (\sqrt{2\pi}/\rho_x)P_m(K, \phi) \tag{46}$$

where $P_m(K, \phi) = (\sqrt{2\pi}/L_y)S_m(\vec{K}_r)$ is the spectrum of the reflectivity modulation signal appropriately normalized by the average power envelope so that the integral of P_m over all wavenumbers yields the total echo power modulation variance μ^2 for any particular antenna azimuth ϕ. Writing the modulation spectrum in terms of the frequency ω, this normalization may be expressed as (e.g., for infinite bandwidth)

$$\mu^2(\phi) \equiv \frac{<[W(\tau, t) - W_o(\tau)]^2>}{[W_o(\tau)]^2} \equiv \int_{-\infty}^{\infty} P'_m(\omega, \phi)d\omega \tag{47}$$

where $W_o(\tau) = <W(\tau, t)>$ is the average power envelope and where $P'_m \equiv P_m/c'$. The dependence of SNR_o on the bandwidth ω_p or, equivalently, on the reciprocal of the range resolution, we note, is quite different from the conventional definition of the SNR as the ratio of the average echo signal power to thermal noise power (S/N) which scales in inverse proportion to the pulse or receiver bandwidth. Because of thermal noise, one cannot arbitrarily increase the system bandwidth to increase the above SNR. For bandwidths such that S/N = 1, thermal noise and the clutter noise contribute equally to the total output SNR of the SP system, and any increase in the bandwidth beyond this point will only serve to decrease the total output SNR. However, as shown by Jackson et al. (1985a) and Jackson (1987b), S/N ratios in excess of + 5 dB are easily achievable at satellite altitudes with available satellite radar altimeter hardware (e.g., Seasat and Geosat altimeters) when incidence angles near 10° and antenna apertures of about 1 m^2 are considered, and so thermal noise is not as a rule a significant factor for near-vertical incidence wave spectrum measurements with SP systems at satellite altitudes.

By way of further interpretation of the SNR results for the SP and DF measurement techniques, we note that both nonimaging radar techniques share a common $1/L_y$ dependence on the azimuth beamwidth. In the case of the DF technique, the $1/A_o$ dependence can be interpreted as the result of sensing the reflectivity modulation spectrum through a narrow spectral window of width

$$\delta\vec{K}_r = \delta K_x \delta K_y = 2\pi/L_x L_y = 2\pi^2/A_o \tag{48}$$

In the SP case where the K_x analysis bandwidth is arbitrary, the SNR depends only on frequency spread of the fast time fading spectrum. However, the K_y analysis bandwidth is the same as that in the DF case, i.e., $\delta K_y = \sqrt{2\pi}/L_y$. Since the directional resolution $\delta\phi \sim \delta K_y/K$, it follows that in the case of both SP and DF nonimaging radar techniques, the higher the directional resolution, the lower the SNR. In the case of the SAR technique to be discussed below, we note that the measurement SNR is not similarly related to the directional resolution, since the spectral analysis is performed on the detected two dimensional image, usually digitally, and therefore the loss of signal power variance with increasing spectral resolution can always be compensated for by increasing the precision of the digital analysis.

As in the case of the DF technique, the SP SNR can be further increased by 'slow-time' filtering the received signal. The most straightforward implementation of the slow time filtering is, as illustrated in Figure 3b, to integrate a number of detected pulses before passing the signal to the (fast time) spectrum analyzer. Because of the rapid motion of the platform, this integration must be performed on the sampled backscattered pulses in range bins that are suitably adjusted so as to compensate for the relative motion of the surface. If the antenna beamwidth is sufficiently narrow (about 1° - 2°), wavefront curvature can be neglected, and the motion compensation can be accomplished simply by varying the delay on the trigger signal to the analog-to-digital (A/D) converter according to an input relative velocity vector. If N_p independent pulses are integrated in such a manner, the fading signal variance will be reduced by a factor of N_p, and accordingly, the SNR for the output of the system depicted in Figure 3b will become

$$\mathrm{SNR} = N_p \ \mathrm{SNR}_o \tag{49}$$

where SNR_o is the SNR in the unfiltered case represented by Eq. (45). The major system design and analysis problem for the nonimaging SP technique reduces to determining the appropriate integration time and the corresponding number of independent pulses. Generally, the PRF will be less than the Doppler spread $\Delta f_{dop} = \Delta\Omega_{dop}/2\pi$ except near $\phi = 0$ and 180 degrees (cf., Eq. 35), and so consecutive pulses will be independent vis-a-vis the Rayleigh fading statistics. The main factor determining the maximum number of pulses N_p that can be integrated is then the limit on integration time imposed by wavefront curvature effects. (This is the same factor that limits the DF SNR via the finite width of the resonance line $\Delta\Omega_{res}$). For typical space geometries, the wavefront curvature (in the horizontal plane) will limit the integration time to about 0.05 s (Jackson, 1987b). If the PRF is, for example, 1 KHz (Seasat class altimeter), then $N_p = 50$. For typical pulse bandwidths (e.g., 300 MHz for Seasat), typical SP SNRs of the order of 10 dB result.

3.2.3. *Synthetic Aperture Radar.* As indicated in the previous discussion, both the dual-frequency and short-pulse systems yield measurements of the spatial spectrum of the radar cross section of the ocean surface, which can be related to the wave height spectrum through the radar modulation transfer function. In both cases, the spectrum is obtained at spatial wavenumbers lying in the plane of incidence of the radar. Thus, in order to measure the full two-dimensional wave spectrum, the antenna must be rotated through 360°. An alternative method for obtaining two-dimensional spectra with a fixed antenna is provided by the synthetic aperture technique. The geometry for this type of system is shown in Figure 2, where the antenna look direction is held perpendicular to the platform velocity vector.

In a synthetic aperture radar (SAR), conventional ranging techniques are used to provide resolution in the plane of incidence. Usually, pulse compression methods are used with a linear FM coding of the transmitted pulses, as discussed in the previous section. Thus, Eqs. (36) and (38) describe the form of the transmitted and received fields for this case as well. The difference is that in the SAR case, the pulse-to-pulse variation in the phase of the received signal, due to the motion of the platform relative to the scene, is used to provide resolution in the along-track direction as well. As discussed in the previous section, it is useful to view the received signal as a two-dimensional function, where one dimension is the 'fast' time τ which describes the temporal variations of the signal received from a given pulse, and the other dimension is the 'slow' time t which describes the temporal variations of the signal from one pulse to the next.

The pulse-to-pulse variation in the signal received from a surface element at the position $\vec{x}$ in the coordinate system shown in Figure 2 can be calculated by including the next-order term in the expansion of the range distance R, i.e.

$$R(\vec{x}, t) = R_o + x\sin\theta + (y - Vt)^2/2R_o \tag{50}$$

where θ is the incidence angle and R_o is the range distance to the origin of the $x - y$ coordinate system, as previously defined, and $\vec{V}$ is the platform velocity, which is assumed to be in the y-direction (i.e. perpendicular to the plane of incidence). Substituting this expansion into Eq. (38) results in a quadratic phase variation in the slow time t due to the platform motion, in addition to the quadratic phase variation in the fast time τ due to the linear frequency modulation of the transmitted pulse. Thus, a pulse compression in the slow time coordinate can be performed in exactly the same manner as is done for the fast time, in order to produce a two-dimensional image of the surface. The resulting image can then be Fourier transformed in both dimensions to estimate the two-dimensional wave spectrum.

If the surface reflectivity is independent of time, this process is capable of producing images with comparable resolutions in both dimensions. However, because the ocean surface is continually in motion, the temporal variations in the surface reflectivity can not usually be neglected. The effects of these motions can, however, be modeled without great difficulty, as discussed below.

The pulse compression in the fast time coordinate may be accomplished with a matched filter operation involving the convolution of the received signal with a replica of the transmitted pulse as described in the previous section. Alternatively, in the case of a large time-bandwidth product (TBP) chirped waveform such as given by Eq. (37), the pulse may be compressed nearly optimally by convolving the received signal with the complex conjugate of the FM chirp signal alone, i.e., by convolving the signal with the phasor $\exp(j\dot{\omega}\tau^2/2)$. Similarly, the azimuth compression in the slow time coordinate may be achieved by convolving the range-compressed signal with an azimuth reference function having the same linear FM form of the fast time chirp,

$$E_a(t) = \exp(jkV^2t^2/R_o) \tag{51}$$

where $k = \omega_o/c$ is the radar wavenumber. The result of this set of operations may be written as

$$E_c(x, y) = \iint F_r'(x - x')a(\vec{x}', t)r(\vec{x}', t)\, e^{-j(\Omega_y t + \phi)} d\vec{x}'\, dt \tag{52}$$

where

$$F_r'(x - x') = \int p_o(\tau)e^{j\dot{\omega}(x - x')\tau/c'}\, d\tau \tag{53}$$

is the range point target response (PTR) function (for the field strength), $c' = c/(2\sin\theta)$; $\Omega_y = 2kV(y - y')/R_o$; and ϕ is a function of $\bar{x}$ and $\bar{x}'$ which is irrelevant for present purposes, since it drops out when the magnitude of the compressed signal is taken.

The partial expectation of the image intensity $I(\bar{x})$, which is proportional to the detected power $W \propto |E_c|^2$, may now be calculated from Eq. (52) by making use of the delta-correlation property of the reflectivity, Eq. (1). For concreteness, assume that the antenna pattern is given by the separable Gaussian pattern Eq. (30), and let $a_x(x)$ and $a_y(y)$ represent respectively the range and azimuth pattern factors. Then one obtains, to within a proportionality factor

$$<I(\bar{x})>_p =$$

$$\iint a_x^2(x')\,|F_r'(x - x')|^2 F_a'(y - y';\bar{x}',t)\,\sigma_o(\bar{x}',t)\,d\bar{x}'\,dt \tag{54}$$

where F_a' is the (time-dependent) azimuth response function given by

$$F_a' = a_y(y' - VT) \int a_y[y' - V(t - \Delta t)]\,\rho(\bar{x}', t;\Delta t)\, e^{-j\Omega_y \Delta t}\, d(\Delta t) \tag{55}$$

The SAR response problem is considerably simplified if it can be assumed that both ρ and σ_o are independent of the 'center time' coordinate t. In the case of satellite measurements, rather large areas (e.g., 10 km X 10 km) can be mapped out in relatively short times (of the order of 1 s) compared to the wave periods of interest (typically 10 s), and so the motion of the field of $\sigma_o(\bar{x}, t)$ during the 'scene painting' time can always be ignored. For aircraft measurements, longer scene-painting times are needed in order to generate long enough along-track records for minimum desired spectral stability and resolution, and so as a rule scene motion must be accounted for in the reduction of the image data (Raney and Vachon, 1988). As mentioned in section 2.1.4, there is presently some uncertainty about the nature of and relative importance of the time-dependence of the autocorrelation function in SAR ocean imaging; we note also, that in our simplified model, Eq. (20), both the t and t' dependence of the autocorrelation function have not been accurately modelled except for the shortest observation times. Accordingly, we will for the moment ignore any possible center time dependence of the autocorrelation function (tacitly assuming it to be small) and simply set $t = 0$ in both ρ and σ_o. This allows the t-integration in Eq. (54) to be performed immediately, yielding the simpler result,

$$<I(\bar{x})>_p = \int F_r(x, x')F_a(y, y')\sigma_o(\vec{x}', 0)d\bar{x}' \tag{56}$$

where we have let $F_r(x, x') \equiv a_x^2(x)\,|F_r'(x - x')|^2$ represent the range power PTR and F_a the azimuth response,

$$F_a(y, y') = V^{-1} \int A_y(V\Delta t)\,\rho(\bar{x}', 0;\Delta t)e^{-j\Omega_y \Delta t} d(\Delta t) \tag{57}$$

where $A_y = a_y \otimes a_y$ is the convolved azimuth gain pattern.

The azimuth response function $F_a(y, y')$ can be calculated readily on assuming the Gaussian forms, Eqs. (30) and (20), for the azimuthal antenna pattern a_y and the temporal autocorrelation function ρ respectively, and the result expressed as

$$F_a(y, y') = \pi T_a T_a^* \exp[-(y - y' - \Delta y)^2 / 2\rho_y^{*2}] \tag{58}$$

where $T_a = \sqrt{2} L_y / V$ is the beam transit-time, or integration time parameter (one sigma value for power); $T_a^* = [1/T_a^2 + 1/T_c^2]^{-1/2}$ is the effective coherent integration time, including the effect of the finite scene coherence time T_c; $\Delta y = (R_o/V)\bar{V}_r(\bar{x}', 0)$ is the azimuthal pixel displacement and where $\rho_y^* = R_o/(2kVT_a^*)$ is the effective azimuthal resolution. In the case of no surface motion, $\bar{V}_r = 0$ and $T_c = \infty$, in which case F_a is symmetrical about the expected pixel location at the nominal azimuth position $y = Vt$ and the maximum synthetic aperture resolution, $\rho_y^* = \rho_{ya} = R_o/2kVT_a$ is obtained. This limiting resolution is comparable to the physical length of the antenna aperture in the along-track direction as shown, for example, by Brown (1967). In practice, however, this limit is usually not achievable when imaging the ocean. In the case of satellite ocean imaging, the azimuth resolution is determined almost entirely by the scene coherence time T_c. Accordingly, the resolution is given approximately by ρ_y^* approximately equal to ρ_{yc}

$$\rho_{yc} = R/2kVT_c = (R/V)\sigma_v \tag{59}$$

where here for notational convenience the subscript-zero on the slant range has been dropped. As an example of the magnitude of this effect, assume that $\sigma_v = 0.50$ m/s, $R = 700$ km and $V = 7$ km/s. Then $\rho_y^* = 50$ m, which is rather large for good imaging of ocean waves. Corresponding to this degradation of azimuthal resolution caused by the random scatterer motion (small coherence time T_c) is a loss of spectral response at high azimuth wavenumbers, which has been termed the 'azimuth fall-off effect' (Beal et al., 1983). The azimuthal image shifts Δy caused by the net velocity of a surface resolution element leads to the so-called 'velocity bunching' effect (Alpers et al., 1981). This effect can cause image intensity modulations even in the absence of variations in the radar cross section, and is frequently the dominant mechanism for imaging azimuth travelling waves. To illustrate this effect let $\sigma_o \equiv$ constant; for convenience, suppress the range (x) dependence in (56). Then, let $\xi = y + \Delta y = y' + (R/V)\, V_r(y')$, so that (58) becomes $F_a(y, y') \equiv f_a(\xi - y)$, where f_a stands for the exponential in the specified argument. With this transformation, it follows that the image intensity is given by

$$\langle I(y) \rangle_p = \sigma_o \int f_a(\xi - y)\, |d\xi/dy'|^{-1} d\xi \tag{60}$$

providing the Jacobian of the transformation exists. Now if the width ρ_y^* of the azimuth PTR and the image shift Δy are both assumed to be small compared to the wave scale, then (60) reduces immediately to the well-known linear velocity bunching result

$$\langle I(y) \rangle_p = \sigma_o [1 - (R/V)(\partial \bar{V}_r / \partial y)] \tag{61}$$

This result predicts modulations, or image contrasts, that are proportional to the radial surface velocity component gradient in the azimuth direction. For small incidence angles, and for waves travelling in the positive y direction, (61) predicts image intensity modulations that are a maximum in the wave troughs, i.e., it predicts a modulation that is 180° out of phase with the roughness modulation m_h, and 90° out of phase with the tilt modulation m_t. For large R/V ratios, the velocity bunching may become strongly nonlinear. The modulation is then comparatively stronger; maximum contrast occurs when the Jacobian goes to zero. Driving the SAR response any more nonlinear results in the breakdown of the 1:1 mapping from object to image plane (Alpers et al., 1981; Hasselmann et al., 1985; Lyzenga, 1988).

The Fourier transform of the image (i.e., partially averaged over the speckle) in the general center-time-dependent case can be calculated from Eq. (54) using the integral form, Eq. (55), for F_a' and the result expressed as

$$<F_I(\vec{K})>_p = (R/2kV)\tilde{F}_r(K_x) \int a_x^2(x)\tilde{F}_a'(K_y;\vec{x}, t)\sigma_o(\vec{x}, t)e^{-j\vec{K}\cdot\vec{x}}d\vec{x}dt \tag{62}$$

where $\tilde{F}_r$ is the spatial transform of $|F_r'|^2$ and

$$\tilde{F}_a' = a_y(y - Vt)a_y[y - V(t - t_y)]\rho(\vec{x}, t;t_y) \tag{63}$$

where $t_y = K_yR/2kV$. Now, if we again neglect the scene motion and surface velocity changes (accelerations) and set $t = 0$ in ρ and σ_o, we get on performing the t-integration the simpler result,

$$<F_I(\vec{K})>_p = (R/2kV)\tilde{F}_r(K_x)\,\tilde{G}_a(K_y)\int a_x^2(x)\rho(\vec{x},0;t_y)\sigma_o(\vec{x}, 0)e^{-j\vec{K}\cdot\vec{x}}d\vec{x} \tag{64}$$

where we have let $\tilde{G}_a(K_y)$ be identically equal to $A_y(Vt_y)$ represent the azimuth response for a stationary surface. Expressions for the image transform similar to these have been derived by Lyzenga (1988) using the SIFT algorithm of Hasselmann (1980) and Martin (1981). It is evident from Eq. (64) that for stationary surfaces ($\rho = 1$), the Fourier transform of the image is identical to the finite-resolution Fourier transform of the cross section (i.e., the transform as seen through the antenna pattern data window) i.e., in the range direction. Since we have not applied an azimuth data window, the AZIMUTH resolution in Eq. 64 appears to be infinite. For moving surfaces, however, the aforementioned azimuth fall-off and velocity bunching effects come into play via the time and space dependence of the temporal correlation. Following Eq. (10), the radial velocity field of the large-scale sea surface (scale larger than the SAR resolution) at the nominal scene observation time of $t = 0$ may be written as

$$\bar{V}_r(\vec{x}, 0) = \mathrm{Im}\int \Omega_k\; g(\theta, \phi)F_\eta(\vec{K})e^{j\vec{K}\cdot\vec{x}}d\vec{x} \tag{65}$$

where $g(\theta, \phi) = \cos\theta + j\sin\theta\sin\phi$, $\sin\phi = K_x/K$, and $K = |\vec{K}|$. Similarly, if the subresolution scatterer velocities are considered to be primarily a result of the orbital velocity advection of Bragg scattering patches, the rms velocity parameter σ_v may be expressed as

$$\sigma_v^2 = \int_{K_{sar}}^{K_{emh}} |g|^2\,\Omega_K^2 S_\eta(\vec{K})d\vec{K} \tag{66}$$

where K_{sar} corresponds to the SAR resolution limit, or Nyquist frequency, and K_{emh} represents the wavenumber corresponding to the scale separation between the scattering elements and the longer tilting and straining waves, a scale separation which is typically on the order of several em wavelengths (e.g., Hasselmann et al., 1985). If the expression, Eq. (65), for the large-scale radial velocity is now substituted into Eq. (20) for the temporal autocorrelation function, and the result linearized assuming $2k\bar{V}_r t_y << 1$, then the following expression for the expected value of the normalized image spectrum may be derived (on neglecting the antenna pattern term and the pattern spectral window function assuming it is narrow)

$$< | <F_I(\vec{K})>_p |^2> = |R_{sar}(\vec{K})|^2 < |F_\eta(\vec{K})|^2>$$

$$+ |R_{sar}(-\vec{K})|^2 < |F_\eta(-\vec{K})|^2> \quad (67)$$

where

$$R_{sar}(\vec{K}) = R_{p_{tr}}(\vec{K})[R_m(\vec{K}) + R_v(\vec{K})] \exp[-(\rho_{yc} K_y)^2/2] \quad (68)$$

is the net SAR transfer function, in which $R_{p_{tr}}(\vec{K}) \equiv (R/2kV)\tilde{F}_r(K_y)\tilde{G}_a(K_y)$ is the SAR stationary point target spectral response function; $R_m = mK_x$ is the combined tilting/hydrodynamic modulation transfer function (MTF) for wave height, m being the MTF with respect to wave slope as defined in section 3.1; R_v is the SAR-unique 'velocity-bunching' MTF, $R_v = -(R/V)g(\theta,\phi)\Omega_k K_y$; and where the exponential function expresses the so-called azimuth fall-off effect. It is seen that the SAR image spectrum here expressed in the absence of speckle noise, even in the simple linear approximation, is not simply related to the wave height variance spectrum defined by Eq. (11). However, some simplification results if it can be assumed that $R_{sar}(\vec{K})$ is symmetrical with respect to a polar reflection. In this case, one can express the symmetrical image spectrum $S_I(\vec{K})$ in terms of the symmetrical wave height spectrum $S_\eta(\vec{K})$ defined by Eq. (11). If the broad background clutter spectrum $S_c(\vec{K})$ is included, then

$$S_I(\vec{K}) = |R_{sar}(\vec{K})|^2 S_\eta(\vec{K}) + S_c(\vec{K}) \quad (69)$$

For 'one-look' processing, the clutter background S_c is inversely proportional to the product of the fast- and slow-time signal bandwidths, or the same, directly proportional to the full resolution pixel area $\rho_x \rho_y$ (Hasselmann et al., 1985). Accordingly, the SNR in a SAR measurement will be greater than the equivalent DF SNR$_o$ (unfiltered case, Eq. 31) by a factor equal to the number of pixels in the illuminated area. (By equivalent, we mean assuming that the azimuth fall-off is negligible and the waves are range-travelling). Similarly, the SNR advantage over the SP SNR$_o$ (unfiltered case, Eq. 45) will be equal to number of azimuthal samples. In the filtered SP case, Eq. (49), however, the SP SNR is comparable to the SAR SNR, provided the echo signals have the same time-bandwidth products. This might be expected, since there is necessarily a close relationship between the maximum number of independent pulses N_p entering into the SP system integration and the number of azimuthal pixels in the SAR image (i.e., which are contained in the antenna azimuthal footprint dimension).

$R_{sar}(\vec{K})$ is symmetrical in the case where tilt modulation dominates over hydrodynamic modulation, but not otherwise. As mentioned in section 3.A.2, tilting effects dominate over hydrodynamic modulation effects for small nadir angles, and so in the case of near-nadir operation, one may expect Eq. (69) to be applicable generally. Figure 5 provides sample calculations of $R_{sar}(\vec{K})$ in the symmetrical, tilting-dominated MTF case.

As mentioned in section III.A.4, the simple advective orbital velocity model Eq. (66) for the coherence time is not physically realistic for near nadir, quasi-specular scatter. Nevertheless, the advective model predicts values of T_c that are not very different from the values of T_c computed from an exact quasi-specular scattering solution. For example, if we assume a simple Phillips' isotropic, equilibrium-range model spectrum $S_\eta(\vec{K}) = (B/2\pi)|\vec{K}|^{-4}$, $B = 0.005$, $K > K_o$ (= peak wavenumber), and we assume a value of $K_{sar} = 2\pi/50$ m appropriate to the Seasat SAR, then Eq. (66) gives (for $g(\theta, \phi) \cong 1$)

$$\sigma_v \cong [gB/K_{sar}]^{1/2} = 0.63 \text{ m/s} \tag{70}$$

which implies a value of $T_c = (2k\sigma_v)^{-1} = 0.029$ s for L-band ($k = 27$ m^{-1}) which is fairly close to the value of $T_c = 0.037$ s obtained by Winebrenner and Hasselmann (1988) for the isotropic spreading case (cf. their Table 1 and note that their $T_s = 2T_c$). Generally, however, σ_v and T_c may be expected to be environmentally dependent. In particular, one may expect the velocity dispersion and coherence time to be wind speed dependent through the wind speed dependence of the rear-face portion of the equilibrium range of the wave spectrum (cf., M. Donelan, this volume). As suggested by Jackson and Peng (1985), this dependence, rather than being a nuisance, may provide a means for inferring the wind speed, or air-sea momentum flux, from measurements of the azimuth fall-off in SAR image spectra.

Surface accelerations, which have for the sake of simplicity been neglected in the foregoing analysis may further act to degrade the image or distort the image spectrum. Hasselmann et al. (1985) showed that while the small-scale surface accelerations have a negligible effect on the SAR image formation process, large-scale surface accelerations can affect the SAR response by producing an azimuth fall-off that is proportional to the integration time T_a. Alpers and Bruening (1986), on the other hand, find that the higher-order, large-scale surface motions tend more to smear the image spectrum rather than the image itself and so do not contribute to the azimuth fall-off. It would be interesting to extend Lyzenga's (1988) analytical model, for example, so as to include acceleration effects in order to verify the Alpers and Bruening (1986) Monte Carlo results.

A quasi-empirical model for the azimuth falloff effect has been given by Beal et al. (1983) and Beal (1987). For the two-sigma point of the Gaussian azimuth falloff, defined in terms of the equivalent 'minimum detectable' azimuth component wavelength, λ_y (min), Beal and his co-workers find the relationship

$$\lambda_y(\text{min}) \sim 2(R/V)H_s^{1/2} \tag{71}$$

where $H_s = 4<\eta^2>^{1/2}$ is the significant wave height and the units are mks. Since, for wind seas, H_s varies between being linearly and quadratically dependent on wind speed, this model implicitly contains a wind speed as well as a peak wavelength dependence. As an example of the order of magnitude of the azimuth falloff predicted by Eq. (71) consider R/V ratios of 100

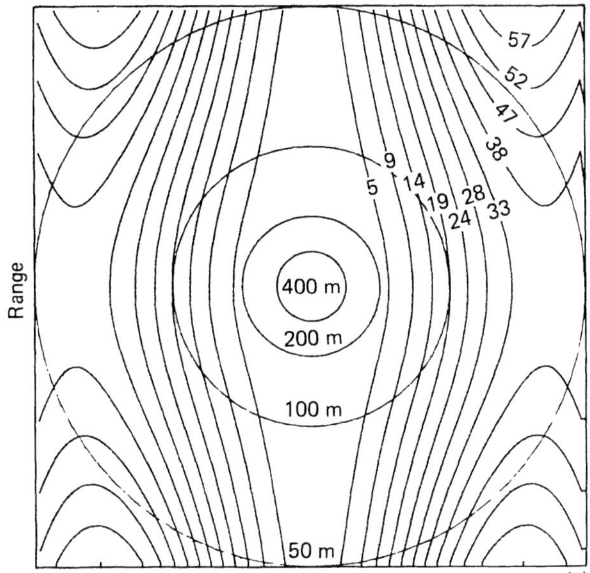

(a)

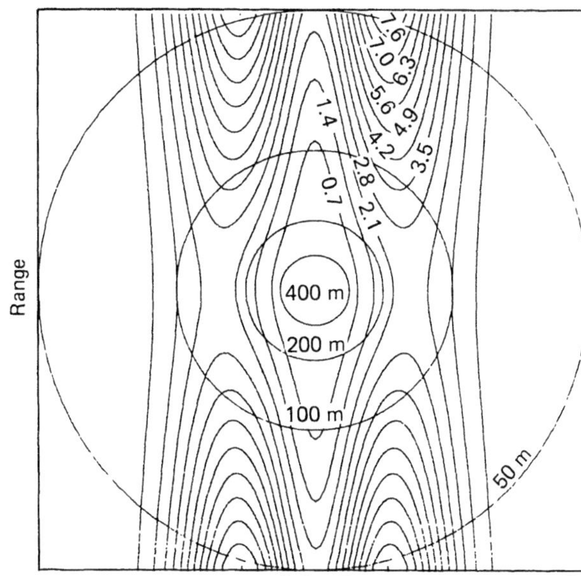

(b)

Fig. 5 — Modulus-squared of $R_{sar}(\vec{K})$ in the symmetrical, tilt-dominated MTF case for an R/V ratio of 30 s. In (a), a two-sigma azimuth falloff wavenumber of $2\pi/50$ m is prescribed while in (b) a two-sigma point of $2\pi/125$ m is assumed. [From Monaldo and Lyzenga (1988)].

s and 30 s, which are characteristic of 'free-flyer' and low earth orbit platform altitudes respectively. Then, for a 4 m sea, $\lambda_y(\min)=400$ m and 120 m for the high and low altitude cases respectively.

As discussed previously, at high R/V ratios, velocity bunching can become nonlinear, and the mapping of surface scattering centers into the image plane can cease to be 1:1. The velocity bunching mechanism, which acts to promote the imaging of waves with a large azimuthal component at moderate R/V ratios, then acts to degrade the image. The combination of severe azimuth fall-off and strongly nonlinear velocity bunching effects may render SAR impractical for conventional 'free-flyer' application, but these effects are mild enough at lower earth-orbit altitudes (e.g., Space Shuttle altitudes) that SAR should be capable of providing useful estimates of the directional wave spectrum for the energy-containing waves of most interest (wavelengths > 100 m). The results of the Shuttle Imaging Radar-B (SIR-B) experiment (Beal et al., 1986), which will be discussed in the following section, confirm this expectation.

While the problems associated with SAR ocean-wave imaging—particularly, those problems stemming from surface-motion effects—are obviated in the case of the nonimaging radar techniques discussed in the previous sections, SAR nevertheless possesses a considerable advantage as an imaging device in allowing two-dimensional spatial variations of the wave field to be observed within the imaged area. This capability makes possible certain types of investigations such as wave/current interaction (e.g, refraction) studies and envelope evolution (e.g., groupiness) studies. However, even with a low earth-orbiting SAR system, we will still need to better understand the SAR response as a function of wavenumber and azimuth relative to the SAR look direction and as a function of environmental parameters (such as wind speed) in order to produce reliable spectra from the imagery, and to provide a solid basis for determining the optimum parameter values of a spaceborne SAR system for measuring wind-wave spectra. These parameter values may be quite different from those which are best suited for SAR detection of other ocean-surface phenomena such as internal wave signatures.

4. Example Results from Airborne and Spaceborne Systems

Numerous examples of wave spectral estimates from synthetic aperture radar data have been reported (see references in section 2). However, most of these have been limited to estimates of the dominant wavelength and direction, and few have included independent measurements of the wave spectrum with sufficient resolution to adequately assess the accuracy of the SAR estimates. Indeed, few alternative techniques exist for measuring wave spectra with resolutions comparable to those obtained from radar measurements. Consequently, the comparisons discussed here are mainly between different types of radar measurements.

The most straight-forward type of radar measurement of wave spectra is that which involves a form of direct surface range measurement such as employed by the Surface Contour Radar (Kenney et al., 1979). The credibility and usefulness of the Surface Contour Radar (SCR) for making wave spectrum measurements has been established by Walsh et al (1985a) by means of an analysis of the consistency of the directional information obtained from this device as compared with pitch and roll buoy data. Figure 6 depicts the SCR measurement geometry. The swath width, which determines the directional resolution of the SCR, is presently limited to a range of 200 m to 400 m, depending on the aircraft altitude and the desired spatial resolution (antenna spot size). Because of the limited swath width and contamination of the spectra by range noise, particularly at the larger off-nadir angles, the spectra

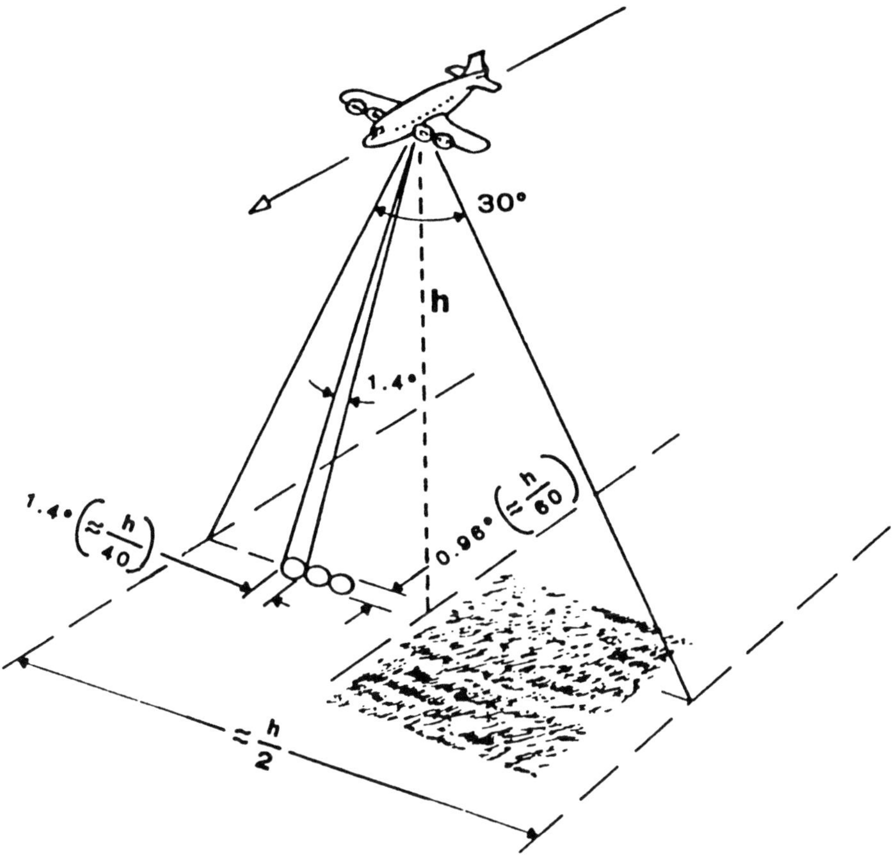

Fig. 6 — Surface Contour Radar (SCR) measurement geometry. The operating altitude of the SCR is normally between 400 m and 800 m, and so the corresponding swath width ranges between 200 and 400 m approximately.

presently produced by the SCR often exhibit a significant aircraft heading dependence and high background noise level, and so some care must be taken in the interpretation of the spectra (Walsh et al., 1985b). A new, upgraded version of the SCR is presently being developed that will be capable of operating at higher altitudes with correspondingly higher directional resolution, and with significantly reduced range noise levels, and so these problems with the present system will be virtually eliminated (E. Walsh, personal communication).

Because of its finite scene-mapping time, the SCR is capable of providing the true direction of wave travel. SCR wavenumber spectra, produced from the polar-symmetric encounter frequency spectra by appropriately Doppler shifting the spectra, accordingly contain the true spectrum in the direction toward which the waves are actually travelling and, in addition, an 'artifact' spectrum in the opposite half space that is produced by Doppler shifting the encounter spectrum in the wrong sense. By overlaying two SCR spectra from aircraft tracks 180° or so apart in direction, one can usually distinguish the true spectrum from the artifact, since the two artifact spectral lobes will be displaced one from the other while the 'true' spectra in the proper half space will coincide, as illustrated by the example of Figure 7. Figure 6 illustrates the utility of the SCR in mapping a wave field that exhibits a strong spatial variabilty. Here, waves emanating from the Delaware Bay off the U. S. East Coast are seen to be refracted about Cape May, producing a wave field north of the cape which is travelling nearly parallel to the New Jersey coast.

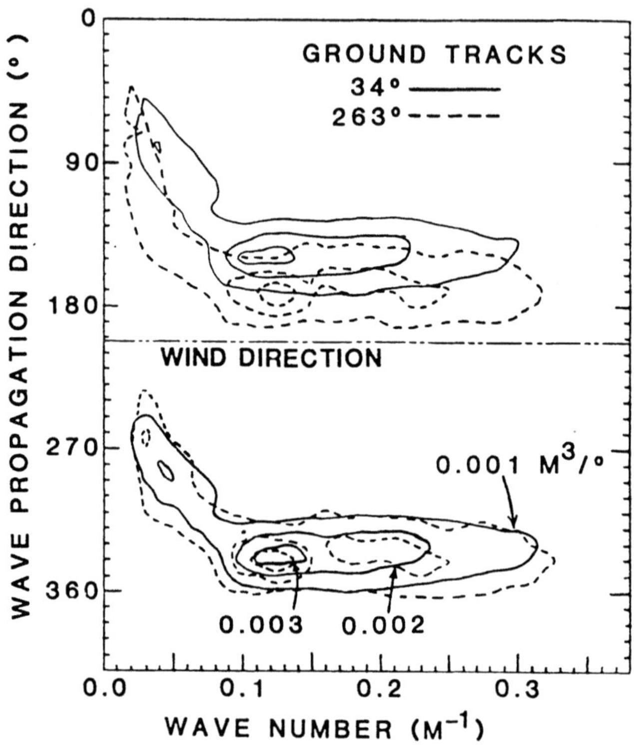

Fig. 7 — Overlays of SCR Doppler-corrected height spectra (wavenumber in radians/m) from two different ground tracks. The spectra are processed assuming that both positive and negative frequencies in the symmetrical encounter-frequency spectrum are real, while in fact only one or the other corresponds to the real direction of wave travel. This results in 'artifact' spectral lobes which are displaced from each other (upper half space) and 'true' spectral lobes which are seen to coalesce (lower half space).

The K_u-band frequency (14 GHz) Radar Ocean Wave Spectrometer (ROWS), which operates on the basis of the short-pulse technique discussed in the previous section, has been validated by comparison with both *in situ* and SCR measurements (Jackson et al., 1985a,b; Jackson, 1987b). As a result of these validation experiments, the ROWS has come to be

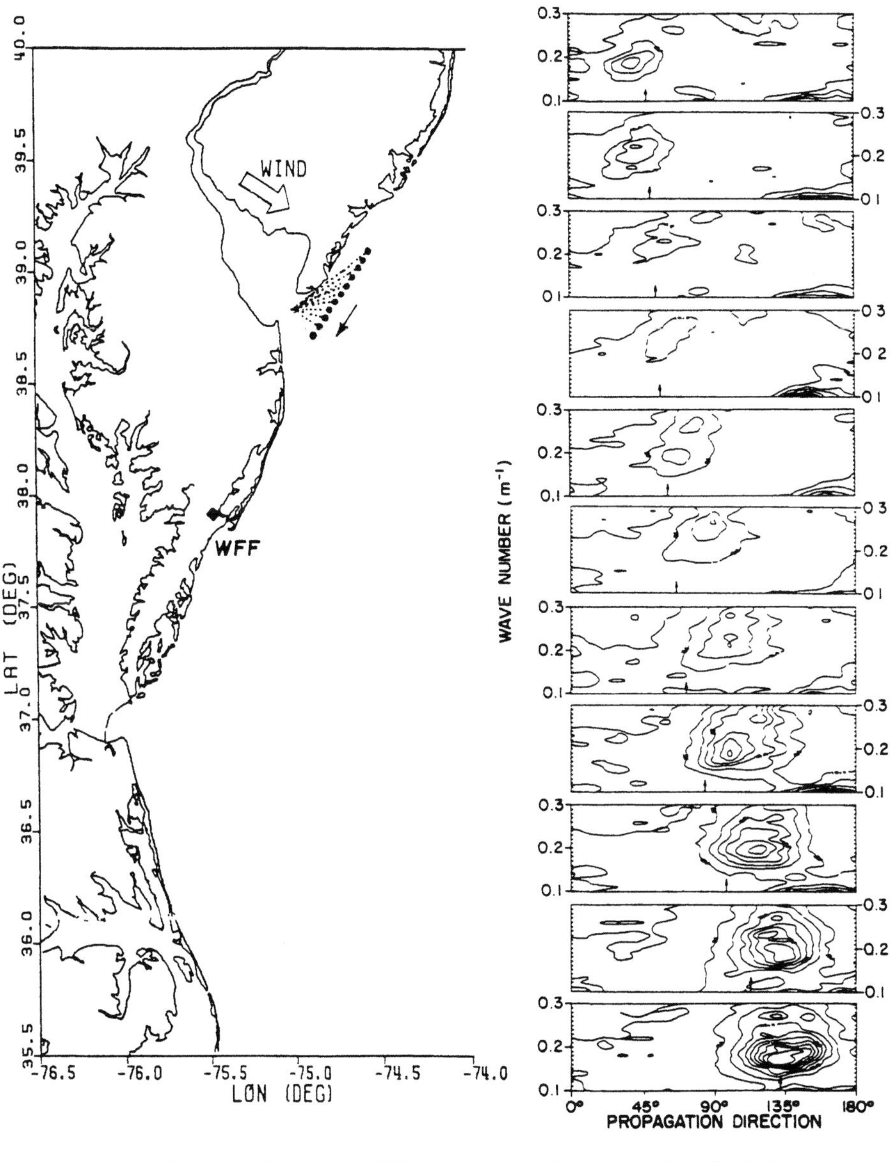

Fig. 8 — Sequence of SCR directional height spectra obtained off the mouth of the Delaware Bay showing how waves emanating from the bay are refracted around Cape May, New Jersey

regarded along with the SCR as a standard of comparison for high-resolution wave spectrum measurements, despite the fact that it is a less direct measurement than the SCR. In fact, since the ROWS responds more directly to the wave slope than to the wave height, it appears to be somewhat more sensitive to low-amplitude short-wavelength components than the SCR. The ROWS also has the advantage of being capable in principle of operating from spacecraft platforms, since it is a pulse-limited rather than a beam-limited device. Figure 9 shows the measurement geometry for the airborne ROWS system. The present ROWS platform, an Orion P-3A aircraft, limits the ROWS to 6 or 7 km altitudes and the antenna footprint dimensions accordingly to ca. 500 m in azimuth and 1000 m in surface range, and so the directional resolution is somewhat limited for the longer wavelengths (minimum resolution of 50° for a 400 m wave compared to 30° for a 200 m water wavelength). Since the system is capable of being operated from much higher altitudes, plans are underway to put the ROWS instrument on a higher-flying (and less expensive) aircraft in order to improve the spectral resolution and increase the sample area covered per flight hour/dollar.

ROWS directional wave spectra (DWS) with approximately 40 degrees of freedom per 0.001 cpm wavenumber and 15° direction band are produced from minimum record lengths of 100 s, consisting of 10 complete rotations of the conically scanning antenna. Estimates of the reflectivity modulation spectra $P_m(K, \phi)$, as defined below Eq. (46), for 15° azimuth bins ($N_p = 40$ pulses) are computed by FFT from rewindowed reflectivity modulation data versus surface range (wavefront sphericity effects being corrected for in transforming from fast time bins to surface range bins of 12 m), and height spectra are computed from the measured $P_m(K, \phi)$ according to the linear tilt model described in section 3.A assuming an incidence angle of 13°, which is the nominal angle for the peak power return within the 10° elevation beamwidth. The mean square slope parameter entering into the tilt MTF (Eq. 15) is derived from least-squares fitting the average near nadir cross section data from the ROWS off-nadir, rotating 'spectrometer mode' beam and an additional, broad, nadir-directed 'altimeter mode' beam with a Gaussian, specular-point model cross section. The nadir-directed beam provides the total mean square slope parameter, while the off-nadir beam provides the relative azimuthal dependence of $<\eta_x^2>$ (i.e., it provides the principal components ratio). Final estimates of the polar-symmetric height spectrum $S_\eta(\vec{K})$, $\vec{K} = K(\cos\phi, \sin\phi)$ are obtained by averaging looks 180° apart (this serves to double the degrees of freedom).

Figure 10 is a dramatic example of how the airborne ROWS system can be used to map a wave field over a large area. The data were obtained on a 1978 NASA CV-990 aircraft mission at an altitude of 10 km. Storm seas, with wave heights up to 10 m, were generated off the coast of northern Norway by an intense extratropical cyclone (Figure 11). The DWS of the wave field shown in Figure 10 exhibit a bimodal structure that is probably characteristic of wave fields lying in the wake of such fast-moving cyclones. A long, 330 m system is seen travelling along the coast, and a shorter 200 m system is seen travelling toward the coast at nearly right angles to the first system. These remote observations, together with Waverider buoy observations located near the ROWS file labeled 'A' have been used to test the performance of two numerical wave hindcast models (study to be published). Figure 12a shows a hindcast DWS from V. Cardone's (Oceanweather, Inc.) 'first generation' ODGP model compared to the ROWS observation at file location 'A'. Figure 12b provides a three-way comparison between the hindcast, ROWS-observed, and buoy nondirectional height spectra. The comparison shows excellent agreement between the ROWS and buoy spectra. The best agreement with the hindcast is obtained if the hindcast spectrum for nine hours earlier is used. This indicates the difficulty of modelling such transient events with proper phasing.

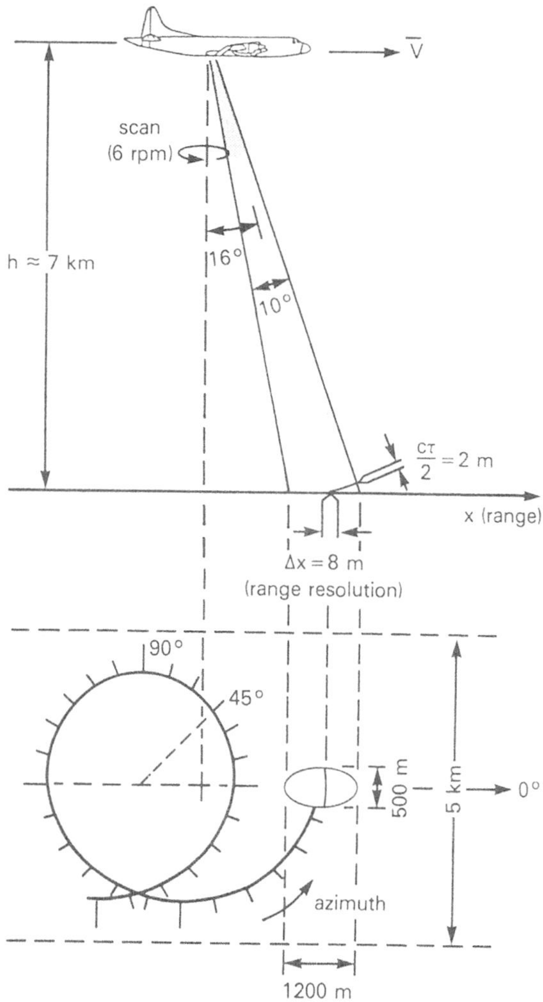

Fig. 9 — Radar Ocean Wave Spectrometer (ROWS) measurement geometry. The 7 km altitude indicated is close to the nominal cruising altitude of the NASA P-3 aircraft.

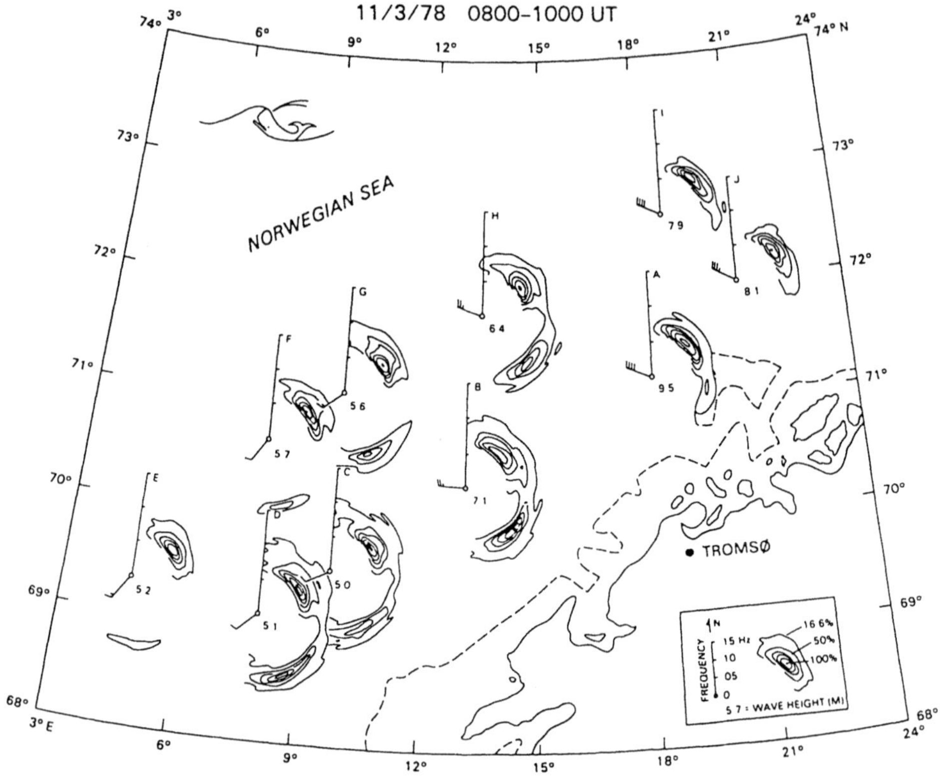

Fig. 10 — Spatial field of ROWS directional height spectra (here transformed to the frequency domain) obtained on the NASA CV-990 aircraft at 10 km altitude just after the passage of an intense storm in the Norwegian Sea (see Fig. 11 for weather map). The northernmost files are still in a region of high winds behind the cyclone whose center is now to the north of the flight box, and a local wind sea can be seen in the spectrum of file 'A'.

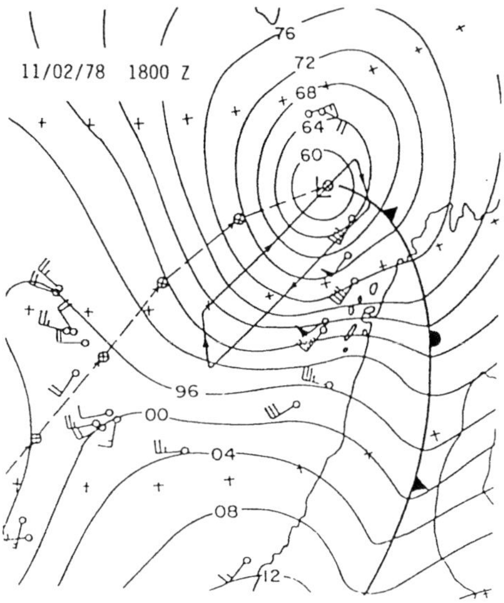

Fig. 11 — Surface chart for November 2, 1978 showing an intense storm in the Norwegian Sea and the generating region for the S-E travelling wave component observed in Fig. 10

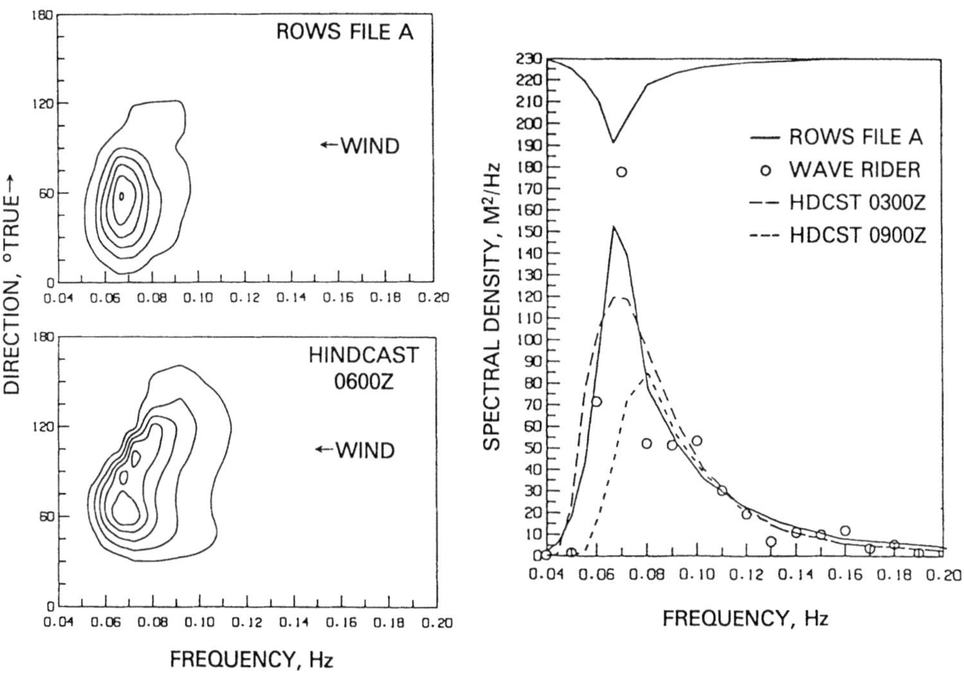

Fig. 12 — ROWS-inferred height spectra, file location 'A' (cf. Fig. 10) compared to model hindcast and Waverider buoy spectra at nearby locations: (a) ROWS DWS (directional wave spectrum) measured at 0810Z compared to hindcast DWS for 0600Z; (b) ROWS-inferred nondirectional height spectrum compared to Waverider buoy spectrum at 0830Z and model outputs for 0900Z and for 0000Z the night before. The upper curve is the ROWS 95% confidence interval.

A comparison of the wave measurement capabilities of a synthetic aperture radar (SAR) system operating from low earth orbit with those of the ROWS and SCR has been presented by Beal et al (1986). The data for this intercomparison were collected during the Shuttle Imaging Radar-B (SIR-B) mission by means of a series of aircraft underflights off the coast of southern Chile. Figure 13 provides a four-way comparison of DWS (wavenumber height spectra in m^4) between the two aircraft sensors, the SCR and ROWS on the NASA P-3 (operating at 0.8 km and 6 km altitudes, respectively), the SIR-B on the Space Shuttle (operating at ca. 240 km altitude), and additionally, the U. S. Navy operational wave for forecast product (for $\tau = 0$) over four consecutive days. All spectra were temporally and spatially colocated to within 1 hour and 100 km. SAR height spectra were obtained from ca. 30 m resolution imagery of 6 km square sample areas, with SIR-B nadir angles varying between 37° (flight day 2) to 19° (flight day 5), after subtracting the measured noise background S_c from the image intensity spectrum, by applying the linear SAR response model outlined above in section 3.B.3 for the tilt-dominated MTF case for which Eq. (69) is applicable and with an azimuth roll-off specified according to Eq. (71). (See Monaldo and Lyzenga, 1988, for details).

The comparison shows a remarkably good overall agreement between the different remote sensors. Seas and swells with wavelengths varying between 80 m and 400 m and heights H_s varying between between 1.7 m and 4.6 m were accurately measured by both the ROWS and SCR on all flight days, as evidenced by the generally excellent agreement shown among the DWS and absolute wave heights inferred from the two radar sensors and ancillary laser profiler data. SIR-B spectra are in fairly good agreement with the aircraft DWS on days 2, 4 and 5, but on day 3 the SAR failed to image the short, steep wind-sea. The broadly angularly spread spectrum of the sea evident in the aircraft spectra lies mostly without the two sigma point of the SAR's azimuth fall-off as determined by Eq. (71), and so it could not be recovered from the image spectra. On day 5, the DWS of the nearly azimuth-travelling 400 m swell system was accurately measured in a relative sense, but the inferred wave height disagreed with the aircraft estimates by a factor of two (Monaldo, personal communication; Monaldo and Lyzenga, 1988), indicating that there are some problems with the velocity bunching MTF that is presumably responsible for wave visibility in azimuth-travelling wave cases. On the other two days, the swells and seas were more or less range-travelling, and the absolute as well as relative levels of swell and sea components measured by SIR-B are in rather good agreement with the aircraft observations, although some rotation of the SAR spectrum into the range direction is evident on day 2. The U. S. Navy forecast, by and large, did quite well considering the paucity of surface wind data in the southern hemisphere. The consistent clockwise rotation of the forecast model DWS outputs with respect to the observations evident in Figure 13 is believed to be due to a mispositioning of the major synoptic storm centers during the mission period.

The SIR-B experiment demonstrates that useful wave spectra measurements can be obtained with a SAR in a low earth orbit. For R/V ratios much higher than SIR-B's, surface motions can so disrupt the synthetic aperture formation as to make SAR a poor candidate for global wave spectrum measurements. The marked contrast between the low-altitude SIR-B (R/V ~30 s) and the high-altitude Seasat SAR (R/V ~130 s) spectrum examples shown in Beal (1987) well-illustrates the truth of this statement.*

*Note in proof: Additional SIR-B experiment results have been published recently in a special issue of the *Journal of Geophysical Research* (Vol. 93, No. C12, 1988), to which the reader is referred. We note, particularly, the work of Bruening et al. in this issue, in which it is shown that even at the relatively low SIR-B altitude, severe spectral distortion can occur in the *range* direction due to the behavior of the velocity bunching MTF. Since $R_v \alpha K_y$, the image spectrum may actually bifurcate about $K_y = 0$ in cases where $|R_v| >> |R_m|$ (cf. Figure 5b). Such bifurcated spectra were observed by Bruening et al. in the SIR-B data collected over the North Sea.

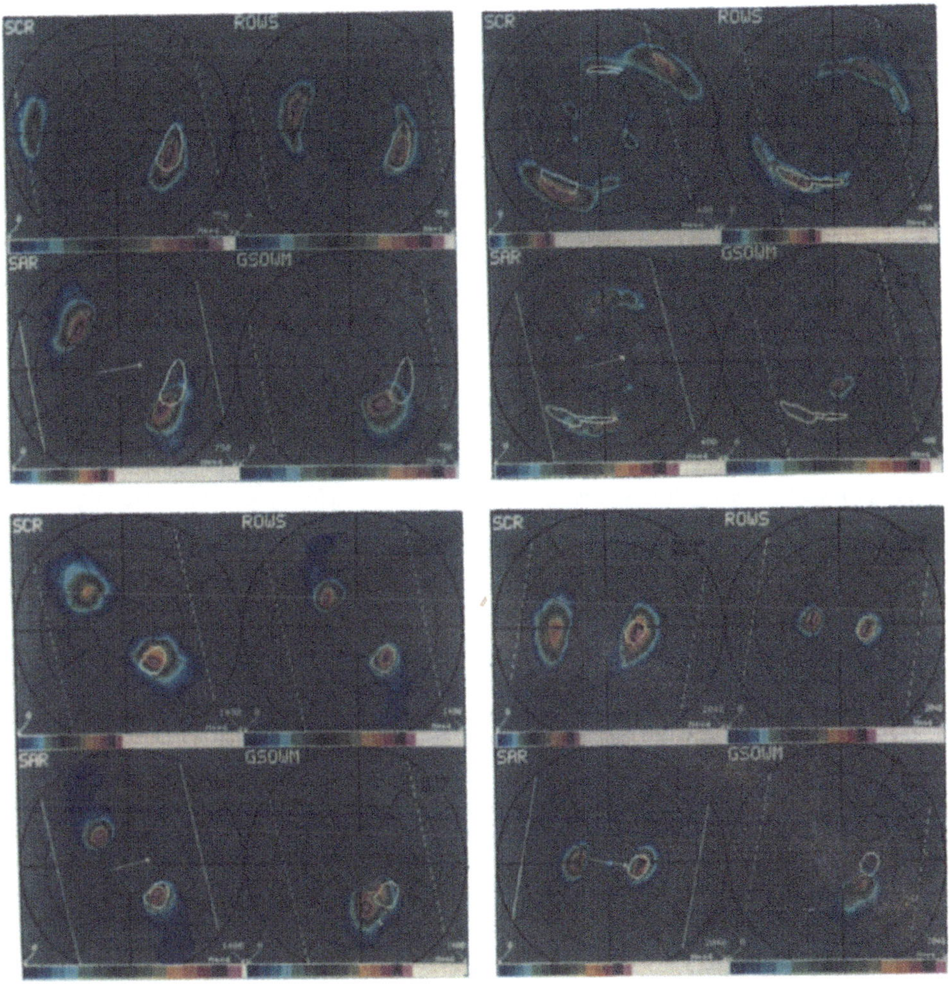

Fig. 13 — Four way, four day comparison between SCR, ROWS, SIR-B, and U.S. Navy operational forecast (for $\tau = 0$) DWS obtained off the coast of southern Chile in October, 1984 (units in m^4). Arrows indicate the headings for the Space Shuttle and NASA P-3. White dashed lines represent the expected two sigma azimuth fall-off contour for the SIR-B spectra according to Eq. (71) in the text, where the wave height H_s was determined from the aircraft data. The white contour superimposed on each spectrum for a particular day is the ROWS 3-dB contour level, which serves as a reference contour. Note that the outermost wavenumber ring is $2\pi/100$ m for all days but day 3, for which it is $2\pi/50$ m. Day 2 is top left, day 3 top right; day 4 bottom left, and day 5 is bottom right.

5. Conclusions

Several types of radar systems have been demonstrated to be capable of measuring ocean wave spectra from surface or aircraft platforms. A subset of these techniques also appears to be capable of operating from spacecraft platforms, and thus to be potentially able to provide synoptic global observations of ocean wave conditions. To date, only one of these (a synthetic aperture radar) has actually been operated in space. However, experience gained from aircraft measurements using other techniques indicates that these should also be usable in space. Consequently, the goal of making global estimates of directional ocean wave spectra appears to be technologically feasible.

6. Acknowledgements

We thank W. J. Plant for his useful comments on an early draft of this report and for his encouragement throughout the course of this work. J. Johnson, F. Monaldo, R. Beal and E. Walsh provided valuable discussions of their work and graciously provided their figures for reproduction.

7. References

Allen, T.D. (ed.), *Satellite Microwave Remote Sensing*, Ellis Horwood Ltd., Chichester, UK and John Wiley and Sons, USA, 1983.

Alpers, W. and K. Hasselmann, The Two-frequency Microwave Technique for Measuring Ocean Surface Wave Spectra from an Airplane or Satellite, *Boundary Layer Meteorol., Vol. 13*, pp. 215-230, 1978.

Alpers, W.R., D.B. Ross and C.L. Rufenach, On the Detectability of Ocean Surface Waves by Real and Synthetic Aperture Radar, *J. Geophys. Res., Vol. 86*, pp. 6481-6498, 1981.

Alpers, W.R. and C. Bruening, On the Relative Importance of Motion-Related Contributions to the SAR Imaging Mechanism of Ocean Surface Waves, *IEEE Trans. Geosci. Remote Sensing, Vol. GE-24*, pp. 873-885, 1986.

Atanassov, V.W. Rosenthal, and F. Ziemer, Removal of Ambiguity in Two-Dimensional Power Spectra Obtained by Processing Ship Radar Images of Ocean Waves, *J. Geophys. Res., Vol. 90*, pp. 1061-1068, 1985.

Atlas, S.R.C. Beal, R.A. Brown, P. De Mey, R.K. Moore, C.G. Rapley, and S.T. Swift, Problems and Future Directions in Remote Sensing of the Oceans and Troposphere: A Workshop Report, *J. Geophys. Res., Vol. 91*, pp. 2525-2548, 1986.

Barrick, D.E., Remote sensing of sea state by radar, in *Remote Sensing of the Troposphere*, V. E. Derr, ed., U.S. Govt. Printing Office, 1972.

Barrick, D.E. and B.J. Lipa, A Compact Transportable HF Radar System for Directional Coastal Wave Field Measurements, in *Ocean Wave Climate*, ed. by M.D. Earle and A. Malahoff, pp. 153-209, Plenum, New York, 1979.

Beal, R.C. (ed), *Measuring Ocean Waves from Space*, Johns Hopkins APL Tech. Digest Vol. 8, No. 1, 1987.

Beal, R.C., D.G. Tilley, and F.M. Monaldo, Large- and Small-Scale Spatial Evolution of Digitally Processed Ocean Wave Spectra from the Seasat Synthetic Aperture Radar, *J. Geophys. Res., Vol. 88*, pp. 1761-1778, 1983.

Beal, R.C., F.M. Monaldo, D.G. Tilley, D.E. Irvine, E.J. Walsh, F.C. Jackson, D.W. Hancock, D.E. Hines, R.N. Swift, F.I. Gonzalez, D.R. Lyzenga, and L.F. Zambresky, A Comparison of SIR-B Directional Ocean Wave Spectra with Aircraft Scanning Radar Spectra and Global Spectral Ocean Wave Model Predictions, *Science, Vol. 232*, pp. 1531-1535, 1986.

Beckmann, P. and A. Spizzichino, *The Scattering of Electromagnetic Waves from Rough Surfaces*, Macmillan, New York, 503 pp. + vii, 1963.

Brown, W.M., Synthetic Aperture Radar, *IEEE Trans. Aerosp. Elect. AES-3*, pp. 217-229, 1967.

Cote, L.J., J.O. Davis, W. Marks, R.J. McGough, E. Mehr, W.J. Pierson, Jr., J.F. Ropek, G. Stephenson, and R.C. Vetter, The Directional Spectrum of a Wind Generated Sea as Determined from Data Obtained by the Stereo Wave Observation Project, *Meteorological Papers, Vol. 2*, No. 6, pp. 1-88, New York University, New York, 1960.

Crombie, D.D., and J.M. Watts, Observations of Coherent Backscatter of 2-10 MHz Radio Surface Waves from the Sea, *Deep Sea Res., Vol. 15*, pp. 81-87, 1968.

Dean, R.G., Directional Wave Spectra: Some Applications and Storage, in *Ocean Wave Measurements and Analysis*, American Society of Civil Engineers, New York, 1975.

Dean, R. (ed)., *NRC Workshop on Waves Measurement Technology*, Marine Board, National Academy of Sciences, Wash., DC, 1983.

Deley, G.W., Waveform Design, Chapter 3 in *Radar Handbook*, ed. by M. Skolnik, pp. 301-347, McGraw-Hill, 1970.

Gonzalez, F.I., T.W. Thompson, W.E. Brown, P.S. DeLeonibus, J.W. Sherman III, J.F.R. Gower, D. Lichy, D.B. Ross, C.L. Rufenach, and R.A. Shuchman, Seasat Synthetic Aperture Radar: Ocean Wave Detection Capabilities, *Science, Vol. 204*, pp. 1418-1421, 1979.

Hasselmann, K., A Simple Algorithm for the Direct Extraction of the Two-Dimensional Surface Image Spectrum from the Return Signal of a Synthetic Aperture Radar, *Int. J. Remote Sensing, Vol. 1*, pp. 219-240, 1980.

Hasselmann, K., R.K. Raney, W.J. Plant, W. Alpers, R.A. Shuchman, D.R. Lyzenga, C.L. Rufenach, and M.J. Tucker, Theory of Synthetic Aperture Radar Ocean Imaging: A MARSEN View, *J. Geophys. Res., Vol. 90*, pp. 4659-4686, 1985.

Holliday, D., G. St-Cyr, and N.E. Woods, A Radar Ocean Imaging Model for Small to Moderate Incidence Angles, *Int. J. Remote Sensing, Vol. 7*, pp. 1809-1834, 1986.

Holthuijsen, L.H., Observations of the Directional Distribution of Ocean Wave Energy in Fetch-Limited Conditions. *J. Phys. Oceanogr., Vol. 13*, pp. 191-207, 1983.

Jackson, F.C., An Analysis of Short Pulse and Dual Frequency Radar Techniques for Measuring Ocean Wave Spectra from Satellites, *Radio Science, Vol. 16*, pp. 1385-1400, 1981.

Jackson, F.C., W.T. Walton, and P.L. Baker, Aircraft and Satellite Measurement of Ocean Wave Directional Spectra Using Scanning-Beam Microwave Radars, *J. Geophys. Res., Vol. 90*, pp. 987-1004, 1985a.

Jackson, F.C., W.T. Walton, and C.Y. Peng, A Comparison of In Situ and Airborne Radar Observations of Ocean Wave Directionality, *J. Geophys. Res., Vol. 90*, pp. 1005-1018, 1985b.

Jackson, F.C. and C.Y. Peng, Comment on 'Imaging Radar Observations of Directional Properties of Ocean Waves' by M. McLeish and D. Ross, *J. Geophys. Res., Vol. 90*, pp. 767-7370, 1985.

Jackson, F. C., The Physical Basis for Estimating Wave Energy Spectra with the Radar Ocean Wave Spectrometer, in *Measuring Ocean Waves from Space*, edited by R. Beal, pp. 70-73, Johns Hopkins APL Tech. Digest, Vol. 8, No. 1, 1987a.

Jackson, F.C., The Radar Ocean Wave Spectrometer, in *Measuring Ocean Waves from Space*, edited by R. Beal, pp. 116-127, Johns Hopkins APL Tech. Digest, Vol. 8, No. 1, 1987b.

Jackson, F.C., A Physically Based Model of the Altimeter Wind Speed Response, in *WOCE/NASA Algorithm Workshop*, Corvallis, Oregon. Aug. 1987, D. Chelton, ed., 1988.

Jackson, F.C., W.T. Walton, B.A. Walter, and C.Y. Peng, Sea Surface Mean-Square Slope from Ku-band Backscatter Data, *J. Geophys. Res.*, in press, 1989.

Johnson, J.W. and D.E. Weissman, Two-Frequency Microwave Resonance Measurements from an Aircraft: A Quantitative Estimate of the Directional Ocean Surface Spectrum, *Radio Science, Vol. 19*, pp. 841-854, 1984.

Keller, W.C., W.J. Plant, and D.E. Weissman, The Dependence of X-band Microwave Sea Return on Atmospheric Stability and Sea State, *J. Geophys. Res., Vol. 90*, pp. 1019-1030, 1985.

Kenney, J.E., E.A. Uliana, and E.J. Walsh, The Surface Contour Radar: A Unique Remote Sensing Instrument, *IEEE Trans. Microwave Theory Tech., Vol. MTT-27*, pp. 1080-1092, 1979.

Lipa, B.J., D.E. Barrick, and J.W. Maresca, HF Radar Measurements of Long Ocean Waves, *J. Geophys. Res., Vol. 86*, pp. 4089-4102, 1981.

Lyzenga, D.R., Numerical Simulation of Synthetic Aperture Radar Image Spectra for Ocean Waves, *IEEE Trans. Geosci. Remote Sensing*, *Vol. GE-24*, pp. 863-872, 1986.

Lyzenga, D.R., An Analytic Representation of the Synthetic Aperture Radar Image Spectrum for Ocean Waves, *J. Geophys. Res.*, *Vol. 93*, pp. 15374-15388, 1988.

Martin, P.J., Direct Determination of the Two-Dimensional Image Spectrum from Raw Synthetic Aperture Radar Data, *IEEE Trans. Geosci. Remote Sensing, Vol. GE-19*, pp. 194-203, 1981.

McLeish, W. and D.B. Ross, Imaging Radar Observations of Directional Properties of Ocean Waves, *J. Geophys. Res., Vol. 88*, pp. 4407-4419, 1983.

Monaldo, F.M. and R.S. Kasevich, Daylight Imagery of Ocean Surface Waves for Wave Spectra, *J. Phys. Oceanogr., Vol. 11*, pp.272-283, 1981.

Monaldo, F.M. and D.R. Lyzenga, On the Estimation of Wave Slope- and Height-Variance Spectra from SAR Imagery, *IEEE Trans. Geoscience and Remote Sensing, Vol. GE-24*, pp. 543-551, 1986.

Monaldo, F.M. and D.R. Lyzenga, Comparison of SIR-B Ocean Wave Image Spectra with Linear Model Predictions Based on Aircraft Measurements, *J. Geophys. Res.*, *Vol. 93*, pp. 15,374-15,388, 1988.

Phillips, O.M. and K. Hasselmann, *Wave Dynamics and Radio Probing of the Ocean Surface*, Plenum Publishing Corp., 1986.

Plant, W.J., Studies of Backscattered Sea Return with a CW, Dual-Frequency, X-Band Radar, *IEEE Trans. Antennas Propag., Vol. AP-25*, pp. 28-36, 1977.

Plant, W.J. and D.L. Schuler, Remote Sensing of the Sea Surface using One- and Two-Frequency Microwave Techniques, *Radio Science, Vol. 15*, pp. 605-615, 1980.

Plant, W.J., W.C. Keller, and A. Cross, Parametric Dependence of Ocean Wave - Radar Modulation Transfer Functions, *J. Geophys. Res., Vol. 88*, pp. 9747-9756, 1983.

Plant, W.J. and A.B. Reeves, The Dual-Frequency Scatterometer Re-Examined, *NASA Conf. Pub. 2303, Frontiers of Remote Sensing of the Oceans and Troposphere from Air and Space Platforms*, pp. 193-198, 1984.

Plant, W.J., W.C. Keller, A.B. Reeves, U.A. Uliana, and J.W. Johnson, Airborne Microwave Doppler Measurements of Ocean Wave Directional Spectra, *Int. J. Remote Sensing, Vol. 8*, pp. 315-330, 1987.

Raney, R.K. and P.W. Vachan, SAR Imaging Ocean Waves from an Airborne Platform: Focus and Tracking Issues, *J. Geophys. Res.*, *Vol. 93*, pp. 12475-12486, 1988.

Reed, I.S., On a Moment Theorem for Complex Gaussian Processes, *IRE Trans. Information Theory, Vol. IT-8*, pp. 194-195, 1962.

Schroter, J.F. Feindt, W. Alpers, and W.C. Keller, Measurement of the ocean wave-radar modulation transfer function at 4.3 GHz, *J. Geophys. Res.*, *Vol. 91*, pp. 923-932, 1986.

Shuchman, R.A., J.D. Lyden, and D.R. Lyzenga, Estimates of Ocean Wavelength and Direction from X- and L-band Synthetic Aperture Radar Data Collected during the Marineland Experiment, *IEEE J. Oceanic Eng.*, *Vol. OE-8*, pp. 90-96, 1983.

Shuler, D.L., W.J. Plant, A.B. Reeves, and W.P. Eng, Removal of Clutter Background Limitations in Dual-Frequency Scattering From the Ocean, *Int. J. Remote Sensing, Vol. 6*, pp. 1091-1112, 1985.

Stilwell, D., Jr., Directional Energy Spectra of the Sea from Photographs, *J. Geophys. Res.*, *Vol. 74*, pp. 1974-1986, 1969.

Trizna, D.B., J.M. Headrick, R.W. Bogle, and J.C. Moore, Directional Sea Spectrum Determination Using HF Doppler Radar Techniques, *IEEE Trans. Antennas Propagat.*, *Vol. AP-25*, pp. 4-11, 1977.

Trizna, D.B., R.W. Bogle, J.C. Moore, and C.M. Howe, Observation by HF Radar of the Phillips Resonance Mechanism for Generation of Wind Waves, *J. Geophys. Res.*, *Vol. 85*, pp. 4946-4956, 1980.

Valenzuela, G.R., Theories for the Interaction of Electromagnetic and Oceanic Waves—A Review, *Boundary-Layer Meteorology, Vol. 13*, pp. 61-85, 1978.

Vesecky, J.F. and R.H. Stewart, The observation of ocean surface phenomena using imagery from the SEASAT synthetic aperture radar: an assessment, *J. Geoph. Res.*, *Vol 87*, pp. 3397-3403, 1982.

Walsh, E.J., D.W. Hancock, D.E. Hines, R.N. Swift, and J.F. Scott, Directional Wave Spectra Measured with the Surface Contour Radar, *J. Phys. Oceanogr.*, *Vol. 15*, pp. 566-592, 1985a.

Walsh, E.J., D.W. Hancock, D.E. Hines, and J.F. Scott, Elimination of Directional Wave Spectrum Contamination from Noise in Elevation Measurements, *J. Oceanic. Eng.*, *Vol. OE-10*, pp. 376-381, 1985b.

Walsh, E.J., D.W. Hancock, D.E. Hines, R.N. Swift, and J.F. Scott, Wave Measurement Capabilities of the Surface Contour Radar and the Airborne Oceanographic Lidar, *Johns Hopkins APL Technical Digest, Vol. 8*, pp. 74-81, 1987.

Weissman, D.E. and J.W. Johnson, Measurements of Ocean Wave Spectra and Modulation Transfer Function with the Airborne Two-Frequency Scatterometer, *J. Geophys. Res.*, *Vol. 91*, pp. 2450-2460, 1986.

Wetzel, L.B., Models for Electromagnetic Scattering from the Sea at Extremely Low Grazing Angles, NRL Memo. Rpt. 6098, Naval Research Lab., Dec., 1987.

Wiegel, R.L. (ed.), *Directional Wave Spectra Applications*, *Proc. of a Conference*, 14-16 September, 1981, Univ. of California, Berkeley, American Society of Civil Engineers, New York, p. 495.

Winebrenner, D.P., and K. Hasselmann, Specular Point Scattering Contribution to the Mean SAR Image of the Ocean Surface, *J. Geophys. Res.*, *Vo. 93*, pp. 9281-9294, 1988.

Young, I.R., W. Rosenthal, and F. Ziemer, A Three-Dimensional Analysis of Marine Radar Images for the Determination of Ocean Wave Directionality and Surface Currents, *J. Geophys. Res.*, *Vol. 90*, pp. 1049-1060, 1985.

Zamarayev, R.D., and A.I. Kalmykov, On the Possibility of Determining the Spatial Structure of an Agitated Ocean Surface by Means of Radar, *Izv. Atmos. and Ocean Physics*, *Vol. 5*, pp. 124-127, 1969.

Chapter 15

PASSIVE MICROWAVE REMOTE SENSING OF OCEAN SURFACE WIND SPEED

CALVIN T. SWIFT
Department of Electrical and Computer Engineering
University of Massachusetts
Amherst, Massachusetts 01003

1. Introduction

The interest in passive microwave remote sensing of the ocean developed in the mid 1960's when the analytical work of Stogryn indicated that radiometric emission from the ocean depends upon the degree of surface roughness. Because the ocean surface becomes rougher with increasing wind speed, a potential technique became available for the global monitoring of surface winds from earth orbiting satellites. The qualitative nature of the theoretical results were verified in a series of experiments conducted by Hollinger. At the same time, Nordberg and colleagues collected data from an aircraft which were tens of Kelvin more responsive to ocean surface winds than predicted by the Stogryn theory. The difference between theory and experiment was correctly attributed to increased emission resulting from the wind generation of surface foam. Subsequent to these exploratory measurements, Webster et al. experimentally derived linear coefficients relating excess radiometric emission to increased wind speed as a function of instrument viewing angle, polarization, and electromagnetic frequency. Finally Jones, Black and colleagues demonstrated the capability of a multi-frequency C-band microwave radiometer to measure ocean surface wind speed. This sensor was used in connection with hurricane research, and has measured wind speeds up to 140 kts through rain rates exceeding 50 mm/hr.

Although this chapter concentrates on these research milestones, the intent is to present these and other results to illustrate the underlying physics and problems in measuring ocean surface wind speed with microwave radiometers. Section 2 discusses total emissions from the earth, including problems presented in separating atmospheric effects. Section 3 focuses on thermal emissions from an undisturbed ocean surface. Topics discussed here include effects of viewing angle and polarization which are considered in quantitative terms. The section forms a baseline for Section 4, which discusses the physical mechanisms which allow a microwave radiometer to measure ocean surface wind speed. The tie between the two sections is that the apparent radiating temperature of calm water is very cold due to the low emissivity, whereas, radiometric warming occurs as the wind disturbs the surface. Since several 10's of °K warming can occur, relatively accurate measurements of wind speed can be made over a large dynamic range of wind speeds, which is the subject of the material presented in Section 5. The chapter concludes with a discussion of present and potential future directions of research in this area.

G. L. Geernaert and W. J. Plant (eds.), Surface Waves and Fluxes, Volume II, 265–292.

2. Composite Radiation Received by Earth Viewing Microwave Radiometers

Before we discuss details of the remote sensing of surface phenomena by microwave radiometer, we first consider all thermal radiation components received by an Earth-viewing radiometer in space. Figure 1 shows such a combined thermal signal received by a radiometer. Descriptively, this combined signal, which we call the brightness temperature T_B can be expressed in words, as follows:

BRIGHTNESS TEMPERATURE =

(Surface Emissivity) X(True Surface Temp) X(Atmospheric Absorption)	+	(Upwelling Atmospheric Radiation)	+	(Reflected Downwelling Atmospheric Radiation) X(Atmospheric Absorption)	+	(Reflected SPACE Radiation) X(Double Atmospheric Absorption)

Or, translating words into a mathematical expression, we have:

$$T_B = eT\exp(-\tau) + T_{UP} + (1 - e)T_{DOWN}\exp(-\tau) + (1-e)T_{SP}\exp(-2\tau) \quad (1)$$

Where:

T_B	is the total brightness temperature, or combined signal
e	is the emissivity of the surface
$\exp(-\tau)$	is the atmospheric attenuation term
$T_{UP} = T_{DOWN}$	is the total atmospheric radiation
T_{SP}	is the 2.7K radiation residual from the big bang
T	is the surface temperature

The quantity $(1 - e)$ is the reflectivity of the surface, and the following simplified expression adequately expresses the total atmospheric radiation:

$$T_{UP} = \langle T \rangle [1 - \exp(-\tau)] \quad (2)$$

The quantity τ is defined as the atmospheric opacity, and functional forms exist for oxygen, water vapor, clouds, and rain. The term $\langle T \rangle$ is the average true temperature of the atmospheric column. At this point, we can illustrate the problem that the atmosphere can cause when one wishes to derive the surface parameters of interest. If the opacity is very large, which it can be if we do our observations in the presence of rain, the component of surface emission received by the surface measurement becomes severely compromised because a change in surface emission of many degrees is received as a proportionally smaller change as a direct result of the exponential weighting in equation (1). Indeed, for large opacity, the bulk of the received radiation is T_{UP}. If the opacity is infinite, then we never see the surface. On the other hand, as the opacity becomes vanishingly small, as it does at very long microwave wavelengths, then all we see is emission from the surface. In this case, the atmosphere is transparent, regardless of meteorological conditions.

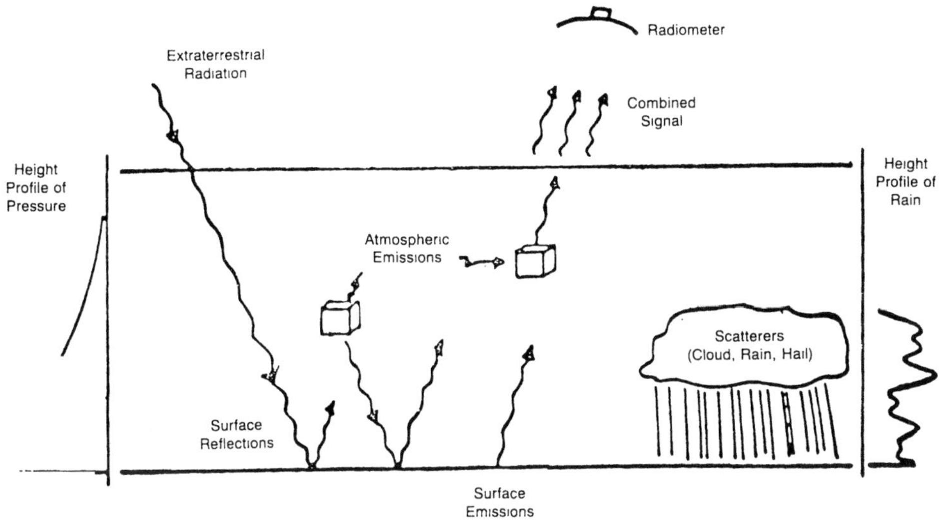

Fig. 1 — Sources of thermal radiation received by a microwave radiometer. (Originally prepared by L.D. Staton, NASA Langley Research Center.)

For optical and infrared remote sensing, the atmosphere presents a severe problem for remote sensing of the earth's surface. The principal problem in this part of the electromagnetic spectrum is attenuation due to cloud cover. At these wavelengths, the cloud drop size is large compared to a wavelength so that the transmitted radiation that originates at the surface scatters in all directions, and therefore severely limits our capability to observe surface phenomena. As the electromagnetic wavelength becomes longer, scattering by clouds becomes weaker. If, indeed, the cloud drop radius is very small in terms of a wavelength, the scattering strength diminishes as λ^{-4}. In principle, therefore, clouds will be invisible at sufficiently long electromagnetic wavelengths. The fact that microwaves are able to penetrate clouds has continued to keep interest alive, even though the spatial resolution of microwave sensors has been coarse compared to resolution achieved by its visible and infrared cousins. An exception has been the Synthetic Aperture Radar (SAR), where signal processing techniques are used to overcome the diffraction limit of conventional antenna systems.

Figure 2 is a plot of atmospheric attenuation over six decades of the electromagnetic spectrum extending from near microwaves to the visible portion of the spectrum. As we consider this figure, we focus attention first to attenuation by clouds. We note that clouds are at least translucent within the conventional microwave band, defined as those frequencies less that 100 GHz (or wavelength of greater than 3mm). At frequencies less than 4 GHz, clouds are virtually invisible, and only relatively small corrections are needed when measuring surface emissions up to 10 GHz (3 cm). The other nonresonant form of atmospheric attenuation shown in this curve is due to rain. The basic attenuation mechanism is essentially the same as for clouds, except that the drop size is much larger than that in clouds, and attenuation effects therefore extend to the longer electromagnetic wavelengths. Attenuation also increases with rain rate (measured in mm/hr) because the number of attenuating particles per unit volume

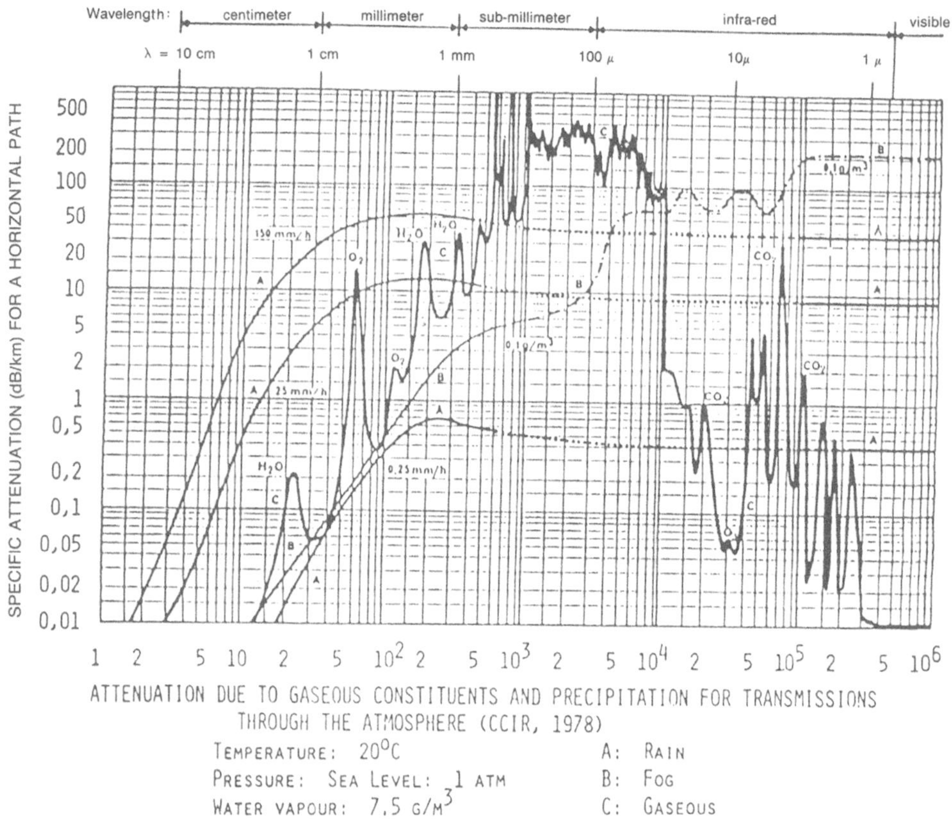

Fig. 2 — Attenuation due to atmospheric constituents as a function of electromagnetic frequency. (Received from M. Maxwell, NASA Goddard Space Flight Center.)

increases with rain rate. Note, however, that even rain tends to be translucent at frequencies less than 10 GHz, except for exceptionally heavy rainfall.

Although rain appreciably attenuates microwave signals, the occurrence of rain is relatively infrequent compared to cloud cover and the probability of very heavy rainfall is even smaller. Relative occurrence of cloud cover and rain rate for various parts of the world are accessible in the literature.

The final atmospheric contribution shown in this figure is attenuation due to atmospheric gases. Gaseous attenuation is not a monotonic function because the absorption is of a resonant nature due to electric and magnetic dipoles associated with the principal constituents; namely, oxygen, carbon dioxide, and water vapor. Indeed, we note that the submillimeter wave band extending roughly from 1 mm to 30 micrometers is essentially an atmospheric wall for surface observations. The attenuation is severe not only because the wings of molecular vibrational modes begin to merge, but also because rotational lines are strong for the submillimeter wavelength. It is also of passing interest to note that technology is demanding in this band because conventional system components are difficult to manufacture. Therefore, atmospheric visibility constraints favor spectral regimes at wavelengths less than 30 micrometers or greater than 3 mm (100 GHz).

At frequencies less than 100 GHz, there is a reasonable operating band (window) in the 80-95 GHz range; however, a strong oxygen line occurs in the 50-70 GHz, which, does not favor surface observations. At frequencies less than 50 GHz, there is only a relatively weak water vapor line to contend with that is centered at 22 GHz. We will focus the remainder of the discussion of atmospheric attenuation to frequencies below 50 GHz.

Figure 3 shows a plot of the apparent temperature that an upward looking radiometer would measure over the microwave band extending up to 35 GHz. Here, we see that even though the 22 GHz water vapor line is relatively weak, it can contribute almost 100°K on very humid days (approximately seven cm of condensed water). Only on perfectly dry days are we left with a residual bias of less than 5°K due to radiation from the low frequency wing of the 60 GHz oxygen resonance. Also shown in this curve is a quantitative calculation of the total atmospheric radiation that would be expected from a very thick cumulus cloud prior to the formation of rain drops. This calculation essentially gives an upper bound on the atmospheric contribution due to non-precipitable clouds. Figure 4 shows the results of a calculation of radiation due to rain over the microwave band for two different rain rates of 4 mm/hr and 10 mm/hr. These two curves represent relatively light rainfall. We note that the apparent temperature is a strong function of the choice of electromagnetic frequency, and we have therefore split the scale to show the contribution in more detail at the lower microwave frequencies. These two figures again illustrate the point we have been making concerning the choice of frequency; i.e., the atmosphere almost becomes transparent if we go low enough in frequency. It has been said that the atmosphere is a 4% problem for surface observations at 5 GHz, whereas, it is a 90% problem at infrared wavelengths.

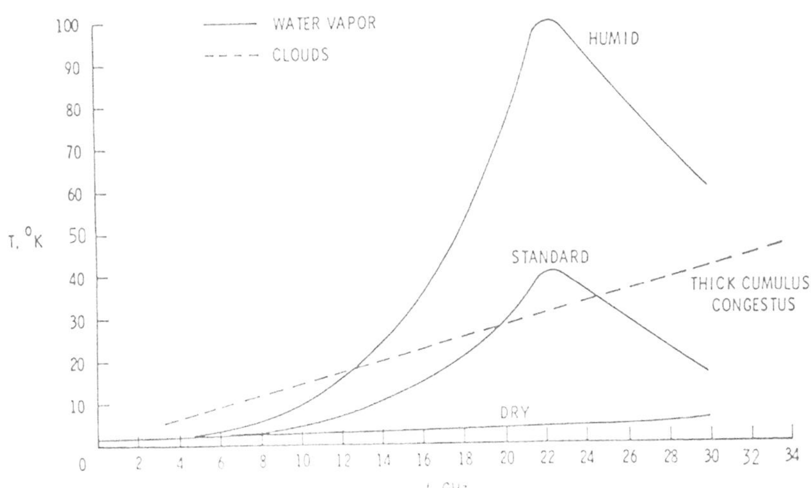

Fig. 3 — Apparent temperature observed by a microwave radiometer viewing a non-precipitable atmosphere for electromagnetic frequencies below 35 GHz

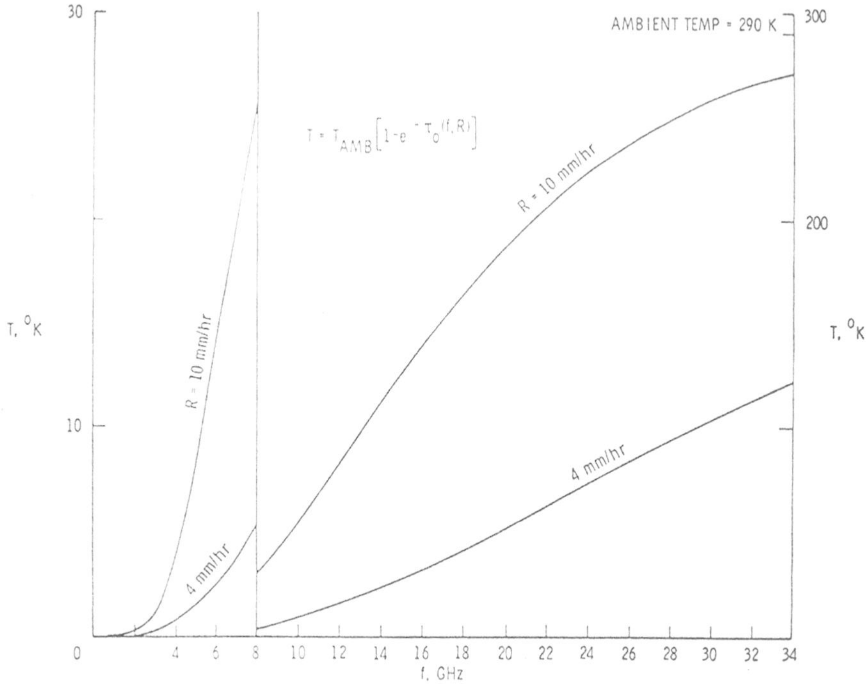

Fig. 4 — Apparent temperature observed by a microwave radiometer viewing precipitation for electromagnetic frequencies below 35 GHz

One is therefore tempted to blindly choose an appropriate frequency less than 10 GHz for surface observations. There are, however, a number of reasons for considering a choice of frequencies up to 35 GHz, and also frequencies in the "dirty" atmosphere windows at 94, 140, and 215 GHz. First, some geophysical observations require higher frequencies. For example, discrimination of new sea ice from old sea ice (defined as that which has survived at least one melt season) favors a choice of both a low microwave frequency (less that 18 GHz) and a frequency at least as high as 35 GHz. A second over-riding reason for selecting higher frequencies, at least in this stage of spacecraft microwave radiometry, is a pressing need for improved spatial resolution from orbital altitudes. For example, the spatial resolution Δ of a microwave radiometer at orbital altitude H, connected to an antenna of diameter D, and operating at a wavelength λ, is given by:

$$\Delta = \frac{\lambda H}{D} \tag{3}$$

The orbital altitude is fixed, and there is usually an upper bound to the size of the antenna that can be integrated into the spacecraft. The remaining variable that determines spatial resolution is therefore the freedom to choose wavelength. Therefore, for a fixed antenna size, one will achieve a ten-fold improvement in spatial resolution by choosing 35 GHz as opposed to 3.5 GHz. However, at 35 GHz, the accuracy of surface observations will be seriously degraded by atmospheric attenuation. For example, a sudden increase in apparent temperature, may be due to the surface, clouds, or an increase in water vapor. A way out of this dilemma is to observe the earth at a number of frequencies. Since the frequency response of clouds and water vapor are known functions of wavelength, the use of three frequencies, say 18, 22, and 35 GHz, can be used to correct for atmospheric effects, while a third frequency can be used to infer surface observations. Additional degrees of freedom are offered if the measurements are performed at both polarizations. Thus, obvious penalties that are paid for observing the earth at the higher frequencies are: a) at least a three-fold increase in measurement hardware, and b) added algorithm complexity (to eliminate atmospheric effects). We should further note that the atmospheric bias becomes more appreciable as we move up in frequency and operate at the "dirty" windows. Cloud contributions to attenuation continues to increase with frequency and the tails of various high frequency oxygen and water vapor resonance distributions add significant attenuation as evidenced in Figure 2. An important consideration when using multiple frequencies is the mismatch in spatial resolution. For example, the hardware designer will invariably make maximum use of the available antenna aperture. If all the frequencies use separate feeds but a common reflector, the beamwidth of the higher frequencies will be narrower than that of lower frequencies. Thus, an 18 GHz channel will have coarser spatial resolution than a 37 GHz channel by approximately a factor of two. The mismatch in spatial resolution can cause difficulties in the interpretation of the derived geophysical results, particularly if lateral gradients in the upwelling brightness temperature occur. This problem could easily be resolved merely by spoiling the beam of the high frequency channel to match that of the lowest frequency channel, but this means sacrificing spatial resolution at the high frequencies, which users are reluctant to do. Thus, both the design engineer and the geophysicist resist the remote sensing specialist on this issue.

An advantage of operating at the higher frequencies is the avoidance of radio frequency interference. Frequencies below 10 GHz are used for a variety of civilian and military applications relating to communications, navigation, and radar. As point sources, these signals are orders of magnitude stronger than natural emissions from the earth. The severity of the problem depends upon the specific operating frequency and geography. Experience with data collected from space at 6.6 GHz, indicates that strong interference occurs only near populated areas of the world. There are also a number of protected bands that one can choose. For example, 1.413 GHz is protected for use by radio astronomy.

We have discussed strategies for dealing with atmospheric attenuation, including choosing low frequencies to avoid the atmospheric problems altogether. We now consider a lower bound on selecting frequencies. Figure 5 shows the apparent temperature of the sky at frequency bands extending from 100 MHz to 50 GHz. The curve shows that the sky becomes warmer with decreasing frequency below 1 GHz. The source is due to non-thermal synchrotron radiation from the Milky Way, which is much stronger than the natural emission from the earth. The ionosphere also begins to pose problems at these lower frequencies. Even though the ionosphere is relatively transparent at 1 GHz, the earth's magnetic field causes Faraday rotation, which means that the linear polarized component received in space is not necessarily that emitted from the earth. It is also difficult to predict the degree of rotation because the amount depends upon the electron density in the ionosphere and the earth's magnetic field, both of which are highly variable. Finally, at extremely low frequencies, below

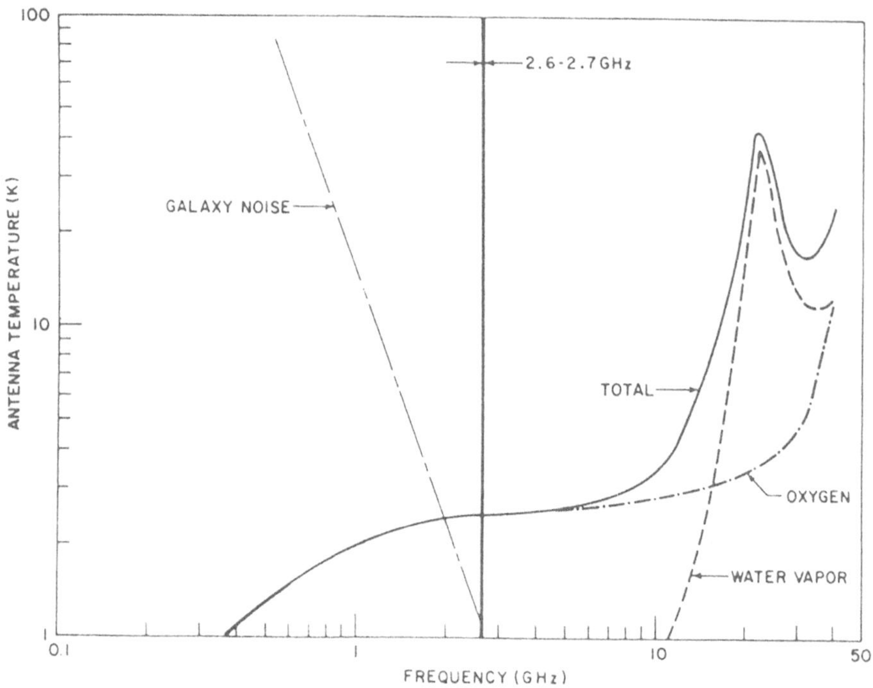

Fig. 5 — Apparent temperature observed by a microwave radiometer viewing clear sky over an electromagnetic frequency band extending from 100 MHz to 50 GHz. (This figure appears in NASA Contract Report CR-1960 submitted as a final report by the Rockwell Science Center.)

10 MHz, the ionosphere is opaque, and observations of the earth's surface are no longer possible from points above 50 km. Such observations are not practical also because of the enormous size of antennas required. Not shown in this figure is the apparent temperature of the sun as a function of frequency. The sun is an extremely hot source, as it is in the visible band, therefore satellite geometry that favors a bistatic reflection of sunlight into the receiver should be avoided. This criterion is independent of the remote sensing band used, including visible and infrared.

We conclude this section on the atmosphere by showing an example of experimental data which should serve to illustrate the role of passive microwave remote sensing. Figure 6 shows a tracing of analog data that were simultaneously collected from a 2.65 GHz microwave radiometer and a 10.6 micrometer infrared radiometer from an aircraft flying several thousand feet above the ocean. Both sensors viewed the ocean in the nadir direction, and the data trace represents a 20 minute record. The upper trace represents data collected by the microwave radiometer. Although the data exhibit a peak-to-peak noise fluctuation of about one-third of a degree, the mean value is constant over the entire 20 minute flight over smooth and isothermal sea water. The lower trace is the corresponding data received by the infrared radiometer. The record is reasonably noise free during the first eight minutes; however, the mean value is not constant. This observed variation of more than one-half degree correlates with changes in the aircraft altitude and corresponding changes in the column depth

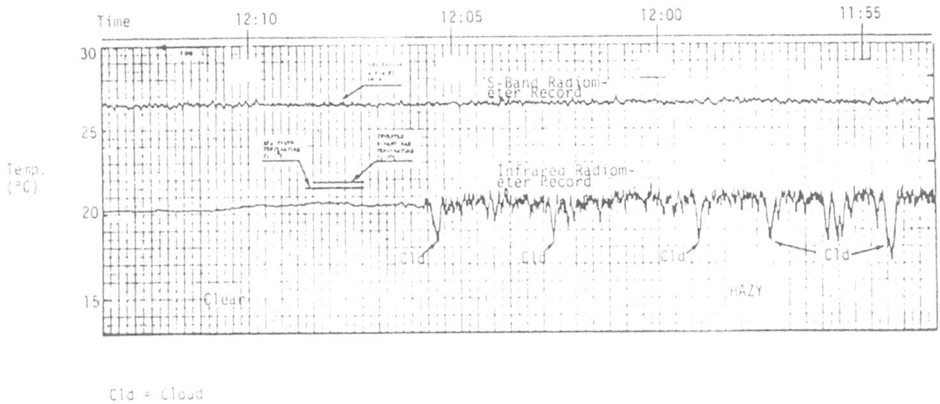

Fig. 6 — S-band and infrared radiometer data collected from an aircraft platform near Bermuda. (Figure prepared by H.J.C Blume, NASA Langley Research Center.)

of the atmosphere. The last twelve minutes of the infrared record shows a marked increase in the noise level, and periods which resemble data drop-outs. The increase in the infrared noise level is associated with overflight of a haze layer, which scatters the upwelling radiation from the ocean. The cooler dips in the record are attributed to radiation and absorption from intervening clouds.

3. Fundamentals of Surface Microwave Emission from an Undisturbed Ocean Surface

The fundamentals of surface emission can best be detailed through a discussion of Figure 7. Here, an antenna is viewing a flat surface, and the user scientist has control of the viewing angle Θ, the operating wavelength λ of the receiver, and the state of polarization. It is conventional to define the two independent states of polarization as follows: we have horizontal (perpendicular) polarization if the antenna collects that component of the thermal radiation whose electric vector is perpendicular to the figure. Likewise, we have vertical (parallel) polarization if the antenna is rotated 90 degrees to observe the incident radiation with the electric field oriented parallel to the plane of the figure. Spacecraft radiometers usually collect both polarization components at a fixed viewing angle.

The amount of noise power received by the radiometer is given by the following equation:

$$P = kT_B B \tag{4}$$

where P is received power in Watts, B is the receiver bandwidth in Hz, and T_B is the brightness temperature in °K. The remaining factor k is the Boltzmann constant. The receiver bandwidth and the Boltzmann constant are almost always discarded, so that only brightness temperature is displayed as data. As we have previously discussed, the brightness temperature is the product of the emissivity and the physical (or true) surface temperature. Moreover, if the surface is in thermal equilibrium, then emission is equal to absorption; otherwise

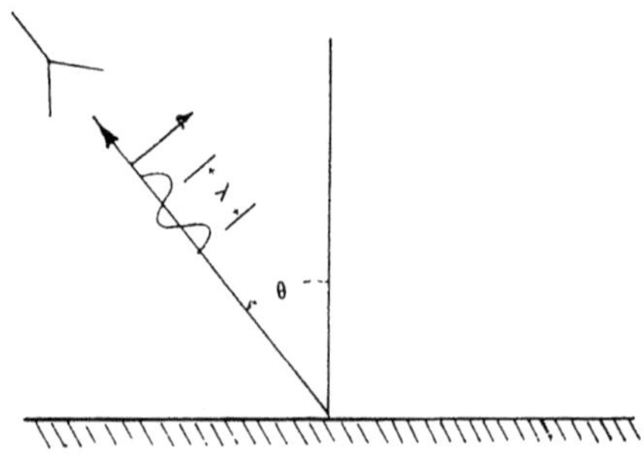

$P_R = \kappa T_B B = A\kappa TB = (1-R)\kappa TB$

B = Receiver bandwidth

T = True surface temperature

T_B = Brightness temperature

A = Attenuation coefficient (emissivity)

R = Reflection coefficient (function of λ, θ, polarization)

Fig. 7 — Fundamental parameters describing thermal emission from smooth radiating half space

the total volume will undergo continuous heating or cooling at a rate much greater than that associated with normal climatic changes. In other words, for every photon absorbed by the medium, one must be emitted; therefore

$$\text{power emission} = \text{power absorption}$$

This is a powerful relationship, because it allows us to develop a quantitative relationship between received brightness temperature and the medium that we are observing. Since the surface is flat and the material below it is assumed to be infinitely deep, the absorption A is simply related to the surface reflection coefficient as follows:

$$A = 1 - R \tag{5}$$

The reflection coefficient is obtained by solving the boundary-value problem of a plane wave incident upon a flat half space. The absorption A, or, equivalently, the emissivity is then found by subtracting the result from unity. In words, we can say all power not reflected is absorbed. The solution to the boundary value problem is found in most basic texts on electromagnetic theory, and the results are:

Vertical polarization:

$$R_V = \left| \frac{\epsilon_r \cos\Theta - \sqrt{\epsilon_r - \sin^2\Theta}}{\epsilon_r \cos\Theta + \sqrt{\epsilon_r - \sin^2\Theta}} \right|^2 \tag{6}$$

Horizontal polarization:

$$R_H = \left| \frac{\cos\Theta - \sqrt{\epsilon_r - \sin^2\Theta}}{\cos\Theta + \sqrt{\epsilon_r - \sin^2\Theta}} \right|^2 \tag{7}$$

Where ϵ_r is the relative dielectric constant of the medium, ϵ/ϵ_0. We have now illustrated two important concepts at this point. First, we have shown why the radiation received is dependent upon polarization. Second, the above equations indicate that the reflection coefficients depend upon the dielectric constant of the material that we are observing (the dielectric constant is the square of the index of refraction). The measurement of the dielectric constant establishes the remote sensing connection with geophysics.

Before proceeding, we should also point out that the dielectric constant ϵ_r is a complex number with the imaginary part of being directly proportional to the ohmic conductivity of the medium and inversely proportional to electromagnetic frequency. The imaginary part introduces the attenuation necessary for thermal emission and serves to justify our approximation that the medium is infinitely deep. The actual penetration depth, which can be considered as the layer where most of the radiation occurs can also be quantified, as follows:

$$\delta = \frac{2\pi}{\lambda_0} \mathrm{Im}(\sqrt{\epsilon_r}) \tag{8}$$

where δ is the 1/e depth of penetration, λ is the wavelength, and $\mathrm{Im}(\sqrt{\epsilon_r})$ is the imaginary part of the complex index of refraction. We can illustrate the depth of penetration by considering sea water. In coastal zones of the earth, the conductivity of water can range from near zero at estuary sources to 4 mhos/m as the water mass mixes with the ocean. We would therefore expect to observe a smaller depth of penetration as the water becomes saltier. Figure 8 shows that this is indeed the case. This curve shows that at a frequency of 1.43 GHz the skin depth is approximately 10 cm for fresh water, and drops abruptly as small volume amounts of salt are introduced. We note that the penetration depth of sea water is approximately 1 cm at 1.43 GHz and will become even smaller as the microwave frequency increases. Thus, oceanic observations can be considered as surface observations. However, we should point out that even though emission at microwave frequencies occurs in only a fraction of a centimeter, the microwave depth is over an order of magnitude greater than the infrared depth. This is an important difference, since some oceanographers claim that substantial thermal gradients can occur within a millimeter of the surface; hence microwave observations may provide a better average of the true radiating properties of the ocean. As a last word on the depth of penetration of water, we note that the penetration depth is finite for zero conductivity as a result of a residual imaginary part of the dielectric constant of water that is associated with molecular collisions.

We will now use sea water as an example to illustrate the variation of microwave emissivity with viewing angle and polarization. Water provides an interesting example because it

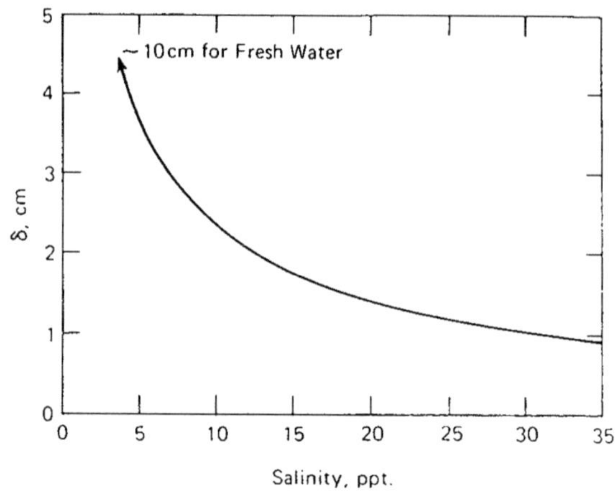

Fig. 8 — Electromagnetic depth of penetration vs ocean salinity

has a high dielectric constant with a corresponding low reflectivity. For example if we substitute a real dielectric constant of 80, which is the approximate value for water, into the formula for the reflection coefficients, we find that for nadir viewing:

$$R_V = R_H = 0.64 \tag{9}$$

If the ambient water temperature is 290K, then the measured brightness temperature will be:

$$T_B = (1 - R)T = 104.9K \tag{10}$$

Therefore, water is a cold medium at microwave frequencies because of its low emissivity. If we now calculate the reflectivities as a function of angle, we construct the two curves shown in Figure 9. Here we have plotted the emissivities for both vertical and horizontal polarization as a function of viewing angle as measured from nadir. Zero angle is looking straight down and 90 degrees is looking at the horizon. We note that the emissivity at vertical polarization increases to unity and then abruptly drops to zero. The angle where the emissivity is unity is called the Brewster angle, which occurs when there is total transmission into the medium. We note that if the surface is viewed at this angle and polarization, we have an unambiguous measurement of the surface temperature. Unfortunately, it is not practical to choose this angle because of severe attenuation caused by the long propagation path through the atmosphere. The emission at horizontal polarization is relatively uncomplicated. In this case, the emissivity, hence the brightness temperature, monotonically decreases to zero.

We now compare the calculations with actual experimental data. A number of years ago, a microwave radiometer system was installed on a bridge to collect basic data on the radiating properties of sea water. Four radiometers were installed on a rotating carriage with an antenna system configured to receive both polarizations. The rotating carriage allowed measurements to be taken over selected frequencies ranging from 750 MHz to 7.5 GHz. Actual

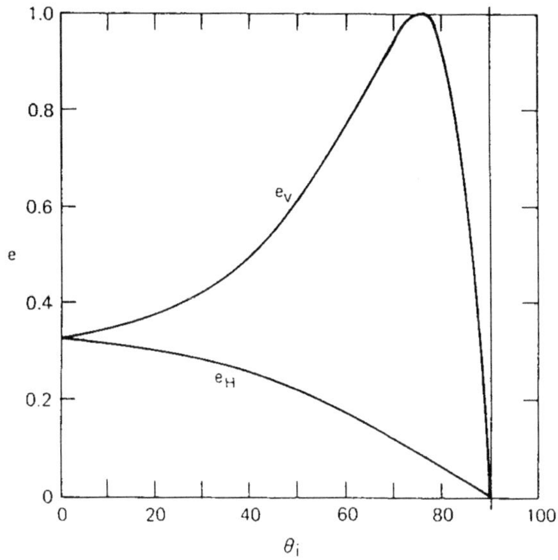

Fig. 9 — Radiometric emissively of a flat ocean surface as a function of viewing angle for horizontal and vertical polarization

experimental data are shown in Figure 10, and compared with the theoretical calculations. In the experiment the viewing angle could be varied from −25 degrees (under and behind the bridge) to +150 degrees (sky, near zenith in front of the bridge). Thus, the radiometer scanned through nadir to the horizon, and then through the atmospheric column. As shown in this figure, there is a range of viewing angles where there is reasonable agreement between theory and experiment. Also, there are some angles where the agreement is poor, but there are good reasons for the discrepancies. We first discuss the discrepancy at nadir. The measured 35K excess brightness temperature over the calculated value occurs because the water emits radiation which is reflected by the bridge, and then reflected again by the water and into the radiometer—thus causing a measurement error. This phenomena serves to illustrate the limitations imposed by measurements taken from platforms. Even if aircraft are used as platforms, the aircraft should fly at altitudes of at least 1000 feet if the radiometer views nadir. The theory and experiment also disagree at angles near the horizon. In the case of vertical polarization, the measured brightness temperature is much less than the calculated value. This is a direct result of the finite beamwidth (field-of-view) of the antenna, in this case, 8 degrees. Near grazing, the vertically polarized antenna receives the weighted average of the relatively hot water and the relatively cold sky, thus resulting in a much colder measured value than the calculation predicts. In the case of horizontal polarization, the opposite is true. Here, we have the weighted average of a relatively cold surface and a warm atmosphere, resulting in an impulse that essentially traces out the antenna pattern.

The data also shows a typical characteristic of sky radiation. This curve shows that the apparent sky temperature is very cold at zenith, but becomes appreciably warm when observations are taken at the horizon. This occurs because of the increased optical depth of the atmosphere when the horizon is viewed. This is the reason we can view beautiful sunsets.

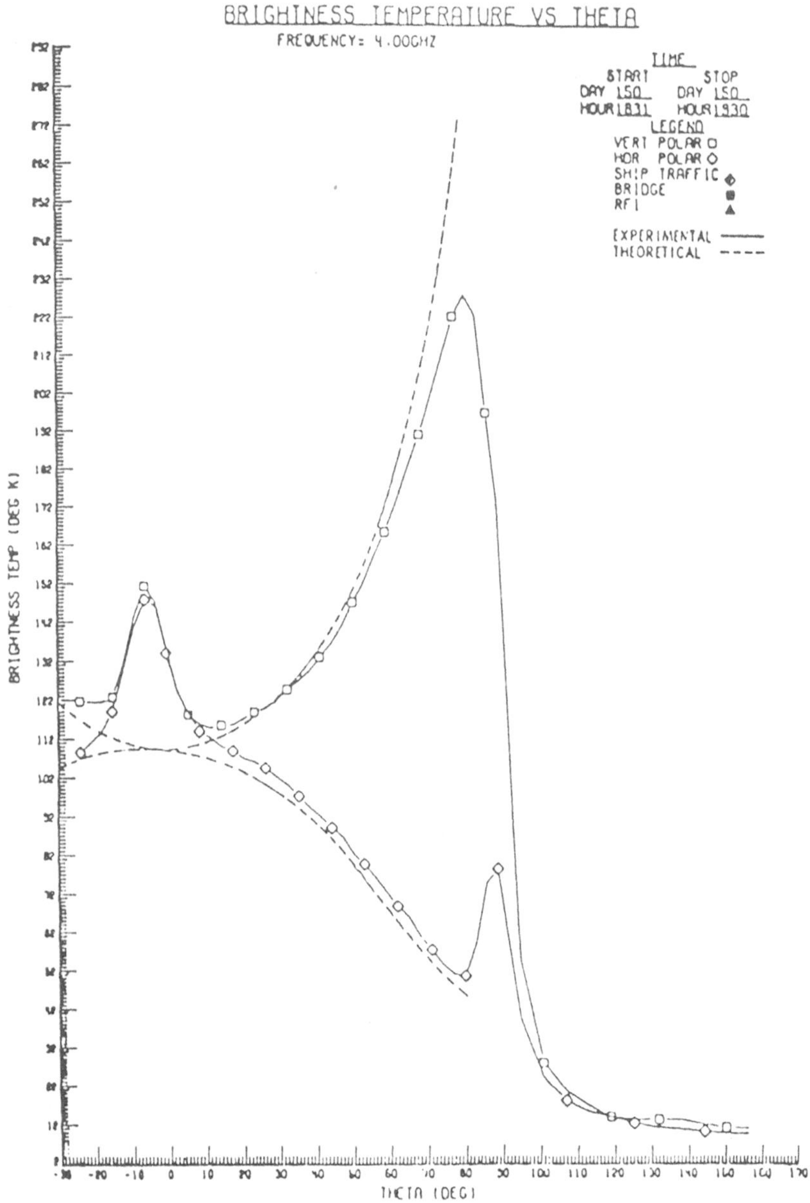

Fig. 10 — Measured radiometric brightness temperature of saline water as a function of viewing angle and polarization at an electromagnetic frequency of 4 GHz

To be more quantitative, the opacity τ varies as $\tau = \tau_o \sec \Theta$, where τ_o is the opacity at zenith, and Θ is the angle measured from zenith. If we include angle, the total sky temperature that one would measure is given by the following formula:

$$T_{\mathrm{DOWN}} = \langle T \rangle [1 - \exp(-\tau_o \sec \Theta)] \tag{11}$$

The experimental example shown here, indicates that the simple models that were presented provide a first order interpretation of experimental data; however, accurate interpretation involves much more effort, such as including the sky component and the averaging effects of the antenna pattern. In practice, this is done, and excellent agreement between theory and experiment is achieved with well calibrated instruments.

4. Fundamentals of Microwave Emission from a Wind-Driven Ocean Surface

Figure 11 is a schematic representation of additional factors to be considered when viewing the earth, in this case, the ocean. As eluded to in the previous section, the atmospheric component must be included, and this is done by utilizing the radiative transfer equation. One additional factor is surface roughness that occurs with the generation of waves by the wind. The ocean becomes rougher as the wind speed increases and we expect to observe a change in the brightness temperature. A theoretical understanding of the surface roughness effects has not been fully developed. Calculations show that surface roughness is a small effect under the assumptions used in existing theories. Although these models seem to be adequate in interpreting radar backscatter, they do not explain radiometric emission with sufficient precision. This is because 1/2 dB is an extremely accurate radar measurement, whereas 0.005 dB accuracy is needed for some applications of microwave radiometry. Consequently, more accurate modelling is required. Because of the present shortcomings in the theory, empirical corrections are made based on experimentally derived relationships between brightness temperature and wind speed. The generation of surface foam is another factor that impacts the surface brightness temperature. As it so happens, foam causes a major perturbation in surface emission. This occurs because foam consists of a mixture of water and a substantial amount of air. Because we now have water mixed with air, the average dielectric constant of foam is much less than 80, the value associated with water. Since the dielectric constant is lower than that of water, the reflectivity is also lower, therefore, the brightness temperature is higher—indeed, foam appears to behave like a blackbody; i.e., a material with an emissivity of one. The composite emission of foam and water can be modeled quite easily. If F_f is the area fraction of foam within the field of view of the radiometer, then the total brightness temperature received by the radiometer is:

$$T_{BT} = (1 - F_f)(1 - R)T + F_f T \tag{12}$$

where R is the reflectivity of sea water, T is the physical temperature of the water and foam, and T_{BT} is the total brightness measured by the radiometer. Thus, if there is no foam on the surface, we measure only the brightness temperature of water—which at nadir is on the order of 110K, but if the ocean surface is 100% foam covered, then we measure the true water temperature, of say 290K. Thus, there is a 180K swing in brightness temperature, which is substantial. Since the percent foam coverage increases with surface wind speed, we would

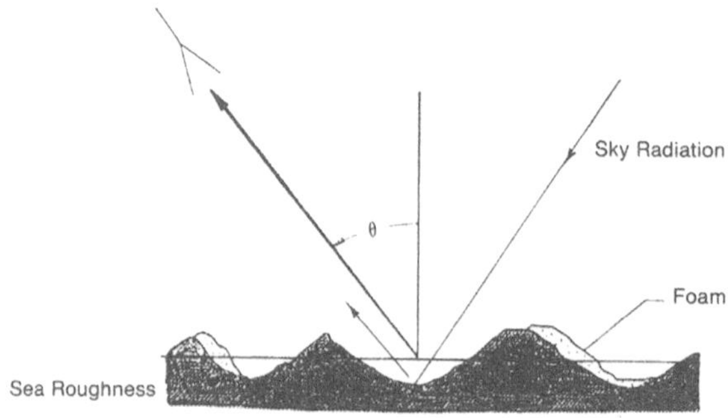

Fig. 11 — Schematic of additional factors that influence the thermal emission from an ocean surface

expect the brightness temperature to increase accordingly. Indeed, this phenomena is used to measure wind speed over the ocean. Again, the present wind speed algorithms are based upon empirical correlations with other data sources such as buoys. In this case an empirical relationship was needed because the theory of foam generation (whitecaps and streaks) is not that well developed. Indeed, it is of interest to note that radiometric measurements are presently being used to advance understanding of foam generation.

We now consider some examples relating to brightness temperature changes as a result of surface roughness and foaming. Figure 12 shows experimental results of the effect of surface roughness. The open squares and circles represent data collected from the radiometer system mounted on the bridge. These are three plots of brightness temperature vs viewing angle at frequencies of 1.4, 4.0, and 7.5 GHz. An equipment problem did not allow data to be at vertical polarization for the 1.4 GHz channel, and the 750 MHz data were not useful because of strong interference from UHF television stations. The other experimental data (dots and crosses) are results first reported by J. P. Hollinger of the Naval Research Laboratory (Hollinger, 1971). The open squares are data that were taken under low wind conditions, and, hence a smooth surface. The open circles represent data collected when high winds were blowing parallel to the plane of incidence, and against the tide, which resulted in a very rough surface. The water surface was foam free. The solid lines in the figure are calculations of emission from a smooth water surface, and the dashed lines are calculated values of emission from a rough surface based upon a geometrical optics model. The roughness parameter in the model is the rms tilt of the scattering facets, and this parameter is allowed to vary from zero (the smooth surface) to 15 degrees. An upper bound for the rms slope of ocean waves is less than 10 degrees as a result of wave breaking. The theoretical results show only a small increase in brightness temperature at near-nadir angles (approximately 2°K). The theory further predicts that the horizontally polarized brightness temperature for rough surfaces always shows an increase over the smooth values, regardless of the viewing angle. On the other hand, the vertically polarized component shows a very interesting effect. For viewing angles less than approximately 50 degrees, surface roughness causes an increase in brightness temperature, but at viewing angles greater than 50 degrees, it causes a decrease. In other words, there is a viewing angle one can select where the emission is invariant with surface roughness. The value of this observation is that surface roughness uncertainties will not cause

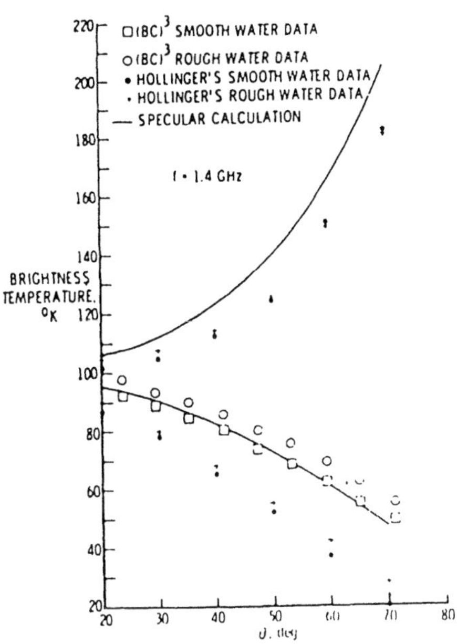

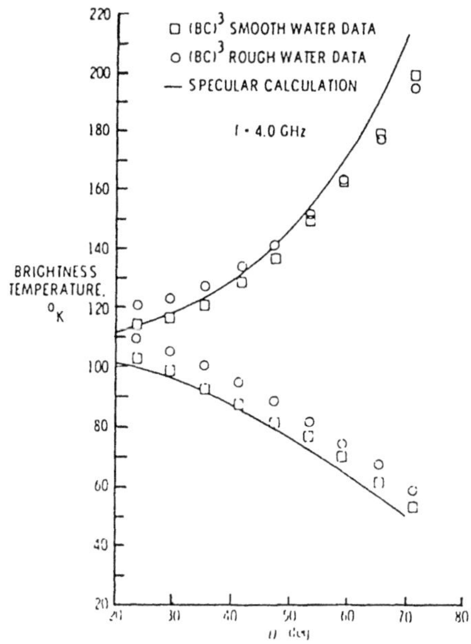

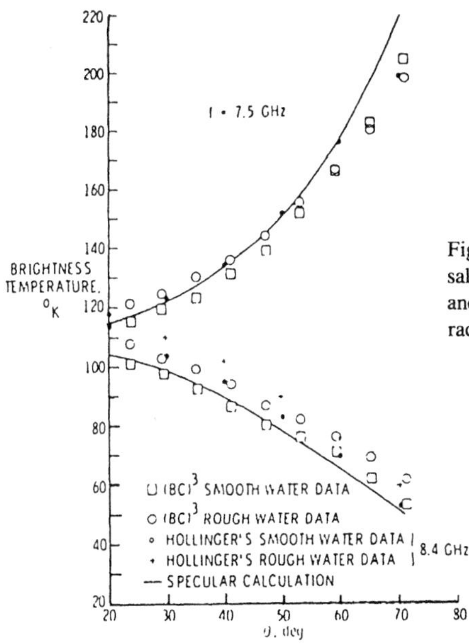

Fig. 12 — Measured radiometric brightness temperature of saline water as a function of viewing angle, polarization, and electromagnetic frequency. These results compare radiometric emission of smooth and rough surfaces.

errors in determining other geophysical parameters of interest, if data are collected at this viewing angle and polarization. This property of rough surface emission was first reported in a 1967 paper by A. Stogryn and is the reason that spacecraft radiometers designed to collect earth surface measurements view the ocean surface at an angle of 50 degrees.

Although the theory has been successful in providing quantitative descriptions of the emission from rough surfaces, it has thus far been insufficient if we desire accurate interpretations. For example, a close comparison of theory with experiment, reveals some detailed differences. Measured increases exceed the theoretical values (except for unrealistic slopes of greater than 10 degrees), and the experimental angle of roughness invariance is more like 60 degrees instead of 50 degrees. There is also an indication from the figure that the emission depends on electromagnetic frequency, yet the geometric optics theory predicts that the results are independent of frequency. Figure 13 shows experimental results reported a number of years ago by Hollinger. This is a plot of increased brightness temperature due to wind speed, hence surface roughness as a function of frequency up to 19 GHz. As we have noted before, the brightness temperature increases with increasing wind speed, but the expanded frequency base of data also clearly shows that the brightness temperature indeed depends upon the choice of electromagnetic frequency, which is not consistent with the rough surface theory. As a result, all wind speed algorithms were empirical relationships.

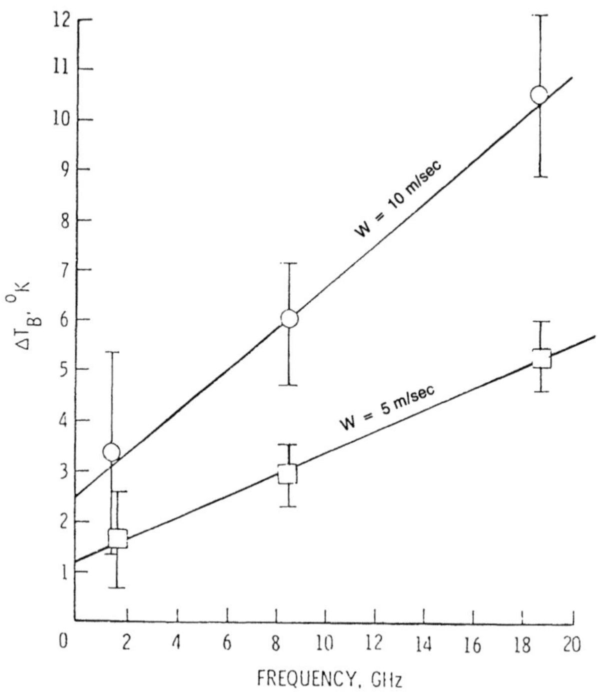

Fig. 13 — Measured increase in brightness temperature vs. electromagnetic frequency for wind speeds of 5 and 20 m/sec. (A cross-plot of results published by J.P. Hollinger, U.S. Naval Research Laboratory.)

We now turn the discussion toward effects of surface foam. Figure 14 shows a plot of percent foam coverage, as compiled by Stogryn (1972) based upon data reported by four oceanographic studies. The least squares fit through the data points results in an increasing percentage of foam with increasing surface wind speed. The contribution of this work was the development of power law relationship between percent foam and wind speed. In this case a relationship that is slightly more than cubic in wind speed is derived from the data. If this cubic relationship is correct, there is an upper bound of approximately 40 m/s on the remote measurement of wind speed, because percent foam cannot exceed 100%. More recent experimental work indicates a wind speed squared relationship, which fortunately extends the dynamic range of the measurement of wind speed. Indeed, an airborne microwave radiometer has remotely measured wind speeds up to 70 m/s (140 knots). Although one can quibble with the specific power law relationship, the point of this discussion is that percent surface foam does increase with wind speed, and we would expect to observe corresponding increases in brightness temperature.

Figure 15 shows some early observations of increased ocean brightness temperature as a function of wind speed at a frequency of 19 GHz, a 50 degree viewing angle, and horizontal polarization. Nadir measurements are also shown for reference. The excess brightness temperature for wind speeds less than 10 m/s is principally due to surface roughness. At higher wind speeds, foaming is the predominant process resulting in excess brightness temperature of more than 30K. The breakpoint in the solid curve occurs because the physics of emission is different at high wind speeds. Recent work has indicated that the wind speed relationship is linear at the low wind speeds, and follows a power law at the higher wind speeds. Figure 16 shows the linear sensitivity of emitted radiation as a function of frequency and polarization. In this figure, the vertical axis is a linear slope constant that relates increased brightness temperature to wind speed. For example, if the constant is one, a 1K increase in brightness temperature represents a 1 m/s increase in wind speed. The figure reinforces many of the points we have raised: the wind speed response depends upon frequency, polarization, and the viewing angel. At an angle of 50 degrees, vertically polarized emission exhibits a much weaker response than does the horizontally polarized component. The results also show that the emission becomes insensitive to wind speed at low microwave frequencies. This is probably associated with two effects. First, the surface roughness contribution becomes very weak as a result of the observed frequency dependence, and second, the longer wavelengths have a tendency to penetrate through the foam to more readily observe water.

5. Airborne Microwave Radiometer Measurements of Ocean Surface Wind Speed Through Heavy Rain

In 1974, NASA Langley Research Center began the design and construction of a Stepped Frequency Microwave Radiometer (SFMR) which would operate in the 4-6 GHz band. The SFMR was essentially a microwave spectral radiometer, whereby, the experimenter could program the machine to sequentially collect data at any number of center frequencies within this broad band of operation. One practical engineering reason for the development of this sensor was to avoid corruption of data due to Radio Frequency Interference (RFI) that was expected within this widely used band. For example, if the mission plan specified an operating frequency of 4.0 GHz, and the operator observed RFI on the output display, he would then tune

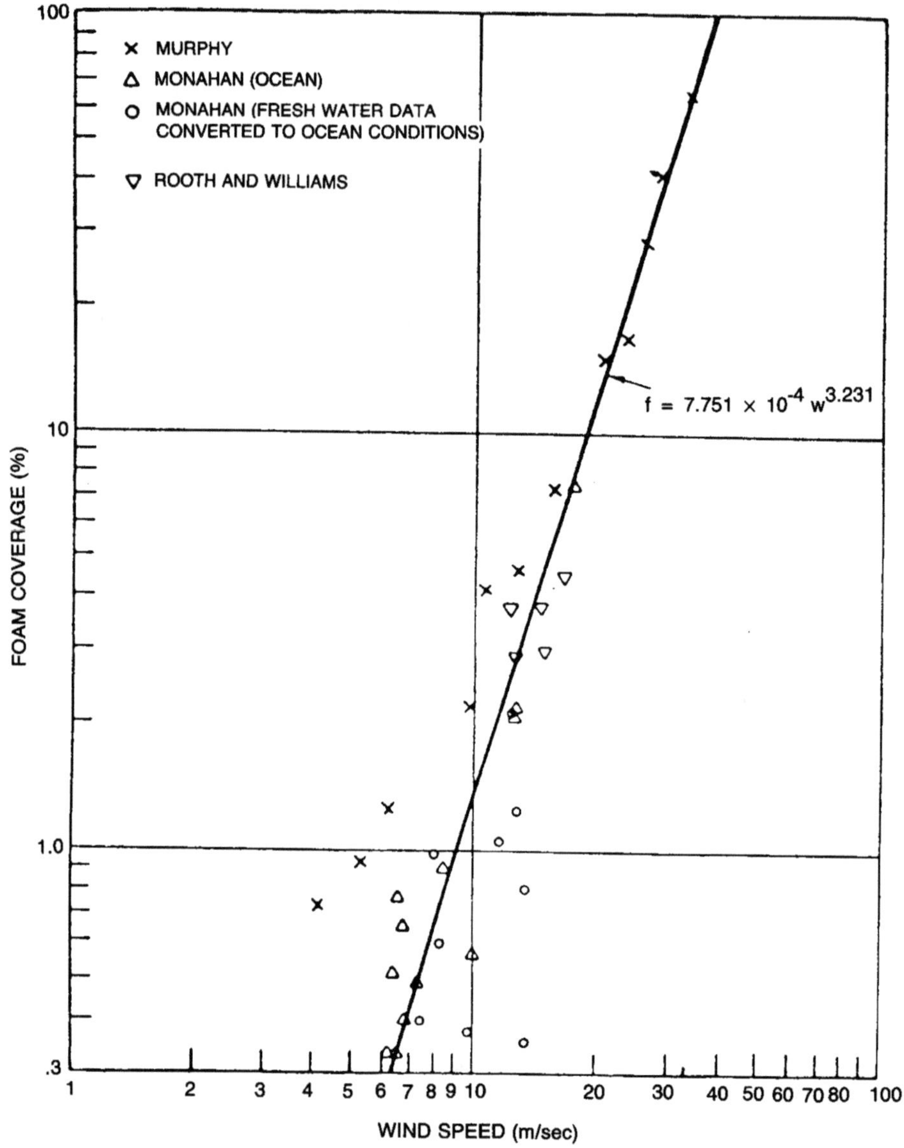

Fig. 14 — Percent foam coverage over the ocean as a function of surface wind speed. (Figure prepared by A. Stogryn, Aerojet-General Corporate [Ref. 9].)

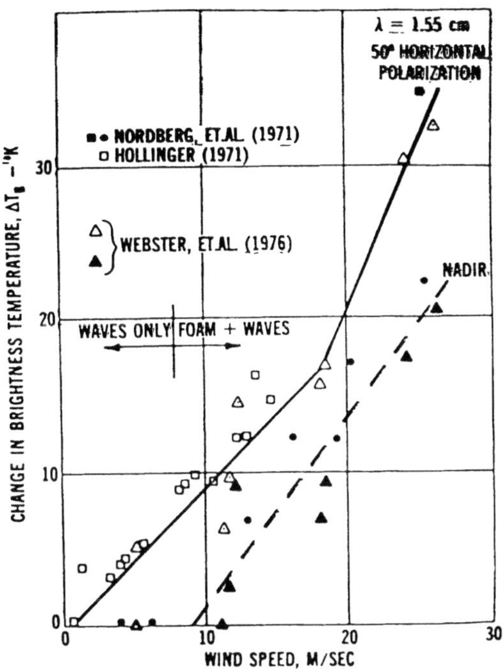

Fig. 15 — Measured increase in brightness temperature as a function of wind speed for an electromagnetic frequency of 19 GHz. (Figure received from P. Gloerson, NASA Goddard Space Flight Center.)

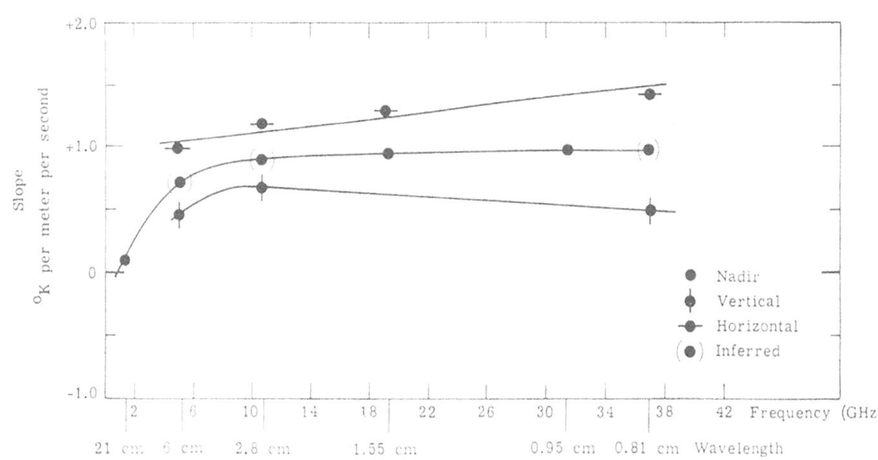

Fig. 16 — Constant of proportionality between wind speed and increased brightness temperature as a function of electromagnetic frequency. (This figure was taken from Ref. 12.)

around this specified frequency to find a quiet band. Since surface phenomena are slowly varying functions of frequency, the fact that data may have been collected at 4.15 GHz instead of 4.00 GHz would have no impact on the analysis and reporting. In fact if the data had been collected at 4.0 GHz, it could have been useless because of severe RFI contamination.

One of the scientific reasons for building the sensor was to demonstrate that the ocean surface wind speed could be measured through heavy rainfall. To illustrate the idea behind the concept, reference to Figure 16 shows that the brightness temperature response vs wind speed is only a weak function of frequency within the design band of the SFMR—and even weaker if we further restrict the band of operation from 4.5 GHz to 6.6 GHz. On the other hand, if we refer to Figure 4 we see that when the radiometer is observing rain, the change in brightness temperature is a very strong function of frequency in the 4.5 GHz - 6.6 GHz band.

These two curves, therefore, suggest that if two or more frequencies are used, an algorithm can be developed to measure both wind speed and rain rate. Such an algorithm was developed, and the SFMR was installed aboard a Department of Commerce aircraft in the summer of 1980 to participate in that year's hurricane mission.

The SFMR is also a nadir viewing instrument, and a 20 minute record of the data is shown in Figure 17 on a flight over smooth ocean on a clear day. The vertical scale of brightness temperature ranges from 100 K to 260 K. The SFMR was programmed to step in frequency in the sequence, 4.5 GHz, 5.0 GHz, 5.6 GHz, 6.6 GHz and then repeat the sequence every five seconds. We note that the mean value of the brightness temperature is constant, with only a slight separation among channels as a result of the variation of the dielectric constant of water with frequency.

The situation is completely different when the wind speed increases and when there is a column of rain between the ocean surface and the aircraft. Figure 18 shows a plot of brightness temperature vs time when the aircraft flew over a hurricane. In this pass over the hurricane, the aircraft first flew over the north eye wall, then over the eye, then over the south eye wall and away from the storm. While the aircraft was flying over the north eye wall, each of the frequencies measured an increase in brightness temperature, with the higher frequencies recording a greater brightness temperature than the lower ones. From discussion of the previous figures, we can associate the general increase for each frequency with an increase in wind speed, and the frequency spread among channels with emission from rain. Indeed, just by glancing at the figure, one can tell where rain occurs, and whether the rain is relatively heavy or light. As the aircraft passes over the eye, all the channels converge, indicating that there is no rain. Also, at the center of the eye, the brightness temperature falls below 120 K, indicating low wind speeds.

The algorithm that was generated was applied to the data of Figure 18, and the resulting wind speed values are shown in Figure 19. This figure indeed shows that high winds (up to 70 m/s) are associated with the north wall, and that they drop almost to zero within the eye. The circles in the figure represent the winds that were measured in-situ by an aircraft flying at an altitude of 1500 meters in tandem with the high altitude aircraft that was used by the SFMR. The agreement between the in-situ and the remote sensing results is quite good, and the detailed differences may be due to a combination of residual calibration errors in the SFMR, and the distinct possibility that the in-situ data collected at 1500 m is not a completely accurate representation of the true surface winds.

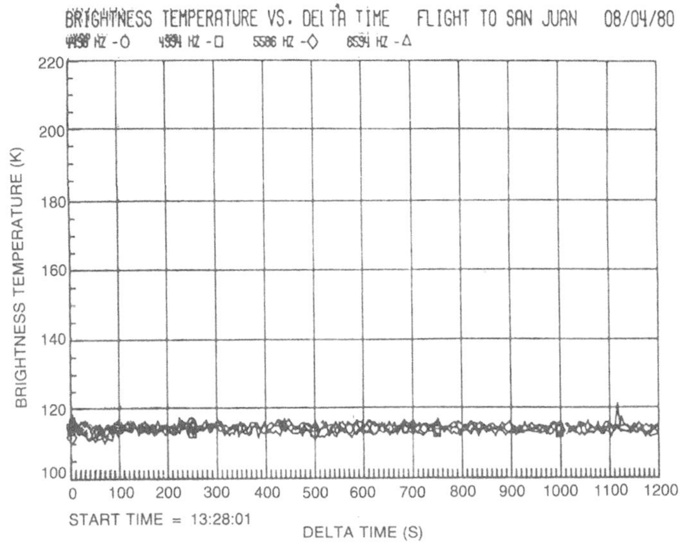

Fig. 17 — Brightness temperature vs. time for four C-Band frequencies observing calm sea conditions. (The remote sensing data were reduced by R.F. Harrington, NASA Langley Research Center, and the insitu data was suppled by P.G. Black, NOAA Hurricane Research Division. Much of the figure material appears in Jones, et al. (1981).)

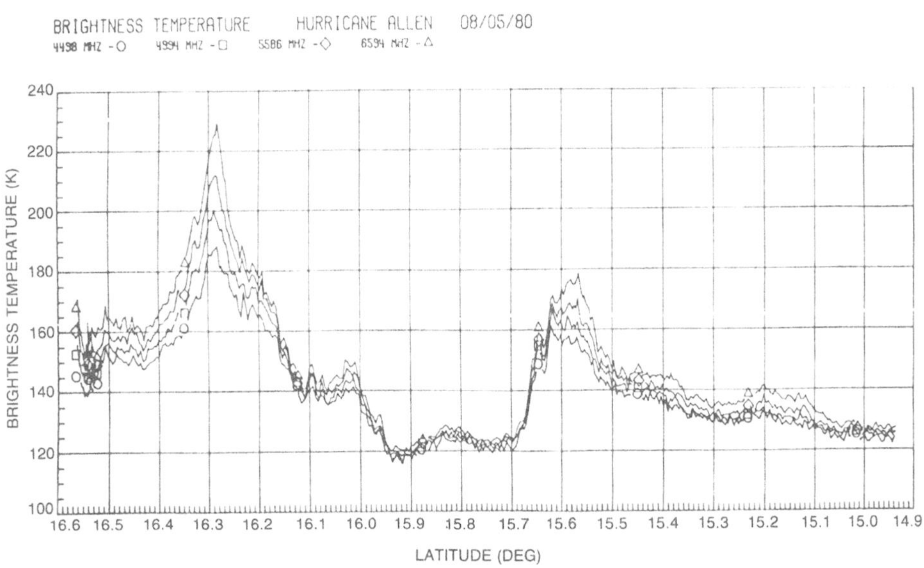

Fig. 18 — Brightness temperature vs time for C-Band frequencies observing rain and high wind speeds associated with hurricanes. (The remote sensing data were reduced by R.F. Harrington, NASA Langley Research Center, and the insitu data was suppled by P.G. Black, NOAA Hurricane Research Division. Much of the figure material appears in Jones, et al. (1981).)

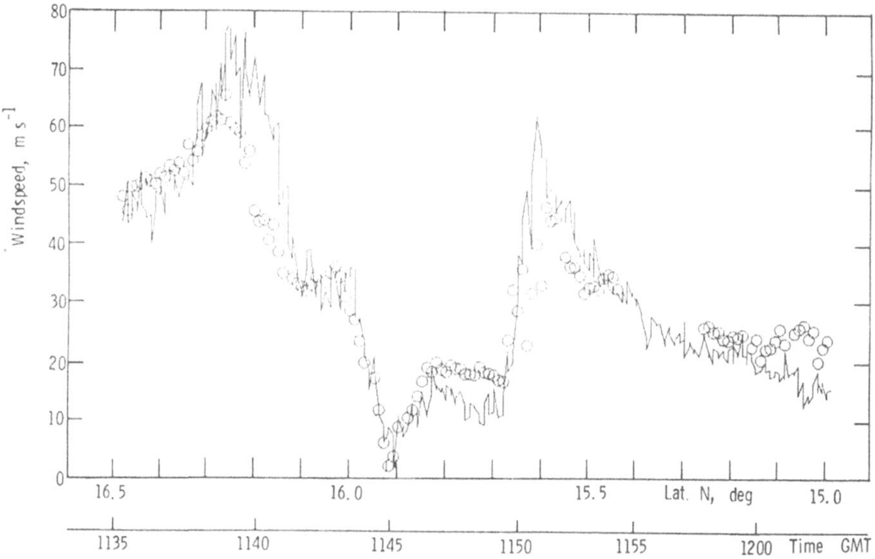

Fig. 19 — Ocean surface wind speed derived from the C-Band data shown in Fig. 18 (continuous line). The circles represent insitu measurements of wind speed obtained from a separate aircraft flying at a lower altitude. (The remote sensing data were reduced by R.F. Harrington, NASA Langley Research Center, and the insitu data was suppled by P.G. Black, NOAA Hurricane Research Division. Much of the figure material appears in Jones, et al. (1981).)

The corresponding rain rate is shown in Figure 20. As anticipated from observing the raw brightness temperature data, heavy rain is associated with the northern eye wall, and lighter rain with the south eye wall. The eye itself is rain free. The circles represent ''in-situ'' values that were inferred from the weather radar on board the aircraft. The rain rate algorithm is still being evaluated, but it is of interest to note that the radiometer provides reliable measurements of wind speed even through very heavy rain. At these frequencies, the noise level, or minimum detectable rain rate is 10 mm/hr. Because higher frequencies are used by the NIMBUS satellite radiometer system, wind speed cannot be measured when the rain rate exceeds 3 mm/hr because of severe attenuation that occurs at frequencies above 18 GHz. The attenuation with frequency was illustrated in Figure 4.

The present status of the capability of the SFMR to measure wind speed is summarized in Figure 21, where we have plotted excess brightness temperature referenced to the smooth water vs wind speed measured directly with a coincident low flying aircraft. The data are cross-plotted on log paper as a means to establish a power law relationship between wind speed and increased brightness temperature. The least squares fit through the data points indicates that the brightness temperature increases with wind speed to the 1.88 power for wind speeds greater than 20 m/s. The standard deviation about this power law fit is 2 m/s. Although we have derived a different power law than that shown previously in Figure 14, the power law relationship is strong evidence that the presence of foam is the physical mechanism for the passive microwave measurements of wind speed. More recent flights into hurricanes indicates that the power law begins to fail when the wind speed falls below 20 m/s, and the

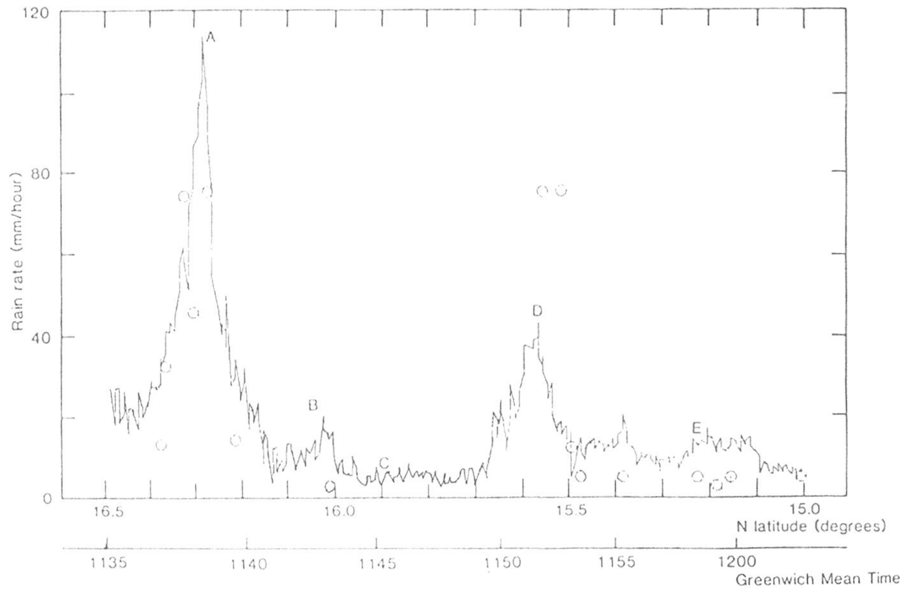

Fig. 20 — Rain rate derived from the C-Band data shown in Fig. 18 (continuous lines). The circles represent rain rate derived from the aircraft radar. (The remote sensing data were reduced by R.F. Harrington, NASA Langley Research Center, and the insitu data was suppled by P.G. Black, NOAA Hurricane Research Division. Much of the figure material appears in Jones, et al. (1981).)

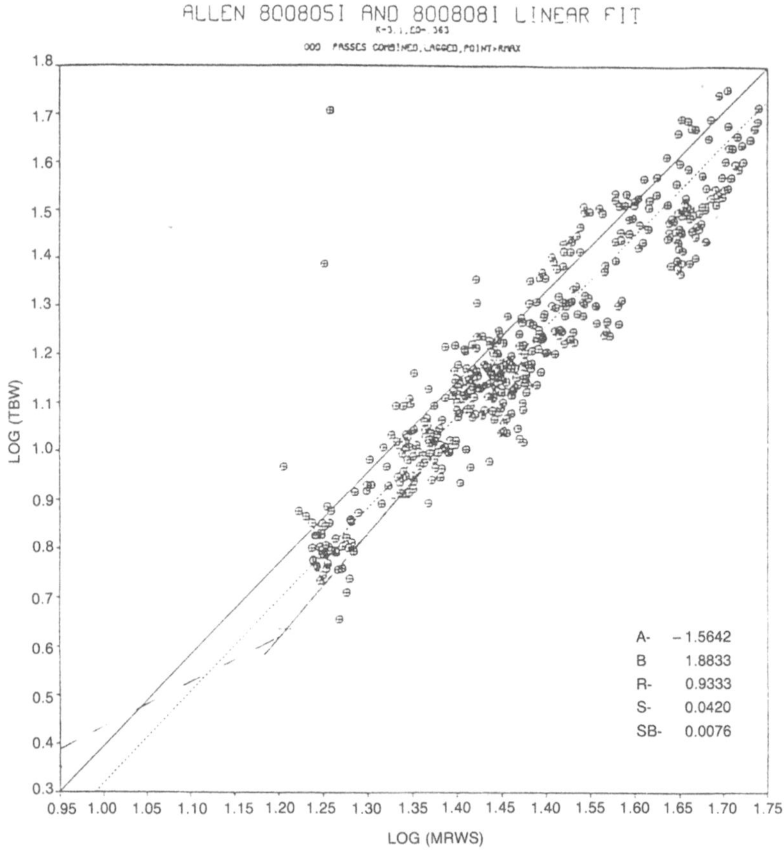

Fig. 21 — Log-log plot of increased brightness temperature vs insitu measurements of wind speed measured from a low-flying aircraft (The remote sensing data were reduced by R.F. Harrington, NASA Langley Research Center, and the insitu data was suppled by P.G. Black, NOAA Hurricane Research Division. Much of the figure material appears in Jones, et al. (1981).)

curve assumes a linear relationship. The physical reason for this is that rough surface effects play a major role in the emission process of low wind speeds where foam generation begins to fall off.

The value of these results is that direct measurements of surface winds can be made safely at high altitudes above severe storms. Aircraft flying 500 m above the surface are in the turbulent boundary layer, and as such, experience forces of up to two or three g's. The microwave radiometer also adds capability to observe the storm at night time, since the emission process does not depend on the time of day. This sensor therefore at least doubles the available observational time.

6. Present Status and Potential Future Directions for Microwave Radiometer Measurements of Ocean Surface Wind Speed

The C-Band Stepped Frequency Microwave Radiometer (SFMR) discussed in Section 5 can be considered a success for the application of remote sensing. A version of the original SFMR has served as a NOAA research instrument during the hurricane seasons of 1984 through 1987. Each year has resulted in improvements in both the hardware and algorithm performance. Typically, it is installed on a Department of Commerce P-3 aircraft in July of each year, and is not removed until the end of the hurricane season in September. There have been no instrument failures, and the derived wind speeds show no anomolies; i.e., the brightness temperature has always increased in proportion to the square of the wind speed. Further detailed comparisons with in situ measurements have resulted in a standard deviation of 1.5 ms^{-1}. The algorithm has been integrated into the software of the aircraft mini-computer in order to provide on-board users with real-time graphics of surface wind speed and rain rate. Furthermore, the instrument is self-operating except for periodic calibrations that are done prior to take-off and over smooth water.

Until 1987, satellite measurements of ocean surface wind speed have been somewhat disappointing. The Scanning Multi-channel Microwave Radiometer (SMMR) was part of the instrument complement on both SeaSat and NIMBUS 7. Both satellites were launched in 1978. Although SeaSat failed after 100 days of operation, the NIMBUS 7 SMMR continued to collect data until the scan mechanism failed in 1987. Initial assessment of the SMMR wind speed algorithm was reported, principally by Wentz et al. (1982). However, global products were never released on a routine basis. Our own experience with NIMBUS data indicates that the wind speed retrievals present significant problems, particularly in the derived rms accuracy. The underlying reasons for the problem were not specifically identified, but it was well recognized that the instrument experienced calibration problems. In June, 1987, the Department of Defense launched a multi-frequency microwave radiometer system called the Special Sensor Microwave/Imager (SSM/I). Design changes were incorporated which resulted in significant improvements in the instrument calibration. An initial assessment has been completed for DoD, and these preliminary results indicate that the sensor can measure wind speed to an accuracy of ± 2 ms^{-1}. It is anticipated that results of the evaluation will appear in the scientific literature, possibly co-incident with the publication of this book.

It is envisioned that the SSM/I will provide synoptic measurements of ocean surface wind speed into the 21st century both on DoD and NASA satellites. In addition, NASA has emphasized a new program which they refer to as the Earth Observing System (EOS). The principal component of this program is a number of space "platforms" in polar orbit. These platforms will carry several new radiometer systems which are collectively referred to as the High-Resolution Multifrequency Microwave Radiometer (HMMR). The new thrust in the design of these systems is a push toward better spatial resolution which will require the use of larger antennas. In addition, the space station itself will probably carry earth viewing sensors, and microwave radiometers are likely candidates for this purpose.

7. Acknowledgments

The nature of review papers is that the work contains research results, and thoughts generated by other scientist and engineers, particularly the figure material. In some cases, these individuals have not published the work themselves. I would like to identify those researchers in connection with the figures that appear in the chapter.

8. References

Black, P.G. and Swift, C.T.: 1984, "Airborne Stepped Frequency Microwave Radiometer Measurements of Rainfall Rate and Surface Wind Speed in Hurricanes, Proc. 22nd Conf. on Radar Meteor., AMS Publication, 433-438.

Cox, C. and Munk, W.: 1954, "Measurements of the Roughness of the Sea Surface from Photographs of the Sun's Glitter," *J Opt. Soc. Am.* **44**, 838-850.

Hollinger, J.P.: 1971, "Passive Microwave Measurements of the Sea Surface," *IEEE Trans. Geo Science Elect.*, GE-91, 165-169.

Jones, W.L., Black, P.G., Delnore, V.E., and Swift, C.T.: 1981, "Airborne Microwave Remote Sensing Measurements of Hurricane Allen," Science, **214**, 279-280.

Klein, L.A. and Swift, C.T.: 1977, "An Improved Model for the Dielectric Constant of Sea Water at Microwave Frequencies," *IEEE Trans. Antennas and Propag.*, AP-25, 104-111.

NASA: 1987, "HMMR High-Resolution Multifrequency Microwave Radiometer Earth Observing System.

Nordberg, W., Conway, J., and Thaddeus, P.: 1969, "Microwave Observations of Sea State from Aircraft," *Quart. J.R. Meteoral. Soc.* **95**, 408-413.

Peake, W.H.: 1959, "Interaction of Electromagnetic Waves with Some Natural Surfaces," *IEEE Trans. Antennas Propag.*, (Special Suppl.), AP-7, 5324-5329.

Stogryn, A.: 1972, "Radiometric Emission from a Rough Sea Surface," Aerojet—General Corporation, Report No. 3000 R-1.

Stogryn, A.: 1967, "The Apparent Temperature of the Sea at Microwave Frequencies," *IEEE Trans. Antennas and Propag.*, AP-15, 278-286.

Swift, C.T.: 1974, "Microwave Radiometer Measurements of the Cape Cod Canal," *Radio Sci.* **9**, 641-653.

Webster, W.J., Wilheit. T.T., Ross, D.B., and Gloerson, P.: 1976, "Spectral Characteristics of the Microwave Emission From a Wind-Driven Foam-Covered Sea, *J. Geophys. Res.* **81**, 3095-3099.

Wentz, F.J. Cardone, V.J., and Fedor, L.S.: 1982, "Intercomparison of Windspeeds Inferred by the SASS, Altimeter, and SMMR," *J. Geophys. Res.* **87**, (C5), 3378-3384.

Chapter 16

REMOTE SENSING OF SURFACE TURBULENCE HEAT FLUX

W. T. LIU
Jet Propulsion Laboratory
California Institute of Technology
Pasadena, CA 91109

1. Introduction

The ocean interacts with the atmosphere through exchanges of momentum, heat and moisture. The differential heating of the atmosphere by the ocean fuels atmospheric circulation which, in turn, drives ocean currents and redistributes the fuel. Short-term climate changes are believed to be strongly influenced by such exchanges. Locally, the ocean absorbs heat in the summer and releases it in the winter; globally, heat accumulated in the tropics is transported poleward by ocean currents. The redistribution of heat reduces the extreme temperature contrasts that otherwise would exist. The heat and moisture fluxes change both the horizontal and the vertical density gradients (baroclinicity and stability) of the atmosphere and the ocean. They, in turn, modify the wind and current shears. The moisture flux connects the energy cycle to the hydrological cycle. In the atmosphere, condensation provides not only the means for precipitation but also much of the diabatic heating. In the ocean, evaporation cools the upper layer and increases the salinity.

The ocean exchanges heat with the atmosphere through radiative and turbulence processes. The radiative fluxes are discussed in both the preceding and following chapters. Both the sensible heat and moisture fluxes are largely transported by turbulence in the atmosphere. The moisture flux carries latent heat which is much larger than the sensible heat, both in the mean and the variability over much of the ocean. In the past, large-scale fluxes were computed using meteorological reports from ships of opportunity. In the tropical and southern oceans, ship observations are sparse and the temporal and spatial variabilities of fluxes cannot be determined accurately using ship data alone. Spaceborne sensors can provide repeated global coverage. This chapter will be largely devoted to the determination of latent heat flux variability using spaceborne sensors, particularly the microwave sensors which have all weather capability. The present status will be summarized and the potential future advancements will be explored.

2. Bulk Parameterization

The conventional method of computing large-scale wind-stress (τ), sensible heat flux (H) and moisture flux (E) between the ocean and the atmosphere is by using bulk parameterization formulae which relate the fluxes to measurements made routinely on board ships.

G. L. Geernaert and W. J. Plant (eds.), Surface Waves and Fluxes, Volume II, 293–309.

$$\tau = \rho C_D U^2 \tag{1a}$$

$$H = \rho c C_H U(T_s - T) \tag{1b}$$

$$E = \rho C_E U(Q_s - Q) \tag{1c}$$

where ρ is the surface air density, c is the isobaric specific heat. The latent heat flux (LE) is the product of E and the latent heat of vaporization. Using this bulk technique, the measurements required include the sea surface temperature (T_s), the wind speed (U), the air temperature (T), and the humidity mixing ratio (Q) measured on board ships. The mixing ratio at the air-sea interface (Q_s), is generally taken to be the saturation value at T_s, and Q can be derived from the dew-point temperature measured. The transfer coefficients (C_D, C_H, and C_E) have been determined by analyzing point measurements (10 min to 1 hr time average at a fixed location). Since the bulk aerodynamic methods are presented in a preceding chapter, only some often overlooked errors with respect to global flux determination will be discussed in this section.

Since U, T, and Q change with height and their vertical gradients depend on atmospheric stratification, the transfer coefficient should be functions of height and stratification. The transfer coefficients are generally determined from measurements at 10 m and near neutral conditions, and they can be adjusted with flux-profile relations based on similarity theory (Deardorff, 1968). As the flux-profile relations, initially developed over land, are extended over water, the effects of humidity fluctuations on buoyancy is often overlooked. Bunker (1976), for example, tabulated the values of the transfer coefficients according to classes of U and $\Delta T \equiv T_s - T$, but not $\Delta Q \equiv Q_s - Q$. In the extratropical oceans, the effects of humidity fluctuations may be small compared with the effects of temperature fluctuations. But in the tropical oceans, due to the rapid increase of saturation humidity at high temperature (Clausius-Clapeyron Equation), humidity fluctuations can have significant effects on atmospheric stability and the variability of the transfer coefficient. The transfer coefficient should depend on the stability parameter ($\zeta \equiv z/L$), where z is the height and L is the Monin-Obukhov length. The parameters, ζ, represents the ratio of turbulence production due to buoyance to production by shear and it can be approximated (Deardorff, 1968) by

$$\zeta = \frac{kgz(\Delta T + 0.61\,\theta \Delta Q)}{\theta U^2 \sqrt{C}} \tag{2a}$$

assuming $C_D = C_H = C_Q = C$. Over land, the term including ΔQ is usually omitted and (2a) reduces to

$$\zeta_* = \frac{kgz\,\Delta T}{\sqrt{C}\,\theta U^2} \tag{2b}$$

where θ is the average absolute temperature. Figure 1 shows the ratio of ζ/ζ_* at various T_s and two values of ΔT, assuming $C = 1.3 \times 10^{-3}, U = 7$ m/s and a relative humidity of 80%. It is obvious that the error for omitting ΔQ can be larger than 50% over warm water (> 25 ° C). In Fig. 2 from Liu (1988), the values of C_E as defined by (1c) at various U and ΔQ typical of the tropical oceans were shown. They were computed using the model of Liu et al. (1979) and assuming $T_s = 25$ ° C, a relative humidity of 80%, and a reference height of 10 m. The thick curve represents the value at neutral stability ($\zeta = 0$). In the western tropical Pacific and eastern Indian Ocean, with a typical wind speed of 4 m/s, and a typical

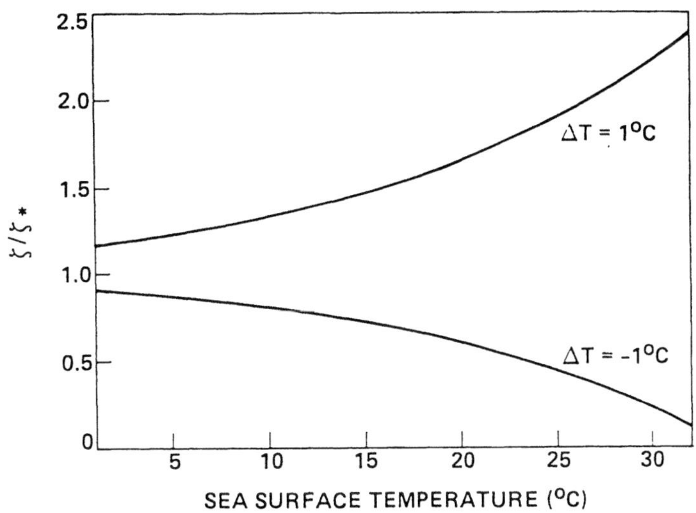

Fig. 1 — The ratio of the stability parameter including humidity effects to the stability parameter excluding humidity effects as a function of temperature for two cases of sea-air temperature differences ΔT.

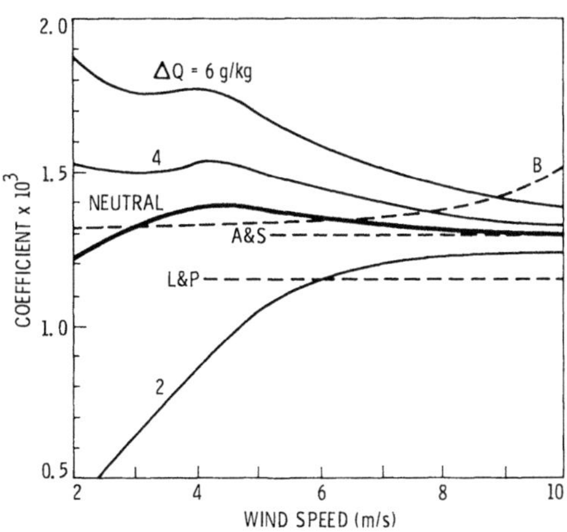

Fig. 2 — The variability of the moisture transfer coefficients with wind speed and sea-air humidity difference ΔQ, compared with the model of Liu et al. (1979), with the thick curve representing values at neutral stability. The near neutral values of Bunker (1976) (B), Anderson and Smith (1981) (A&S), and Large and Pond (1982) (L&P) are shown as dash-lines. (From Liu, 1988)

sea-air humidity difference of 6 g/kg (Weare et al., 1980), the coefficient is approximately 1.8×10^{-3} which will give a latent heat flux approximately 40% higher than the value given by a neutral coefficient of 1.3×10^{-3}. The typical ΔT may be small and the temperature induced buoyancy alone cannot adequately account for the stability effect on C_E.

In the model of Liu et al. (1979), which is a physical approach to surface flux bulk parameterizations, the three flux-profile relations based on similarity theory are solved simultaneously. The relations are,

$$U/U_* = 2.5(z/z_o - \psi_U). \tag{3a}$$

$$(T - T_s)/T_* = 2.2(z/z_T - \psi_T). \tag{3b}$$

$$(Q - Q_s)/Q_* = 2.2(z/z_Q - \psi_Q). \tag{3c}$$

By definition, U_*, T_*, and Q_* are function of τ, H and E. The ψ_U, ψ_T, and ψ_Q are function of ζ and can be expressed in terms of the three fluxes. The lower boundary parameters z_o, z_T, and z_Q are functions of τ and fluid properties. The three unknowns τ, H, and E can be determined by solving the three implicit equations. This method is similar to that used by Deardorff to account for the effects of stability on the values of the transfer coefficient and is equivalent to using (1) with variable coefficients. The neutral values of C_E given by Anderson and Smith (1981), Large and Pond (1982), and Bunker (1976) are also shown in Fig. 2 for comparison. It is generally assumed that the surface roughness represented by z_o increases with U in the fully rough regime (with U above a critical value) and so does C_D. If z_Q similarly increases with U, C_E should also increase with U, as the values given by Bunker indicate. Liu (1979) postulated that while momentum can be transported by form drag at the interface, the transport of moisture is constrainted by molecular diffusion and therefore C_E would not increase with U, as C_D does. This postulate appears to be supported by available observations for a moderate range of winds speeds.

Most of observations used to derive or validate the transfer coefficients in the bulk formulae were taken in moderate wind conditions (4-20 m/s). When breaking waves disrupt the continuous material surface at high winds and when the rough surface changes into a smooth one at low winds, we have very little flux observations. At low wind speeds, the stability parameter becomes very large and the turbulence transport under this near free convection condition is not very well known. The Humidity Exchange over the Sea (HEXOS) experiment (Katsaros et al., 1987), conducted in the North Sea, was designed to study the transfer processes in high winds and field experiments are being planned in the western tropical Pacific (S.J. Godfrey, 1988, personal communication) to study the processes in low winds. The bulk parameterization method of Liu et al. (1979) was verified with hourly measurements. Esbensen and Reynolds (1981), using weather ship data in sub-tropical and midlatitude oceans, demonstrated that the methods can be applied to monthly mean parameters with insignificant errors and Liu (1988) extended their results to the near equatorial oceans.

Spaceborne sensors can measure U and T_s, but not T or Q. Microwave radiometers, however, can measure the precipitable water (W) which is the column-integrated water vapor in the atmosphere. If we can relate Q to W, (1c) can be used to determine E. Since T has only secondary effects on E through the stability term, it can be approximated by assuming a constant relative humidity and the model of Liu et al. (Eqn 3abc) may be applied. In the next section, the method of relating Q to W will be discussed.

3. Retrieval of Surface Layer Humidity

Spaceborne microwave radiometers can measure W to the same accuracy as radiosondes, approximately 0.2 g/kg (Katsaros et al., 1982: Alishouse, 1983, and Chang et al., 1984). Atmospheric scientists have long been interested in the relation between the precipitable water and surface humidity. Statistical relations between W and some forms of surface humidity have been derived from radiosonde soundings at various locations over land for different time scales (e.g. Reitan, 1963; Smith, 1966). Liu (1986) established a global relation between monthly mean W and Q, using 17 years of radiosonde reports from 46 stations distributed over global oceans. Figure 3 shows the global relation (curve) compared with the 5567 monthly means in the data set of Liu (1986). The root-mean-square difference between the predicted values and observations is 0.73 g/kg and the correlation coefficient is 0.99. The relation is found to be valid for all oceans and adequate in describing seasonal and interannual variabilities except in the region around the Bering Sea during summer. The failure in the Bering Sea region was attributed to the frequent saturation of warm and moist continental air advecting over the cold ocean surface. There are no other obvious seasonal or regional biases.

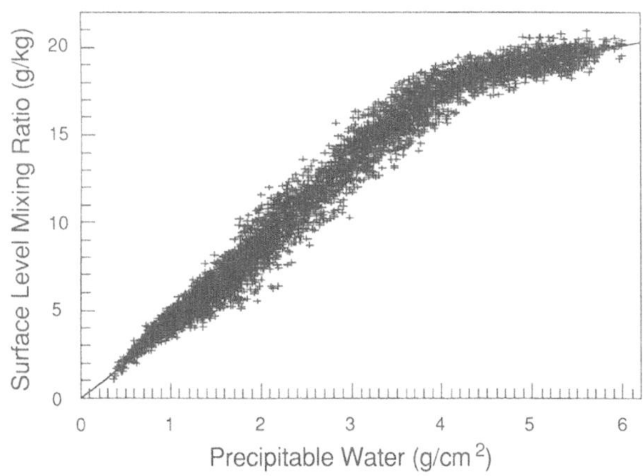

Fig. 3 — Scatter plot of Q verus W for 46 stations distributed over global oceans taken over a period of 17 years. Each point is a monthly average. The solid line represents the global relation by Liu (1986).

The Q-W relation was first applied by Liu and Niller (1984) on data from the Scanning Multichannel Microwave Radiometer (SMMR) on Seasat. Liu (1988) derived Q from SMMR data on Nimbus-7 and compared them with ground truth measurements at 27 stations over the global tropical oceans for 46 months starting in January 1980. Examples of the comparison at four stations are shown in Fig. 4. An intense climate anomaly, the El Niño Southern Oscillation (ENSO) episode, took place during 1982-1983. The variability of Q derived from

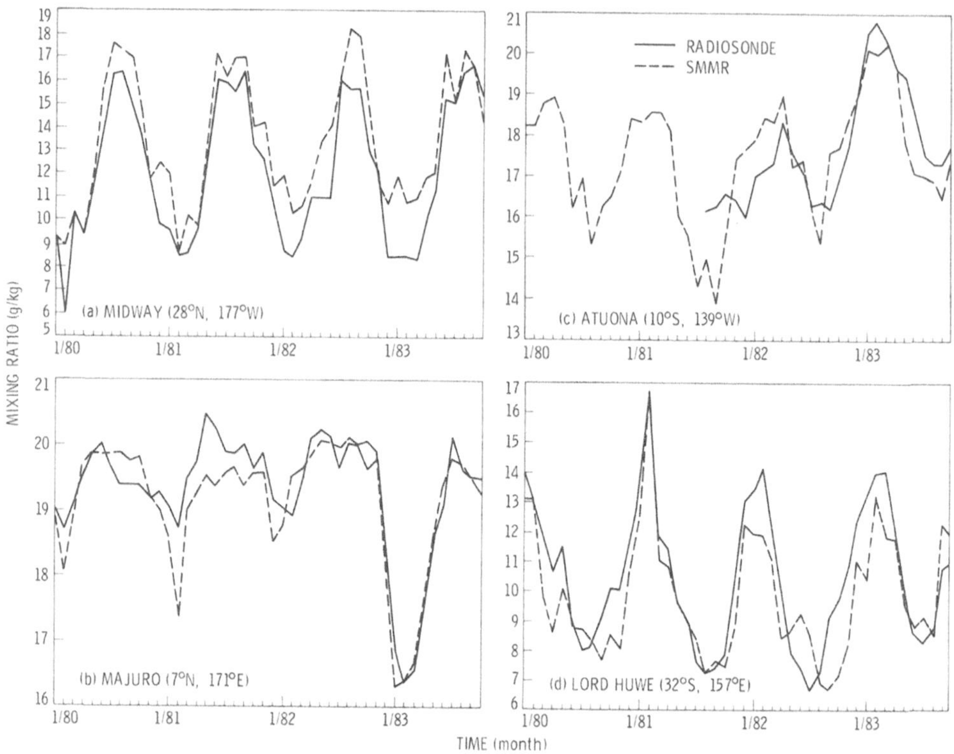

Fig. 4 — Comparison of the temporal variations of surface-level mixing ratio derived from Nimbus/SMMR and from in situ measurements at four stations in the Pacific.

SMMR W closely follows both the seasonal cycles and the ENSO anomalies indicated by the in-situ measurements. During non-ENSO periods, Majuro and Atuona are located, repectively, under the ascending and descending branches of the zonal Walker circulation, with Majuro more humid than Atuona. During the ENSO episode, the system is dislocated with a Q-deficit at Majuro and a Q surplus at Atuona. Midway and Lord Huwe are in midlatitude sub-tropical oceans and are influenced by the subsidence of the meridional Hadley circulation. The comparison in Fig. 4 demonstrates that the relation of Liu (1986) is valid for a variety of atmospheric regimes over oceans.

The relation between Q and W depends on the variability of the vertical distribution of water vapor. The empirical orthogonal functions (EOF) are the most efficient way of representing uncorrelated modes of the variance (Lorenz, 1977). The number of EOFs required to explain a given total variance is the fewest of any set of orthogonal functions. The EOF of humidity soundings for 4 years at 3 stations are shown in Fig. 5. The soundings are semi-daily for Wake and St. Paul but only daily soundings are available at Majuro. The first EOFs accounts for 77.5%, 61.7% and 56.7% of the variance respectively. When monthly-mean soundings between 1964-1984 are used, the variances accounted for by the first EOFs increase to 96%, 89%, and 82% respectively. In order to explain 70% of the daily variance at Majuro, for example, at least two independent predictors are needed. But to

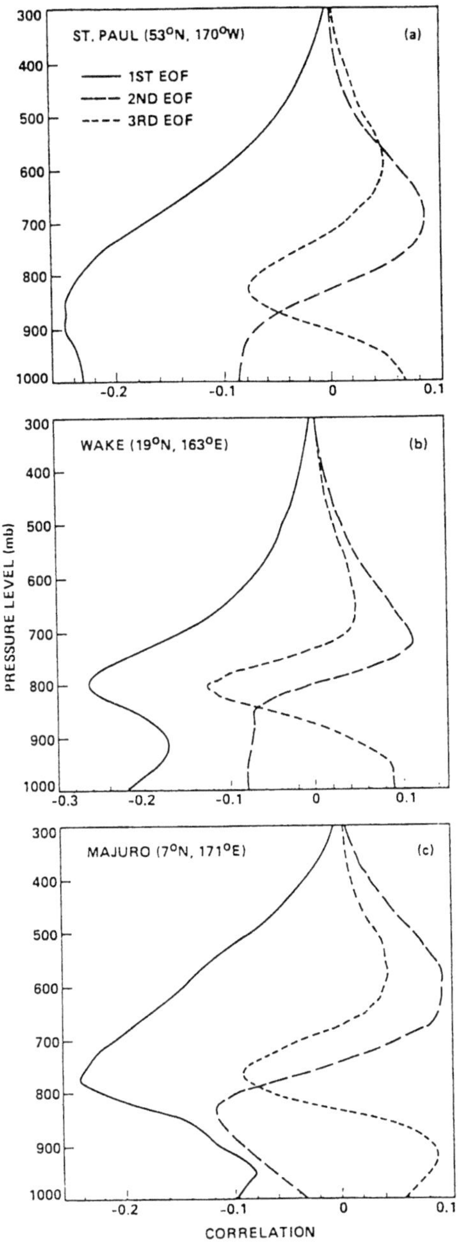

Fig. 5 — Three empiral othogonal functions accounting for most variance of four years of individaual humidity soundings at three stations in the Pacific.

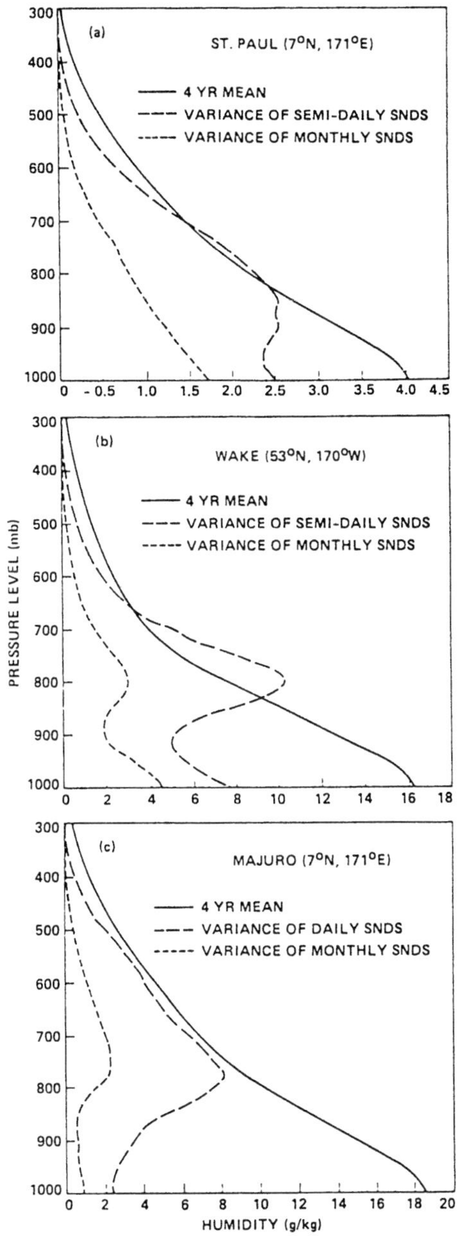

Fig. 6 — Vertical profiles of the four year mean humidity soundings and the variance profiles of individual soundings and the monthly mean soundings.

explain the same amount of variance on the monthly scale, one predictor is more than enough. This illustrates the reason for the success of determining Q from W (one parameter prediction) on the monthly time scale. On synoptic time scales, additional predictors are needed to account for the same amount of variance. In Fig. 6, the variance profiles of individual soundings and the monthly-mean are compared. The maximum variance is found between 900 and 800 mb, i.e., at the top of the planetary boundary layer. Most of the variance at this level has time scales less than a month. To determine Q on synoptic scales, an additional predictor providing information at this level is preferable. A likely choice would be the radiance measured by one of the water vapor channels of atmospheric sounders whose weighting function peaks between 900 mb to 800 mb. The weighting functions of the water vapor channels of the Advanced Microwave Sounding Unit (AMSU) to be flown on the next operational NOAA polar orbiter is shown in Fig. 7 (R. Kakar, 1989, personal communication). They represent the contributions to the radiance as a function of height and are defined as

$$\partial \lambda / \partial \ln P$$

where P is the pressure and λ is the transmittence of the respective water vapor channel. To determine the monthly Q, measurements by microwave radiometers are adequate; for synoptic scales, additional measurements by atmospheric sounders may help.

NOAA operationally provides the average humidity between 700 mb and the surface derived from data of the TIROS-N Operational Vertical Sounder (TOVS) (Smith et al., 1979). Similar retrieval from the VISSR Atmospheric Sounder (VAS) is discussed by Smith (1983). The vertical resolution of the lowest level humidity is too coarse to be useful in the bulk formula. Unlike microwave sensors, these infrared sensors cannot penetrate clouds and the quoted accuracies in W measurements are lower than those for microwave sensors.

4. Retrieval of Wind Speed and Interfacial Humidity

The surface wind vector can be measured by active microwave sensors, such as the scatterometer on Seasat (SASS), as discussed in the previous Chapters. The error of SASS wind speed when compared with individual in situ measurements is often found to be approximately 1.5 m/s (e.g., Jones et al., 1982). Assuming random error and a decorrelation time scale of 3 days, there would be 10 independent measurements in a month and the error in monthly means is likely to be 0.5 m/s. Liu (1988) compared SMMR wind speeds with measurements at two equatorial moorings in the eastern and central pacific. Mooring measurements within the hour of satellite overpass were scalar averaged and compared with SMMR averages in 2° latitude by 2° longitude monthly bins centered at mooring locations. The standard deviations of the differences were found to be 0.64 and 0.69 m/s for 22 months and 31 months at the two locations respectively.

The interfacial humidity Q_s is the saturation humidity at T_s. In comparing individual ship measurements of T_s with Seasat/SMMR data, Liu (1983) found the standard deviation of the difference to be 1.7°C. A similar magnitude of the standard deviation was also found by Gloersen et al. (1984) in comparing Nimbus/SMMR data with individual buoy measurements. Other systematic errors were also suggested by Milman and Wilheit (1985). If the errors are random, they could be reduced by time averaging. Bernstein and Morris (1983) found a standard deviation of 0.5°C in comparing monthly means T_s from Seasat/SMMR and from ship.

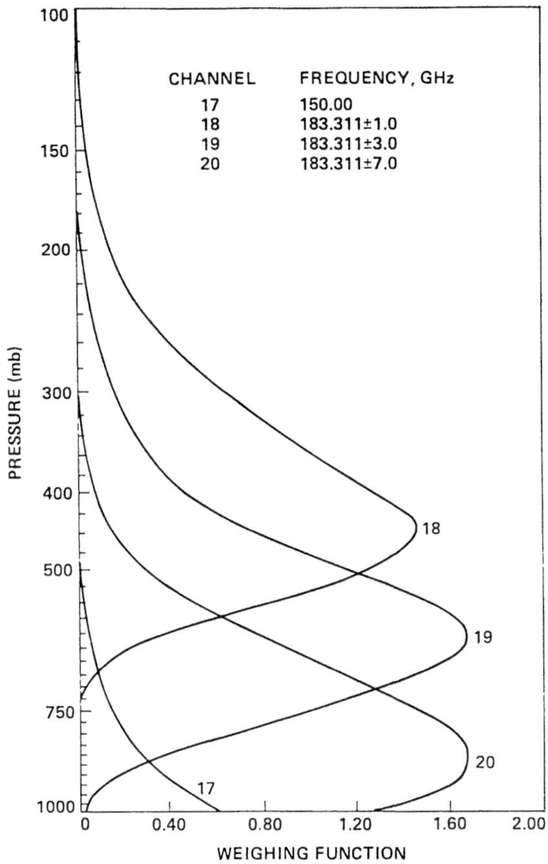

Fig. 7 — Sketch of the weighting functions of various humidity channels of the Advanced Microwave Sounding Unit (AMSU).

Liu (1988) found the standard deviation of difference between measurements at an equatorial mooring and 2° latitude by 2° longitude monthly average of satellite data to be 0.8°C. But the error may include spatial variation because the location has strong zonal gradients. More discussion on the retrieval of sea surface temperature is presented in a preceding chapter. According to the Clausius-Clapeyron Equation (Fleagle and Businger, 1963), we have:

$$\frac{\partial Q_s}{\partial \theta} = \frac{Q_s L}{R \theta^2} \tag{4}$$

At $\theta = 298$ °K, an error of 0.5 °C in T_s corresponds to an error of 0.6 g/kg in Q_s.

5. Retrieval of Latent Heat Flux

5.1. INDIRECT RETRIEVAL

With U, W, and T_s measured by SMMR and the global $Q - W$ relation by Liu (1986), monthly mean E can be computed over global oceans using the bulk parameterization methods described in Section 2. Other indirect methods of retrieving LE from a combination of satellite and in situ measurements have been suggested (e.g., Chou and Atlas, 1982) but they are useful only under limited conditions. Surface fluxes have recently become operational output products of operational General Circulation Models (Simonot and Le Trent, 1987) but much improvement is needed to provide realistic variabilities in data-poor regions (Liu, 1988).

The evaluation of E derived from satellite data is hampered by the lack of extensive quality of in situ standards. Liu (1988) estimated the random error in monthly mean LE derived from Nimbus/SMMR to be 26 W/m^2 based on the propagation of errors in U, Q, and T_s determined in comparision with measurements at tropical mooring and weather stations. An error of similar magnitude was found in comparing SMMR LE with those from ship reports, taking into consideration the random noise in ship data. An advantage of this technique is that the parameters required can be retrieved from different spaceborne sensors. For example, the LE in August 1978 has been computed using U from the SASS and W and T_s from the Seasat SMMR. They are compared with LE computed from the ship reports of the Comprehensive Ocean-Atmosphere Data Set (COADS) in Fig. 8. Each point is a monthly 2° latitude by 2° longitude average. Only those bins with more than 10 ship reports, between 40°N and 40°S and 600 km from land are included. The random errors are similar in magnitude to those derived from SMMR data alone.

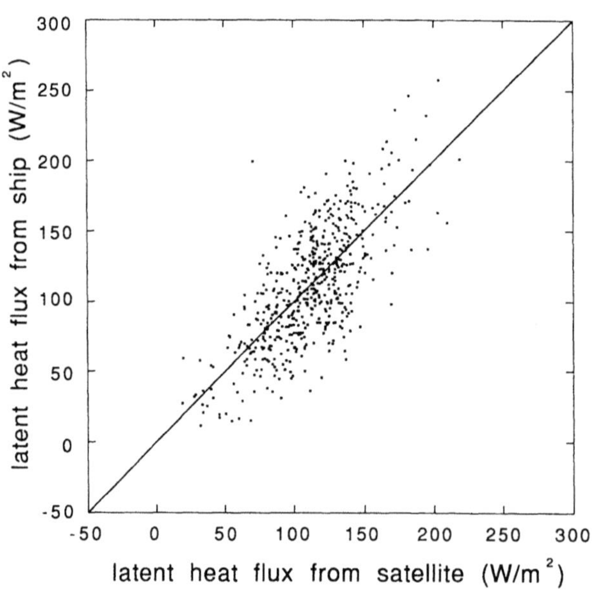

Fig. 8 — Comparison of latent heat fluxes from Seasat/SMMR and from ship reports during August 1978. Each point is a 2° latitude by 2° longitude monthly mean. Only those bins with more than 10 reports, between 40°N and 40°S and 600 km from land are included.

In Fig. 9, the global map of the LE is shown in the maps of dealiased surface wind vectors (Baker et al., 1984) from SASS and W from SMMR. The latent heat flux is strong over the Tradewind regions, particularly in the winter hemisphere, where U is strong. Despite high U, the LE is low over the upwelling regions along the equator in the eastern Pacific and Atlantic and in the Arabian Sea where T_s and Q_s are low. Over the major convergence zones, the Intertropical Convergence Zone (ITCZ) located just north of the Equator in the Pacific and the Atlantic and the South Pacific Convergence Zone (SPCZ) running southeast from New Guinea, W is high but LE is low due both to low Q and low U. It is apparent that the high water content in the atmosphere over this regions is not due to local evaporation but advected by surface winds. In the Indian Ocean, the evaporation is much higher south of the equator than in the Arabian Sea. The rainfall on the subcontinent must be dependent on the moisture advected across the equator by the Monsoon. The highest W is found over the warm ocean in the western tropical Pacific and eastern Indian ocean. Much information on the hydrologic balance can be revealed by a combination of spaceborne sensors. Discussion on the seasonal variation and ENSO anomalies of remotely sensed LE and other parameters can be found in Liu (1988). The study demonstrates that satellite data can provide a basin-wide multi-year look at the temporal and spatial variabilities of latent heat flux and other related parameters.

5.2. DIRECT RETRIEVAL

The bulk parameterization methods described in Section 2 and applied in Section 5A are designed to use measurements made on ships. But what is actually measured by SMMR are microwave radiances at 10 channels (6.6H, 6.6V, 10.7H, 10.7V, 18H, 18V, 21H, 21V, 37H, 37V), i.e., frequencies of 6.6, 10.7, 18, 21, 37 GHz for both horizontal and vertical polarizations. Geophysical parameters were derived using algorithms (Wilheit and Chang, 1980) in the general form of

$$S_i = a_i + \sum_{j=1}^{j=10} b_{ij} B_j \tag{5}$$

where B_j is a simple function of the brightness temperature, the radiance expressed as an equivalent blackbody temperature of the jth channel, and S_i is the ith geophysical parameter (U, W, or T_s). If LE is the product required, it may be desirable to relate it directly to the brightness temperature rather than spending the effort to retrieve the intermediate parameters U,W, and T_s. A direct relation may also bypass the uncertainties related to bulk parameterization. Perhaps not all the channels are sensitive to the variabilities of LE and some channels may be sensitive to more than one geophysical parameter. A direct relation may use an optimal set of channels and eliminate redundancy.

As a simple demonstration, stepwise multivariate regression (Draper and Smith, 1981) of the form

$$\mathrm{LE} = c_o + \sum_{i=0}^{i=10} c_i B_i \tag{6}$$

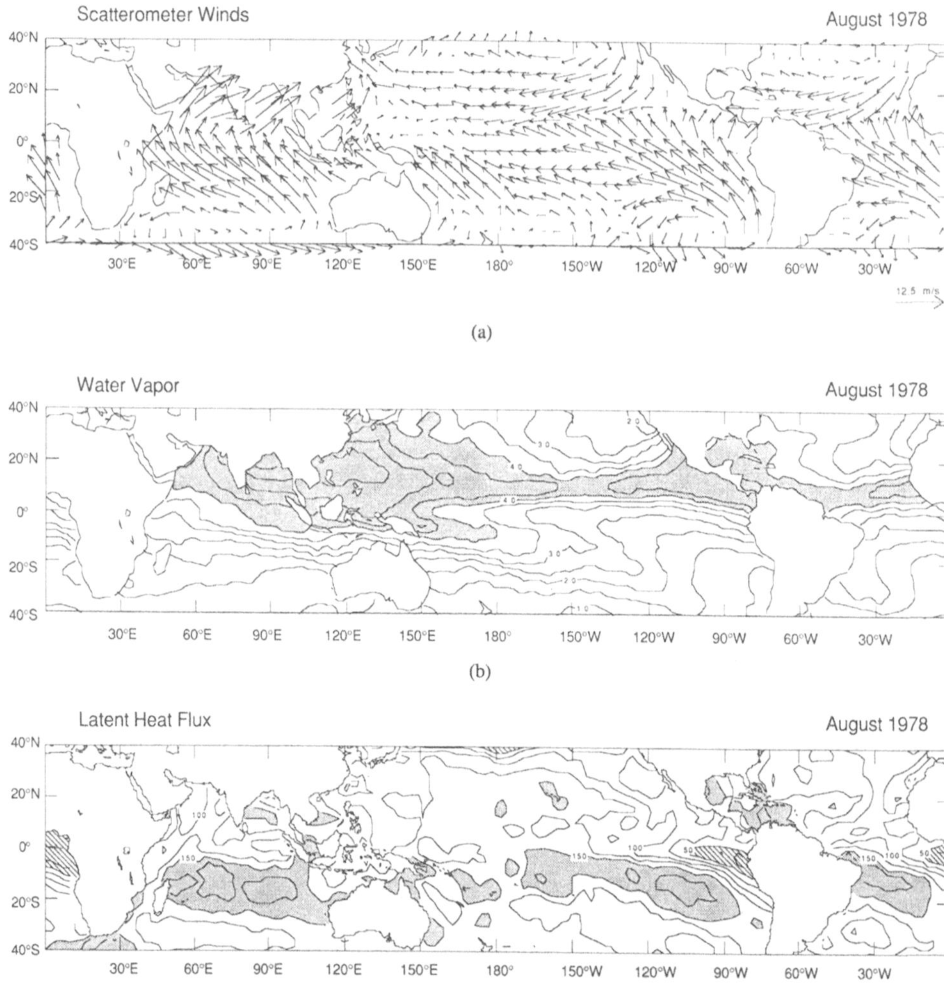

Fig. 9 — Distribution of surface wind vector from Seasat/SASS (a), of precipitable water from Seasat/SMMR (b), and of latent heat flux from both Seasat/SASS and Seatsat/SMMR (c).

was used to relate LE to the relevant brightness temperatures. The coefficients were derived from four months (January, April, July and October) data in 1982. Ship reports were used to derive LE, and B was taken to be the brightness temperature observed by Nimbus/SMMR. Monthly 2° by 2° averages for those bins with more than 10 ship reports and 600 km from land were used. Only the brightness temperature at six channels were found to pass the 5% significant level test. They are 6.6H, 10.7H, 10.7V, 21H, 21V, and 37H. Using the coefficients derived in (6), the LE over oceans in the Northern Hemisphere was computed for all twelve months of 1982 and the yearly zonal means are compared with those derived from ship data in Fig. 10.

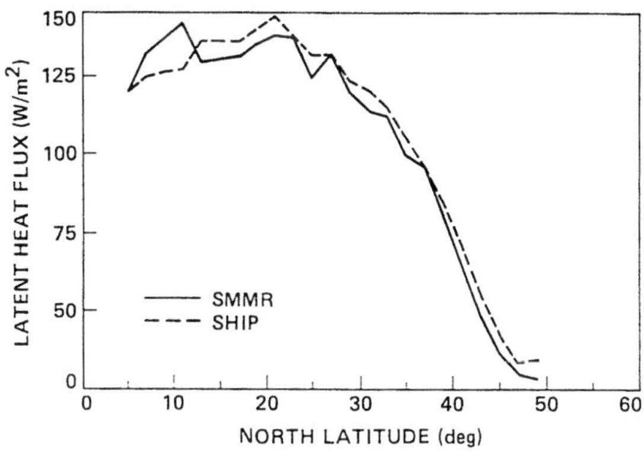

Fig. 10 — Comparing zonal mean latent heat flux derived from ship reports and from Nimbus/SMMR brightness temperatures from 1982.

6. Retrieval of Sensible Heat Flux

There were attempts of deriving T from Q, or vice versa, by assuming a certain relative humidity. This is similar to using a Bowen ratio (LE/H). In Fig. 11, the 12 yr time series of Q at Ocean Weather Ship I (59°N, 19°W) is shown with the saturation mixing ratio ($\hat{Q}$) at air temperature. They are typical of most high and mid-latitude stations. The temporal variation of temperature and humidity appear to be coherent and in phase giving support to the use of the constant relative humidity. A similar comparison using Truk atoll (7°N, 152°E) data is shown in Fig. 12. There are temporal variations of temperature that are not present in the humidity. A more vigorous study is conducted by Liu and Niller (1989).

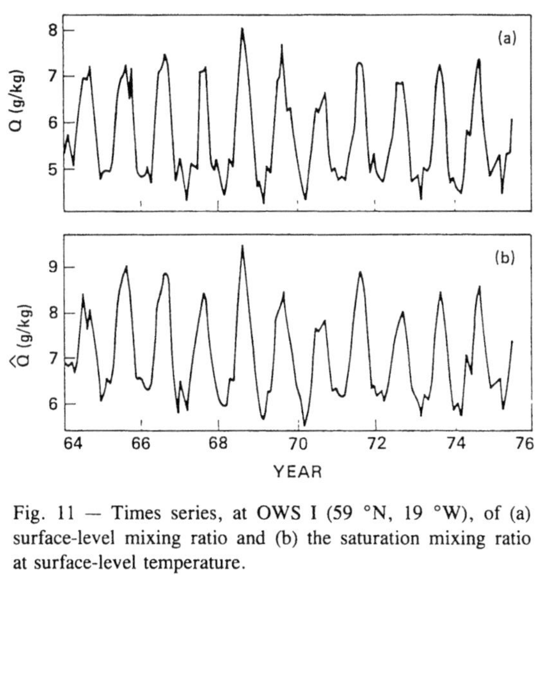

Fig. 11 — Times series, at OWS I (59 °N, 19 °W), of (a) surface-level mixing ratio and (b) the saturation mixing ratio at surface-level temperature.

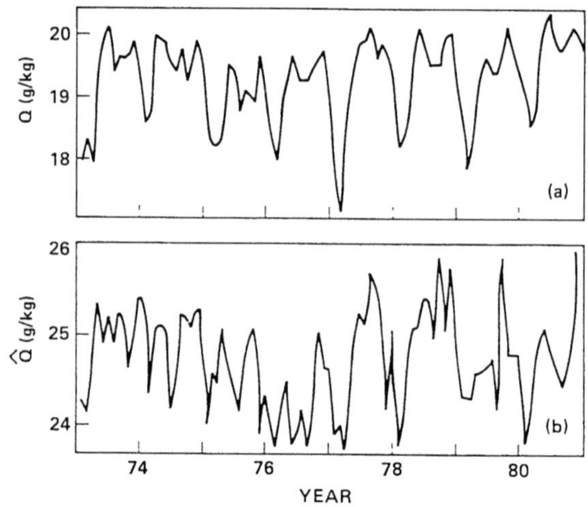

Fig. 12 — Time series, at Truk atoll (7°N, 150°E), of (a) surface-level mixing ratio, and (b) the satuation mixing ratio at surface-level temperature.

With the improvement of vertical resolution of the atmospheric sounders, such as the Atmospheric Infrared Sounders (AIRS) planned for the polar platform of the Earth Observing System (Eos), the potential of measuring a boundary layer (1 km) temperature exists. There is also potential of measuring boundary layer height by Lasa altimeter. The parameterization schemes discussed in Section 2 apply only to parameters measured in the surface layer (~50 m deep). Boundary layer parameterization models that include the Coriolis force and the baroclinicity (e.g. Arya, 1977; Brown and Liu, 1982) need to be carefully studied to make use of the new technology in determining the surface sensible heat flux.

7. Future Sensors Synergism

There has not been an operating scatterometer in space since the demise of Seasat in 1978. A similar Ku-band scatterometer was approved by NASA (NSCAT) but the space platform and launching date are uncertain at present. The Active Microwave Instrument (AMI) to be launched on the European Remote Sensing Satellite (ERS-1) is a C-band radar capable of operating as a wind scatterometer. It will have a more serious sampling problem because it will sample on only one side compared with the two-side sampling of Seasat/SASS and it will operate in the scatterometer mode only part of the time. Two scatterometers are planned for the polar platforms of Eos.

To compute turbulence heat fluxes, only wind speed is required; currently, remote measurements with a microwave radiometer can provide the measurement. With the deterioration of Nimbus/SMMR, there will be no microwave radiometer in the near future that will measure the same suite of frequencies as did SMMR. The Special Sensor Microwave Imager (SSMI) on the operational Defense Meteorological Space Program (DMSP) can measure W and possible U, but it lacks the low frequency channels which are sensitive to T_s variability. The next microwave radiometer having the same full range of frequencies as SMMR is the Japanese Advanced Microwave Scanning Radiometer (AMSR) planned for the Eos in the mid-1990s. Until then, turbulent heat flux has to be computed from the bulk parameterization methods using U and W from SSMI and T_s from operational Advanced Very High Resolution Radiometer (AVHRR). Better T_s measurements may be provided by the Advanced Medium Resolution Imaging Radiometer (AMRIR) and the Moderate Resolution Imaging Spectrometer (MODIS), planned for Eos.

Improved sounder technology is expected in the next NOAA operational satellite, such as the AMSU. With the additional information of the vertical humidity structure, Q at shorter time scales may be determined. Eos sensors, such as AIRS, may provide information for the retrieval of sensible heat flux. However, simultaneous and coincident coverages are more important for computing heat fluxes at shorter time scales, where calculations are based on measurements from a combination of sensors. There are urgent needs to design payload scenarios of the Eos platforms for optimal synergism and to establish a data management system tailored for high level products like the heat fluxes.

8. Acknowledgment

This study was performed at the Jet Propulsion Laboratory, California Institute of Technology, under contract with the National Aeronautics and Space Administration (NASA). It was partially supported by the NASA Scatterometer (NSCAT) Project and partially by the Ocean Processes Branch of NASA. The Seasat radiometer data were obtained from NASA Ocean Data System, the Seasat scatterometer wind vectors from the NSCAT Project, and the Nimbus/SMMR geophysical parameters from the National Space Science Data Center. The radiosonde reports were provided by the National Climatic Data Center and the ship reports by the National Center for Atmospheric Research. Dr. Eni Njoku kindly made the Nimbus/SMMR brightness temperatures available and Professor Pearn P. Niiler initially suggested using empirical orthogonal functions to study the humidity variability.

9. References

Alishouse, J.C., 1983: Total precipitable water and rainfall determinations form the SEASAT scanning multichannel microwave radiometer. *J. Geophys. Res.*, *88*, 1929-1935.

Arya, S.P.S., 1977: Suggest revisions to certain boundary layer parameterization schemes used in atmospheric circulation models, *J. Atmos. Sci.*, *105*, 215-227.

Anderson, R.J. and S.D. Smith, Evaporation coefficient for the sea surface from eddy flux measurements, *J. Geophys. Res.*, *86*, 449-459, 1981.

Baker, W.E., R. Atlas, E. Kalnay, M. Halem, P.M. Woiceshyn, and D. Edelman, 1894: Large-scale analysis and forecast experiments with wind data from the Seasat-A scatterometer, *J. Geophys Res.*, *80*, 4927-4936.

Bernstein, R.L. and J.H. Morris, 1983: Tropical and mid-latitude North Pacific sea surface temperature variability from Seasat SMMR, *J. Geophys. Res.*, *88*, 1877-1891.

Brown, R.A., and W.T. Liu, 1982: An operational large-scale marine planetary boundary layer model, *J. Appl. Meteor.*, *21*, 261-269.

Bunker, A.F., 1976: Computations of surface energy flux and annual air-sea interaction cycles of the North Atlantic Ocean, *Mon. Wea. Rev.*, *104*, 1122-1140.

Chang, H.D., P.H. Hwang, T.T. Wilheit, A.T.C. Chang, D.H. Staelin, and P.W. Rosenkranz, 1984: Monthly distributions of precipitable water from Nimbus-7 SMMR data, *J. Geophys. Res.*, *89*, 5328-5334.

Chou, S.H. and D. Atlas, 1982: Satellite estimates of ocean-air heat fluxes during cold air outbreaks. *Mon. Wea. Rev.*, *110*, 1434-1450.

Deardorff, J.W., 1968: Dependence of air-sea transfer coefficient on bulk stability, *J. Geophys. Res.*, *73* 2549-2557.

Esbensen, S.K. and R.W. Reynolds, 1981: Estimating monthly average air-sea transfer of heat and momentum using the bulk aerodynamic method, *J. Phys. Oceanogr.*, *11*, 457-465.

Fleagle, R.G. and J.A. Businger, 1983: *An Indroduction to Atmospheric Physics*, Academic Press, New York, p. 62.

Gloersen, P., P.J. Cavaleri, A.T.C. Chang, T.T. Wilheit, W.T. Campbell, O.M. Tohannessen, K.B. Katsaros, K.F. Kunzi, D.B. Ross, D. Staelin, E.P.L. Windsor, F.T. Barath, P. Gudmandsen, E. Langhan, and R.O. Ramseirer, 1984: Results from the first Nimbus-7 SMMR observations, *J. Geophys Res.*, *89*, 5335-5344.

Jones, W.L., L.C. Schroeder, D.H. Boggs, E.M. Bracalente, R.A. Brown, G.L. Dome, W.J. Pierson, and F.L. Wentz, 1982: The SEASAT-A satellite scatterometer: the geophysical evaluation of remotely sensed wind vectors over the ocean, *J. Geophys. Res.*, *87*, 3297-3317.

Katsaros, K.B., P.K. Taylor, J.C. Alishouse, and R.G. Lipes, 1982: Quality of SEASAT SMMR (Scanning Multichannel microwave radiometer) atmospheric water determinations. *Oceanography From Space*, J.F.R. Gower (ed.) Plenum Press, New York, 691-706.

Katsaros, K.B., S.D., Smith, and W.A. Oost, 1987: HEXOS — Humidity exchange over the sea, a program or research on water vapor and droplet fluxes from sea to air at moderate to high wind speeds. Bull. Amer. Meteor. Soc., 68, 466-476.

Large, R.G. and S. Pond, 1982: Sensible and latent heat flux measurements over the ocean, *J. Phys. Oceanogr.*, *12*, 464-482.

Liu, C.T., 1983: Tropical Pacific sea surface temperature measured by SEASAT microwave radiometer and by ships, *J. Geophys. Res.*, *88*, 1909-1914.

Liu, W.T., 1986: Statistical relation between monthly mean precipitable water and surface-level humidity over global oceans, *Mon. Wea. Rev.*, *114*, 1591-1602.

Liu, W.T., 1988: Moisture and latent heat flux variabilities in the tropical Pacific derived from satellite data, *J. Geophys. Res.*, in press.

Liu, W.T., K.B. Katsaros, and J.A. Businger, 1979: Bulk parameterization of air-sea exchanges of heat and water vapor including the molecular constraints at the interface. *J. Atmos. Sci.*, *86*, 1722-1735.

Liu, W.T., and P.P. Niller, 1984: Determination of monthly mean humidity in the atmospheric surface layer over oceans from satellite data. *J. Phys. Oceanogr.*, *14*, 1451-1457.

Liu, W.T., and P.P. Niller, 1989: Evaluation of latent heat flux approximations used in ocean circulation models. *J. Geophys. Res.*, to appear.

Lorenz, E.N., 1977: An experiment in nonlinear statistical weather forecasting. *Mon. Wea. Rev.*, *105*, 590-602.

Milman, A.S., and T.T. Wilheit, 1985: Sea surface temperatures from the scanning multichannel microwave radiometer on Nimbus-7, *J. Geophys., Res.*, *90*, 11631-11641.

Reitan, C.H., 1963: Surface dew point and water vapor aloft. *J. Appl. Meteor.*, *2*, 776-778.

Simonot, J.Y. and H. Treunt, 1987: Surface heat fluxes from a numerical weather prediction system. *Climate Dynamics* 2, 11-28.

Smith, W.L., 1966: Note on the relationship between total precipitable water and surface dew point. *J. Appl. Meteor.*, *5*, 726-727.

Smith, W.L., 1983: The retrieval of atmospheric profiles from VAS Geostationary radiance observations, *J. Atmos. Sci.*, *40*, 2025-2035.

Smith, W.L., H.M. Woolf, and C.M. Hayden, D.Q. Wark, and L.M. McMillin, 1979: The TIROS-N Operational Vertical Sounder, *Bull. Amer. Meteor. Soc.*, *60*, 1177-1187.

Wilheit, T.T. and A.T.C. Chang, 1980: An algorithm for retrieval of ocean surface and atmospheric parameters from the observations of the scanning multichannel microwave radiometer, *Radio Sci.*, *15*, 525-544.

Chapter 17

SATELLITE REMOTE SENSING OF SEA SURFACE TEMPERATURE

ENI G. NJOKU
Earth Science and Applications Division
Office of Space Science and Applications
National Aeronautics and Space Administration
Washington, DC

1. Introduction

1.1. BACKGROUND

Since the early days of satellite environmental remote sensing, efforts have been made to measure sea surface temperature (SST) from space with scientifically useful accuracy. The synoptic view from space allows SST to be mapped globally, including regions where ships and in situ measuring devices are sparse or nonexistent. Furthermore, satellites can provide frequent coverage, enabling the motions of current eddies and meanders to be studied in detail. During the past decade much progress has been made in satellite measurement of SST. The satellite instruments and SST retrieval algorithms have improved considerably, scientific applications and requirements are much better understood, and the potential benefits of satellite observations are becoming widely recognized.

The physical basis for measurement of SST from space is the detection by a spaceborne microwave or infrared radiometer of the thermal energy emitted by the ocean surface. The surface radiation is modified in its passage through the atmosphere to the radiometer by processes of atmospheric absorption, emission, and scattering, for which appropriate corrections must be made in order to obtain valid estimates of the surface radiance. The corrected radiances must be converted to surface temperature, which requires knowledge or measurement of the surface emissivity and corrections for reflected solar and atmospheric radiation. Infrared and microwave techniques provide these measurements and corrections in somewhat different ways, each with unique advantages and disadvantages.

Environmental satellites of the early 1970s carried infrared radiometers designed primarily for meteorological purposes (i.e. cloud imaging and atmospheric sounding), but which also included channels in spectral regions of low atmospheric absorption that could be used to observe the ocean surface and hence determine SST (Rao et al., 1972; Brower, et al., 1976). Beginning in late 1981 with the launch of the five-channel AVHRR instrument on the NOAA polar-orbiting satellites, an improved operational sea surface temperature product became available using a multichannel retrieval approach (McClain et al., 1983). This multichannel SST (MCSST) product incorporated an improved atmospheric correction algorithm and cloud

G. L. Geernaert and W. J. Plant (eds.), Surface Waves and Fluxes, Volume II, 311–338.

filtering procedure, and superseded the former less accurate GOSSTCOMP (Global Operational SST Computation) product that had been produced routinely by NOAA. These and other NOAA SST products were directed primarily towards meeting the needs of NOAA's operational users.

Following the launch of the NASA Seasat and Nimbus-7 satellites, and the NOAA Tiros-N generation of satellites, in late 1978, a number of other satellite instruments and retrieval methods were evaluated for research-oriented applications of SST. These instruments including the Scanning Multichannel Microwave Radiometer (SMMR) on the Seasat and Nimbus-7 satellites (Hofer et al., 1981; Wilheit et al., 1984); the High-resolution Infra-Red Sounder and Microwave Sounding Unit (HIRS/MSU) on the Tiros-N and NOAA satellites (Susskind et al., 1984); and the Visible-Infrared Spin Scan Radiometer (VISSR) Atmospheric Sounder (VAS) on the NOAA GOES series of geostationary satellites (Bates and Smith, 1985). Studies have shown that these instruments are capable of measuring SST to an rms accuracy of approximately 1 °C and perhaps significantly better in some cases (Bernstein and Chelton, 1985). Summary characteristics of these sensors are provided in Table 1.

Table 1 — Characteristics of recent satellite sensors used in measurement of SST

Spacecraft	Orbit Type	Sensor	Spectral Channels[||]	Sensor Spatial Resolution, km	Spatial Resolution of Global SST Product, km[§]	Swath Width	Research/ Operational Status of SST Product
NOAA 6-11	Polar, 850 km altitude	AVHRR	Visible, plus 3.7, 11, 12 μm (infrared)*	1	50-250	2500 km[†]	Operational (several products)
NOAA 6-11	Polar, 850 km altitude	HIRS/MSU	Visible, plus 3.7-15 μm (19 infrared channels), 50.3-58.0 GHz (4 microwave channels)	20	60-250	2200 km	Research
Nimbus 7	Polar, 955 km altitude	SMMR[††]	6.6, 10.7, 18, 21, 37 GHz (microwave)	150	150	800 km	Research
GOES 4-7	Geosynchronous, 0°N, 75°W (GOES-East); 0°N, 135°W (GOES-West)	VAS	Visible, plus 3.9, 11, 12.6 μm (infrared)**	14	50-75	Earth disk[†]	Research

* This channel not included on NOAA, 6, 8, and 10
† Useful area reduced by viewing angle contraints
‡ SMMR was also carried on the Seasat satellite
Nominal positions (spacecraft failure may necessitate repositioning of remaining active satellite)
§ Depending on method used to generate product
|| Not al channels used in SST determination
** Additional infrared channels available in dwell-sounding mode

Applications of satellite-derived SSTs are numerous, and can be grouped according to the time and space scales of the phenomena being observed (Njoku et al., 1985). Users of SST products for fisheries research and fishing operations require the highest spatial resolutions available (~1 km) and good relative SST accuracy for locating fronts and eddies (high absolute accuracy is somewhat less critical). At the other end of the scale, absolute accuracy requirements for global climate research are very demanding, especially for the warm regions of the tropical oceans where SST accuracies of 0.3°C are desired. Weekly to monthly averages on space scales of 50 km or greater are sufficient however for climate purposes (WCRP, 1986).

The use of satellite-derived SST data in oceanographic applications is increasing. Evaluation workshops have been organized (Njoku, 1985; WCRP, 1986) to compare the various SST methods and products, the estimate their accuracies, and to plan their use for mesoscale and global scale air-sea interaction studies. There are ongoing efforts to assimilate SST data into ocean models used by the U.S. Navy, NOAA's National Meteorological Center (NMC), and by universities, (e.g. Clancy and Pollack, 1983). These involve both objective analysis of data fields and, eventually, full four-dimensional, space-time continuous data assimilation.

1.2. TECHNIQUES AND SENSORS FOR SST REMOTE SENSING

For SST measurement using infrared radiometry the basic technique is to measure thermal infrared radiation from the ocean at a wavelength corresponding to an atmospheric window, typically in the 3.5 to 4 μm or 10.5 to 12.5 μm range, where the atmospheric absorption is low. In these spectral regions the ocean emissivity is approximately constant and close to unity, thus changes in surface-emitted radiance can be related directly to changes in surface temperature. Improved SST accuracy can be obtained by using spectral information from two or more window channels to correct for the variable component of atmospheric absorption (principally due to water vapor). The AVHRR instrument has three infrared spectral window channels (Table 1) and operates according to this principle. The two visible-wavelength channels are useful for identifying cloudy fields of view where accurate SST retrievals cannot be made due to the opacity of clouds in the infrared. The inherent spatial resolution of the sensor is 1 km, but for global coverage the raw data are sub-sampled to a resolution of 4 km. Several operational AVHRR SST products are routinely generated which, after cloud-filtering algorithms have been applied, are available at resolutions of 20-50 km or greater. A description of the AVHRR instrument, operational SST algorithms, and derived products can be found in the article by McClain et al., (1985).

Other sensors on the NOAA satellites have also been used by researchers to generate SSTs. The motivation has been to improve the quantitative accuracy of SST measurements in regions of variable humidity and cloudiness where the AVHRR capability experiences some limitations. In one such approach, data from selected channels of the 20-channel HIRS and the 4-channel MSU are combined in a retrieval algorithm which corrects for atmospheric attenuation and variable cloudiness. The inherent spatial resolution of the HIRS/MSU measurements (~20 km) is poorer than that of the AVHRR, but the SST retrievals are obtainable under conditions of greater cloudiness. The HIRS/MSU technique has been discussed by Susskind and Reuter (1985). Another useful approach is that of combining selected AVHRR and HIRS infrared channels to obtain an improved atmospheric correction (Schluessel et al., 1987). All infrared techniques experience severe degradations in accuracy when measuring SST under heavily cloud overcast conditions. In these situations, microwave techniques offer the only possibility for satellite SST measurement since absorption of surface radiation by atmospheric gases and clouds is much lower at microwave wavelengths. However, the ocean emissivity is also lower (~0.4-0.8) in the microwave region than in the infrared, and is variably affected by wind-induced roughness and foam. Hence there is a requirement for additional microwave spectral channels to correct for the effects of variable windspeed, in addition to humidity and clouds. In principle this is straightforward, and has been demonstrated using the SMMR instrument which has dual-polarized channels at five frequencies (between 6 and 37 GHz). The spatial resolution of SMMR SST measurements is approximately 150 km. Details of the SMMR technique are given in the article by Milman and Wilheit (1985).

Infrared instruments in geostationary orbit, such as the VAS on the NOAA GOES satellites, have the advantage of frequent observations from a stationary vantage point over a given region of the ocean. The high temporal observation frequency is useful for tracking fast-moving SST features such as eddies and current boundaries, and provides a good probability of cloud-free views of the ocean over short periods of time such as for generating daily or weekly average SST maps. In its multispectral imaging mode the VAS instrument utilizes similar spectral channels and retrieval methods to those of the AVHRR. The inherent spatial resolution of the VAS sensor is 14 km, with SST retrievals after cloud filtering available at approximately 50 km resolution. The VAS technique is described in Bates and Smith (1985).

2. Surface Emission and Radiative Transfer

2.1. RADIATIVE TRANSFER

Satellite measurements of SST are based on passive techniques in which infrared or microwave radiometers detect thermally emitted radiation from the ocean surface. Retrieval of SST information thus requires an understanding of the processes by which electromagnetic radiation is emitted and scattered at the ocean surface, and by which radiation is emitted and attenuated by the intervening atmosphere as it propagates towards the satellite. The propagation of radiation through a nonscattering medium in thermodynamic equilibrium is described mathematically by the radiative transfer equation (Chandrasekhar, 1960):

$$\frac{dI(\nu,\theta)}{ds} = -\{I(\nu,\theta) - B[\nu, T(s)]\}\, k(\nu, s) \tag{1}$$

where: $I(\nu;\theta)$ is the intensity of radiation at frequency ν, and position s, propagating in direction θ. $B[\nu, T(s)]$ is the Planck blackbody emission function at temperature $T(s)$; and $k(\nu, s)$ is the volume absorption coefficient.

For a spaceborne radiometer viewing the Earth, as in the geometry of Fig. 1, boundary conditions are applied so that the radiative transfer equation in its integral form can be expressed as:

$$I(\nu,\theta) = \epsilon(\nu,\theta)B(\nu, T_s)\tau_0(\nu,\theta) + I_a(\nu,\theta)$$

$$+ \tau_0(\nu,\theta)\iint_{2\pi} \gamma(\nu,\theta;\theta')\,[I_d(\nu,\theta') + \tau_0(\nu,\theta')I_h(\nu,\theta')]\,d\theta' \tag{2}$$

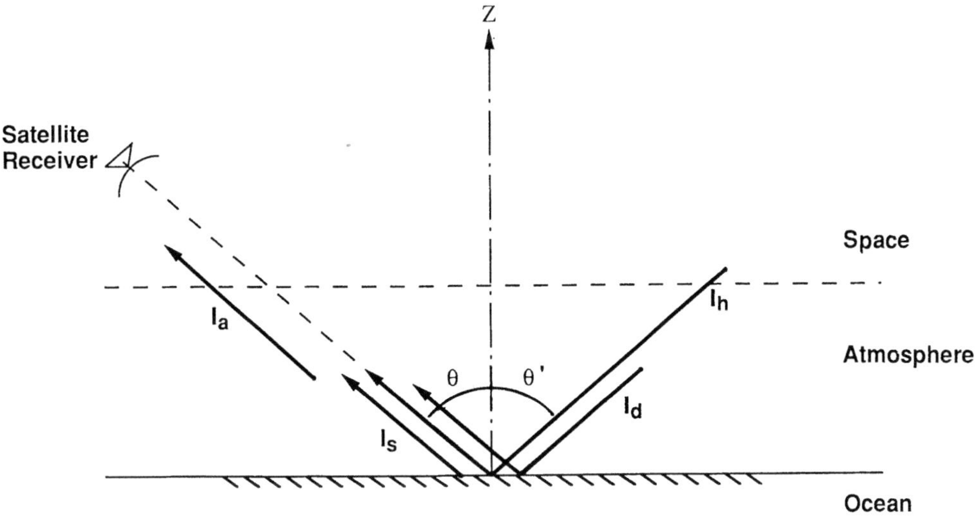

Fig. 1 — Radiation components observed by a satellite radiometer with geometry for emission and reflection at the ocean surface

In this expression, $I(\nu, \theta)$ is the radiometer-observed clear-column radiance assuming that no clouds or aerosols are within the instrument field of view (which would give rise to scattering of the transmitted radiation). The surface emissivity $\epsilon(\nu, \theta)$ $(0 < \epsilon < 1)$ depends on the physical characteristics of the ocean surface; T_s is the ocean surface temperature. The atmospheric spectral transmittance $\tau_0(\nu, \theta)$ along the viewing path from the surface to the satellite is determined by the absorption coefficients of the atmospheric gaseous constituents. The first term in (2) represents the emitted surface radiation I_s reaching the radiometer. The Planck function depends on temperature according to the relation;

$$B\ (\nu, T) = \frac{2h\nu^3}{c^2[\exp\ (h\nu/kT) - 1]} \tag{3}$$

The form of the Planck function is shown in Fig. 2. Note that the Planck function peaks near 10 μm wavelength in the infrared for a typical Earth temperature of ~300°K.

From the radiative transfer equation, the upward component of atmospheric emission at the satellite $I_a(\nu, \theta)$ can be expressed as:

$$I_a(\nu, \theta) = \int_0^{z_0} B\ [\nu, T(z)]\ \frac{\partial \tau\ (\nu, \theta, z)}{\partial z}\ dz \tag{4}$$

Also,

$$\tau_0(\nu, \theta) = \tau\ (\nu, \theta, 0) \tag{5}$$

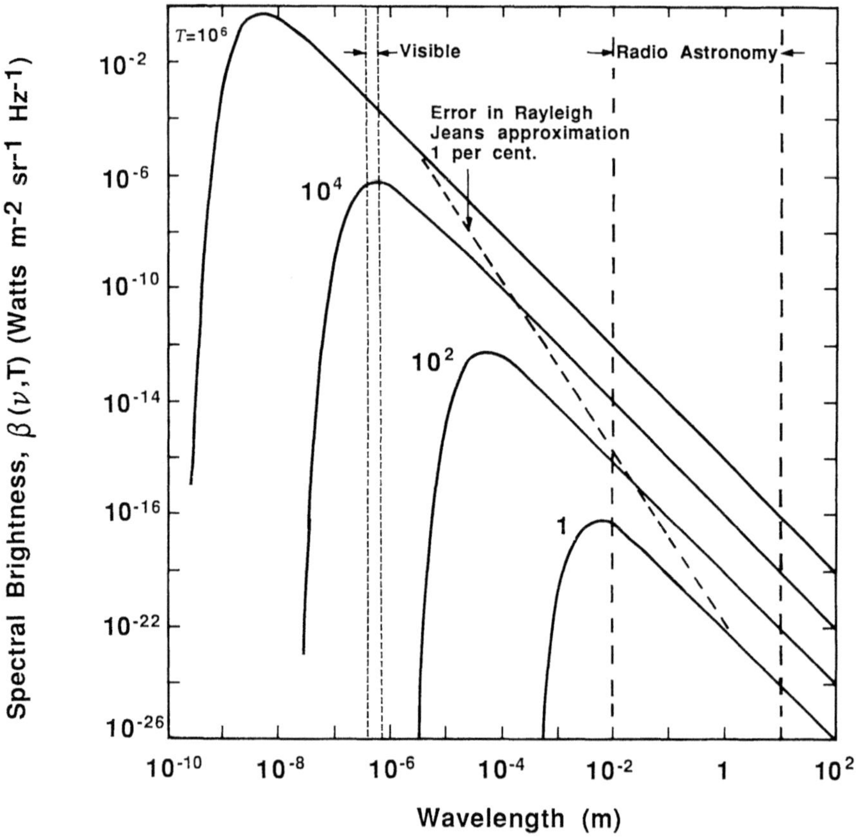

Fig. 2 — Blackbody spectral brightness at various temperatures as determined by the Planck function

where $T(z)$ is the vertical atmospheric temperature profile as a function of height z, and $\tau(\nu, \theta, z)$ is the atmospheric transmittance along the viewing path between height z and the satellite altitude z_0. The downward component of atmospheric emission at the surface, $I_d(\nu, \theta')$, can be expressed in a manner similar to $I_a(\nu, \theta)$. The background space radiation $I_h(\nu, \theta')$ includes contributions from solar radiation and cosmic background radiation. The surface scattering coefficient $\gamma(\nu, \theta; \theta')$ is assumed to be azimuth independent; it depends on the physical characteristics of the ocean surface and is related to the emissivity by the reciprocity relation:

$$\epsilon(\nu, \theta) = 1 - r(\nu, \theta)$$

and

$$r(\nu,\theta) = \iint_{2\pi} \gamma\,(\nu,\, \theta;\, \theta')\, d\theta' \tag{6}$$

where $r(\nu, \theta)$ is the surface reflectivity. Computation of satellite radiances using (2) can be performed by assuming a plane parallel atmosphere with specified vertical profiles of temperature and gaseous constituents, and using an appropriate model for the scattering coefficients at the ocean surface in order to compute the scattering integral (Chahine et al., 1983).

2.2. INFRARED EMISSION AND ATTENUATION

For a cloud-free field of view and a given surface temperature, the radiation received by an ocean-viewing radiometer is determined principally by the ocean emissivity ϵ and the atmospheric transmittance τ_0. In the infrared portion of the spectrum the atmospheric transmittance is affected by a variety of absorbing gaseous constituents such as H_2O, CO_2, O_3, N_2O, and others (Chahine et al., 1983). However, there are several spectral "window" regions in which the transmittance is high (~ 0.8) and which can therefore be used for remote sensing of the surface. Gaseous absorption in the atmosphere is spatially and temporally homogeneous since the absorbing gases are well mixed. An exception is the absorption due to water vapor which is variable due to changes in atmospheric humidity. Figure 3 shows infrared atmospheric transmittance computed for three different model atmospheres each with differing amounts of water vapor (Maul, 1985). The computations were performed using the LOWTRAN-4 code of Selby et al. (1978). It can be seen that the 3.4-4 μm and 8-13 μm spectral windows both have a high transmittance, but at 8-13 μm the effects of variable humidity are significant. In some regions of the tropics less than 40% of the radiation reaching the satellite within this spectral range originates from the surface. (The rest originates from the atmosphere principally due to emission by tropospheric water vapor.) On the other hand, in the 10-12 μm wavelength range the ocean surface emissivity is high (~ 0.99), which is a distinct advantage since the downwelling atmospheric and solar radiation scattered at the surface is almost negligible. At 3.5-4 μm the atmospheric transmittance is less susceptible to variations in atmospheric humidity but the surface emissivity is lower (~ 0.97) and the surface reflectivity correspondingly higher. Reflection of solar radiation at the surface can make the 3.5-4 μm spectral region unsuitable for SST measurement during daylight hours. The wavelength dependence of the infrared reflectance of sea water is shown in Figure 4 and is discussed by Friedman (1969) and Hobson and Williams (1971).

Models for computing infrared radiance as viewed by a satellite radiometer must integrate the effects of several molecular spectral line absorptions (weighted by the filter spectral bandwidth of the radiometer). The computations can be performed line by line, using the spectral line shape and intensity of the individual molecules, which can be highly computer-intensive due to the large number of spectral lines that need to be included. Band approximations are often used to simplify the computations (Chahine et al., 1983). In the atmospheric window regions, absorption from weak spectral lines and wings of strong lines far from the spectral region of interest are contributing factors. Absorption by water vapor dimers (two water molecules joined by a hydrogen bond) is also present in the 10.5-12.5 μm interval and is highly temperature dependent (Bignell, 1970). Uncertainties in modelling this continuum emission can lead to errors in the computed radiances.

Clouds are opaque at infrared wavelengths, hence infrared measurements with clouds in the field of view tend to indicate the cloud temperature rather than that of the underlying ocean surface. To a lesser extent, this is also true of fields of view containing aerosols such as fog and other atmospheric particulates. Scattering of infrared radiation by aerosols is particularly

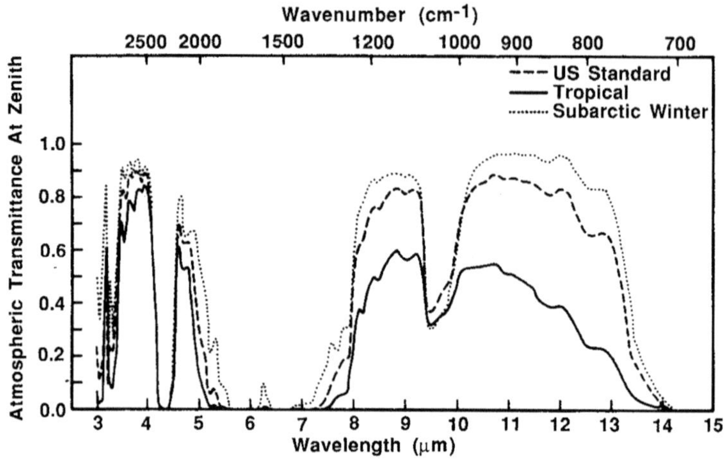

Fig. 3 — Computed zenith atmospheric transmittance for three standard atmospheres with different amounts of water vapor (from Maul, 1985)

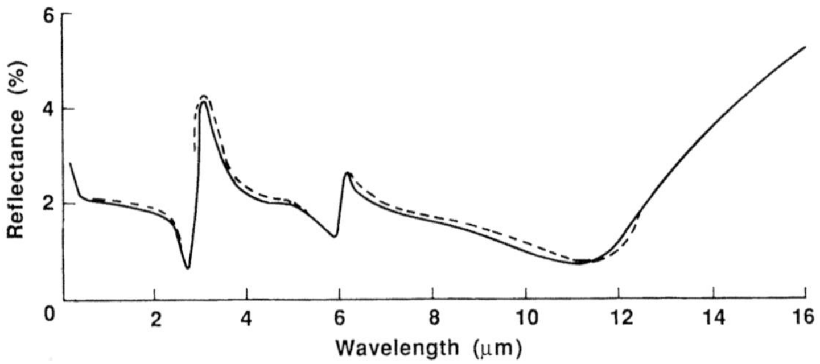

Fig. 4 — Reflectance of pure water (solid line) and sea water (dashed line) for radiation incident normal to a plane water surface (from Stewart, 1985)

significant at the shorter wavelengths (3.5-4 μm). In determining SST from infrared radiances, considerable care must be taken to detect cloud- and aerosol-contaminated fields of view if valid results are to be obtained. Corrections are sometimes possible for fields of view which are only partially cloud covered, although the resulting SSTs are often of lower accuracy and reliability.

2.3. MICROWAVE EMISSION AND ATTENUATION

In the microwave region of the spectrum below ~ 100 GHz the Rayleigh-Jeans approximation is often used to simplify the radiative transfer equation. In this approximation, $h\nu << kT$, and the Planck emission function takes the form: $B(\nu, T) = 2kT\nu^2/c^2$ (see Figure 2). For a given frequency (or wavelength) the emitted radiance is then proportional to temperature and the radiative transfer equation can be expressed in terms of radiometric temperatures or "brightness temperatures." Thus, equation (2) can be written:

$$T_B(\nu,\theta) = \epsilon(\nu,\theta)T_s\,\tau_0(\nu,\theta) + T_a(\nu,\theta)$$

$$+\,\tau_0(\nu,\theta)\iint_{2\pi} \gamma(\nu,\theta;\theta')\,[T_d(\nu,\theta') + \tau_0(\nu,\theta')\,T_h(\nu,\theta')]\,d\theta' \qquad (7)$$

Expressions equivalent to $I_a(\nu, \theta)$ and $I_d(\nu, \theta')$ can be derived for $T_a(\nu, \theta)$ and $T_d(\nu, \theta')$ respectively. In this frequency range (< 100 GHz) the principal gaseous absorbing constituents are oxygen and water vapor. Water droplets in the form of clouds and rain also contribute to atmospheric absorption, while droplets with large diameter-to-wavelength ratios will cause scattering of the transmitted radiation (Chahine et al., 1983). The atmospheric transmittance at frequencies below 10 GHz is generally greater than 95% even in the presence of heavy clouds. For this reason microwave techniques have been viewed as a favorable alternative to infrared for regions where extensive and/or persistent cloudiness precludes reliable infrared measurement of SST. Figure 5 shows the microwave zenith attenuation for various atmospheric water vapor and cloud contents (Liebe, 1983). Note that an attenuation of 1 dB corresponds to a transmittance of ~80%, hence below 20 GHz even thick cumulus congestus clouds with the liquid water amount shown (1.8 mm) are relatively transparent. (Precipitation significantly increases the attenuation however.) These curves are based on measurements and theoretical models which in the case of gaseous absorbers have been documented extensively (e.g. Waters, 1976). The theory for Rayleigh and Mie absorption and scattering by liquid droplets is well understood (Van de Hulst, 1957), but for practical situations modeling the drop-size distributions and geometrical configurations of clouds is difficult since these distributions are not easy to determine a priori and their effects must be averaged over the radiometer field of view.

The ocean emissivity is significantly lower at microwave than at infrared wavelengths and is more variable, principally due to wind-induced surface roughening and foam (Wilheit, 1979). In addition, the emissivity is a function of temperature T_s and salinity S due to the dielectric properties of water at microwave frequencies (Klein and Swift, 1977). In Figure 6 the 2.65 GHz computed brightness temperature at normal incidence for a smooth water surface is shown as a function of surface temperature ($T_B = \epsilon(S, T_s)T_s$) for a range of salinities from 0-38 °/oo. The curves are approximately linear at lower temperatures and the slopes vary with salinity. At frequencies higher than 5 GHz the effects of salinity become negligible and the slopes (sensitivity to T_s) increase to a maximum value near 5-6 GHz.

The effect of incidence (viewing) angle on the observed brightness temperature is shown in Figure 7. The computations were performed using a geometric optics model for the wind-induced roughness, but without considering the generation of foam (Stogryn, 1967). A characteristic of the curves is that for vertical polarization the brightness temperature shows little sensitivity to wind speed, particularly at incidence angles near 50°, whereas the horizontal polarization is highly sensitive. Both polarizations are sensitive to the effects of foam which increases the brightness temperature (Stogryn, 1972). Wilheit (1979) has developed an empirical model, consistent with available experimental data, which incorporates both roughness and foam effects.

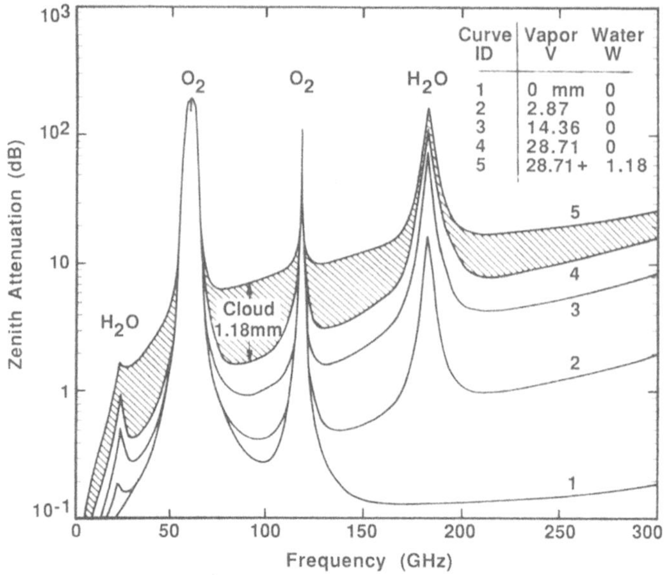

Fig. 5 — Zenith atmospheric attenuation in the microwave and millimeter wave region. Different curves are for increasing amounts of water vapor and cloud (from Liebe, 1983).

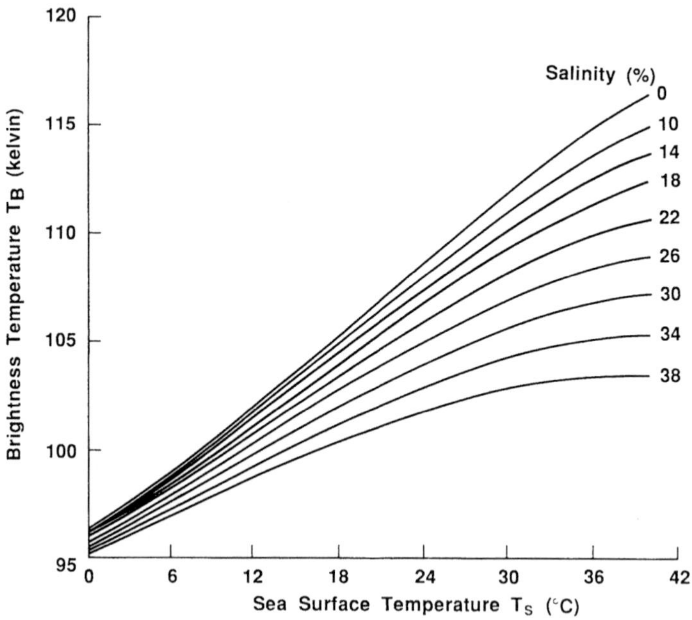

Fig. 6 — Brightness temperature of a smooth water surface at normal incidence and frequency of 2.65 GHz, as a function of temperature and salinity (from Swift, 1980)

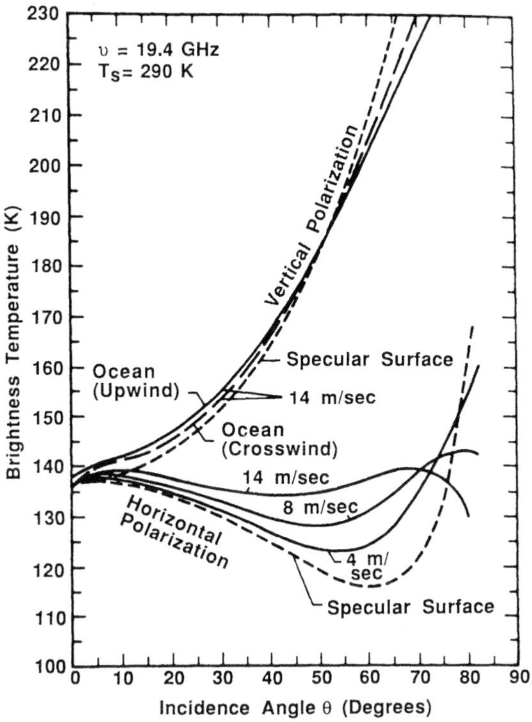

Fig. 7 — Dependence of vertically and horizontally polarized ocean brightness temperature on incidence angle for a smooth (specular) surface, and for different wind speeds, at a frequency of 19.4 GHz (from Maul, 1985)

The relative effects of oceanic and atmospheric parameters on the observed microwave radiance can be determined by computing the normalized sensitivity of the brightness temperature to these parameters. For a radiometer viewing at nadir, such a computation is shown in Figure 8 (Wilheit et al., 1980). In the figure, the parameters P_i; (and typical units) are; ocean salinity (°/ₒₒ), ocean temperature (°C), surface wind speed (m/s), column integrated water vapor (g/cm^2), and column integrated cloud liquid water (mg/m^2). For SST the sensitivity is maximum near 5 GHz, whereas there is little sensitivity at that frequency to salinity, water vapor, or clouds. This ideally makes 5 GHz an optimum frequency for microwave SST measurement. However, the brightness temperature sensitivity to wind speed is also significant in the 5 GHz region. Algorithms which retrieve SST from microwave measurements at this frequency must therefore utilize methods to correct for the effects of wind speed. For the SMMR (on the Seasat and Nimbus-7 satellites) this was accomplished by incorporating five spectral channels which spanned the range of sensitivities to the environmental parameters (in Figure 8 the channel frequencies are indicated by arrows). In addition the SMMR instrument was designed to center receive orthogonal (vertical and horizontal) polarized radiation at each frequency and the antenna was designed to scan conically at a fixed viewing angel of 50° in order to take advantage of the difference in roughness sensitivities of the two polarizations at

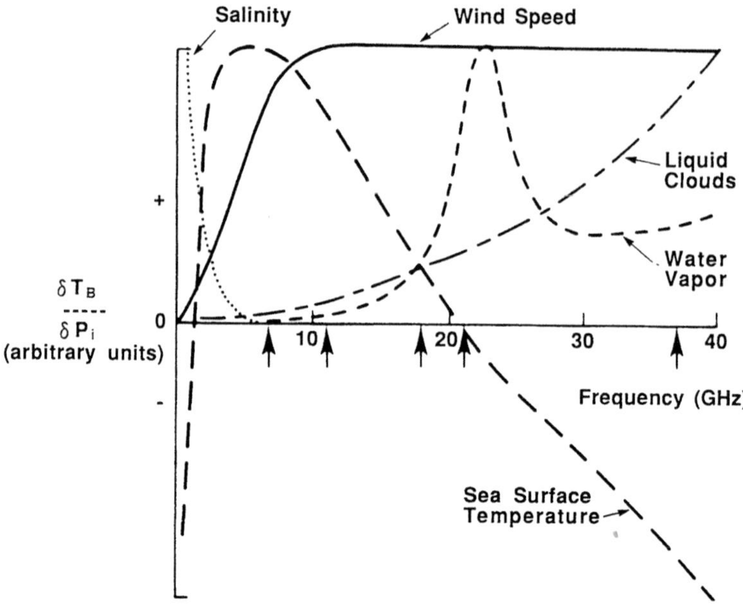

Fig. 8 — Schematic representation of the sensitivity of brightness temperature T_B to various ocean and atmosphere parameters P_i as a function of frequency. The SMMR frequencies are indicated by arrows (from Wilheit et al., 1980).

that angle (Figure 7). Using regression techniques various combinations of SMMR channels were evaluated and used to derive multichannel SST algorithms for both the Seasat and Nimbus-7 instruments (Wilheit and Chang, 1980; Hofer and Njoku, 1981).

It is important in the microwave region, as in the infrared, that physical models developed for computing satellite radiances be accurate and well understood. The main uncertainty in the microwave models lies in the effects of wind-induced roughness and foam on emissivity. Theoretical and experimental studies have been carried out to improve these models (e.g. Wu and Fung, 1972; Webster et al., 1976; Wilheit, 1979; Wentz, 1983). However the physical wind-wave-foam relationships and the electromagnetic radiation interactions at the ocean surface are quite complex, and model uncertainties still contribute a significant source of error to the microwave SST retrievals.

3. Retrieval Methods

3.1. INFRARED SST RETRIEVAL

The atmospheric layer above the ocean in which absorption and re-emission of upwelling infrared radiation takes place is generally cooler than the ocean surface. Thus, the radiometric temperature viewing the ocean from space will in general be lower than the actual ocean temperature and should not be used as a measure of ocean temperature unless corrections for atmospheric effects are made. The temperature difference can be as large as ~10 K in the 10.5-12.5 μm region for tropical (humid) atmospheres, but is less (<4 K) in the 3.5-4 μm region for these cases. Early infrared radiometers had a single window channel and hence made use of independent estimates of water vapor (such as from coincident radiosondes or spaceborne sounding instruments) in order to compute an atmospheric correction. Histogram and discriminant function techniques were used to determine cloud-free pixels for which corrected temperatures could then be mapped into routine products such as NOAA's GOSSTCOMP for operational use (Brower et al., 1976). Analyses of these products by research users subsequently revealed significant large-scale errors (~1-4°C) in the satellite-derived SSTs when compared to high quality in situ measurements, indicating that the methods used for atmospheric correction were inadequate (Barnett et al., 1979).

With the advent of the AVHRR instruments on the TIROS-N/NOAA series of satellites beginning in 1978 additional infrared window channels became available. The current five-channel AVHRR instruments have two spectral channels in the visible range (0.58-0.68 and 0.725-1.1 μm) and three in the infrared (3.55-3.93, 10.3-11.3, and 11.5-12.5 μm). In anticipation of these instruments several simulations and experimental studies using multispectral techniques were performed (e.g. Anding and Kauth, 1970; Prabhakara et al., 1974; McMillin, 1975; Deschamps and Phulpin, 1980; Imbault et al. 1981). These studies indicated that it was possible to obtain accurate SST measurements by using linear combinations of infrared window channel measurements, i.e. the SST could be written as:

$$T_s = a_0 + \sum_{i=1}^{n} a_i T_i \tag{8}$$

where T_s is the SST, and n is the number of channels used. The coefficients a_i depend upon the absorption coefficients k_i at the different wavelengths, and T_i are the radiometric temperatures at wavelengths λ_i. McMillin and Crosby (1984) showed that the coefficients a_i are independent of variations in the temperature and moisture profiles provided that the total atmospheric absorption is relatively small so that the approximations used in deriving (8) remain valid.

Simulations show that linear algorithms of the form given by (8) can generate accurate SSTs under most cloud-free conditions provided that appropriate window channels are used. The coefficients a_i can be derived by statistical regression which also provides rms error estimates that are valid for the ensemble of simulated or real measurements included in the regression (but not necessarily for all conditions encountered in the real environment). The theoretical simulations by Deschamps and Phulpin (1980), for example, considered three infrared window channels centered at 3.7, 11, and 12 μm. Radiances were computed using a theoretical atmospheric transmittance model and a combination of standard atmosphere and radiosonde profile data. For the atmospheres chosen, the theoretically-derived three-channel linear algorithm resulted in an estimated rms SST error of ~0.1 K, indicating the ability of this type of algorithm to provide adequate atmospheric correction. When radiometer noise of 0.2 K was introduced in each channel the theoretical error estimate increased to ~0.4 K. Several authors have performed simulations of this type for a variety of channel combinations and for different viewing angles. Some of these results are discussed by Barton (1983). Differences in algorithm coefficients derived by these authors for similar wavelengths can be ascribed to differences in the models and atmospheric absorption coefficients used, for which there remains some uncertainty particularly in the absorption coefficient due to water vapor. The results reinforce the suggestion that the 3.7 μm channel is particularly useful due to the low atmospheric absorption at this wavelength.

The simulations also show that dual angle techniques, in which measurements of the ocean are made at the same frequency but at two different viewing angles, are potentially capable of improved accuracy since estimation of the wavelength dependence of atmospheric absorption is not required. This dual angle technique will be tested using the Along-track Scanning Radiometer (ATSR) instrument to be launched on the ESA ERS-1 satellite in 1991 (Minnett, et al. 1983).

Before infrared SST retrieval algorithms can be applied to the satellite radiances these radiances must be screened for contamination by clouds. In the NOAA operational MCSST procedure (McClain et al., 1985) several tests are applied to identify cloudy pixels. These include: visible albedo tests, which identify clouds by their higher reflectance (as compared to ocean) in the visible and near infrared spectrum; uniformity tests, in which measurements of adjacent fields of view exhibit high variability when partially cloud-filled; and channel intercomparison tests, in which SSTs derived by different combinations of channels are compared to discriminate uniform low-level clouds. Observed radiances which successfully pass these cloud screening tests are then operated on by the multichannel retrieval algorithms.

In practice, SST measurements obtained from satellite data using theoretically-derived algorithms often show significant biases when compared to in situ measurements. This can be due to a number of causes, among which are inaccuracies in the models used to simulate the satellite measurements, inaccuracies in the atmospheric absorption coefficients, radiometer calibration errors, etc. Empirical approaches are commonly used to remove these observed biases. The satellite measurements are regressed linearly against statistically representative sets of in situ SST measurements to obtain empirical algorithm coefficients. These regressions do not provide insight into the causes of the biases, and the coefficients often require fine-tuning during the satellite mission. If regression coefficients are derived under limited environmental conditions they may give erroneous results when applied globally. Strong and McClain (1984) derived "global" coefficients for the AVHRR by regression of satellite measurements with measurements from a geographically distributed set of fixed and drifting buoys. The following equations were obtained:

$$SST(\text{day}) = 1.0346\, T_{11} + 2.58(T_{11} - T_{12}) - 283.21$$

$$SST(\text{night}) = 1.017\, T_{11} + 0.97(T_{3.7} - T_{12}) - 276.58 \tag{9}$$

(where SST is in °C and T_i is the infrared brightness temperature (K) for channel i). Note that the daytime algorithm does not make use of the 3.7 μm channel due to the high reflectance of incident solar radiation in this channel. Figure 9 (Strong and McClain, 1984) shows statistics for comparisons of AVHRR SSTs derived using the above algorithms with SSTs from an independent set of fixed and drifting buoys. Colocation of the satellite and in situ measurements was made to within 24 hours and 25 km. The rms differences are 0.5 - 0.6 K, with daytime statistics slightly better. Other recent studies using linear multichannel or dual angle algorithms are those of Barton (1985), Castagne, et al. (1986), Ho et al. (1986), Llewellyn-Jones et al. (1984), Maul (1983), McClain et al. (1985), McMillin and Crosby (1984), and Schluessel et al. (1987).

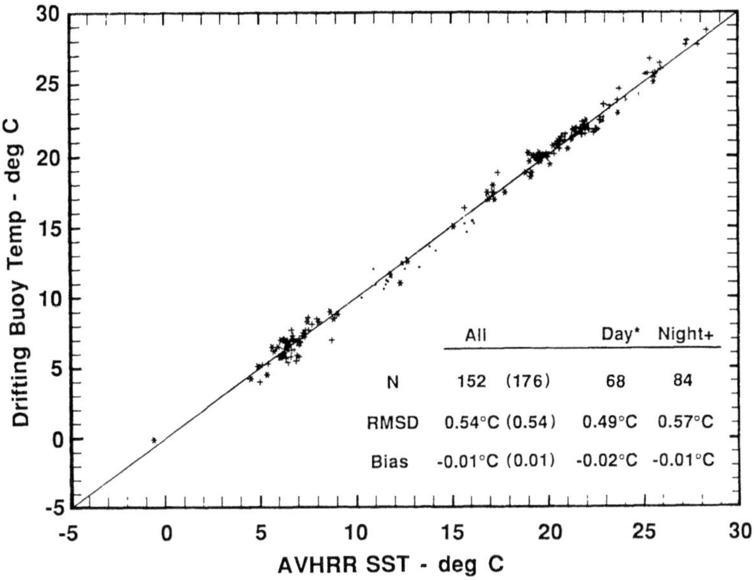

Fig. 9 — Comparisons of satellite-derived (AVHRR) SST measurements with measurements from fixed and drifting buoys. Data were obtained during the period November 1981 to August 1982. (The fixed buoy measurements are shown by dots.) (From Strong and McClain, 1984)

The HIRS and MSU are 20-channel infrared and 4-channel microwave sounders, respectively, which together with the stratospheric sounding unit (SSU) comprise the TIROS operational vertical sounder system (TOVS) flown on the NOAA polar-orbiting satellites. The AVHRR SST operational measurements described earlier do not incorporate TOVS data. However, a research SST product has been developed (Susskind et al., 1984) which uses HIRS and MSU data. In this approach, surface and atmospheric conditions (including clouds) are determined which, when substituted into the radiative transfer (2), result in computed radiances that match the observed radiances to within a specified amount. An iterative scheme is used, starting with first guess values for the meteorological parameters, with computations at each step utilizing those channels most sensitive to the parameters being determined (the HIRS includes window channels at 3.7, 4, and 11 μm). If satisfactory convergence is not obtained the retrievals are flagged as questionable. The method can be computationally intensive due to the iterative calculation of the radiative transfer equation at each retrieval point. A key aspect of the approach is the ability to perform SST retrievals in partially cloud-filled scenes. This is done by reconstructing from the observed radiances in a group of neighboring pixels the clear-column radiances that would have been observed if no clouds were present. SST retrievals are then performed on these derived radiances. The HIRS and MSU have several sounding as well as window channels and SST is just one of a number of parameters that can be obtained simultaneously using the iterative procedure. Other retrievable parameters include fractional cloud cover, cloud-top temperature, cloud-top pressure, atmospheric temperature profile, and land surface temperature and emissivity.

An advantage to using the HIRS instrument is the availability of infrared sounding channels. These inherently can provide corrections for the effects of variable vertical distributions of atmospheric water vapor and temperature on radiances received by the infrared window channels. Schluessel et al., (1987) have examined the benefits obtained by extending the AVHRR multichannel regression technique to include HIRS channels at 11.1, 13.3, and 13.6 μm (which provide information on the temperature profile) and 7.3 μm (which provide additional information on the water vapor structure). Interpolation is required to colocate the differently sized fields of view, and partial cloud filling of the HIRS fields of view is also a problem. However, with careful analysis it appears that some improvement in the SST accuracy for cloud-free scenes can be achieved by incorporating the HIRS data in this manner.

Infrared instruments on geostationary satellites have been used to derive SST using techniques similar to those used for the AVHRR. The VISSR on the early GOES satellites contained a visible spectral channel and a single infrared channel centered at 11.0 μm. Independent methods for atmospheric correction using either in situ data or polar-orbiter multispectral data were employed (Maul, 1981; Zandlo et al., 1982). The VAS, first launched in 1980, contains one visible spectral band and twelve infrared spectral bands. In its multispectral imaging mode (which provides frequent images of the Earth disk) only three infrared channels may be selected, which for SST purposes are usually chosen to be the 3.9, 11.0, and 12.6 μm channels. Bates and Smith (1985) describe an algorithm for deriving SST from these VAS channels, derived by linear regression of satellite and in situ measurements, resulting in equations similar in form to the AVHRR MCSST equations 9). Cloud-filtering is performed using techniques also similar to those for the AVHRR.

3.2. MICROWAVE SST RETRIEVAL

With the launch of the Seasat and Nimbus-7 satellites in 1978, both carrying a scanning multichannel microwave radiometer (SMMR), satellite measurements of SST using microwave techniques became possible for the first time. As with infrared retrievals, theoretical simulations showed that a linear algorithm could be used to estimate SSTs provided the spectral channels were chosen correctly. As previously discussed, in the microwave case (<40 GHz for the SMMR) it is unnecessary to incorporate elaborate cloud-filtering procedures since the microwave radiation penetrates even heavy clouds.

For small excursions around mean values of SST and wind speed, and for low absorption by water vapor and cloud liquid water, the radiative transfer equation (7) is approximately linear. Hence, linear inversion algorithms of the form given by equation (8) can be used. In this case T_i are the brightness temperatures at appropriately chosen microwave wavelengths, and the coefficients a_i are obtained by regression using model simulations or experimental data as in the infrared case. When the range of environmental conditions are such that the radiative transfer equation cannot be considered linear over the whole range, equation (8) can be applied in a piecewise linear manner or alternatively nonlinear terms can be developed for use in place of the brightness temperatures. (For instance logarithmic functions of T_i can be used as suggested by Wilheit and Chang (1980)). Iterative schemes can also be developed to solve for the SST.

In the microwave region the atmospheric absorption windows are broad and the choice of precise measurement frequencies is not critical. However, the frequencies and polarizations must be chosen for sensitivity to SST and for the ability to correct for effects of wind-generated surface roughness and foam, atmospheric water vapor, and clouds. The channels selected for the SMMR instrument are indicated in Figure 8 (6.6, 10.7, 18, 21, and 37 GHz). The 6.6 GHz channel provides sensitivity to SST in a spectral region where atmospheric effects are small. The 21 GHz channel provides water vapor estimation and correction, and the 10.7, 18, and 37 GHz channels provide corrections for the effects of surface roughness and clouds. In addition, each channel is dual-polarized which provides an improved ability to measure wind speed and to correct for the surface roughness.

The linear algorithm coefficients a_i can be derived by using a statistical regression approach. Wilheit and Chang (1980) based their regressions on an ensemble of artificially generated surface and atmospheric conditions, with combinations of values for the geophysical variables (SST, wind speed, water vapor, cloud water content, etc.) spanning the full range expected in a real environment. Brightness temperatures at each frequency are computed for each case using the radiative transfer equation, and algorithm coefficients are derived by multiple linear regression. These theoretically derived algorithms, when first applied to the SMMR data, resulted in significant SST bias and rms errors (Milman and Wilheit, 1985). A number of causes were identified including: (1) radiometer calibration errors; (2) radiative transfer and emission model errors; (3) occasional occurrence of radio-frequency interference (RFI) from ground transmitters; (4) reflected solar radiation from the ocean surface; (5) precipitation; and (6) land contamination from the antenna sidelobes in coastal regions. Careful filtering could remove most errors due to RFI, sunglint, precipitation, and land contamination. However, calibration drifts of the SMMR instrument and uncertainties in the geophysical models (particularly in the effect of wind speed on surface emissivity) remain a problem in obtaining accurate SSTs from the SMMR. Algorithms which compensate for the calibration and model errors have been derived empirically by regressions using external calibration data such as SST climatology and in situ data from ships and buoys (Milman and Wilheit, 1985), As with the infrared case, these empirical corrections mask understanding of the underlying error sources and must be used with care when the required corrections are large.

Another potential source of error is the variability of the antenna beam earth-incidence angle. The SMMR was designed with a conical antenna scan geometry to achieve a constant incidence angle over the cross-track swath (Njoku, 1980). From the slopes of the curves in Figure 7 it can be seen that a small change in incidence angle near 50° can cause a significant change in the observed brightness temperature particularly for vertical polarization. If ignored in the algorithms, incidence angle variations such as might be caused by satellite attitude errors can contribute to the SST error. Later improvements to the post-launch SST algorithms thus included an incidence angle term in addition to calibration terms incorporating instrument temperatures in order to improve the overall calibration stability. Subsequent analyses of SMMR-derived SSTs in the north Pacific ocean still revealed some erroneously high values which appeared to be related to the occurrence of winter storms in the region. These errors were possibly due to inadequate modelling of the change in ocean emissivity caused by wind-induced roughness and foam. Additional nonlinear terms in the SST algorithms were developed to account for this effect. An interesting history of these SMMR SST algorithms and their subsequent modifications is contained in the paper by Milman and Wilheit (1985). From this discussion it is clear that the SMMR was effective as an experimental sensor from which much was learned about the microwave calibration and geophysical model improvements required for next-generation instrument and algorithm designs. The SMMR has provided an opportunity for demonstrating the microwave SST technique, particularly the capability for making measurements in the presence of non-raining clouds.

4. Observational Products and Evaluations

4.1. ACCURACY OF SATELLITE AND IN SITU SST PRODUCTS

For both infrared and microwave instruments, the pre-flight sensor calibration techniques have been found to be in need of improvement if stable in-flight radiometric accuracies of better than 0.5 K are to be achieved (Brown et al., 1985; Milman and Wilheit, 1985). These techniques include better thermal modelling of the instrument, more comprehensive pre-flight thermal vacuum test measurements and analysis, and more stable on-board thermal environments. Inevitably, even with careful attention to the calibration and geophysical algorithms, post-launch comparisons of satellite SSTs with in situ measurements exhibit spatially and temporally varying biases due to residual uncompensated effects. Blended products, in which satellite and in situ SST data are objectively combined, are being studied as a possible way to externally "calibrate" the satellite data using in situ measurements, and to provide the spatial and temporal coverage that in situ data alone cannot provide (Reynolds, 1988). In order to develop optimal blending procedures the error characteristics of the satellite and in situ SST measurements must be known. In the case of satellite measurements however the global error characteristics have been difficult to determine.

Figure 10 shows a typical monthly global distribution of in situ SST measurements from routine ship and buoy reports. It can be seen that large areas of the oceans are sampled infrequently or not at all by these measurements. The ship data are predominantly engine intake temperatures which typically have rms accuracies of $\sim 1°K$, with a small warm bias of $\sim 0.3°K$ (Bernstein and Chelton, 1985). Higher quality in situ SST data are available from research ships and from fixed and drifting buoys which are useful as high quality point measurements but are limited in their geographical distribution.

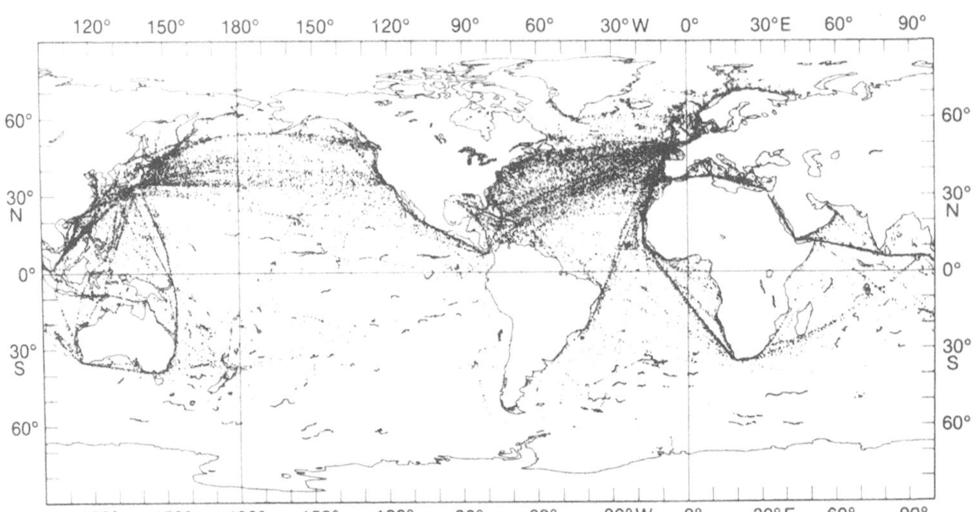

Fig. 10 — Distribution of routine in situ (ship and buoy) SST data reported over the Global Telecommunications System (GTS) for October 1986 (from Reynolds, 1988)

The ship and buoy instruments measure the ocean bulk temperature at 1 to 10 meter depth, whereas satellite instruments measure the radiative skin temperature of the upper few millimeters (microwave) or microns (infrared) of the ocean surface. The skin temperature is normally coller than the bulk temperature due to evaporative cooling (Robinson et al., 1984), and the difference can be several tenths of a degree centigrade in magnitude. (Wind speeds greater than ~10 m/s will create mixing in the upper layer and reduce the skin cooling.) Additionally, under conditions of high insolation and minimal surface mixing (i.e. low wind), diurnal heating can raise the ocean temperature in the upper few meters by as much as a degree or more (Ostapoff and Worthem, 1974). These various effects give rise to skin-bulk temperature differences that show up as sources of "error" in satellite SST comparisons with in situ data, and must be considered when utilizing SST measurements for air-sea interaction studies. For illustration Figure 11 shows a histogram of the differences between bulk and skin temperatures recorded during the cruise of the R/V Meteor in October/November 1984 in the North Atlantic (Schluessel et al., 1987). Bulk temperatures were measured by thermistors at a depth of 4 m and skin temperatures were measured using a ship-borne Barnes PRT5 infrared radiometer. For the entire cruise data set, the skin had a mean cool bias of 0.2°C, with a range of 0.5°C warmer to 0.9°C cooler than the bulk temperature.

Present satellite sensors are considered to have achievable SST accuracies in the range 0.5 to 1.5°C rms with biases (after calibration) in the range ± 0.3°C. Results from the NASA workshops (Njoku, 1985) and subsequent studies have provided some insight into these statistics. The workshops focussed on global data sets from three satellite sensors: AVHRR, HIRS/MSU, and SMMR; and from ships. Measurements from the VAS satellite instrument and in situ XBT data were also evaluated, but on a regional scale only. Global data sets for the workshops were generated by subtracting an interpolated climatology (Reynolds, 1982) from each satellite SST retrieval point and binning the resultant SST "anomalies" into monthly-averaged 2° × 2° latitude-longitude bins (Hilland et al., 1985).

Four months of data were available for the workship studies. Table 2 summarizes the mean, standard deviation, and rms deviation of the differences between the satellite (each sensor) and ship binned anomalies, along with the number of bins containing observations (Bernstein and Chelton, 1985). Also shown are the differences between the ship data and the

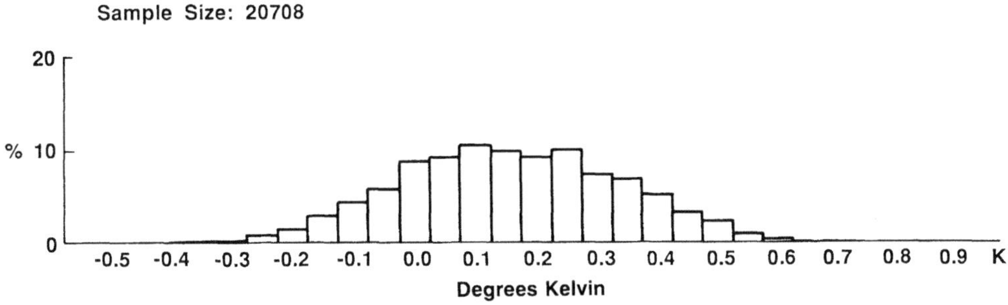

Fig. 11 — Distribution of the differences (bulk minus skin) between bulk water temperature, at a depth of 4 m, and surface skin temperature during the cruise of R/V Meteor, October/November 1984 (From Schluessel et al., 1987)

Table 2 — Global binned difference temperature statistics (°C)
(from Bernstein and Chelton, 1985)

Satellite Sensor	November 1979	December 1981	March 1982	July 1982
			(a) Satellite Minus Ship	
AVHRR				
Mean	0.19 (0.24)	−0.30 (−0.33)	−0.36 (−0.44)	−0.48 (−0.35)
Standard deviation	0.58 (0.35)	0.50 (0.28)	0.51 (0.29)	0.79 (0.52)
rmsd	0.61 (0.42)	0.58 (0.43)	0.62 (0.53)	0.92 (0.63)
Number of observations	723 (324)	729 (235)	795 (368)	644 (274)
HIRS				
Mean	−0.04 (−0.20)	0.13 (0.21)	0.30 (0.29)	−0.07 (0.09)
Standard deviation	1.01 (0.62)	0.88 (0.42)	0.92 (0.31)	0.69 (0.38)
rmsd	1.01 (0.65)	0.89 (0.47)	0.97 (0.42)	0.69 (0.39)
Number of observations	735 (324)	729 (235)	795 (368)	662 (327)
SMMR				
Mean	0.52 (0.72)	0.72 (0.71)	−0.21 (−0.17)	−0.43 (−0.69)
Standard deviation	1.27 (0.81)	1.17 (0.79)	1.11 (0.79)	0.97 (0.60)
rmsd	1.37 (1.08)	1.37 (1.06)	1:13 (0.81)	1.06 (0.91)
Number of observations	395 (152)	677 (226)	690 (300)	522 (230)
			(b) Ships Minus Climatology	
Mean	0.00 (0.00)	−0.03 (−0.08)	−0.09 (−0.13)	−0.46 (−0.70)
Standard deviation	0.80 (0.54)	0.61 (0.38)	0.52 (0.35)	0.74 (0.63)
rmsd	0.80 (0.54)	0.61 (0.39)	0.53 (0.37)	0.87 (0.94)
Number of observations	735 (324)	729 (235)	795 (368)	635 (336)

Only bins with five or more ship observations are included. No bins within 600 km of land are included. Values in parentheses are difference statistics for 3 × 3 spatially smoothed bins.

climatological mean. The ship-climatology differences can be viewed as the anomlay "signal" on scales greater than ~200 km and 1 month (inflated somewhat by ship SST noise). A plot of the statistics derived from Table 2 is shown in Figure 12. Note that the AVHRR standard deviations remain fairly constant at 0.5 - 0.6°C for the first three months, but then climb to nearly 0.8°C in July, 1982. This change in July 1982 is associated with the large-scale stratospheric aerosol cloud generated by the El Chichon volcanic eruption. Inspection of the month-to-month variations in Figure 12 suggests that the three satellite datasets all have time-dependent biases sufficiently large to mask out actual SST variations as measured by the ships. To a large extent this is due to the different algorithms used in deriving the SSTs during the period studied (November 1979 to July, 1982) and points out the necessity for continuous validation and calibration of the satellite data using in situ measurements.

The workshop analyses indicated that, of the satellite data sets evaluated, the AVHRR exhibits the greatest self-consistency as well as the lowest rms errors and bias variations. Nevertheless, a number of unexplained spatial and temporal variations in the biases were observed which could create large errors if assimilated unchecked into ocean and climate prediction models. Recent results indicate that the operational AVHRR MCSST algorithms and validation procedures have improved significantly. Table 3 shows a set of statistics for comparisons between AVHRR SSTs and drifting buoys (WCRP, 1986). The data set has been stratified by region and by atmospheric moisture level to examine the performance of the

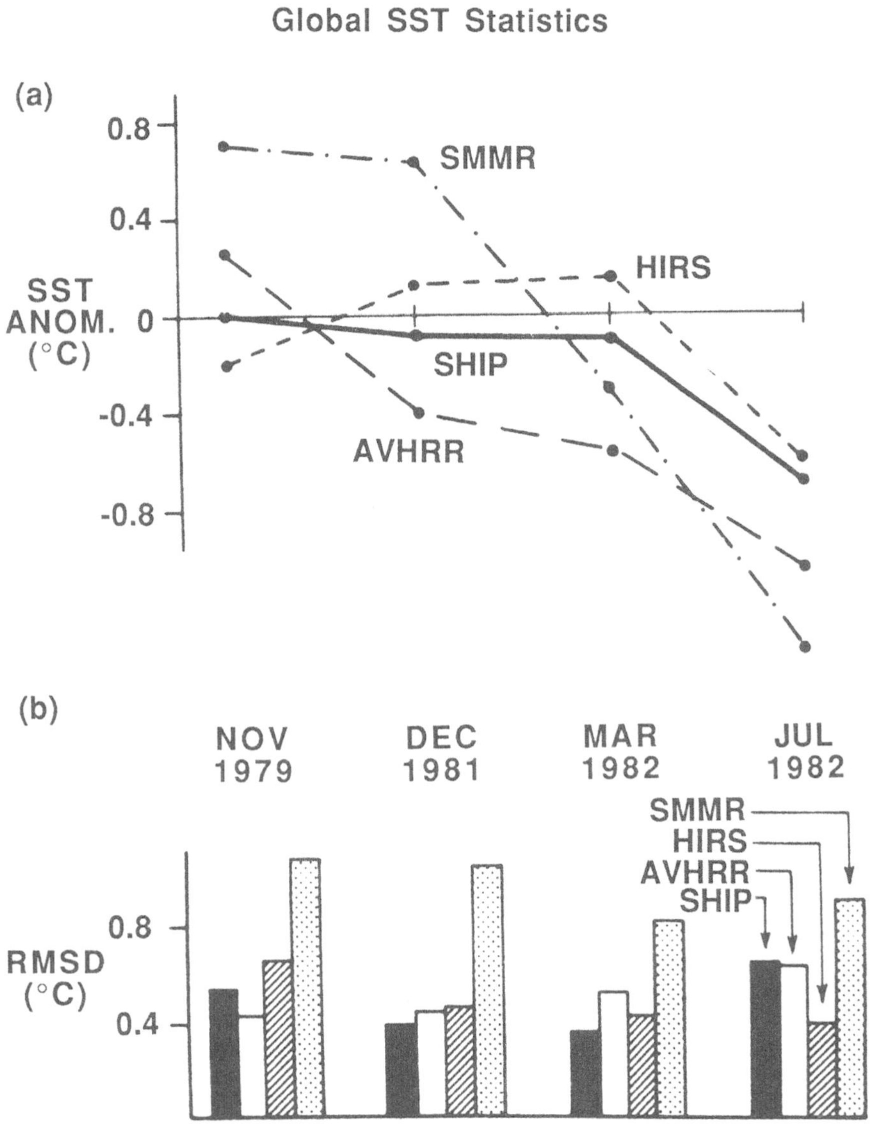

Fig. 12 — Plot of global SST statistics derived from Table 2 (3 × 3 smoothed values): (a) mean difference with respect to climatology, and (b) ship standard deviation, and satellite minus ship rms differences, for each month (from Bernstein and Chelton, 1985)

Table 3 — Analysis of regional and moisture-related variability in AVHRR SST accuracy. Statistics are for differences between colligated satellite and drifting buoy SST measurements acquired during the period February to August 1985. Stratification of the comparisons is by: (a) region; and (b) humidity, as determined by AVHRR channel 4 minus channel 5 brightness temperature difference (from WCRP, 1986)

(a)

REGION	Day				Night			
	No.	Bias	Scat.	RMSD	No.	Bias	Scat.	RMSD
World	1625	+0.01	0.55	0.55	1042	+0.12	0.47	0.48
W. Trop. Pac.	189	−0.02	0.47	0.47	103	+0.22	0.33	0.40
E. Trop. Pac.	433	−0.15	0.53	0.55	290	+0.08	0.41	0.41
Trop. Atl.	424	+0.08	0.50	0.51	277	+0.12	0.04	0.51
Trop. Ind.	277	+0.08	0.50	0.50	163	+0.11	0.45	0.46
N. Atl. & Pac.	208	+0.18	0.57	0.60	115	+0.18	0.52	0.54
N. Pac.	112	+0.25	0.60	0.65	74	+0.16	0.57	0.59
N. Atl.	96	+0.10	0.52	0.53	41	+0.23	0.40	0.46
Southern Ocean	139	+0.08	0.62	0.63	113	+0.11	0.43	0.44

(b)

MOISTURE (Tropical Oceans)	Day				Night			
	No.	Bias	Scat.	RMSD	No.	Bias	Scat.	RMSD
$0° < T_4 - T_5 < 1°$	500	−0.08	0.46	0.47	218	+0.16	0.44	0.47
$1° < T_4 - T_5 < 2°$	694	+0.00	0.56	0.56	553	+0.09	0.45	0.46
$2° < T_4 - T_5 < 3°$	69	−0.01	0.62	0.61	38	−0.01	0.53	0.52

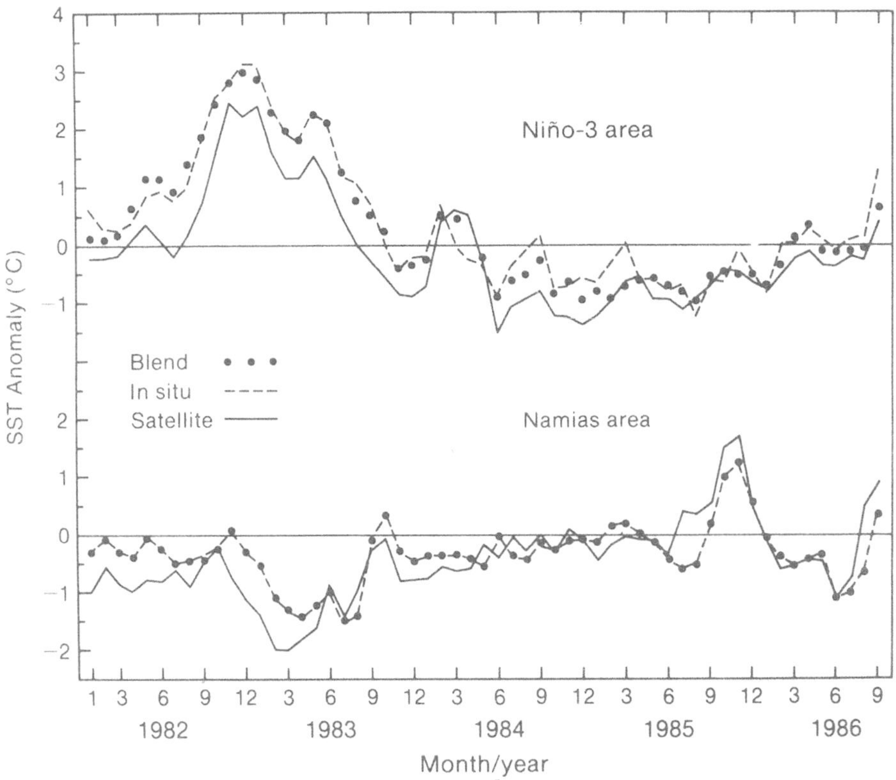

Fig. 13 — Time series of monthly average in situ, satellite (AVHRR), and blended SST anomalies for two regions in the Pacific ocean: (a) ''Nino-3'' (5°N to 5°S; 150 to 90°W); (b) ''Namias'' (30 to 50°N; 165 to 140°W) from Reynolds, 1988

SST algorithm under geographically and environmentally distinct conditions. No systematic trends are evident, and consistent rms accuracies of 0.5 to 0.6°C are obtained, with biases of less than 0.2°C.

The nature of the biases between satellite and in situ SST data can be examined more closely by time series analyses in fixed regions of the ocean. One such analysis is shown in Figure 13 (Reynolds, 1988). Two regions of the Pacific ocean in the tropics and mid-latitudes were selected in which all operational AVHRR SST retrievals and in situ ship SST reports were separately averaged by month. A time series was prepared for the two regions from 1982 through 1986 to compare the satellite and in situ measurements. The satellite data exhibit a cold bias of ~0.5°C when compared to ships during the first part of the period but this difference is not consistent over the entire period. As noted earlier the ship data may be biased slightly warm, and skin-bulk temperature differences introduce a variable component. Also it is noted that the AVHRR algorithms are tuned by regression against drifting buoys and are thus ''calibrated'' to the buoy bulk-temperature level. The effects of stratospheric aerosols from the eruption of El Chicho in March 1982 may have contributed to the cold bias in 1982 and 1983.

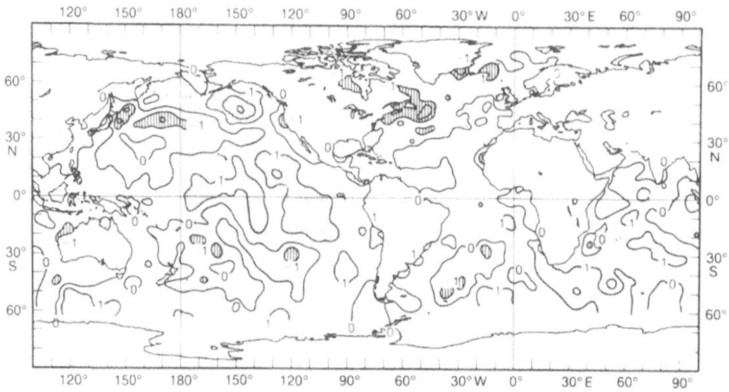

(a)

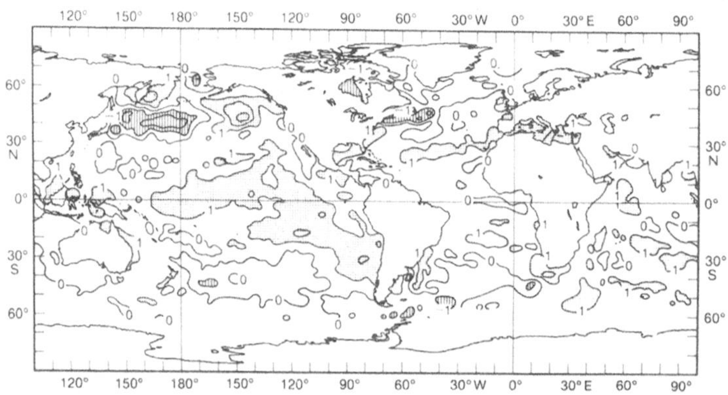

(b)

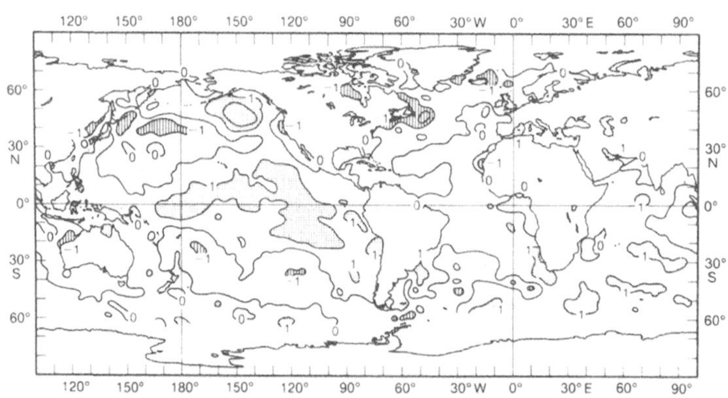

(c)

Fig. 14 — SST anomalies for October 1986. The contour interval is 1° C. Values less than -1° C are shaded; values greater than 1° C are stippled. (a) In situ, (b) AVHRR, (c) Blended product (from Reynolds, 1988).

4.2. COMPOSITE SST PRODUCTS

Assuming that in situ ship SST measurements do not have spatially variable biases they can be used as benchmark SST values, located predominantly along the highly travelled ship routes where the measurements are taken (Figure 10). Satellite SST data can then be "blended" with the in situ data to fill in the shape of the SST field in regions where the ship data are sparse, thus combining the merits of both SST products. A number of automated procedures can be developed to perform the blending. Reynolds (1988) derives blended fields by solving Poisson's equation with appropriate boundary conditions, and has used this approach to produce routine monthly SST anomaly analyses which appear in NOAA's "Oceanographic Monthly Summary" publication (National Weather Service, Washington, D.C.). Figure 14 shows an example of the results obtained by this method. SST anomalies are shown for (a) in situ ship data, (b) AVHRR SST data, and (c) the ship/satellite blended product. The satellite anomalies usually show good large-scale coherence in the tropics and southern Pacific where the sparse ship data cannot represent the SST field adequately. However, the satellite anomalies occasionally differ significantly from the in situ anomalies in regions where both should be adequate, indicating probable errors in the satellite field. The blended analysis preserves the desirable features of both fields.

5. References

Anding, D. and R. Kauth, Estimation of sea surface temperature from space, *Rem, Sens. Environ.1*, 217-220, 1970.

Barnett, T.P., W.C. Patzert, S.C. Webb, and B.R. Bean, Climatological usefulness of satellite determined sea-surface temperatures in the tropical Pacific, *Bull. Am. Meteorol. Soc.* 60, 197-205, 1979.

Barton, I.J., Dual channel satellite measurements of sea surface temperature, *Quart. J. Roy. Met. Soc.*, 109, 365-378, 1983.

Barton, I.F., Transmission model and ground truth investigations of satellite-derived sea surface temperatures, *J. Clim. Appl. Meteorol.*, 24, 508-615, 1985.

Bates, J.J. and W.L. Smith, Sea surface temperature: observations from geostationary satellites, *J. Geophys. Res.*, 90, 11609-11618, 1985.

Bernstein, R.L. and D.B. Chelton, Large-scale sea surface temperature variability from satellite and shipboard measurements, *J. Geophys. Res.*, 90, 11619-11630, 1985.

Bignell, K.J., The water vapor infrared continuum, *Quart. J. Roy Met. Soc.*, 96, 390, 1970.

Brower, R.L., H.S. Gohrband, W.G. Pichel, T.L. Signore, and C.C. Walton, Satellite derived sea surface temperatures from NOAA spacecraft, *NOAA Tech. Memo.*, *NESS* 78 pp., U.S. Dept. of Commerce, Washington, D.C., 1976.

Brown, O.B., J.W. Brown, and R.H. Evans, Calibration of Advanced Very High Resolution Radiometer observations, *J. Geophys. Res.*, 90, 11667-11677, 1985.

Castagne, N., P. LeBorgne, J.L. Vourch, and J.-.P Olry, Operational measurement of sea surface temperature at CMS Lannion from NOAA-7 AVHRR data, *Int. J. Rem. Sens.*, 7, 953-984, 1986.

Chaline, M.T., D.J. McCleese, P.W. Rosenkrane, and D.H. Staelin, Interaction Mechanisms within the Atmosphere, *Manual of Remote Sensing, Vol. 1, no. 5,* (D.S. Simonett and F.T. Ulaky, eds.). American Soc. of Photogrammetry, Falls Church, VA, 1983.

Chandrasekhar, S., *Radiative Transfer,* Dover, New York, 1960.

Clancy, R.M. and K.D. Pollack, A real-time synoptic ocean thermal analysis/forecast system, *Progress in Oceanography,* 12, 383-424, 1983.

Deschamps, P.Y. and T. Phulpin, Atmospheric correction of infrared measurements of sea surface temperature using channels at 3.7, 11, and 12 μm, *Boundary-Layer Meteorol.*, 18. 131-143, 1980.

Friedman, D., Infrared characteristics of ocean water (1.5-15 μm), *Appl. Optics,* 8, 2073-2078, 1969.

Hilland, J.E., D.B. Chelton, and E.G. Njoku, Production of global sea surface temperature fields for the Jet Propulsion Laboratory workshop comparisons, *J. Geophys. Res.*, 90, 11642-11665, 1985.

Ho, D.,A. Asem, and P.Y. Deschamps, Atmospheric correction for the sea surface temperature using NOAA-7 AVHRR and METEOSAT-2 infrared data, *Int. J. Rem. Sens.*, 7, 1323-1333, 1986.

Hobson, D.E. and D. Williams, Infrared reflectance of sea water, *Applied Optics,* 10, 2372-2373, 1971.

Hofer, R. and E.G. Njoku, Regression techniques for oceanographic parameter retrieval using spaceborne microwave radiometry, *IEEE Trans. Geosci. Rem. Sens.*, GE-19, 178-189, 1981.

Hofer, R., E.G. Njoku, and J.W. Waters, Microwave radiometric measurements of sea surface temperature from the Seasat satellite; first results, *Science, 212,* 1385-1387, 1981.

Hollinger, J.P., Remote passive microwave sensing of the ocean surface, *Proc. 7th Int'l. Symp. Rem. Sens. of Environ.* Ann Arbor, MI, 1807-1817, 1971.

Imbault, D., N.A. Scott, and A. Chedin, Multichannel radiometric determination of sea surface temperature: parameterization of the atmospheric correction, *J. Appl. Meteorol.*, 20, 556-564, 1981.

Klein, L.A. and C.T. Swift, An improved model for the dielectric constant of seawater at microwave frequencies, *IEEE Trans. Ant. and Propagat.*, AP-25, 104-111, 1977.

Liebe, H.J., Atmospheric EHF window transparency, *IEEE Trans. Ant. and Propagat.*, AP-31, 127-135, 1983.

Llewellyn-Jones, D.T., P.J. Minnett, R.W. Saunders, and A.M. Zavody, Satellite multichannel infrared measurements of sea surface temperature of the N.E., Atlantic ocean using AVHRR/2, *Quart. J. Roy. Met. Soc.*, 110, 613-631, 1984.

Maul, G.A., Application of GOES visible-infrared data to quantifying mesoscale ocean surface temperature, *J. Geophys. Res.*, 86, 8007-8021, 1981.

Maul, G.A., Zenith angle effects in multichannel infrared sea surface remote sensing, *Rem. Sens. Environ.*, 13, 439-451, 1983.

Maul, G.A., *Introduction to Satellite Oceanography,* Martinus Nijhoff Publishers, Dordrecht, Holland, 1985.

McClain, E.P., W.G. Pichel, C.C. Walton, Z. Ahmad, and J. Sutton, Multi-channel improvements to satellite-derived global sea surface temperatures, *Adv. Space Res.*, 2, 43-47, 1983.

McClain, E.P., W.G. Pichel, and C.C. Walton, Comparative performance of AVHRR-based multichannel sea surface temperatures, *J. Geophys. Res.*, 90, 11587-11601, 1985.

McMillin, L.M., Estimation of sea surface temperature from two infrared window measurements with different absorption. *J. Geophys. Res.*, 80, 5113-5117, 1975.

McMillin, L.M. and D.S. Crosby, Theory and validation of the multi-window sea surface temperature technique, *J. Geophys. Res.*, 89, 3655-3561, 1984.

Milman, A.S., and T.T. Wilheit, Sea surface temperatures from the scanning multichannel microwave radiometer on Nimbus-7, *J. Geophys. Res.*, 90, 11631-11641, 1985.

Minnett, P.J., A.M. Zavody, and D.T. Llewellyn-Jones, Satellite measurements of sea surface temperature for climate research, *Large-scale Oceanographic Experiments and Satellites* (C. Gautier and M. Fieux, eds.), 57-85, D. Reidel Publishing Co., Dordrecht, Holland, 1983.

Njoku, E.G., Antenna pattern correction procedures for the scanning multichannel microwave radiomenter (SMMR), *Boundary-Layer Meteorol.*, 18, 79-98, 1980.

Njoku, E.G., Satellite-derived sea surface temperature: Workshop comparisons, *Bull. Am. Met. Soc.*, 66, 274-281, 1985.

Njoku, E.G., T.P. Barnett, R.M. Laurs, and A.C. Vastano, Advances in satellite sea surface temperature measurement and oceanographic applications, *J. Geophys. Res.*, 90, 11573-11586, 1985.

Ostapoff, C.A. and S. Worthem, The intradiurnal temperature variation in the upper ocean layer, *J. Phys. Oceanog.*, 4, 601-612, 1974.

Prabhakara, C., G. Dalu, and V.G. Kunde, Estimation of sea surface temperature from remote sensing in the 11 to 13 μm window region, *J. Geophys. Res.*, 79, 5039-5044, 1974.

Rao, P.K., W.L. Smith, and R. Koffler, Global sea-surface temperature distribution determined from an environmental satellite, *Mon. Weather Rev.*, 100, 10-14, 1972.

Reynolds, R.W., A monthly averaged climatology of sea surface temperatures, *NOAA Tech. Rep. NWS 31*, NOAA/National Weather Service, Washington, D.C., 35 pp., 1982.

Reynolds, R.W., A real-time global sea surface temperature analysis, *J. Climate*, 1, 75-86.

Robinson, I.S., N. Wells, and H. Charnock, The sea surface thermal boundary layer and its relevance to the measurement of sea surface temperature by airborne and spaceborne radiometers, *Int. J. Rem. Sens.*, 5, 19-45, 1984.

Schluessel, P., H.Y. Shin, W.J. Emery, and H. Grassl, Comparison of satellite-derived sea surface temperatures with in situ skin measurements, *J. Geophys. Res.*, 92, 2859-2874, 1987.

Selby, J.E.A., F.X. Knezys, J.H. Chetwynd, and R.A. McClatchey, Atmospheric transmittance/radiance: Computer code LOWTRAN-4, Rept. # AFGL-TR-78-0053, Air Force Geophysical Laboratory, Hanscomb AFB, MA, 1978.

Stewart, R.H., *Methods of Satellite Oceanography*, University of California Press, Berkeley, CA, 1985.

Stogryn, A., The apparent temperature of the sea at microwave frequencies, *IEEE Trans. Ant. Propagat.*, AP-15, 278-286, 1967.

Stogryn, A., The emissivity of sea foam at microwave frequencies, *J. Geophys. Res.*, 77, 1658-1666, 1972.

Strong, A.E. and E.P. McClain, Improved ocean surface temperatures from space comparisons with drifting buoys, *Bull. Am. Met. Soc.*, 65, 138-142, 1984.

Susskind, J., J. Rosenfield, D. Reuter, and M.T. Chahine, Remote sensing of weather and climate parameters from HIRS2/MSU on TIROS-N, *J. Geophys. Res.*, 89, 4677-4697, 1984.

Susskind, J. and D. Reuter, Retrieval of sea surface temperature from HIRS2/MSU, *J. Geophys, Res.*, 90, 11602-11608, 1985.

Swift, C.T., Passive microwave remote sensing of the ocean—a review, *Boundary-Layer Meteorol.*, 18, 25-54, 1980.

Van de Hulst, H.C., *Light scattering by small particles,* Wiley, New York, 1957.

Waters, J.W., Absorption and emission by atmospheric gases, *Methods of Experimental Physics*, Vol. 12, Part B (M.L. Meeks, ed.), Academic Press, New York, 1976.

WCRP, Report of the COSPAR international workshop on satellite-derived sea surface temperatures for global climate applications, *Report # WCP-110, WMO/TD no. 93,* 65 pp. 1986.

Webster, W.J., T.T. Wilheit, D.B. Ross, and P. Gloersen, Spectral Characteristics of the Microwave Emission from a Wind-driven Foam-covered Sea, *J. Geophys Res., 81*, 3095-3099, 1976.

Wentz, F.J., Model function for ocean microwave brightness temperatures, *J. Geophys. Res.*, 88, 1892-1908, 1983.

Wilheit, T.T., A model for the microwave emissivity of the ocean's surface as a function of wind speed, *IEEE Trans. Geosci. Rem. Sens.*, GE-17, 244-249, 1979.

Wilheit, T.T., A.T.C. Chang, and A.S. Milman, Atmospheric corrections to passive microwave observations of the ocean, *Boundary-Layer Meteorol.*, 18, 65-77, 1980.

Wilheit, T.T. and A.T.C. Chang, An algorithm for retrieval of ocean surface and atmospheric parameters from the observations of the scanning multichannel microwave radiometer (SMMR), *Radio Science,* 15, 525-544, 1980.

Wilheit, T.T., J. Greaves, J. Gatlin, D. Han, B.M. Krupp, A.S. Milman, and E.S. Chang, Retrieval of ocean surface parameters from the scanning multichannel microwave radiometer (SMMR) on the Nimbus-7 satellite, *IEEE Trans. Geosci. Remote Sensing,* GE-22, 133-143, 1984.

Wu, S.T. and A.K. Fung, A noncoherent model for microwave emissions and backscattering from the sea surface, *J. Geophys. Res.*, 77, 5917-5929, 1972.

Zandlo, J.A., W.L. Smith, W.P. Menzel, and C.M. Hayden, Surface temperature determination from an amalgamation of GOES and TIROS-N measurements, *J. Appl. Meteorol.*, 21, 44-50, 1982.

Chapter 18

PARAMETERIZATION SCHEMES AND MODELS FOR ESTIMATING THE SURFACE RADIATION BUDGET

K. B. KATSAROS
Dept. of Atmospheric Sciences AK-40
University of Washington
Seattle, WA 98195

1. Introduction

To understand the processes of air-sea interaction fully we must include the thermodynamic energy exchanges, as well as the transfer of kinetic energy. Jacobs (1942) said it as follows: "For every meteorological change in the atmosphere over the oceans there must be a corresponding change in the energy relationships between sea and atmosphere. Thus, it would seem that any attempts toward a solution of long-period weather problems must be concerned with thermodynamics as well as dynamical factors."

Turbulent fluxes of sensible heat and water vapor, which remove heat from the sea have been discussed in previous chapters. In this chapter the heating and cooling of the sea due to solar and terrestrial radiative energy fluxes will be discussed. Improved evaluation of these fluxes over the global ocean has taken on increased importance in view of the emphasis in the scientific community on global change (e.g., Rasool, 1987). The ERBE (Earth Radiation Budget Experiment, Hartmann, et al., 1986) monitors the radiative energy budget of the planet as a whole from several satellite-borne instruments. To obtain the surface radiation budget (SRB) requires understanding of radiative transfer through the atmosphere, in particular the effects of clouds on this transfer. Estimates of the surface radiation budget have been obtained in the past (e.g., Sellers, 1965; Vonder Haar and Oort, 1973; Budyko, 1974, Stephens et al., 1981). These estimates were based on parameterizations of the flux terms which employ daily averages of standard synoptic ship observations or satellite input data. As our understanding of these parameterizations changes and the flux estimates are revised, other climatic estimates such as the meridional heat transport by the atmosphere or by ocean currents or the local oceanic heat budget have to be changed (e.g., Hastenrath, 1980; Oort and Vonder Haar, 1976; Weare et al., 1981; Weare, 1983; Wyrtki, 1965).

In this chapter, parameterization schemes and models for estimating the terms in the surface radiation budget are discussed. Parameterization schemes are typically simpler, mainly empirical relationships between the radiation term of interest and some readily available or observed parameter such as solar zenith angle, interface temperature, humidity, cloudiness amount - and cloud type, while the models typically are somewhat more sophisticated and require more physical understanding. The models have mainly been used with remotely

G. L. Geernaert and W. J. Plant (eds.), Surface Waves and Fluxes, Volume II, 339–368.

sensed data. However, the parameterization schemes become more physical as our understanding and experience grows and there has been a trend to "parameterize" the models to make evaluations more economical. The distinction is, therefore, not very sharp.

2. Theoretical and Practical Considerations

The components of radiant energy flux to and from the sea surface can be divided into the solar (or shortwave, 0.3 to 3 μm wavelength) and terrestrial (long wave or infrared, 3 to 50 μm wavelength) region designated by subscripts s and l respectively. Hemispheric radiation into unit area of the sea per unit time we call irradiance, E Wm^{-2}, while hemispheric radiation out of the sea we call exitance, M (Raschke, 1978). The net heat flux Q_{net}, at the sea surface due to radiation is then:

$$Q_{net} = E_s + M_s + E_l + M_l \ . \tag{1}$$

The four tems can be measured individually by hemispheric sensors. Pyranometers are used for shortwave irradiance measurements, pyrgeometers for longwave irradiances, and their sum can be measured with a pyrradiometer (e.g., Hinzpeter, 1980; Lind and Katsaros, 1982). The importance of gimballing shortwave sensors on ships and buoys was discussed by Katsaros and DeVault (1986). Effects of instrumental tilt are not as serious for pyrgeometers since the contrast between irradiance from the sky, or rather the atmosphere near the horizon, does not differ greatly from the irradiance from the sea because of the opacity of the atmosphere at most infrared wavelengths for large optical paths. The differences between these heat sensing units is mainly in the material of the hemispheric domes which filter out the undesired portion of the electromagnetic spectrum.

M_s and M_l are difficult to measure correctly from an obstacle such as a ship at sea, and since they can be calculated quite accurately from well proven formulae, when E_s and sea surface temperature (SST) are known, they are seldom measured over water nowadays.

The short and longwave irradiances, E_s and E_l, will be discussed in more detail in this chapter. What parameters do they depend on and how can one infer them without measuring them? The starting point for E_s is the insolation at the surface of the earth under clear sky. E_s for clear sky depends mainly on the irradiance at the top of the atmosphere E_{so}. E_{so} in turn depends on the earth-sun distance and the position of the sun in the sky, and can be calculated from knowledge of the solar constant, the date, time, latitude and longitude. The irradiance at the top of the atmosphere is therefore given. The established formula for E_s under a clear sky, assuming a transmission factor through the atmosphere of 0.7, is given by Seckel and Beaudry (1973). Variations from this value depend on variation in the transmittance through the atmosphere, which depends on the following factors: a) absorption by water vapor and ozone, water vapor being the most important, b) absorption and scattering by atmospheric aerosols, c) reflection, scattering and absorption by clouds in the path of the radiation, and d) over ice-covered ocean (and over most land surfaces) on multiple "reflections" between the surface and overlying clouds and atmosphere (Figure 1).

Atmospheric longwave irradiance at the sea surface depends mainly on the vertical distribution and temperature of the three triatomic gases, carbon dioxide, waper vapor, and ozone. Carbon dioxide is well mixed, so it is mainly the temperature distribution which affects its

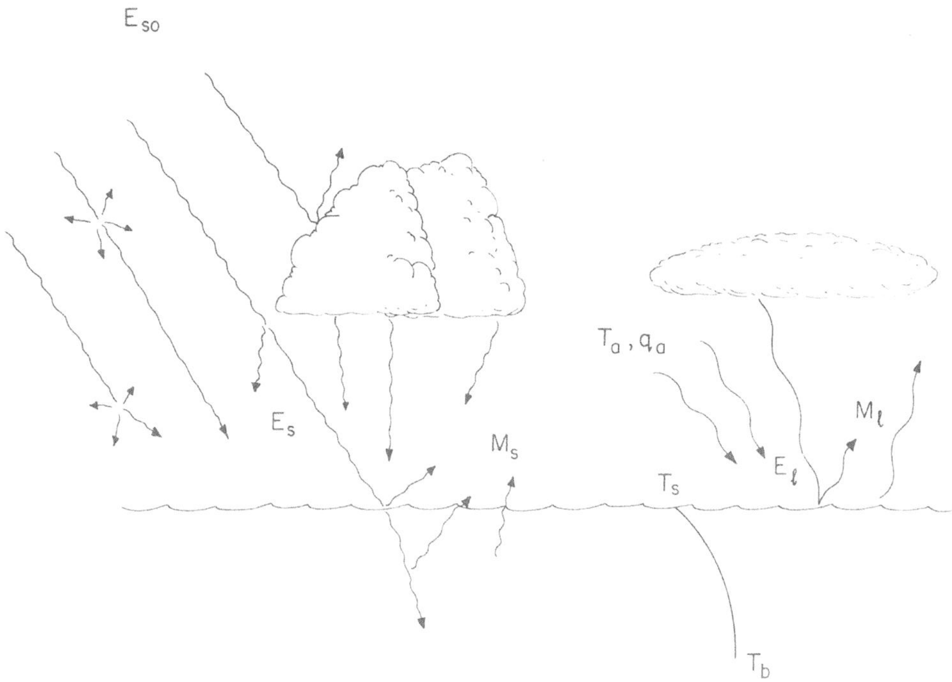

Fig. 1 — Illustration of the four radiative fluxes, which contribute to the net heating at the air-sea interface. To the left we have shortwave irradiance E_{so} impinging on the top of the atmosphere. Most of it is transmitted to the surface, E_s, either directly or after scattering by clouds and atmospheric particles. At the surface some of it is transmitted and some reflected. The transmitted portion is partially scattered in the backwards direction and makes up between 10 to 30% of M_s, the shortwave exitance. Longwave exitance from the sea surface, M_1, is grey body emission dependent on the interfacial temperature T_{so} and has a small contribution due to reflection of the longwave irradiance, E_1, which is made up of contributions from emission by atmospheric gases (mainly carbon dioxide, water vapor, and ozone) and by clouds. T_a is air temperature, q_a a measure of atmospheric humidity.

longwave emission. Concentrations of water vapor and ozone are variable, so their distribution in the vertical should also be known for exact calculations (Schmetz, et al., 1986). Clouds are strong emitters, and low clouds especially are important for longwave emission to the earth's surface. The relative importance of temperature, humidity structure and clouds (especially low clouds) was studied in model calculations by Fung et al. (1984) for several standard atmospheres representative of low, middle, and high latitudes in winter and summer seasons. Their conclusions were that for strongly perturbed conditions under clear skies the net longwave flux varies non-linearly with both temperature and water vapor and that inversions in the atmospheric temperature structure dominate the longwave radiative exchange at all latitudes. In the tropics the net longwave flux can vary from 5 to 70 Wm^{-2} depending on temperature, water vapor profiles and cloud height, but the range is much greater yet in high latitudes showing large seasonal variations. At all latitudes presence of high clouds has little influence, while low level clouds exert a strong influence. Typical errors introduced by inaccurate input data on cloud type or cloud base to the model calculations caused uncertainties in E_l of 10 to 20 Wm^{-2}.

The task for both surface-based and satellite-based parameterization schemes is thus to find readily available parameters and an approximate formulae which can work as a stand-in for the exact downward radiative transfer including all the variables listed above.

None of the existing parameterization schemes for E_s or E_l are exact or likely to be valid for all seasons and locations and most of them have been developed for conditions over land (e.g., Fung et al. 1984; Siegel and Dickey, 1986). Another difficulty arises because the input variables often are poorly measured or observed. It is, therefore, recommended for research projects where accuracy may be especially important, to obtain direct measurements of E_s, E_l, and SST, when possible, and to employ parameterization schemes or models mainly for closing data gaps.

Parameterization schemes and satellite based radiative transfer models for E_s, and E_l will be discussed in Sections 3.1, 3.2, 4.1 and 4.2. M_s and M_l are discussed in Sections 3.3 and 4.3.

To obtain information globally about the surface radiation budget over the oceans we must also consider the sampling limitations of direct measurements, surface based parameterizations and parameterizations based on satellite measurements. Direct measurements are at present very few. The role of direct measurements must, therfore, be as calibration checks for parameterization schemes or model calculations except for research projects, where short term variations may be of interest. Pyranometers could be used on the voluntary observing fleet for direct measurements of E_s but are not presently being used. Such measurements, and parameterization schemes based on surface observations from ships, however, sample the global ocean in a very uneven fashion.

Observations of cloudiness from ships have the advantage that they are obtained from the same perspective as that of a surface radiometer, which is the appropriate perspective for deriving the surface radiation budget. When satellite observations of radiance are employed to calculate surface irradiance, the discrepancy in angular resolution or spectral interval sampled must be taken into account. Justus and Paris (1985) and Ohring et al., (1984) among others discuss these kinds of differences and the problem has been addressed by most SRB modelers who employ satellite data (e.g., Frouin et al., 1988a). Satellite observations from geostationary satellites can be used for sampling every hour or even more frequently from latitudes between 50°S and 50°N. Three-hourly samples are typically employed by Gautier et al., (1980) and Raschke et al., (1987). At higher latitudes, two polar orbiting satellites provide four samples per day, if one includes infrared as well as visible data on cloud cover. A difficulty with using satellite data for estimating the surface radiation budget on the large scale is that it is voluminous and therefore expensive to acquire and employ. Because of nonlinearities in the equations for radiative flux terms, it is evident that whether one calculates fluxes from averages of the meteorological parameters of each synoptic report, typically 4 to 24 per day, or whether one calculates averages of the fluxes from individual flux estimates made for each synoptic report, the averaging time can influence the mean error. The length of the chosen averaging time, one day, one week or a month will also make a difference in the r.m.s. error. Typically, the r.m.s. error is halved when the averaging time is extended from one day to one week.

3. Estimation Schemes for Shortwave Radiation

3.1. SHORTWAVE IRRADIANCE FROM SURFACE OBSERVATIONS

There exists a wealth of parameterization schemes for average daily insolation, some of which were reviewed by Reed (1977), Lind (1981), Lind and Katsaros (1986) and Dobson and Smith (1985;1988). They are of the general form:

$$E_s = E_{s\,\text{clear sky}}\,(1 - aN^b) \tag{2}$$

where $E_{s\,\text{clear sky}}$ is the irradiance from a clear sky, which is modified by a cloud factor. N is the fractional cloud cover (in octas or tenths of the sky) found in the synoptic code; a and b are constants. In some formulations a is dependent on cloud type; b is a power usually equal to 1 (e.g., Kimball, 1928; Tabata, 1964; Reed, 1976), but $b = 2$ was used by Berliand (reference in Budyko, 1974) and $b = 3$ by Laevastu (1967), (Table 1). As is seen in Figure 2 the varying powers do not improve the fit of the data to observations from many oceanic experiments (Table 2). Note also that the various formulations do no agree on the value of E_s for clear skies, and the E_s values normalized with the clear sky irradiance do not go to 1 at C=0 for all parameterization schemes as one would expect. The general shape of the data cloud agrees with observations over land by Kasten and Czeplak (1988). The same six parameterization schemes for daily average shortwave irradiance are compared individually to the measurements from the oceanic field experiments in Figure 3a-f. The experiments listed in Table 2 took place at latitudes from 0° to 60° and at various times of the year, so they cover a wide range of marine conditions.

For shortwave parameterization the cloud type information available from the complete synoptic code was incorporated by Lumb (1964), who devised a formula which evaluates the irradiance for each hourly observation:

$$E_s = 1353 \sin h_s (A + B \sin h_s) \tag{3}$$

where h_s is the solar altitude and A and B are constants. Lumb divided the synoptic code reports of cloud types into 9 categories and evaluated the coefficients A and B by comparison to direct measurements of E_s at weather ships Juliet and Alfa in the North Atlantic. His is the only formula that estimates insolation on an hourly basis. Figure 3g shows daily averages obtained from Lumb's model against data obtained over the sea. Lind and Katsaros (1986) found that all possible synoptic code reports were not covered by Lumb's nine categories, so they broadened the Lumb categories to include the other combinations. The appropriateness of these categorizations were corroborated by Dobson and Smith (1988).

Reed (1977) concluded that his empirical formula for daily mean insolation

$$E_s / E_{so} = 1 - 0.62C + 0.0019\Phi \tag{4}$$

where ϕ is noon solar altitude, appears to be valid for cloud cover from 0.3 to 1.0 and the random error of estimate in monthly mean insolation should be less than ±10% within 95% confidence limit. This formula was tested on data from latitudes near 25°N. By comparison

Table 1: Shortwave irradiance parameterizations (Wm^{-2})

Φ =	Noon solar altitude, degrees	
C =	Daytime cloud cover fraction (0.0 - 1.0)	
Q_0 =	Clear-sky irradiance (Reed, 1977)	
Houghton	(1954) E_S	$= Q_0$ (1.28 - 1.03 C)
Kimball	(1928) E_S	$= Q_0$ (1.00 - 0.71 C)
Laevastu	(1967) E_S	$= Q_0$ (1.00 - 0.60 C^3)
R.K. Reed	(1977) E_S	$= Q_0$ (1.00 - 0.62 C + 0.0019 Φ)
Tabata	(1964) E_S	$= Q_0$ (1.00 - 0.716 C + 0.00252 Φ)
Wyrtki	(1965) E_S	$= Q_0$ (1.00 - 0.38 C - 0.38 C^2)
Lumb	(1964) E_S	= 1350 (A + BS) · S
S	=	Mean sine of solar altitude for each hour
A,B	=	Empirical constants for each of nine cloudiness Categories of Lumb's scheme developed for mid-latitude ship observations

Table 2: Oceanic field experiments where the irradiance data in Figures 2, 3 and 8, and in Tables 3 and 4 were obtained

Experiment Location	Season Year	Duration	Data Available
JASIN 60N, 12W N. Atlantic	Summer 1978	7 weeks	10-minute means two platforms
STREX 50N, 140W N. Pacific	Autumn 1980-	6 weeks	10-minute means
MILDEX 30N, 126W E. Pacific	Autumn 1983	4 weeks	5-minute means two platforms
FASINEX 30N, 70W Bermuda	Spring 1986	5 weeks	10-minute means to platforms
TROPIC HEAT II Hawaii to Equator	Spring 1987	5 weeks	15-minute means
GATE, Phase III Tropical, Atlantic	Summer 1974	3 weeks	hourly means, R/V Researcher (only)

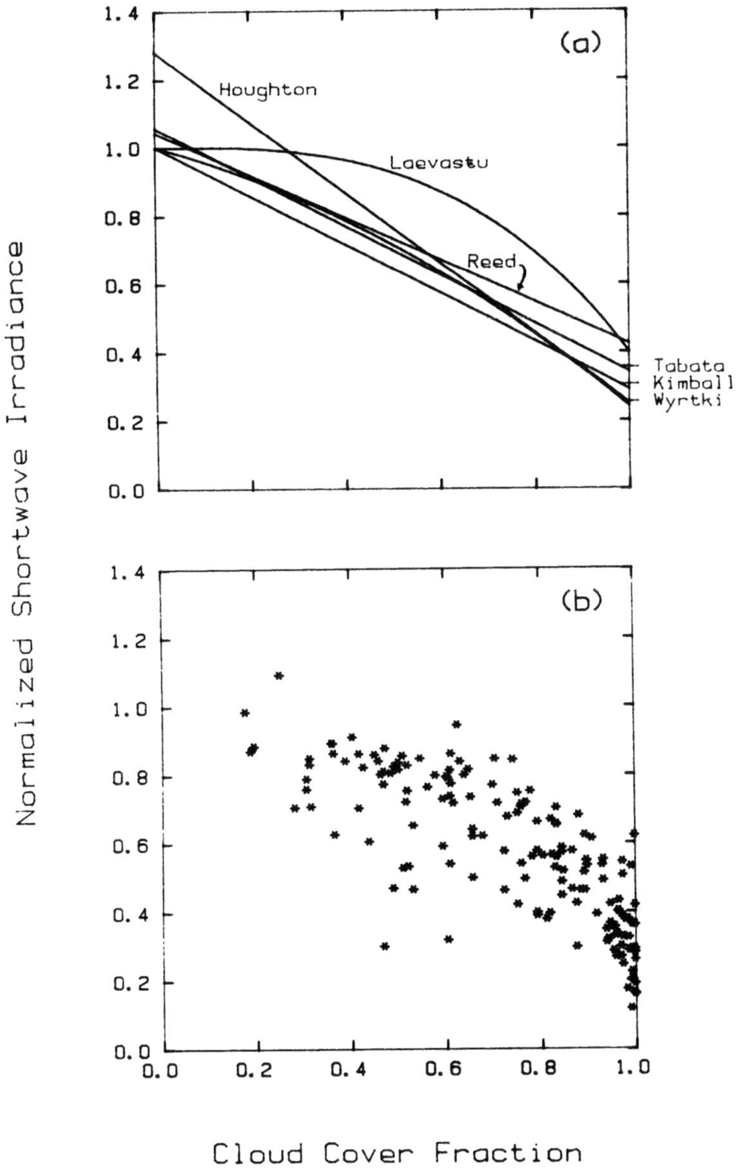

Fig. 2 — Daily averaged normalized shortwave irradiance (ratio of shortwave irradiance to clear sky irradiance) versus daily averaged cloud coverage: (a) predicted from model cloud factors (models listed in Table 1) and (b) measured during oceanic field experiments (listed in Table 1)

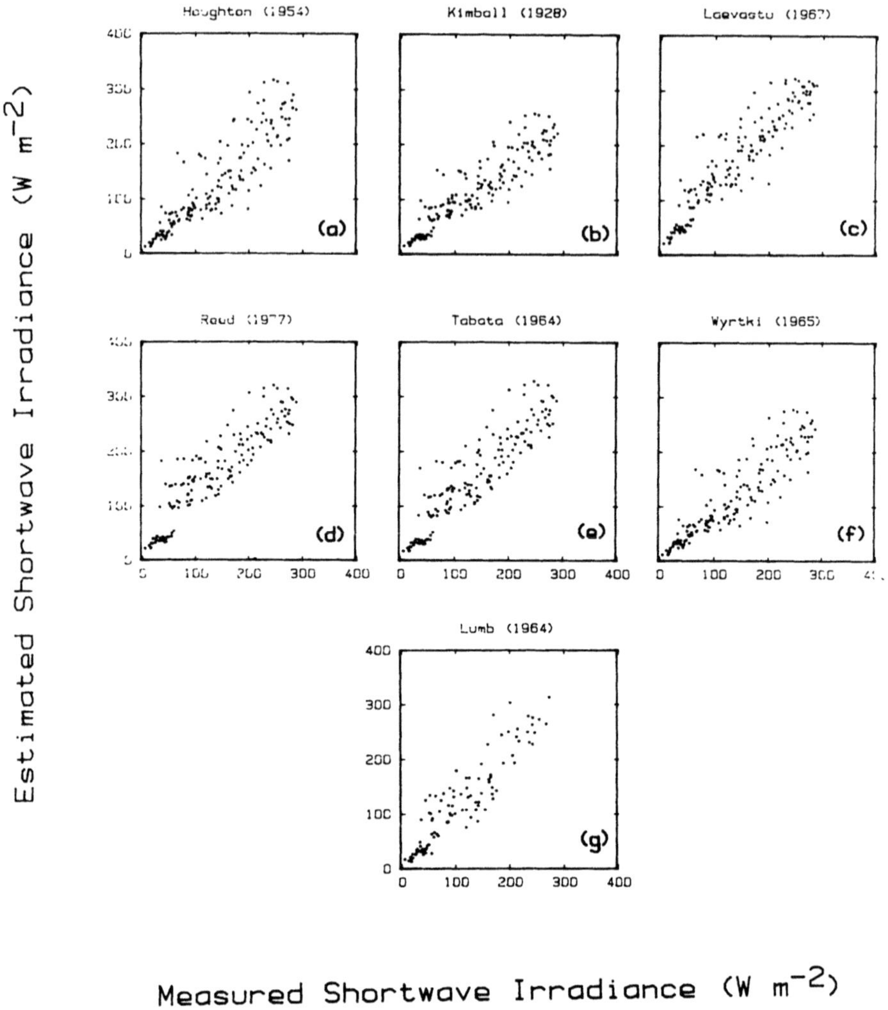

Fig. 3 — Comparison of surface based shortwave irradiance schemes to daily averaged observations: (a) model of Houghton (1954); (b) model by Kimball (1928); (c) model by Laevastu (1967); (d) model by Reed (1977); (e) model by Tabata (1964); (f) model by Wyrtki (1965); and (g) model by Lumb (1964). Comparison data sources are listed in Table 2 and the statistics of the intercomparisons are found in Table 3. Model of Lumb (g) could only be applied to data from GATE,JASIN and STREX Experiments.

with the Smithsonian formula (Seckel and Beaudry, 1973), Reed found for clear skies that Kimball's (1928) and Mosby's (1936) formulae had unacceptably large errors at all solar altitudes. Laevastu's were too large at high solar altitudes, and Berliand's formula generally gave shortwave irradiances 5 to 10% too large. Lumb's (1964) formula performed best. This is in agreement with the conclusions of Lind et al. (1984) who compared models to data obtained on two ships at 60°N during summer in the JASIN (Joint Air-Sea Interaction) Experiment. Frouin, et al. (1988a), however, found by testing five empirical formulae for daily averages of E_S against data at 31°N over the sea, that Mosby's (1936), Kimball's (1928), and Reed's (1977) formulae performed about equally well, while they found in agreement with Reed (1977) that Berliand's and Laevastu's formulae had larger positive biases, 15 and 24 Wm^{-1}, respectively.

Dobson and Smith (1988) compared several models to long term records of shortwave irradiance on midlatitude weather ships, and found that the simplest models which do not include cloud type do about as well as more complex ones. Reed's (1977) model performed best. Our Table 3 indicates, however, that Lumb's model has slightly less bias and r.m.s. errors for our data set. Unfortunately, the complete cloud observations are not usually reported by the voluntary observing fleet, so Lumb's model can only be applied to weather ship data.

In Table 3 error statistics for our selected seven parameterization formulae are compared to measurements at sea. Mean error is defined as E_s (calculated) minus E_s (measured)

$$\text{r.m.s. error} \equiv \left\{ \sum_{1=1}^{N} \frac{\left\{ E_s(calculated) - E_s(measured) \right\}^2}{N} \right\}^{1/2}$$

where N is the number of data points.

Table 3: Comparison statistics for daily averaged insolation computed from empirical formulae and measured during GATE, JASIN, STREX, FASINEX and Tropic Heat II Experiments (shown in Figure 3). Mean error is defined as positive when estimated value exceeds measured value

Reference Formula	Days of Comparison	Mean Error Wm^{-2}	Root-mean Square Error Wm^{-2}	Correlation Coefficient
Houghton	157	−11.4	40.3	0.887
Kimball	157	−23.6	42.7	0.905
Laevastu	157	30.3	45.4	0.923
Reed	157	20.1	40.4	0.906
Tabata	157	12.7	37.3	0.910
Wyrtki	157	−21.2	41.3	0.899
Lumb	103	9.7	33.0	0.918

After Lumb's model, Houghton's and Tabata's formulae perform about equally well. The large biases reported for some formulae are in part a consequence of the original development being carried out against data collected in particular climatic conditions while our test data cover a wide range of conditions. Inspection of Figure 3a, for instance, shows Houghton's formula to be biased high at high values of irradiance, and less so at low to intermediate values.

This somewhat unfair treatment of the parameterization formulae was done intentionally to illustrate that these empirical formulae cannot be applied indiscriminately to conditions different from those for which they were developed. This fact has not been adequately acknowledged or considered by most upper ocean or climate modelers and may severely affect their conclusions of the role of surface radiative fluxes, since the model biases are often large and can be of either sign. In the defense of the numerical modelers one must admit that the literature is confusing and the criteria for selecting one formula over another have not been available. That problem remains acute, because the data base of simultaneous direct measurements and good synoptic code reports is small. A more complete (but still limited) survey of models tested against measurements obtained at sea from polar to equatorial regions is in progress (Katsaros and Lind, 1988).

3.2. SHORTWAVE IRRADIANCE FROM SATELLITE OBSERVATIONS

Several models have been developed for inferring surface irradiance from satellite-based radiance measurements (e.g., Hay and Hanson, 1978; Tarpley, 1979: Gautier et al., 1980; Dedieu et al., 1987; Pinker and Ewing, 1985; Möser and Raschke, 1984). Most remote sensing protocols for meteorological, oceanographic, or climatological quantities acknowledge the need for regular calibration of satellite derived parameters against *in situ* measurements at anchor stations (e.g., Earth System Science, 1988); several authors deplore that shortwave sensors on the GOES (Geostationary Operational Environmental Satellites) are uncalibrated. Complex post-launch procedures have had to be devised in order to derive absolute values (e.g., Frouin and Gautier, 1987; and Gautier and Katsaros, 1984). As mentioned, when employing satellite radiance data to infer surface irradiances, corrections for the angular discrepancy must be considered and such conversions are not exact. A difficulty in comparing surface pyranometer data to satellite estimates lies in the localized sampling by these instruments, which is aggravated for the case of low clouds. Raschke (1985) suggests that averages over a month may be necessary to eliminate the sampling error. Gautier and Katsaros (1984) and Diekmann, et al. (1986) discuss another source of discrepancy between satellite inferred E_s and direct pyranometer measurements, namely the uncertainty of the satellite navigation. Keeping these sources of discrepancy in an intercomparison in mind we can consider the reported r.m.s. errors as upper bounds.

In general, these studies have shown that the surface shortwave irradiance can be estimated from geostationary satellite radiance data to about 10% accuracy for daily averages (e.g. Buriez et al., 1986; Raphael and Hay, 1984; Frouin et al. 1988a; and Gautier and Katsaros, 1984) (see Figures 4, 5 and 6). In the latter two references satellite estimates were compared to direct measurements at sea.

On an hourly basis discrepancies between ground-based measurements and satellite estimates are much larger, typically 20% (e.g., Buriez et al., 1986). As discussed above, they can be attributed in part to uncertainties in satellite navigation and to a mismatch between temporal and spatial resolution of the satellite sensor and the surface pyranometer, which results in different sampling, especially when the cloud cover is variable (e.g., Gautier and Katsaros, 1984). The fundamental uncertainties which remain are due to insufficient knowledge of the state of the atmosphere, particularly in regard to profiles of water vapor, ozone and to concentration, size distribution and optical properties of aerosols in the atmospheric path.

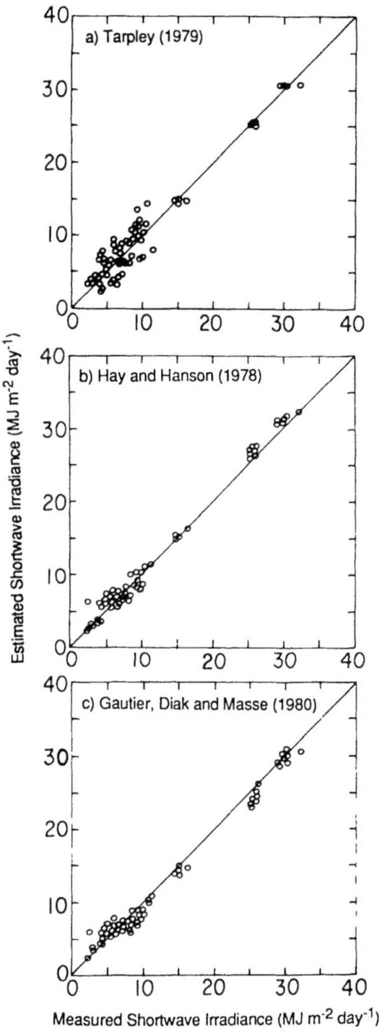

Fig. 4 — Shortwave irradiance schemes based on satellite data compared to measurements: Values are total insolation per day per m^2: (a) Tarpley (1979), (b) Hay and Hanson (1978), and (c) Gautier et al., (1980), (after Raphael and Hay, 1984).

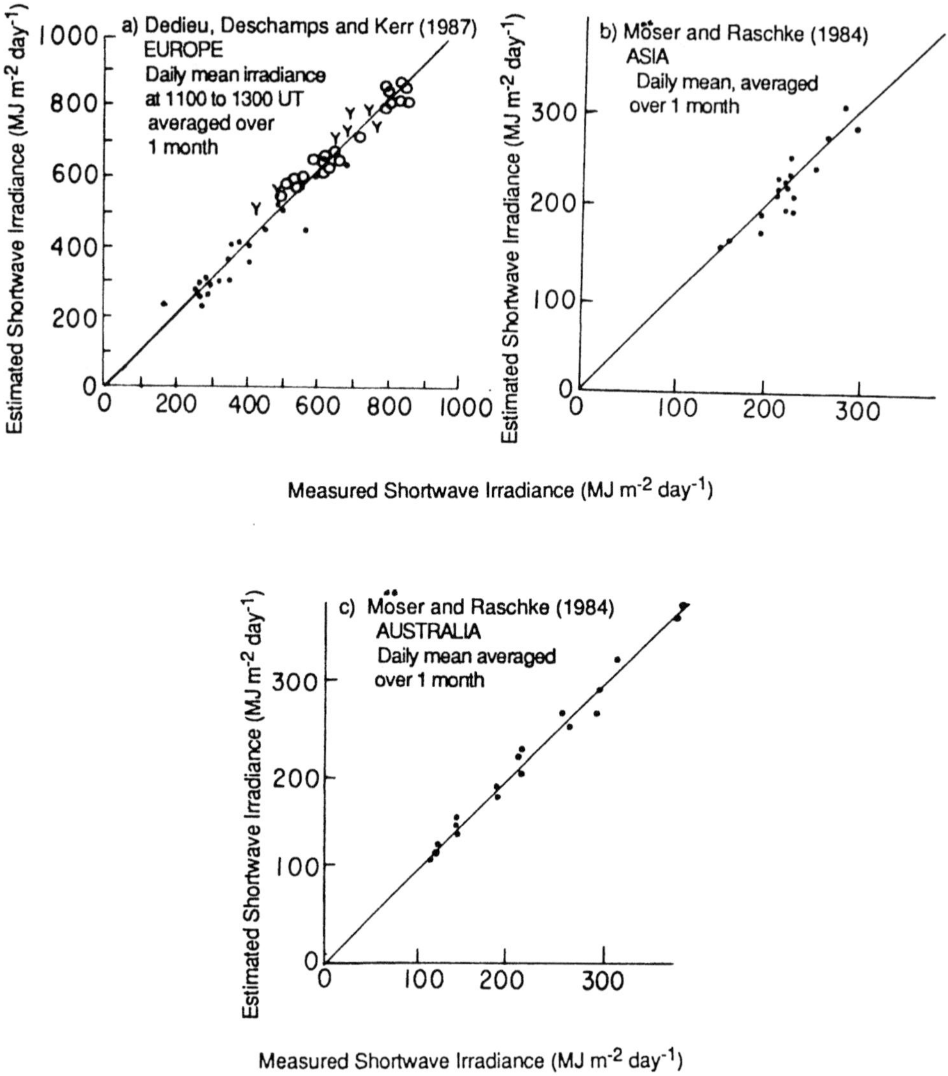

Fig. 5 — Shortwave irradiance schemes based on satellite data compared to measurements. Values are daily means in Wm^{-2} averaged over 1 month. (a) model of Dedieu, Deschamps and Kerr (1987); (b) model of Moser and Raschke (1984) compared to measurements in Asia and (c) model of Moser and Raschke (1984) compared to measurements in Australia. (b and c after Raschke et al., 1986)

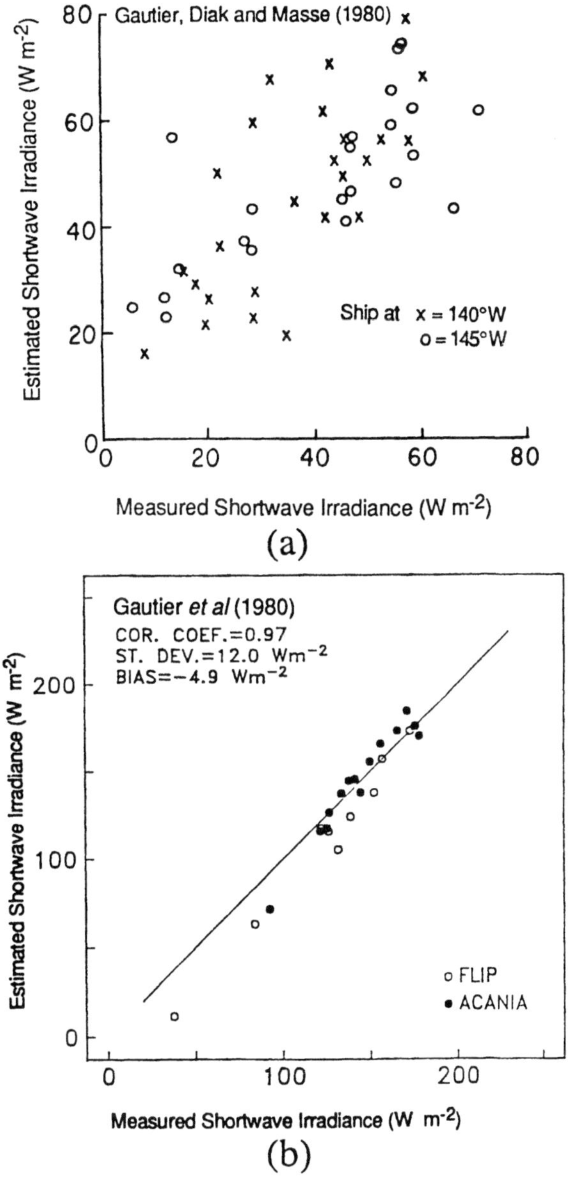

Fig. 6 — Comparison of daily averaged shortwave irradiance estimated with the satellite based model of Gautier et al., (1980) to pyranometer measurements made at sea. (a) in the STREX experiment (after Gautier and Katsaros, 1984) at 50°N and (b) in the MILDEX experiment (after Frouin, et al., 1988a). (See Table 2 for details of the experiments.)

Buriez et al. (1986) performed a sensitivity study of the factors which influence the dependence of the surface shortwave irradiance estimate on the albedo at the top of the atmosphere. They found for instance that the cloud type was not important for estimating the surface irradiance when shortwave radiance values measured from satellites were used to infer the cloud effect. These authors compared satellite schemes based on two different channels of the Advanced Very High Resolution Radiometer's (AVHRR) instruments, i.e., Channel 1 (0.5 to 0.7μm) and Channel 2 (0.7 to 1.1μm), to irradiance measurements over Europe. They attribute biases greater than 10 Wm^{-2} in hourly means (for values of cosine of the solar zenith angle > 0.4) to insufficient knowledge of the atmospheric aerosols.

Using a method of Möser and Raschke (1984) but with the reduced, so called, B3 data sets of the International Satellite Cloud Climatology Project (ISCCP) (Schiffer and Rossow, 1985) from Meteosat and GMS1, sampling Europe and Asia, Raschke et al. (1987) produced maps of monthly averages of irradiance. Comparing to a large number of surface stations (187 in Europe and 40 in Asia), they found r.m.s. differences of 6% for monthly averages, Figure 5. For daily averages, the r.m.s. error was 15% of the mean irradiance. In cloud free conditions, the model estimates are more accurate.

Three European models were compared in a study by Grüter et al. (1986) under the auspices of the European Community Research and Development Program. One is a physical model developed at the University of Cologne (Möser ad Raschke, 1984), the second one is a purely statistical one developed in Stuttgart by Reinhardt et al. (in Grüter et al., 1986), and the third is a French model by Cano and Monget (in Grüter et al., 1986). The Reinhardt et al. model is the only one of these which employs radiance data from polar orbiting satellites as input data; it is, therefore, the only one applicable to latitudes greater than $\pm 50°$.

The general conclusion from comparison of these three models to surface measurements collected from sites all over Europe, after the models had been adjusted to a two-week comparison data set, was that for hourly averages standard deviations of 83 Wm^{-2} were typical, while for daily averages they were 24 Wm^{-2}. (Dobson and Smith (1988) found very similar error magnitudes in their study of schemes based on surface observations.) These models were meant to be operational so detailed analysis of individual satellite scenes were not performed. For the more uniform and predictable albedo of the ocean these models might perform somewhat better.

In Figure 6 comparison of the scheme by Gautier et al. (1980) to direct measurements at sea show the potential of this scheme for improved estimates of surface irradiance over the ocean. This conclusion is based on the small bias in Figures 6a and 6b and on the rather small scatter in Figure 6b. The scatter in the plot from the STREX experiment (Figure 6a) is greater, but it is of the same magnitude along both axes, and even though the scatter looks relatively large in this plot it is absolutely rather small and was obtained in a "worst condition" for the satellite estimates in that STREX occurred in late fall at 50° N, where the sampling from a geostationary satellite is at a large incidence angle, marginal for the Gautier et al. model.

3.3. SHORTWAVE EXITANCE

Shortwave exitance from the sea can be calculated with good accuracy from direct measurements or model-produced estimates of the shortwave irradiance and a value for the albedo of the sea surface. The albedo A is defined as

$$A = -M_s / E_s \tag{5}$$

The albedo in turn can be modeled in terms of the diffuseness of the incident radiation. As the diffuseness of the irradiance increases, the incidence angles of the radiation on the sea surface are decreased on the average for low solar altitudes and increased for high solar altitudes, thereby modifying the albedo (Figure 1 in Katsaros et al., 1985 illustrates this point).

In a seminal study by Payne (1972) direct measurements of albedo from a research platform were related to the diffuseness of the incident radiation. The diffuseness was parameterized in terms of the total transmittance, τ, i.e.,

$$\tau = E_s / E_{so} \tag{6}$$

with E_{so} simply calculated from knowledge of time and location (e.g., Paltridge and Platt, 1976). Thus M_s can be obtained from E_s through

$$M_s = -A(E_s / E_{so})E_s \tag{7}$$

Figure 7 (after Katsaros et al., 1985) shows Payne's (1972) fitted curves with data obtained in two field experiments over over the deep ocean at tropical and high latitude sites.

The agreement between model and observations is generally good in spite of uncertainties in the measurements due to the presence of the ships, from which the measurements were made. Some systematic deviations from Payne's model appear at low transmittances, but most of the differences are probably due to real variations in the hemispheric distribution of the incident irradiance.

The roughness of the sea also contributes to the variation in the incidence angle and therefore to the albedo. It has not been accounted for in Figure 7. This effect is less than ± 20% of the albedo for wind speeds from 4 to 12 ms^{-1}, which translates to 1 to 5% of E_s depending on the value of A. The trend of the dependence of albedo on wind speed, which mainly determines the sea surface roughness was consistent in three independent studies (Payne, 1972; Simpson and Paulson, 1979; Katsaros et al., 1985; see Figure 6 of the latter reference). For wind speeds great enough that wave breaking is occurring, an increase in albedo occurs due to foam cover. Monahan and O'Muircheartaigh (1988) estimate the albedo increase over the low wind speed values to be about 0.006 for winds of 15 ms^{-1} and 0.012 for winds of 20 ms^{-1}.

Thus, for long term energy budget calculations one can simply employ Payne's (1972) empirical fits for different solar altitude bands of the albedo as a function of total transmittance. For instantaneous estimates of albedo, these values may differ from the true value by a factor of 2 according to Figure 7.

4. Estimation Schemes for Longwave Radiation

It will be seen in this section that surface- and satellite-based schemes for estimating long wave irradiance do fairly well but are still much less accurate than shortwave schemes. The main difficulty is that the cloud base temperature and emittance are not readily deduced from standard surface or satellite measurements.

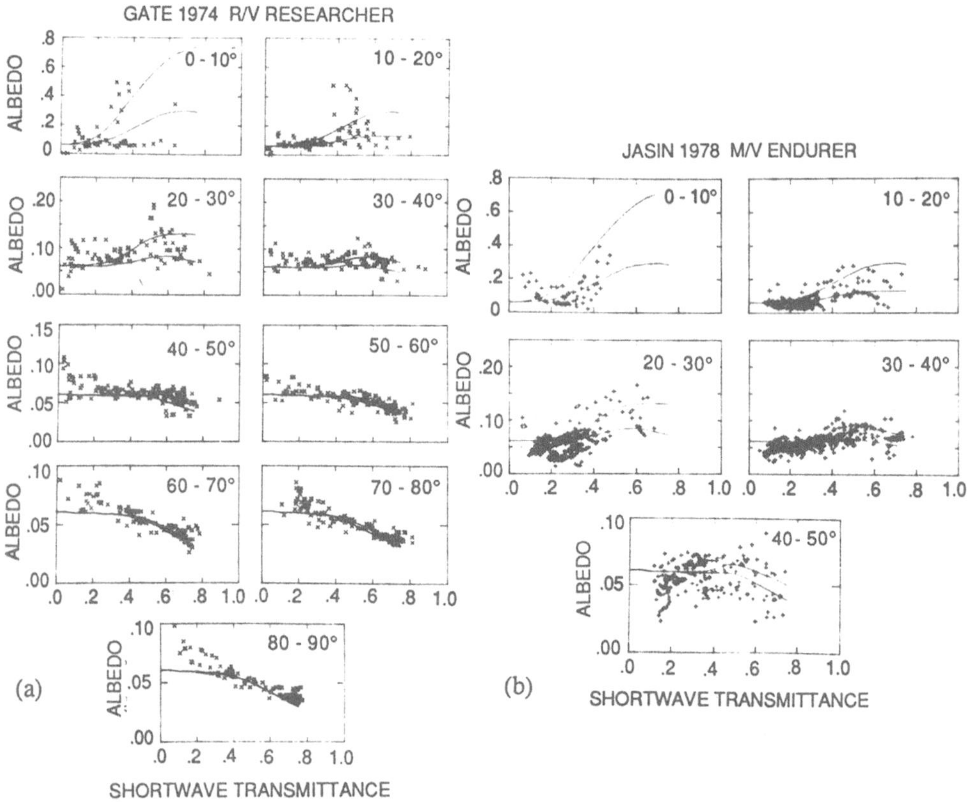

Fig. 7 — Albedo as a function of total shortwave transmittance for 10 bands of solar altitude for 10 minute averages of data from: (a) R/V Researcher in the GARP Atlantic Tropical Experiment, where GARP stands for the Global Atmospheric Research Program, at 7°N, 23°E in Summer 1974 and (b) R/V Endurer in the JASIN experiment at 60°N in Summer 1978 (after Katsaros et al., 1985)

4.1. LONGWAVE IRRADIANCE FROM SURFACE OBSERVATIONS

Parameterizations of E_l based on surface observations have almost as long a history as the ones for E_s. Many of the older schemes seek to estimate the net longwave loss, that is $M_l - E_l$. In several such schemes the two terms are not readily separable, so in order to limit our discussion to E_l, M_l has had to be subtracted from such schemes. The reason for this development is that over land the interface temperature is not easily measured. Over sea, on the other hand, as mentioned above, it is best to simply calculate M_l from sea surface temperature measurements (see Section 4.3). Because of the well-known dependence of E_l on atmospheric gases and clouds a typical parameterization scheme can be written as:

$$E_l = \epsilon \sigma T_x^4 (a - be^p)(1 - dC^n) \tag{8}$$

where ϵ is emittance, σ the Stefan-Boltzmann constant, T_x is either air or surface temperature (T_a or T_s), e is vapor pressure, C is cloud cover fraction, and the power p of the vapor pressure is either 1 or $\frac{1}{2}$; a, b, d, and n are constants. Fung et al. (1984) suggested that $p = \frac{1}{2}$ applies to high humidity conditions prevalent in the tropics, while $p = 1$ applies best in drier conditions, such as are found at high latitudes in winter. Most longwave formulae have the power $n = 1$ except the formulae of Clark et al. (1974), Hastenrath and Lamb (1978), and Adamec and Elsberry (1984). The latter formulae used in evaluating back radiation over the North Pacific is attributed to Gill (1982) by Adamec (personal communication).

Among the longwave irradiance or net irradiance formulae the following have been used frequently: Anderson (1952), Berliand M.E. (cited by Budyko, 1956); Brunt (1932); Clark et al. (1974); Efimova N.A. (cited by Kondratyev, 1969); Hastenrath and Lamb (1978); Idso and Jackson (1969); Reed (1976); and Swinbank (1963). Swinbank's formula is unique in that it depends only on T_a, assuming a correlation between T_a and e_a. This is clearly not a valid assumption for certain oceanic regions, particularly east of continents. Here air masses of both marine and continental origin are frequent and they can have quite a different relationship between T_a and e_a.

A scheme which is based on physical reasoning and for which estimates of E_l are made for each hourly observation in the synoptic code was developed by Lind and Katsaros (1982) inspired by the success of Lumb's (1964) scheme for hourly shortwave irradiance. It requires detailed cloud reports and allows for different radiative temperatures for each cloud deck. In addition to variables reported in the surface code, information about the vertical structure of temperature and humidity is required. This can be obtained from local radiosonde ascents, if they are available. In the general case, the temperature and humidity are read from the synoptic weather charts at the standard levels, i.e., 850, 700, and 500 mb, at the location of interest. Vertical temperature and humidity profiles are constructed from these points. The most difficult aspect of constructing such a model was estimating the cloud emittances as a function of cloud thickness (or liquid water path). Here some subjective judgement and empirical fitting was necessary. A fit between the relationships suggested by Paltridge (1974), and Stephens et al., (1978) was chosen (Lind and Katsaros, 1982). This model estimated longwave irradiance within 3 Wm^{-2} for daily averages and 10 Wm^{-2} for hourly averages when compared to the JASIN (Joint Air-Sea Interaction Experiment) measurements. An error of 1 octa in reported cloud cover or 5°C in estimated cloud bottom temperature for low to middle level clouds resulted in 10 Wm^{-2} error. Since the surface reports from commercial ships seldom include the complete synoptic cloud code, a simplified version of this model which assigns heights to upper level clouds was devised. This version of the model gave 7 Wm^{-2} for daily and 12 Wm^{-2} for hourly averages when comparisons were made to the JASIN data (Lind and Katsaros, 1986).

The statistics of the performance of the longwave parameterizations are found in Table 4 and the scatter plots in Figure 8 with the statistical measures defined similarly as for Table 3. Over the wide range of meteorological conditions encountered in the oceanic experiments used for these statistics the formulae perform almost equally well except for Berliand's formula, which has both the largest bias and the greatest r.m.s. error. The Lind and Katsaros model, for which hourly estimates of E_l are averaged together, has the best statistic for the daily average.

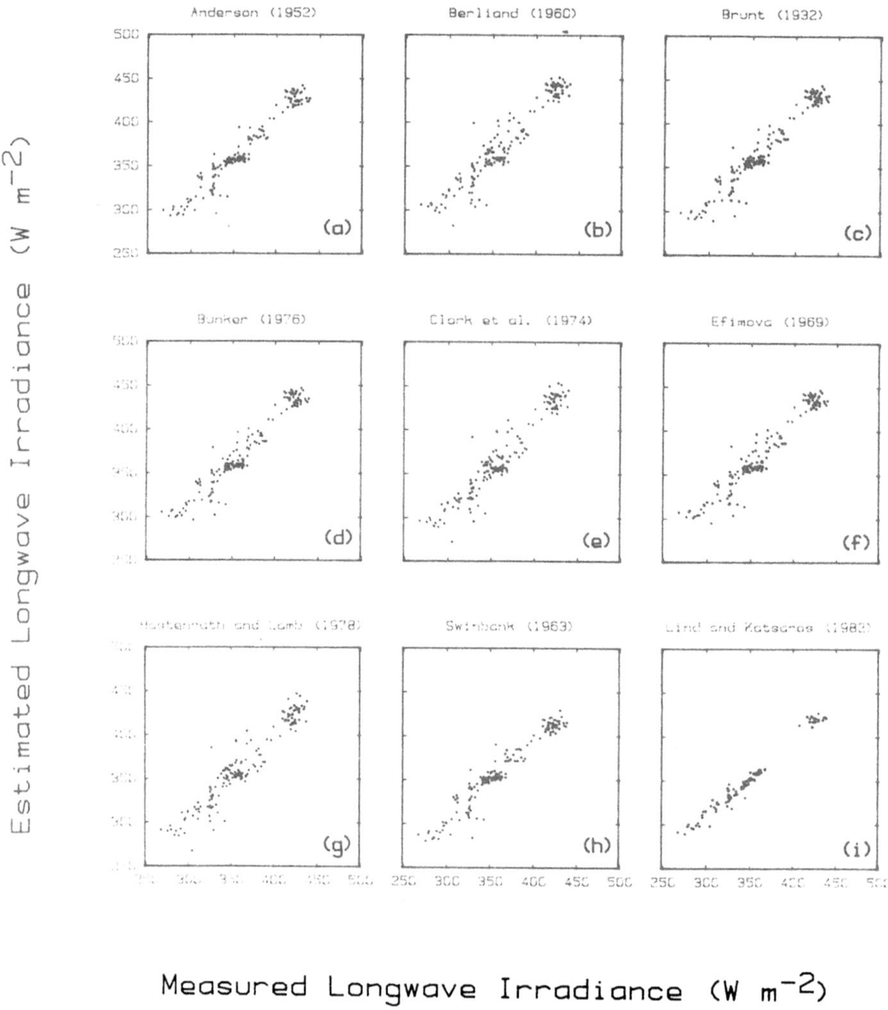

Fig. 8 — Comparison of surface based longwave irradiance schemes to daily averaged observations: (a) model by Anderson (1952); (b) model by Berliand (1960); (c) model by Brunt (1932); (d) model by Bunker (1976); (e) model by Clark et al. (1974); (f) model by Efimova (1969); (g) model by Hastenrath and Lamb (1978); (h) model by Swinbank (1963) and (i) model by Lind and Katsaros (1982). Comparison data sources are listed in Table 2. Model of Lind and Katsaros (i) could only be applied to data from GATE, JASIN and STREX Experiments.

Table 4: Comparison statistics for daily averaged irradiance computed from empirical formulae and measured during GATE, JASIN, STREX, FASINEX and Tropic Heat II Experiments (shown in Figure 8). Mean error is defined as positive when estimated value exceeds measured value

Reference Formula	Days of Comparison	Mean Error Wm^{-2}	Root-mean Square Error Wm^{-2}	Correlation Coefficient
Anderson	157	5.2	12.2	0.978
Berliand	157	11.1	18.2	0.964
Brunt	157	5.7	12.9	0.976
Bunker	157	9.3	15.1	0.974
Clark et al.	157	5.5	15.1	0.967
Efimova	157	10.1	15.8	0.973
Hastenrath and Lamb	157	−0.9	13.8	0.966
Swinbank	157	−4.0	12.5	0.975
Lind and Katsaros	103	1.2	8.6	0.989

4.2. LONGWAVE IRRADIANCE FROM SATELLITE OBSERVATIONS

One of the first attempts to estimate longwave irradiance at the earth's surface from satellite data was by Darnell et al. (1983), who employed sounding data from the TIROS Operational Vertical Sounder (TOVS) with a radiative transfer model (Figure 9). They divided the atmosphere into 50 mb deep layers up to the 250 mb level, their height limit for clouds. The TOVS output provides an effective fractional cloud cover and cloud top pressure. For surface longwave irradiances it is, however, the cloud base temperature and the true cloud cover, which are required. Thus, an estimate of cloud thickness must be provided. These authors chose a standard 50 mb cloud thickness, which is probably acceptable for middle level clouds. They compared the model output for 41 satellite passes to surface based measurements averaged over $\frac{1}{2}$-hour intervals at a ground site in Ann Arbor, Michigan. Their results show a correlation coefficient of 0.87 and a standard error of estimate of 20 Wm^{-2}, which is 6.5% of the mean value. Part of the difference can be attributed to the inherent difference between a point measurement at the surface and the aereal averaging by the footprint of the TOVS which is 300 km along track and 250 to 500 km across track.

A similar model was developed by Frouin et al. (1988b), who also employed TOVS data for the temperature and humidity profiles but hourly GOES (Geostationary Operational Satellite) Visible and Infrared Spin-Scan Radiometer (VISSR) data for the cloud parameters. The cloud base altitude is deduced from the cloud top altitude and the liquid water path interpreted from the visible reflectance. Cloud base temperature can then be obtained from the atmospheric temperature profile obtained from the TOVS data. Other possible methods for obtaining cloud base temperature were also tried, such as fixing the thickness and liquid water path

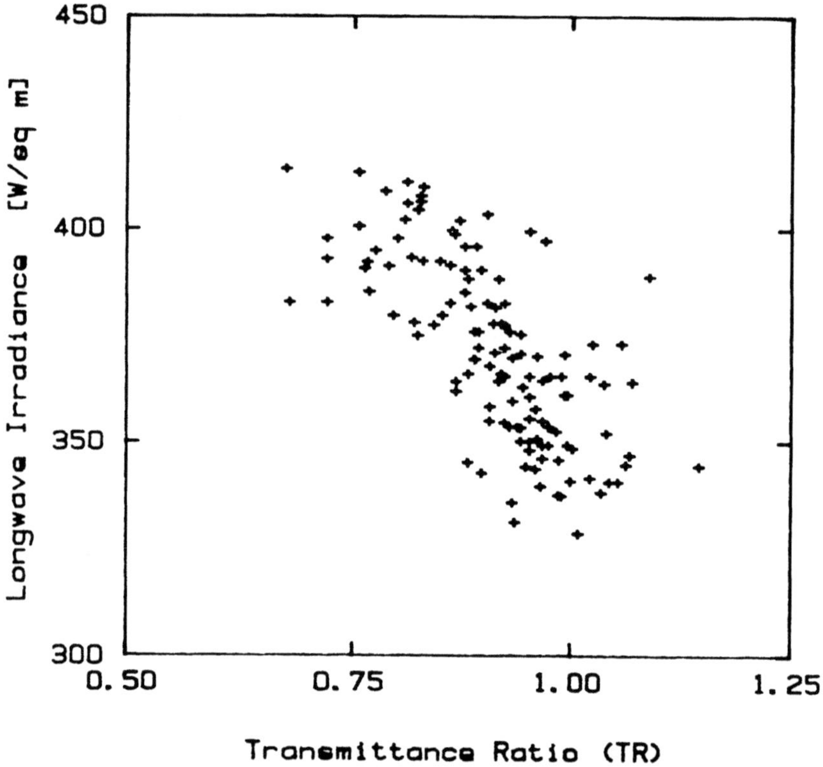

Fig. 9 — Examples of longwave irradiance estimates calculated with satellite data as input compared to pyrgeometer measurements: (a) the most complete model tried by Frouin et al. (1988a) using visible satellite data for the cloud amount. The pyrgeometer data were obtained on R/V Acania and R/P FLIP in the MILDEX experiment by Lind and Katsaros (1987), (after Frouin et al., 1988a) and (b) the model of Schmetz, Schmetz and Raschke (1986) and (c) the model of Darnell, Gupta and Staylor (1983). Note the difference in scales on the ordinate.

to the climatological mean value for the region or relating them to the fractional cloud cover, as was done by Lind and Katsaros (1982) for surface-based parameterizations. The satellite-derived fluxes were then compared to *in situ* pyrgeometer measurements obtained in the MIxed Layer Dynamics EXperiment (MILDEX) conducted in the fall of 1983 off the California Coast (Lind and Katsaros, 1987) (Figure 9).

The radiative transfer models of longwave radiation are quite accurate and perform well when compared to more complete line for line radiative transfer calculations. What is limiting the parameterization of the longwave irradiance at the earth's surface at the present are the limitations in retrievals of temperature and water vapor profiles from TOVS in that the TOVS wavelengths do not penetrate clouds. The spatial resolution is too coarse, the twice-a-day sampling by a single polar orbiting satellite is too infrequent. For the regions which can be observed from the geostationary satellites VAS (VISSR Atmospheric Sounder) offers improved spatial and temporal resolution (e.g., Smith, 1983).

In a recent paper on shortwave irradiance estimates, Frouin et al. (1988a) suggested that employing cloud cover estimates obtained from satellite radiances in bulk parameterization models of the type discussed in Section 3.1 could be an economical and reasonably accurate alternative, since the bulk models are typically limited by subjective cloud observations. Such cloud cover estimates could be tried with empirical longwave irradiance models also.

Another idea, which has not yet been tested, is to use cloud liquid water path estimated from passive microwave data provided by the SMMR (Scanning Multichannel Microwave Radiometers) on Seasat and Nimbus 7 and recently by the SSM/I (Special Sensor Microwave Imager) on a satellite in the Defense Meteorological Satellite Program (DSMP) (e.g., Gloersen and Barath, 1977; McMurdie and Katsaros, 1985; Katsaros and Lewis, 1986). These sensors have typical footprint for cloud water of the order of 50 km x 50 km. However, no official cloud liquid water algorithm exists, partly because the theoretical ones (e.g., Chang and Wilheit, 1979) cannot be readily tested against *in situ* data. The qualitative evidence from case studies (McMurdie and Katsaros, 1985; Katsaros and Lewis, 1986) indicates, however, that the theoretical estimates are reasonable. King (1987) shows that a strong relationship exists between cloud optical thickness (related to liquid water path) and the solar reflection function. However, the albedo of the surface below causes a bias. This problem is, of course, less troublesome over the ocean.

Another possible technique for improving estimates of cloud liquid water path on a smaller spatial scale is to use a so-called "reflectance ratio" from satellite observations or a "transmittance ratio" (TR) from surface observations. The ratios being taken are between the irradiances in two broad bands of the solar spectrum (DeVault and Katsaros, 1983). One band is in the solar near infrared, where cloud water absorbs more than it does in the other band, which is in the solar visible. An example of the correlation between longwave irradiance and the transmittance ratio, both measured on a ship at sea, is found in Figure 10. However, beyond a certain value for the liquid water path (here TR = 1) the longwave exitance becomes independent of the ratio, and almost constant. This occurs when the clouds are thick stratus, which are highly emissive and have a typical bottom temperature which doesn't vary substantially during a short-term experiment.

A very interesting model for estimating E_l which employs geostationary data, is the hybrid model by Schmetz et al. (1986). It uses visible and infrared data from METEOSAT-2 to derive cloud parameters in conjunction with grid point thermodynamic fields. This model calculates the clear sky contribution from empirical formulae (Idso and Jackson, 1969; Idso, 1981) and the cloud contribution from satellite estimates of cloud amount, cloud base height and temperature using a formula after Martin and Berdahl (1984). Surface humidity is not available from these sources, so a climatological mean of 85% relative humidity is substituted. The scheme is limited to daytime applications and latitudes well observed from geostationary satellites. Comparisons to daily averages at four stations in West Germany showed a bias of 3.9 Wm^{-2} and r.m.s. error 11.6 Wm^{-2} (Figure 10). Its accuracy is thus comparable to the accuracies of the best bulk models and is closest to the accuracy of the Lind and Katsaros' (1982) model which also employed hemispheric grid data but with surface observations of cloudiness.

In conclusion, for this section, Frouin et al. (1988b) did not find that the satellite methods for estimating longwave irradiance reduced the standard errors of estimate compared to the empirical bulk formulae for half-hourly or daily averages, but the satellite methods were generally less biased. At present, therefore, the main advantage of satellite techniques for

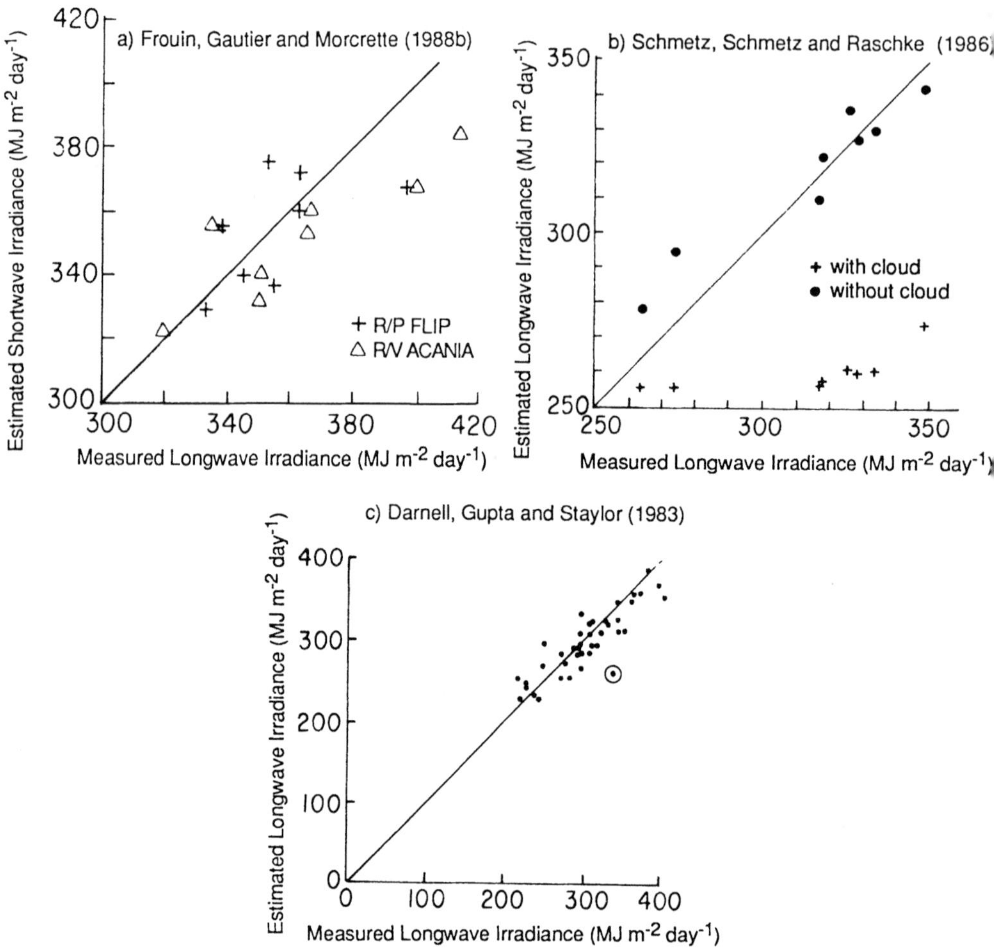

Fig. 10 — Longwave irradiance as a function of the ratio of shortwave irradiance in two bands of the solar spectrum, 0.7 to 2.7 μm and R/V Oceanus in the Frontal Air-Sea Interaction Experiment, February 15 thru March 8, 1986

estimating longwave irradiance at the surface is the global coverage. However, a technique, which depends on the geostationary cloud observations is limited to latitudes < ±50° and, if only visible data are used, to the daylight hours.

4.3. LONGWAVE EXITANCE

Direct measurements of longwave exitance from a ship or research platform requires that the pyrgeometer be extended out from the obstacle by a boom. If the platform can be made to have a small, solid angle both for the skyward and seaward hemispheres, this may be a reasonably accurate measurement. However, as Katsaros et al. (1985) discuss in connection with albedo measurements from the broad side of a ship, the required corrections for blocked solid angles are large. If alternatively, the pyrgeometer or pyrradiometer is extended from the bow, the heaving of the ship will cause substantial fluctuations in the horizon viewed by hemispheric sensors. For obtaining the longwave exitance from the sea, it is therefore usually preferable to use the Stefan-Boltzmann law over attempting to make direct measurements (e.g., Hinzpeter, 1980; Lind and Katsaros, 1986). It is given by:

$$M_l = \epsilon\sigma T_s^4 + (1-\epsilon)E_l \tag{9}$$

where ϵ is emittance, σ the Stefan Boltzmann constant (equal to $5.70 \times 10^{-8} Wm^{-2}K^{-4}$), and T_s is sea surface temperature (SST) in degrees Kelvin. The accuracy of M_l then depends on how well one can estimate ϵ and T_s. By comparison to the large uncertainties in direct measurement of M_l some errors in estimates of ϵ and T_s are acceptable. The average emittance value for the broad range from 3 μm to 50 μm is about 0.98 with a slight dependence on temperature and salinity (see Katsaros, 1980a for relevant references). The uncertainty in ϵ in this broad range of wavelengths is less than about ± 0.01, which introduces an uncertainty in M_l of ± 3.4 Wm^{-2} at 5°C and ± 4.5 Wm^{-2} at 25°C. The appropriate SST to use in Equation 9 is the interface temperature T_{so}, which is only obtainable from measurements with infrared radiometers operating in the atmospheric window between 8 and 12 μm. The surface emission in the 3 to 50 μm range emerges from depths below the air-sea interface of between 25 and 500 μm (McAlister and McLeish, 1969).

Conventional SST's are obtained by immersion buckets, ship's water intake thermometer or buoys from a thermistor, often placed inside the hull of the buoy. These temperatures represent the water temperature at depths of 1 to 4 m, which can differ from the interface temperature by typically from a few tenths of a °C up to several degrees, positively or negatively (e.g., Saunders, 1967; Clauss et al., 1970; Hasse, 1971; Grassl, 1976; Katsaros, 1977; Katsaros, 1980b; Paulson and Simpson, 1981; Schluessel et al., 1987). More and more, and especially for vast expanses of the oceans, where ships rarely travel, we rely on satellite determinations of SST. The accuracy of the retrieved SST's are now of the order of 0.5°C (Bernstein and Chelton, 1985; Robinson et al., 1984; Schluessel et al., 1987). In the latter reference, the satellite retrieved SST's were compared to skin temperatures measured with an infrared radiometer on board a ship. After incidence angle and atmospheric path corrections were made, the deviations between remotely sensed T_s from the satellite and the ship were of the order of 0.3°C for the best case of multiple satellite sensors, but 0.73°C was found for the standard AVHRR (Advanced Very High Resolution Radiometer) SST procedure. The operational satellite retrieval schemes for SST have been fitted to bulk T_s measurements made primarily from buoys (McClain et al., 1985) which should be slightly biased in view of the prevalence of a cool surface film of 0.2°C to 0.5°C deviation (e.g., Katsaros, 1980a, 1980b; Schluessel et al., 1987). For these reasons, there exists an uncertainty of at least ± 0.5°C in the SST values available from ships or satellite retrievals, and likely a bias of at least 0.2°C in the T_s needed for Equation 9. The result is ± 2.5 Wm^{-2} uncertainty in M_l and a negative bias of 1 Wm^{-2} over most of the ocean. Neither of these numbers is significant. Since E_l calculated from surface or satellite observations represents the downward irradiance just above the sea surface and since this radiation is also subject to reflection from the sea with a reflectivity $r = (1-\epsilon)$, one can simplify the estimations of both E_l, absorbed, and M_l by letting $\epsilon = 1$. There is a slight error committed in doing this, since E_l and M_l have their peak radiation at different wavelengths, so that the reflectivity for E_l does not exactly equal $(1-\epsilon)$ where ϵ is the emittance contained in M_l, but this error is very small.

5. Summary and Discussion

The importance of the surface radiation budget for many sub-disciplines of meteorology and oceanography is well established. Therefore, estimates of the surface energy budget, of which the SRB often is the dominating component, are produced in numerous contexts, especially when studying upper ocean mixing, genesis and evolution of sea surface temperature anomalies or in calculations of the meridional heat transport. This review has aimed to establish the accuracies and uncertainties which exist in the various models employed to estimate the four terms in the SRB.

Unfortunately, the energy budgets produced by the model calculations are likely to be greatly in error (e.g. Talley, 1984; Weare, 1983; Smith and Dobson, 1984). The reason is two-fold. First, the models or parameterization schemes have often not been well tested for the region of the world ocean being modeled (a good physical scheme should not be location dependent, but most schemes still contain empirical fits, which for the cloudiness effects on irradiances is especially likely to remain part of the recipe for some time to come). Second, the input data to the models are often not representative of the time and area considered or, they may often be of marginal accuracy.

Concentrating here only on summarizing what is known about the terms in the surface radiation budget, we can say the following:

a. *Shortwave irradiance.* Satellite or surface based estimates of shortwave irradiance can give monthly averages with an accuracy of 5-15% (Raschke, 1985). For direct pyranometer measurements accuracies of 5% are claimed. Thus the best currently available models and schemes for shortwave irradiances are comparable to direct measurements, which is a good state of affairs, since the shortwave irradiance dominates the SRB in most circumstances. However, it is not quite good enough, because the accuracy needed in the net surface energy budget for long term monitoring of climate change is of the order of 5 to 10 Wm^{-2}. We note that 5% of typical total irradiance magnitudes of the order of 300 to 400 Wm^{-2} corresponds to 15 to 20 Wm^{-2}, and this is only one term in the net budget. Thus, it would seem important to try to reduce this error and particularly examine the systematic biases tied to location or season. This can probably be done by inclusion of real time atmospheric profiles of temperature, humidity and aerosol effects. Such data are potentially available, and the know-how for their utilization exists; it is mostly inertia plus the complexity and great expense of including all the information on the same computer, which prevents it from being done routinely. Frouin et al. (1988a) includes such a complete scheme among their many schemes tested, and it is superior to the others.

b. *Shortwave exitance-albedo.* Fortunately, over the ocean the albedo and, therefore, M_s is predictable to good accuracy in contrast to the difficulties that this term causes over land, when attempting to evaluate net shortwave irradiance from space (e.g., Raschke, 1985). Uncertainties still remain in short term estimates due to turbidity of the ocean and surface roughness. This variability could be included in the parameterization if measurements of turbidity and wind speed (or sea state) were made part of the algorithm. Such co-located information is potentially available from multi-wavelength optical systems and scatterometers, respectively, which are part of the suites of measurements being planned for future satellite earth observing systems (e.g., Earth System Science, 1988).

c. *Longwave Irradiance.* This term is generally less variable than the shortwave irradiance, which is fortuitous, since longwave irradiance is the term whose parameterization has had least testing over the sea to date. In low latitudes, this term is typically almost constant due to high surface humidity and only small variations in air temperature; in high latitudes, however, particularly in the wintertime, it is highly variable and is often the dominant term in the SRB. Appropriate schemes for estimating E_l from satellite measurements are only recently being developed, while a plethora of surface based estimation formulae exists. Great improvement will probably only come after we have gained another substantial measure of understanding of cloud physics, so that we can infer the bottom temperature and emittance of a cloud from cloud top temperature and an estimate of its liquid water content or depth and have learned how to handle multilayered clouds. Alternatively, numerical model data, which exist for every grid box on the globe at all times (or at least on 6-hourly intervals) could be folded into the schemes as suggested by Schmetz et al., (1986) and by Lind and Katsaros (1982). The difficulty again is due to inertia, the complexity and expense of joining disparate data sets obtained at different times. Such problems are, however, not insurmountable in due time. Errors of 5% or 15 Wm^{-2}, whichever is greater, seem unavoidable at present.

d. *Longwave Exitance.* This term is rapidly being improved due to improved estimates of SST. Uncertainties of $\pm 0.3°C$ in SST are probably consistent with the optimum accuracy obtainable in the foreseeable future. Errors about twice as large are likely to be typical of the operational ship and buoy SST products for some time to come. This latter uncertainty only results in about ± 5 Wm^{-2} in M_l in any case, which is insignificant in comparison to the errors inherent in E_s and E_l at present.

6. Acknowledgement

Acknowledgement and thanks are due to R.J. Lind for valuable discussions and for calculations presented in Figures 2, 3, 8 and Tables 3 and 4. Ms. Iftekhar Bhatti and Ms. Kim Kawada who contributed to the literature searches, Ms. Kay Dewar who drafted the figures and Ms. Janet Meadows who typed the final manuscript with great care. This review was written with support from the Office of Naval Research under Grant N00014-84-C-0111, PO0017.

7. References

Adamec, D. and R.L. Elsberry, 1984: The Use of Mean atmospheric Forcing in an Ocean Mixed-Layer Model. *J. Phys. Oceanogr.*, **14**, 1670-1676.

Anderson, E.R., 1952: Energy Budget Studies, Water-Loss Investigations: Lake Hefner studies. *U.S. Geol. Surv. Circ.*, **229**, 71-88.

Atwater, M.A. and J.T. Ball, 1978: A Numerical Solar Radiation Model Based on Standard Meteorological Observations. *Solar Energy,* **21,** 163-170.

Bernstein, R.L. and D.B. Chelton, 1985: Large-scale sea surface temperature variability from satellite and shipboard measurements, *J. Geophys. Res.*, **90**, 11,619-11,630.

Brunt, D., 1932: Notes on Radiation in the Atmosphere. *Quart. J. Roy. Met. Soc.*, **58**, 389-420.

Budyko, M.I., 1956: *The Heat Balance of the Earth's Surface* (in Russian), Gidrometeorologicheskeo, Leningrad. (English translation, N.A. Stepanova, U.S. Weather Bureau, Wash., DC 1958.)

Budyko, M.I., 1974: *Climate and Life.* Academic Press, 508 pp.

Bunker, A.F., 1976: Computations of Surface Energy Flux and Annual Air-Sea Interaction Cycles of the North Atlantic Ocean. *Mon. Wea. Rev.*, **104**, 1122-1140.

Buricz, J.C., B. Bonnel and Y. Fouquart, 1986: Theoretical and Experimental Sensitivity Study of the Derivation of the Solar Irradiance at the Earth's Surface from Satellite Data. *Beitr. Phys. Atmosph.*, **59**, 263-281.

Chang, A.T.C. and T.T. Wilheit, 1979: Remote Sensing of Atmospheric Water Vapor, Liquid Water, and Wind Speed at the Ocean Surface by Passive Microwave Techniques from the Nimbus 5 Satellite. *Radio Science,* **14** 793-802.

Clark, N.E., L. Eber, R.M. Laurs, J.A. Renner and F.F.T. Saur, 1974: Heat Exchange Between Ocean and Atmosphere in the Eastern North Pacific for 1961-71. *NOAA Tech. rep. NMFS SSRF-682*, U.S. Dept. of Commerce, Wash., DC.

Clauss, E., H. Hinzpeter and J. Mueller-Glewe, 1970: Messungen zur Temperaturstruktur im Wasser an der Grenzflaeche Ozean-Atmosphere. Meteor Forschungsergeb. Reihe B, 5, 90-94.

Darnell, W.L., S.K. Gupta and W.F. Staylor, 1983: Downward Longwave Radiation at the Surface from Satellite Measurements. *J. Clim. Appl. Met.*, **22**, 1956-1960.

Dedieu, G., P.X. Deschamps and Y.H. Kerr, 1987: Satellite Estimation of Solar Irradiance at the Earth's Surface and of Surface Albedo using a Physical Model Applied to Meteosat Data. *J. Clim. Appl. Met.*, **26**, 79-87.

DeVault, J.E. and K.B. Katsaros, 1983: Remote Determination of Cloud Liquid Water Path from Bandwidth-Limited Shortwave Measurements. *J. Atmos. Sci.*, **40**, 665-685.

Diekmann, F.J., E. Ruschke, and H. Woick, 1986: Operational mapping of global radiation over Central Europe, *Meteor Rund.*, **39**, 59-63.

Dobson, F.W. and S.D. Smith, 1985: Estimation of Solar Radiation at Sea. In *The Ocean Surface,* Y. Toba and H. Mitsuyasu, eds. D. Reidel Publishing Company, 525-533.

Dobson, F.W. and S.D. Smith, 1988: Bulk Models of Solar Radiation at Sea. *Quart. J. Roy. Met. Soc.*, **114**, 165 -182.

Earth System Science, 1988: A Program for Global Change, A Closer View. Report of the Earth System Sciences Committee, NASA Advisory Council, NASA, Wash., D.C., pp. 208.

Frouin, R. and C. Gautier, 1987: Calibration of NOAA-7 AVHRR, GOES-5 and GOES-6 VISSR/VAS Solar Channels. *Remote Sens. Env.*, (in press).

Frouin, R., C. Gautier, Katsaros, K.B. and R.J. Lind, 1988a: A comparison of Satellite and Empirical Formula Techniques for Estimating Insolation over the Oceans. *J. Clim. Appl. Met.*, 97, 1016-1023.

Frouin, R., C. Gautier and J-J. Morcrette, 1988b: Downward Longwave Irradiance at the Ocean Surface from Satellite Data: Methodology and In-Situ Validation. *J. Geophys. Res.*, (in press).

Fung, I.Y., D.E. Harrison and A.A. Lacis, 1984: On the Variability of the New Longwave Radiation at the Ocean Surface. *Reviews in Geophys.*, **22**, 177-193.

Gautier, C., G. Diak and S. Masse, 1980: A Simple Physical Model to Estimate Incident Solar Radiation at the Surface from GOES Satellite Data. *J. Appl. Meteor.*, **19**, 1005-1012.

Gautier, C. and K.B. Katsaros, 1984: Insolation during STREX, 1, Comparison between Surface Measurements and Satellite Estimates. *J. Geophys. Res.*, **39**, 11,778-11,788.

Gill, A.E., 1982: Atmospheric Ocean Dynamics International Geophysics Series, Vol. 30, Academic Press, N.Y., pp. 662.

Gloersen, P. and F.T. Barath, 1977: A Scanning Multichannel Microwave Radiometer for Nimbus 6 and Seasat-A, *IEEE J. Oceanic Engineering*, **OE-2,** 172-178.

Grassl, H., 1976: The Dependence of the Measured Cool Skin of the Ocean on Wind Stress and Total Heat Flux. *Bound. Layer Meteorol.*, **10**, 465-474.

Grüter, W., H. Guillard, W. Möser, J.M. Monget, W. Palz, E. Raschke, R.E. Reinhardt, P. Schwarzmann and L. Wald, 1986: *Solar Radiation Data from Satellite Images: Determination of Solar Radiation at Ground Level from Images of the Earth Transmitted by Meteorological Satellites.* D. Reidel Publishing Company, 100 pp.

Hartmann, D.L., V. Ramanathan, A. Berroir and G.E. Hunt, 1986: Earth Radiation Budget Data and Climate Research. *Rev. Geophys. Space Phys.*, **24**, 2, 439-468.

Hasse, L., 1971: The Sea Surface Temperature Deviation and the Heat Flow at the Sea-Air Interface. *Bound. Layer Meteorol.*, 368-379.

Hastenrath, S., 1980: Heat Budget of Tropical Ocean and Atmosphere. *J. Phys. Oceanogr.*, **10**, 159-170.

Hastenrath, S. and P.J. Lamb, 1978: *Heat Budget Atlas of the Tropical Atlantic and Eastern Pacific Oceans.* The University of Wisconsin Press, Madison, WI., 90 pp.

Hay, J.E. and K.J. Hanson, 1978: A Satellite-Based Methodology for Determining Solar Irradiance at the Ocean Surface during GATE. *Bull. Amer. Meteor. Soc.*, 59, 1549.

Hinzpeter, H., 1980: Atmospheric Radiation Instruments, In *Air-Sea Interaction Instruments and Methods,* F. Dobson, L. Hasse and R. Davis, eds. Plenum Press, New York, 491-507.

Houghton, H.G., 1954: On the Annual Heat Balance of the Northern Hemisphere, *J. Meteor.*, **11**, 1-9.

Idso, B.S., 1981: On the Systematic Nature of Diurnal Patterns of Differences between Calculations and Measurements of Clear Sky Atmospheric Thermal Radiation. *Quart. J. Roy. Meteor. Soc.*, **107**, 737-741.

Idso, B.S., and R.D. Jackson, 1969: Thermal radiation from the atmosphere, *J. Geoph. Res.*, **74**, 5397-5403.

Jacobs, W.C., 1942: On the Energy Exchange Between Sea and Atmosphere. *J. Marine Res.*, **5**, 36-66.

Justus, C.G. and M.V. Paris, 1985: A Model for Solar Spectral Irradiance and Radiance at the Bottom and Top of a Cloudless Atmosphere. *J. Clim. Appl. Met.*, **24**, 193-205.

Kasten, F. and G. Czeplak, 1980: Solar and Terrestrial Radiation Dependent on the Amount and Type of Cloud. *Solar Energy*, **24**, 177-189.

Katsaros, K.B., 1977: The Sea Surface Temperature Deviation at Very Low Wind Speeds: Is There a Limit? *Tellus*, **29**, 229-39.

Katsaros, K.B., 1980a: Radiative Sensing of Sea Surface Temperatures. In *Instruments and Methods in Air-Sea Interaction*, F. Dobson, L. Hasse and R. Davies, eds., Plenum Publishing Corp., New York, 293-317.

Katsaros, K.B., 1980b: The Aqueous Thermal Boundary Layer. *Bound. Layer Meteorol.*, **18**, 107-127.

Katsaros, K.B. and J.E. DeVault, 1986: On Irradiance Measurement Errors at Sea Due to Tilt of Pyranometers. *Atmos. and Oceanic Techn.*, **3**, 740-745.

Katsaros, K.B. and R.M. Lewis, 1986: Mesoscale and Synoptic Scale Features of North Pacific Weather Systems Observed With the Scanning Multichannel Microwave Radiometer on Nimbus 7. *J. Geophys. Res.*, **91**, 2321-2330.

Katsaros, K.B. and R.J. Lind, 1988: Estimates of Shortwave and Longwave Irradiances over the Oceans Compared to Measurements from Equatorial to Polar Regions, (in preparation).

Katsaros, K.B., R.J. Lind and J.E. DeVault, 1988: Surface Radiation Measurements in FASINEX. Preprint, 7th Conference on Atmosphere-Ocean Interaction, Amer. Meteor. Soc., Anaheim, CA., Feb. 1-5.

Katsaros, K.B., L.A. McMurdie, R.J. Lind and J.E. DeVault, 1985: Albedo of a Water Surface, Spectral Variation, Effects of Atmospheric Transmittance, Sun Angle, and Wind Speed. *J. Geophys. Res.*, **90**, 7313-7321.

Kimball, H.H., 1928: Amount of Solar Radiation that Reaches the Surface of the Earth on the Land and on the Sea and Methods by which it is Measured. *Mon. Wea. Rev.*, **56**, 393-399.

King, M.D., 1987: Determination of the Scaled Optical Thickness of Clouds from Reflected Solar Radiation Measurements, *J. Atmos. Sci.*, *44, No.*, 13, 1734-1751.

Kondratyev, K.Ya, 1969: *Radiation in the Atmosphere.* Academic Press, Orlando, FLA., 432-440.

Laevastu, T., 1967: Cloud Factor in Long-wave Radiation Formulae. *J. Geophys. Res.*, 72, 4277.

*Lind, R.J., 1981: Models of Long and Short Wave Irradiance with Radiation Estimates from JASIN 1978. M. Sc. Thesis, Dept. of Atmospheric Sciences, University of Washington, Seattle, WA 98195, 129.

Lind, R.J. and K.B. Katsaros, 1982: A Model of Longwave Irradiance for use with Surface Observations. *J. Appl. Meteorol.*, **12**, 1015-1023.

Lind, R.J. and K.B. Katsaros, 1986: Radiation Measurements and Model Results from R/V *Oceanographer* During Strex 1980. *J. Geophys. Res.*, **91**, 13,308-13,314.

*Lind, R.J. and K.B. Katsaros, 1987: Radiation Measurements from R/V FLIP and R/V Acania during the Mixed Layer Dynamics Experiment (MILDEX). Technical Report, Dept. of Atmospheric Sciences, University of Washington, Seattle WA 98195, 44 pp.

Lind, R.J., K.B. Katsaros and M. Gube, 1984: Radiation Budget Components and their Parameterization in JASIN. *Quart. J. Roy. Meteor. Soc.*, **110**, 1061-1071.

Lumb, F.E., 1964: The Influence of Cloud on Hourly Amount of Total Solar Radiation at the Sea Surface. *Quart. J. Roy. Meteor. Soc.*, **90**, 43-56.
Martin, M. and P. Berdahl, 1984: Characteristics of Infrared Sky Radiation in the United States. *Solar Energy,* **33**, 321-336.
McAlister, E.D. and W. McLeish, 1969: Heat Transfer in the Top Millimeter of the Ocean. *J. Geophys. Res.*, **74**, 3408-3414.
McClain, E.P., W.G. Pichel and C.C. Walton, 1985: Comparative Performance of AVHRR-Based Multichannel Sea Surface Temperatures. *J. Geophys. Res.*, **90**, 11,587-11,6f01.
McMurdie, L.A. and K.B. Katsaros, 1985: Atmospheric Water Distribution in a Midlatitude Cyclone Observed by the Seasat Scanning Multichannel Microwave Radiometer. *Mon. Wea. Rev.*, **113**, 4, 584-598.
Monahan, E.C. and I.B. O'Muircheartaigh, 1988: Comments on "Albedos and Glitter Patterns of a Wind-roughened Sea Surface". *J. Phys. Oceanogr.*, (in press).
Mosby, H., 1936: Verdunstung und Strahlung auf dem Meere. *Ann. Meteor, 64*, 281-286.
Möser, W. and E. Raschke, 1984: Incident Solar Radiation over Europe Estimated from METEOSAT Data. *J. Clim. Appl. Met.*, **23**, 166-170.
Ohring, G., A. Gruber and R. Ellingson, 1984: Satellite Determinations of the Relationship between Total Longwave Radiation Flux and Infrared Window Radiance. *J. Clim. Appl. Met.*, **23**, 416-425.
Oort, A. and T.H. Vonder Haar, 1976: On the Observed Annual Cycle in the Ocean-Atmosphere Heat Balance Over the Northern Hemisphere. *J. Phys. Oceanogr.*, **6**, 781-800.
Paltridge, G.W., 1974: Infrared Emissivity, Short Wave Albedo and Microphysics of Stratiform Water Clouds. *J. Geophys. Res.*, **79**, 4053-3058.
Paltridge, G.W. and C.M.R. Platt, 1976: *Radiative Processes in Meteorology and Climatology. Elsevier, New York, pp. 318.*
Paulson, C.A., and J.J. Simpson, 1981: The Temperature Difference Across the Cool Skin of the Ocean., *J. Geophys. Res., 86, 11.044-11.054.*
Payne, R.E., 1972: Albedo of the Sea Surface. *J. Atmos. Sci.*, **29**, 959-970.
Pinker, R.T., and J.A. Ewing, 1985: Modeling Surface Solar Radiation: model Formulation and Validation. *J. Clim. Appl. Met., 24, No. 5,* 389-401.
Pinker, R.T., J.A. Ewing and J.D. Tarpley, 1985: The Relationship Between the Planetary and Surface Net Radiation, *J. Clim. Appl. Met., 24, No. 11,* 1262-1268.
Raphael, C. and J.E. Hay, 1984: An Assessment of Models which use Satellite Data to Estimate Solar Irradiance at the Earth's Surface. *J. Clim. Appl. Met., 23,* 832-844.
Raschke, E., 1978: *Terminology and Units of Radiation Quantities and Measurements.* In Reports by the IAMAP Radiation Commission, ed. J. Lenoble. Available from Sec. General, NCAR, P.O. Box 3000, Boulder, CO 80307.
Raschke, E., 1985: On the Derivation of Radiation Budget Parameters at the Surface from Satellite Measurements. *Adv. Space Res.*, **5**, 319-327.
Rasckhe, E.A. Gratzki and M. Rieland, 1987: Estimates of Global Radiation at the Ground from the Reduced Data Sets of the International Satellite Cloud Climatology Project. *J. Climatol,* **7**, 205-213.
Rasool, S.I., 1987: Understanding the Global Change: An Opportunity to Seize. *EOS. Transactions, American Geophysical Union,* **68**, 1609-6111.
Reed, R.K., 1976: On Estimation of Net Long-Wave Radiation From the Oceans. *J. Geo-*

phys. Res., **81**, 5793-5794.

Reed, R.K., 1977: On Estimating Insolation over the Ocean. *J. Phys. Oceanogr.*, **7**, 482-485.

Robinson, I.S., N.C. Wells and H. Charnock, 1984: The Sea Surface Thermal Boundary Layer and its Relevance to the Measurement of Sea Surface Temperature by Airborne and Spaceborne Radiometers. *Int. J. Remote Sens.*, **5(1)** 19-45.

Saunders, P.M., 1967: The Temperature at the Ocean-Air Interface. *J. Atmos. Sci.*, **24** 269-273.

Schiffer, R.A. and W.B. Rossow, 1985: ISCCP Global Radiance Data Set: A New Resource for Climate Research. *Bull. Amer. Meteor. Soc.*, **66**, 1498-1505

Schluessel, P., H.-Y. Shin, W.J. Emery and H. Grassl, 1987: Comparison of Satellite-Derived Sea Surface Temperatures with In Situ Skin Measurements. *J. Geophys. Res.*, **92**, 2859-2874.

Schmetz, P., J. Schmetz and E. Raschke, 1986: Estimation of Daytime Downward Longwave Radiation at the Surface from Satellite and Grid Point Data. *Theor. Appl. Climatol*, **37**, 136-149.

Seckel, G.R. and F.H. Beaudry, 1973: The Radiation from Sun and Sky over the North Pacific Ocean (abstr.). *Trans. Amer. Geophys. Union*, **54**, 1114.

Sellers, W.D., 1965: Physical Climatology, *Univ. of Chicago Press*, Chicago, Ill.

Siegel, D.A. and T.D. Dickey, 1986: Variability of Net Longwave Radiation Over the Eastern North Pacific Ocean. *J. Geophys. Res.*,**91**, 7657-7666.

Simpson, J.J. and C.A. Paulson, 1979: Mid-ocean Observations of Atmospheric Radiation. *Quart. J. Roy. Meteor. Soc.*, **105**, 487-502.

Smith, S.D. and F.W. Dobson, 1984: The Heat Budget at Ocean Weather Station Bravo. *Atmos-Ocean*, **22**, 1-22.

Smith, W.L., 1983: The Retrieval of Atmospheric Profiles from VAS Geostationary Radiance Observations. *J. Atmos. Sci.*, **40**, 2025-2035.

Stephens, G.L., G.G. Campbell and T.H. Vonder Haar, 1981: Earth Radiation Budgets. *J. Geophys. Res.*, **86**, 9739-9760.

Stephens, G.L., G.W. Paltridge and C.M.R. Platt, 1978: Radiation Profiles in Extended Water Clouds. III: Observations., *J. Atmos. Sci.*, *35*, 2133-2141.

Swinbank, W.C., 1963: Long-Wave Radiation from Clear Skies. *Quart. J. Roy. Meteor. Soc.*, **89**, 339-348.

Tabata, S., 1964: Insolation in Relation to Cloud Amount and Sun's Altitude (in *Studies on Oceanography*, edited by K. Yoshida. University of Tokyo, Tokyo, pp. 202-210.

Talley, L.D., 1984: Meridional Heat Transport in the Pacific Ocean. *J. Phys. Oceanogr.*, **14**.

Tarpley, J.D., 1979: Estimating Incident Solar Radiation at the Surface from Geostationary Satellite Data. *J. Appl. Meteor.*, **18**, 1172-1181.

Vonder Haar, T.H. and A.H. Oort, 1973: New Estimate of Annual Poleward Energy Transport by Northern Hemisphere Oceans. *J. Phys. Oceanogr.*,**2**, 169-172.

Weare, B.C., P.T. Strub and M.D. Samuel, 1981: Annual Mean Surface Heat Fluxes in the Tropical Pacific Ocean. *J. Phys. Oceanogr.*, **11**, 705-717.

Weare, B.C., 1983: Interannual Variation in Net Heating at the Surface of the Tropical Pacific Ocean. Bull. Am. Meteor. Soc., 64, 873-885.

Wyrtki, K., 1965: The Average Annual Heat Balance of the North Pacific Ocean and its Relation to ocean Circulation. *J. Geophys. Res.*, **70**, 4547-4559.

INDEX

ENVIRONMENTAL FLUID MECHANICS

Publications:

1. W. Brutsaert: *Evaporation into the Atmosphere.* Theory, History, and Applications. 1982; Reprint 1988 ISBN 90-277-1247-6
2. G.T. Csanady: *Circulation in the Coastal Ocean.* 1982 ISBN 90-277-1400-2
3. A.S. Monin and R.V. Ozmidov: *Turbulence in the Ocean.* 1985 ISBN 90-277-1735-4
4. S. Panchev: *Dynamic Meteorology.* 1985 ISBN 90-277-1744-3
5. V.M. Kamenkovich, M.N. Koshlyakov and A.S. Monin (eds.): *Synoptic Eddies in the Ocean.* 1986 ISBN 90-277-1925-X
6. A.S. Monin: *Theoretical Geophysical Fluid Dynamics.* 1990 ISBN 0-7923-0426-8
7. G.L. Geernaert and W.J. Plant (eds.): *Surface Waves and Fluxes.* Volume I – Current Theory. 1990 ISBN 0-7923-0809-3
   ISBN 0-7923-0806-9 (Set, vols. I + II)
8. G.L. Geernaert and W.J. Plant (eds.): *Surface Waves and Fluxes.* Volume II – Remote Sensing. 1990 ISBN 0-7923-0805-0
   ISBN 0-7923-0806-9 (Set, vols. I + II)

KLUWER ACADEMIC PUBLISHERS – DORDRECHT / BOSTON / LONDON

GPSR Compliance

The European Union's (EU) General Product Safety Regulation (GPSR) is a set of rules that requires consumer products to be safe and our obligations to ensure this.

If you have any concerns about our products, you can contact us on ProductSafety@springernature.com

In case Publisher is established outside the EU, the EU authorized representative is:

Springer Nature Customer Service Center GmbH
Europaplatz 3
69115 Heidelberg, Germany

Batch number: 10607220

Printed by Printforce, the Netherlands